ROTORDYNAMICS '92

ROTORDYNAMICS '92

**Proceedings of the International Conference
on Rotating Machine Dynamics
Hotel des Bains, Venice, 28–30 April 1992**

Edited by

Michael J. Goodwin

Springer-Verlag
London Berlin Heidelberg New York
Paris Tokyo Hong Kong
Barcelona Budapest

Michael J. Goodwin BSc, PhD, CEng, MIMechE
Department of Mechanical and Computer-Aided Engineering, Staffordshire Polytechnic,
Beaconside, Stafford ST18 0AD, UK

ISBN-13: 978-3-540-19754-6   e-ISBN-13: 978-1-4471-1979-1
DOI: 10.1007/978-1-4471-1979-1

British Library Cataloguing in Publication Data
Rotordynamics '92: Proceedings of the International Conference
on Rotating Machine Dynamics, Hotel des Bains, Venice,
28–30 April 1992
  I. Goodwin, M. J.
  621.8
ISBN-13: 978-3-540-19754-6

Library of Congress Cataloging-in-Publication Data
A catalog record for this book is available from the Library of Congress

Printed by Antony Rowe Ltd., Chippenham, Wiltshire
69/3830–543210 Printed on acid-free paper

Proceedings of the International Conference on Rotating Machine Dynamics
ROTORDYNAMICS '92
Hotel des Bains, Venice, 28–30 April 1992

Contents

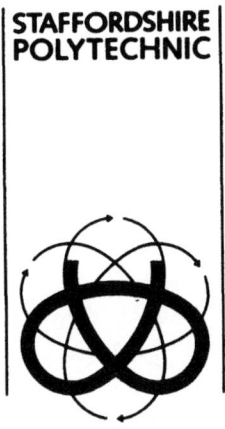

Organised by **Staffordshire Polytechnic**

Proceedings of the International Conference on Rotating Machine Dynamics
ROTORDYNAMICS '92
held at the Hotel des Bains, Venice, 28–30 April 1992

Organised by Staffordshire Polytechnic

Sponsored by:

Staffordshire Polytechnic

The Canadian Society for Mechanical Engineers

The Japan Society of Mechanical Engineers

The Vibration Institute

The Theory of Lubrication and Bearings Institute

Federazione delle Associazioni Nazionali dell'Industria Meccanica Varia ed Affine
(ANIMA)

The Institution of Diagnostic Engineers

The Australian Academy of Science

Associazione Elettrotecnica ed elettronica Italiana

Proceedings of the International Conference on Rotating Machine Dynamics
ROTORDYNAMICS '92
held at the Hotel des Bains, Venice, 28–30 April 1992

Organised by Staffordshire Polytechnic

Foreword

Rotating machinery is one of the largest classes of machinery used, and designers and operators of such equipment have to deal with many related potential problems associated with machine vibration and wear. The use of sophisticated design aids, and of new monitoring techniques has developed as a consequence of the world's increasing demands for quieter machine operation, longer machine life, and greater efficiency of operation. With improved machine design, and greater operating experience, the expectations of the engineer concerned with using machines becomes yet more demanding, and so research into rotating machinery continues at a substantial pace.

The international conference, rotordynamics '92, has drawn together the ideas and practices of engineers concerned with rotating machinery, from throughout the world. The topics discussed included balancing, torsion, cracked structures, bladed systems, seals, and bearings, as well as methods used in analysis and monitoring. Reference to practical experience in the operation of machinery was also made. This book is a record of the contributions to the conference, made by experts from throughout the world. It should provide stimulating reading to all concerned with rotating machinery.

Happy reading!

Michael Goodwin

Proceedings of the International Conference on Rotating Machine Dynamics
ROTORDYNAMICS '92
held at the Hotel des Bains, Venice, 28–30 April 1992

Organised by Staffordshire Polytechnic

List of Papers Presented

List of Contributors

R. J. Alfredson
Monash University
Clayton
Victoria
Australia

P. E. Allaire
University of Virginia
U.S.A.

D. K. Anand
University of Maryland at College Park
Department of Mechanical Engineering
College Park Campus
Maryland, 20742
U.S.A.

N. Bachschmid
Dept Mechanical Engineering
Poltecnica de Milano
Piazza Leonardo da Vinci 32
20132 Milano
Italy

R. W. Baines
Staffordshire Polytechnic
Beaconside
Stafford ST18 0AD
U.K.

A. D. S. Barr
University of Aberdeen
U.K.

H. L. Berger
Siemens Engineering
Besuchsadresse
Wiesenstraße, 35
Germany

L. Biao
Senior Engineer
Xuzhou Mine Equipment Plant
Xuzhou Jiangsu
221007, China

T. J. Bievenue
G E Schenectady NY
U.S.A.

R. Bogacz
Inst. Fundamental Tech. Research
Warsaw
Poland

B. Bou-Said
Laboratoire de Macanique des Contacts
INSA
Villeurbanne
France

R. D. Brown
Heriot Watt University
Department of Mechanical Engineering
James Nasmyth Building
Ricarton
Edinburgh EH14 4AS
U.K.

J. S. Burdess
University of Newcastle-upon-Tyne
U.K.

M. D. Butler
Staffordshire Polytechnic
MCAE
Beaconside
Stafford ST18 0AD
U.K.

R. Cardinali
University of Campines
Brazil

K. L. Cavalca
Politecnico di Milano
Italy

G. M. Chapman
School of Engineering and Manufacture
Leicester Polytechnic
PO Box 143
Leicester LE1 9BH
U.K.

F. Cheli
Politechnico di Milano
Italy

Z. Chen
Group 405
Department of Propulsion
Beijing University of Aero & Astro
Beijin 100083
China

A. Collina
Dept Mechanical Engineering
Poltecnica de Milano
Piazza Leonardo da Vinci 32
20132 Milano
Italy

I. A. Craighead
University of Strathclyde
Department of Mechanical Engineering
75 Montrose Street
Glasgow, GL 1XJ
U.K.

A. Curami
Politecnico di Milano
Italy

S. Daixia
Thermal Power Research Institute
Ministry of Energy
China

M. S. Darlow
Technion
Israel Institute of Technology
Faculty of Mechanical Engineering
Technion City
Haifa 3200
Israel

H. R. T. de Azevedo
Cepel
Caixa Postal 2754
Rio de Janero
Brazil

G. Diana
Politecnico de Milan
Italy

A. D. Dimarogonas
Department of Mechanical Engineering
School of Engineering and Applied
Science
Washington University in St Louis
MO 63130
U.S.A.

J. Ding
University of Melbourne
Australia

Dong Ling
1st Department of Mechanical
Engineering
Chong ging University
Shi Chuan
China

R. A. Driver
NEI Parsons
Heaton Works
Newcastle-upon-Tyne
U.K.

K. L. Fan
Department of Mechanical and
Computer-Aided Engineering
Staffordshire Polytechnic
Beaconside
Stafford ST18 0AD
U.K.

Y. Fang
Department of Mechanical and
Computer-Aided Engineering
Staffordshire Polytechnic
Beaconside
Stafford ST18 0AD
U.K.

B. Fantino
INSA LYON
20 Avenue Albert Enstein
69621 Villeubranne Cedex
France

Fang Feng-Zhou
Dept Mechanical Engineering
HeiLongjang Institute
Jixi City HeiLongJiang
China

C. Frigeri
ENEL Automatica Research Centre
Cologno
Monzese
Italy

D. Gonsalves
University of Aberdeen
Department of Engineering
Kings College
Aberdeen AB9 2UE
U.K.

M. J. Goodwin
Staffordshire Polytechnic
Beaconside
Stafford ST18 OAD
U.K.

I. Green
Georgia Institute of Technology
Mechanical Engineering
Atlanta
Georgia 30332
U.S.A.

Z. Guanghui
Chongqing University
Congquing 63004
China

L. He-Sheng
Dept of Mechanical Engineering
Zhejiang University
Hangzhol
Zhejiang 310027
China

C. Heyou
Shijiazhuang Railway Institute
Shijiazhuang City
China

D. P. W. Horrigan
Department of Mechanical and
Manufacturing Engineering
University of Melbourne
Parkville, Victoria
3052 Australia

L. Hui
Chongqing University
Chongqing 63004
PR China

R. R. Humphris
University of Virginia
U.S.A.

H. Irretier
Inst. of Mechanics
Kassel University
Germany

H. Jiaju
The Gansu HUAYU Automatic
Water Feeder Plant
11 Suenjiati Qilihe District
Lanzhou, China

Z. Jian-Yuan
Department of Power Engineering
Wuhan University of Water
Transportation Engineering
Wuhan Hubel 430063
China

L. Kai
Shaanxi Institute of Mechanical
Engineering 305
Shaanxi
Xi'an
710048 China

J. A. Kirk
University of Maryland
College Park
Maryland 20742
U.S.A.

R. G. Kirk
Mechanical Engineering Department
Virginia Polytechnic Institute and State
University
Blacksburg
Virginia 24061
U.S.A.

C. R. Knospe
University of Virginia
United States of America

A. Kollias
School of Engineering and Applied
Science
Washington University in St. Louis
U.S.A.

S. Kouidri
ENSAM
Boulevard de l'Hopital
Paris
France

J. M. Krodkiewski
The University of Melbourne
Parkville
Victoria 3052
Australia

S. T. Kulig
Siemens AG
Mulheim a d Ruhr
Germany

G. L. Lapini
CISE
SpA
Segrate
Italy

A. S. Lee
Georgia Inst. Tech
Atlanta
Georgia
U.S.A.

P. S. Leung
Newcastle Polytechnic
Faculty of Engineering Science &
Technology
Ellison Building
Newcastle-upon-Tyne NE1 8ST
U.K.

L. K. Lim
Department of Mechanical and
Computer-Aided Engineering
Staffordshire Polytechnic
Beaconside
Stafford ST18 0AD
U.K.

R. P. Lisner
Monash University
Clayton
Victoria
Australia

Huang Ya Luo
Central China Electric Power
Administration
China

Z. H. Luo
Shanghai Jiao Tong University
PR China

G. Marenco
Pompe Gabbioneta
Milano
Italy

E. H. Maslen
University of Virginia
U.S.A.

F. Massouh
ENSAM
Boulevard de l'Hopital
Paris
France

J. F. Mayer
Universitat Stuttgart
7 Stuttgart 80
(Vaihingen)
Pfaffenwaldring 6
Germany

A. V. Metcalfe
University of Newcastle-upon-Tyne
Department of Engineering and
Mathematics
Stephenson Building
Newcastle-upon-Tyne NE1 7RU
U.K.

S. Michimura
Takushoku University
Tokyo
Japan

D. Muster
Department of Mechanical Engineering
University of Houston
Houston
Texas 77204–4792
U.S.A.

S. Narayanan
Indian Inst. of Tech.
Madras
India

R. D. Neilson
Department of Engineering
University of Aberdeen
King's College
Aberdeen AB9 2UE
U.K.

D. Nelias
Laboratoire de Mecanique des
Contacts
INSA
Villeurbanne
France

R. Noguera
ENSAM
151 boulevard de l'Hopital
75013 Paris
France

R. Nordmann
University of Kaiserslautern
Germany

P. J. Ogrodnik
Staffordshire Polytechnic
Beaconside
Stafford ST18 0AD
U.K.

A. Okada
Nippon Vulkan Co Ltd
Japan

Z. A. Parszewski
University of Melbourne
Australia

E. C. Pawtowski
Virginia Polytechnic Inst. and State
University
U.S.A.

J. E. T. Penny
Aston University
Birmingham
U.K.

F. Petrone
Instituto di Macchine
Uno
deli studi di Catania
Italy

B. Pizzigoni
University of Pavia
Italy

A. Rajamani
Vibration Laboratory
Corporate Research &
Development Research
Vikasnagar
Hyderabad
500 593 India

K. Ramakrishna
Vibration Laboratory
BHEL
Vikasnager
Hyderabad
India

R. Rey
ENSAM
Boulevard de l'Hopital
Paris
France

D. D. L. Risk
Staffordshire Polytechnic
Beaconside
Stafford ST18 0AD
U.K.

M. P. Roach
Department of Mechanical and
Computer-Aided Engineering
Staffordshire Polytechnic
Stafford
U.K.

E. Rodriquez
Goddard Space Flight Center
Greenbelt
Maryland
U.S.A.

M. Scali
Pompe Gabioneta
Viale Casiraghi 68
20099 Sesto S
Giovanni
Milano, Italy

H. Stetter
University of Stuttgart
Germany

Y. Suzuki
Takushoku University
Tokyo
Japan

E. Swain
Loughborough University
U.K.

T. Szolc
Polic Academy of Sciences
Institute of Fundamental Technical
Research
Swietokrzyska 21; 00 049
Warsaw
Poland

A. Tamura
Department of Mechanical Systems
Engineering
Takushoku University
815-1 Tate-machi hachioji
Tokyo
193 Japan

Lin Tan
Theory of Lubrication and Bearing
Institute
Xi'an Jiaotong Unversity
Xi'an
710049 China

E. Tanzi
Politecnico di Milano
Italy

G. Thieleke
ITSM
Pfattenwaldring 6
Postfach 80 11 40
7000 Stuttgart 80
Germany

M. Tonsi
ENEL Automatica Research Centre
Cologno
Monzese
Italy

M. Typrin
Virginia Polytechnic Inst. and State
University
U.S.A.

A. Vallini
ENEL DPT-VDT-STE
Pisa
Italy

Y. Wang
Beijing Satellite Institute of
Environmental Engineering
PO Box 785
Beijing 100029
China

X. Wang
Loughborough University
U.K.

Z. H. Wang
Department of Mechanical and
Production
Shandong Polytechnic University
Jinan
Shandong
China

J. Wauer
University of Karlsruhe
Mechanical Department
7500 Karlsruhe 1
Kaiserstraße 12
Postfach 6380 Germany

D. Weber
University of Campinas
Geprom – laboratorio de Projecto
Mecanico
PO Box 6051
13081 Campinas Brazil

H. I. Weber
Campinas State University
Sao Paulo
Brazil

J. Wei
Northwest Institute of Textile Science
and Technology
Department of Mechanical Engineering
Jin Hua Road
Xi'an 710048
China

M. F. White
NTH
Department of Marine Technology/
Engineering
The University of Trondheim
The Norwegian Institute of Technology
N 7034 Trondheim
Norway

T. S. Wilkinson
Parsons Turbine Generators
NEI Parsons Limited
Heaton Works
Newcastle-upon-Tyne NE6 2YL
U.K.

T. S. Wilkinson
NEI Parsons
Heaton Works
Newcastle-upon-Tyne
U.K.

J. F. Williams
University of Melbourne
Australia

S. B. Xia
Department of Power Engineering
Harbin Institute of Technology
Harbin 150006
China

W. Xinhua
Department of Power Engineering
Harbin Institute of Technology
Harbin 150006
China

H. Xiuzhu
Thermal Power Research Institute
Ministry of Energy
China

Z. Xueyan
Thermal Power Research Institute
Ministry of Energy
China

N. Yamagishi
Nigata Converter Co Ltd
Japan

S. Yanabe
Nagaoka Univ. of Technology
Dept. of Mechanical Engineering
Nagaoka-shi Kamitomioka
940-21
Japan

M. Yang
Loughborough University
U.K.

L. Yu-bin
Liao Yang Pharmacy Machinery
Plant 2
Section 1 Sheng Li Road
Liao Yang City
Liao Ning
China

G. A. Zanetta
ENEL Automatica Research Centre
Cologno
Monzese
Italy

M. Zhao
Vibration Shock and Noise Institute
Shanghai Jiao Tong University
1954 Hua Shan Road
Shanghai 200030
China

Z. X. Zhao
Staffordshire Polytechnic
Beaconside
Stafford ST18 0AD
U.K.

W. Zhihong
Shandong Polytechnic University
Jinan
China

Z.W. Zhong
Materials Fabrication Laboratory
RIKEN
The Institute of Physical and Chemical
Research
Wako Saitama 351-01
Japan

M. Zippo
CISE SpA
Segrate
Italy

R. Zmood
RMIT
Melbourne
Australia

A Brief History of Rotor Dynamics

A.D. Dimarogonas

School of Engineering and Applied Science, Washington University in St Louis

ABSTRACT

Rotor Dynamics was initiated in the last quarter of the 19th Century due to the problems associated with the high speed turbine of Gustaf de Laval who invented the elastically supported rotor, called de Laval Rotor, and observed its supercritical operation. Foeppl explained analytically the dynamic behavior of the de Laval rotor and Stodola pioneered the turbomachinery rotor dynamics studies. In the 1920's there was substantial development in the design of turbomachinery and most of the present-day rotor dynamics problems were identified.

THE ORIGINS OF VIBRATION THEORY

The development of vibration theory, as a subdivision of mechanics, came as a natural result of the development of the basic sciences it draws from, mathematics and mechanics. These sciences were founded in the middle of the first millennium B.C. by the ancient Greek philosophers. Of course, people were using the underlying principles in their every-day life long before that, sometimes in a systematic way. Sumerians and Egyptians, for example, have developed primitive arithmetic and geometry to deal with needs of the trade and the ownership of the land.

The term "vibration" was used from the Aeschylos times (Lidell & Scott 1879). Pythagoras of Samos (ca. 570-497 BC) conducted several vibration experiments with hammers, strings, pipes and shells. He established the first vibration research laboratory, the first known man-made research laboratory. Moreover, he invented the monochord (Theon of Smyrna, 2nd Century AD), a purely scientific instrument to conduct experimental research in the vibrations of taut strings and to set a standard for vibration measurements.

That for a (linear) system there are frequencies at which the system can perform harmonic motion was known to musicians but it was stated as a law of nature for vibrating systems by Pythagoras. Moreover, he proved with his hammer experiments that natural frequencies are system properties and do not depend on the magnitude of the excitation (Dimarogonas 1990, Dimarogonas, Haddad 1992).

It is generally believed that the wheel was developed in Mesopotamia at about 4,500 BC. The first wheels, for tranportation or for pottery making, were heavy and rigid. For transportation on flat and sandy lands, the heavy wheels worked well. The low operating speeds, resulted in subcritical operation. Therefore, dynamics of the shaft-wheel system was of no importance. Later-on, when fast chariots were used in upper Egypt, Greece and Rome, where the ground is hard and rough, the wheels and the supporting shafts became light and flexible to obtain high vibration isolation and lubricants were used to minimize friction and allow for higher speeds. In Homer's "Odyssey" Telemachos parked his chariot vertically against a wall to avoid creep of the wheels and the shaft due to its very flexible design that apparently resulted in very high stresses and creep.

Some kinematic aspects of rotor dynamics were observed very early. Aristotles

(384-322 BC who also developed a primitive set of laws of motion similar with Newton's Laws) in his "Mechanical Problems" discused several kinematic problems of the wheel motion. Diodorus the Cicilian (first Century BC) described the whirling motion of the shaft in the bearing clearance. Astronomers Hipparchos (second sentury BC) and Ptolemeos (second century AD) observed the whirl and precession of heavenly bodies.

In the ancient world, there was substantial progress in vibration theory and an extended understanding of the basic principles of natural frequency, vibration isolation, vibration measurements, resonance and sympathetic vibrations (Dimarogonas 1990). This body of knowledge had very limited use in engineering, however, due to the low level of production technology and machinery speeds. Moreover, many branches of mathematics were already extensively developed but calculus and mechanics were at their very early stages to allow for analytical treatment of vibration.

THE ERA OF CONTINUUM MECHANICS.

The beginning of this Era is marked with the works of Galileo and Newton. Moreover, it is marked with the early stages of mechanization and the industrial revolution. The utilization of chemical energy with the associated high power per unit machinery volume, introduced numerous vibration problems. Together with the development of calculus and continuous mechanics led to the rapid development of vibration theory by the mid-19th century. At the time of Galileo and Newton, Physics and Mechanics were much more developed than it is generally assumed. Their fundamental contribution is that they have revived and redefined these sciences at times that were demanding progress in natural science. There were several previous attempts for revival of physics and mechanics in the first half of the second millennium AD. The time, however, was not yet appropriate.

As stated above, extensive experimental results were available for the vibrating strings since the Pythagoras times and further results were obtained by Galileo and Marinus Mersenne (1588-1648), a Franciscan Friar. The vibration modes and the nodal points were observed by J. Sauver (1653-1716), who identified also the *fundamental natural frequency* and the *harmonic tones*. Daniel Bernoulli (1750) explained the experimental results by the *principle of superposition* of the harmonics and introduced the idea of expressing the response as a sum of the simple harmonics. The problem of the vibrating string was solved mathematically first by Lagrange (1759) considering it as sequence of small masses.

The wave equation was introduced by D' Alembert in a memoir to the Berlin Academy (1750). He used it in his memoir also for longitudinal vibration of air columns in pipe organs. Experimental results for the same problem were obtained by Pythagoras.

The solution of the string equation is due to Daniel Bernoulli, D' Alembert and Euler, though Joseph Sauveur (1653-1716) and Brook Taylor (1685-1731) have previously obtained approximate solutions.

Euler (1744) obtained the differential equation for the lateral vibration of bars and he determined the functions that we now call *normal functions* and the equation that we now call *frequency equation* for beams with free, clamped or simply supported ends, while Daniel Bernoulli supplied him with experimental verification. Chladni (1802) investigated experimentally these vibration problems and also longitudinal and torsional vibration of bars.

Euler and James Bernoulli attempted to solve the problem of vibrating plates and shells analytically. Euler (1767) considered an elastic membrane to consist of two systems of stretched strings perpendicular to each other and he obtained the membrane differential equation. Jacques Bernoulli (1759-1789) obtained the differential equation of the vibrating plate considering it as consisting of two systems of beams perpendicular to each other.

Chladni further investigated experimentally plates and shells, bells in particular. His work stirred great interest on the subject. The French Academy, at Napoleon's suggestion who was impressed by a demonstration of Chladni's experiments, proposed as a subject of a prize the investigation of vibrating plates. Sophie Germain (1776-1831) won the prize after a lengthy procedure and controversy (Timoshenko 1953). She derived the correct differential equation but there was concern with the rigor of the proof and the correctness of the boundary conditions. Further improvement was made by Poisson and Kirchhoff but it was Navier (1821) who gave a rigorous theory of bending vibration of plates. He further investigated the general equations of equilibrium and vibration of elastic solids. He formed an expression for the work done in a small relative displacement by all forces and obtained the differential equations by way of the calculus of variations.

The solution of the differential equations of motion for an elastic solid was treated by Poisson (1829) and Clebsch (1862) who founded the general theory of vibrations. The theory of vibration of thin rods was brought under the general equations of vibration of elastic solids by Poisson (1829), in particular the theory of torsional vibration.

The first systematic treatise on vibration was written by Lord Rayleigh (Lord Rayleigh, 1889). He formalized the idea of normal functions, as introduced by Daniel Bernoulli and Clebsch, and introduced the ideas of generalized forces and generalized coordinates. He further introduced systematically the energy and approximate methods in vibration analysis, without solving differential equations. This idea was further developed by W. Ritz (1909). Rayleigh introduced a correction to the lateral vibration of beams due to rotatory inertia and Timoshenko (1916) the correction due to shear deformation.

THE DE LAVAL ROTOR.

At the end of the 19th century. the theory of vibration was very extensively developed. At the same time, there was rapid progress in machinery building, in particular the development of locomotives and steam turbines. The rotating speed of shaft since the invention of the wheel was generally below 1000 RPM. In the 1870's Dr. Gustaf Patric de Laval, a sweedish engineer, invented the milk separator that had to work at 6,000 to 10,000 RPM. De Laval first units were hand- or horse-driven with geared step-up of the speed. He soon saw the need for higher speeds and the steam turbine was born. He was experimenting with 30,000 to 42,000 RPM turbine rotors and he had to overcome the problem of incomplete balancing that, at these speeds, would lead to prohibitively high centrifugal forces with stiff shafts. He then invented the flexible shaft with low critical speed (7 times lower than the speed of rotation in his turbines). He noticed that he could accelerate through the critical speed, and that the operation at speeds way above the critical was very smooth.

Whirling of shafts was studied first by W.A. Rankine (1869). He made the assumption that when the centrifugal force due to a virtual deflection of the shaft is less than the restoring force there is no shaft whirling. This reasoning led him to a limit of speed above which the operation will be unstable. This speed is what later was called critical speed and his conclusion was that operation was possible only below that speed. He computed the limit speed for a simply-supported and for an overhang shaft using appropriate beam formulas.

Extensive analytical and experimental investigations for more complex shafts were reported by Greenhill (1883) and Dunkerley (1894) who mentioned in the paper that the analysis was given by O. Reynolds. Dunkerley found the natural frequencies, which he called *critical speeds*, for one mass rotor on different configuration shafts or for continuous shafts of various geometries and boundary conditions. He noticed that for compound shafts he could compute the compound

4

shaft natural frequency using the natural frequencies computed with the different single or continuous masses with his formula which he found intuitively. Thus he developed a rational numerical method to compute the lowest natural frequency of compound engineering rotors. His formula was proved theoretically later-on by Jeffcott (1919). Further discussion on the critical speeds and the Dunkerley method for compound shafts was presented by Chree (1904) who also introduced the gyroscopic effects due to large diameter disks. Morley (1909) introduced engineering approximations for the computation of critical speeds of compound rotors and introduced the Rayleigh's correction for the rotatory inertia. He seems to be the first to apply the Rayleigh's method for the computation of the critical speeds of a stepped rotor. The same method was further applied by Kerr (1916). In the same paper there is a remarkable discussion where Stodola presented an analysis for the critical speed (approximately half the critical speed) due to the rotor's weight and the Coriolis term and the equations of motion in a rotating coordinate system. More remarkable in this paper is the extensive discussion on critical speeds by Chree, Jeffcott, Stodola, Carter and Barkley. Tha latter discusser presented a surprizing summary of rotor dynamic problems, such as bearing clearance effects, bearing and coupling misalignment, looseness of rotor parts, internal damping, and thermal strains.

MODERN ROTOR DYNAMICS

The whirling problem was solved analytically by A. Foeppl (1895) who was able to explain the quiet operation of the de Laval rotor above the critical speed with the inversion of the dynamic amplitude. An important question is related with the extent that de Laval used analytical methods since he seems to have encountered and solved most of the rotor dynamics problems that were studied later. Stodola in his classic work "Steam and Gas Turbines" (1927) indicated that the rotor dynamics analyses of Rankine, Dunkerley, Reynolds and Foeppl were developed independently by de Laval. It is quite possible, in view of the successful solution of the rotor dynamics problems by de Laval, that he has developed many of the analyses reported later but he published very little on it, apparently considering the information as proprietary.

Stodola (1916) introduced bearing damping in rotor dynamics. He proved that damping limits the amplitude due to the unbalance at the critical speed, observed the decrease of the critical speed due to damping and computed the phase angle with bearing damping. The same author (1918) introduced the analysis for the rotor acceleration through the critical speed.

Prandtl (1918) used the equations of motion of the rotor in a rotating coordinate system and observed the critical speed due to rotor's weight, explaining observations of Guembel (1917, 1918), Foeppl (1916, 1918) and Stodola (1917, 1918).

Balancing of rotors was an early concern (Stodola 1904, 24, Akimoff 1918). Multiplane balancing analyses were introduced by Bishop and Gladwell (1959) and Goodman (1964).

Jeffcott (1919) introduced a damping constant into the Foeppl's analysis to explain the finite vibration amplitude at the critical speed. This paper caused some confusion and the de Laval rotor, named after its inventor by Foeppl and Stodola, is called sometimes "Jeffcott Rotor". This started with a state of the art syrvey in a report by Rieger (1965) where the author, apparently unaware of the relevant German literature, stated that Jeffcott resolved the problem of limiting the amplitude at the critical speed introducing viscous damping in the bearings. However, the damping (already discussed extensively by Stodola in 1916) introduced in Jeffcott's paper is unspecified external viscous damping and otherwise the analysis is identical with Foeppl's. Further, Gunther (1966) in his dissertation (also published as a NASA report) used the term "Jeffcott Rotor" on the basis of the above reference and this became a standard designation in USA. It is interesting that Jeffcott himself used the term "de Laval turbine rotor" in his paper.

Rotor stability problems were identified very early. A. Hurwitz in his important paper (1895) on the stability of linear systems expressed by differential equations with complex coefficients reported that the paper was written in conjunction with A. Stodola. It must be noticed, incidentally, that in applying Hurwitz stability criterion the original reference should be consulted because it has been wrongly applied in the literarture.

The first reports on self-excited rotor instability appeared in the 1920's. At that time the turbine industry designed machines to operate at substantially higher loads and at speeds above the lowest critical speed and thus the modern-day rotor dynamics problems appeared. Some of the earliest works of the great mechanicians of this century, such as Prandtl, A. Foeppl, Stodola, von Karman, Hohenemser, Prager, were during that period when most of the important vibration concepts have been identified.

Newkirk (1924) presented an analysis of shaft whipping, meaning unstable vibration due to different effects, such as interface friction, internal damping in the shaft material, bearings and flexible pedestals. Kimball (1924, 1925) presented further analytical and experimental results for internal and interface damping which cause instability at speeds higher than the critical speed and can be eliminated by sufficient external damping. Dimentberg (1959) and Tondl (1965) applied non-linear models for internal and interface damping.

The influence of fluid bearings on rotor behaviour was introduced by A. Stodola (1916 and 1925), B.L. Newkirk and H.D. Taylor (1925) who observed the "oil whip" phenomenon, instability due to the oil motion in the bearing clearance at shaft speeds two-times the critical speed or higher. Robertson (1933) introduced the Reynold's equation for the bearings into the rotor motion. Further work on the modelling and linearization of bearings as they affect the rotor's behavior was reported by Haag & Sankey (1958) and Sternlicht (1959).

Rotor instability due to thermal strains caused by rubbing was observed by Newkirk (1926) and Kroon (1940). Dimarogonas (1973) developed an analytical model to find that stability exists, generally, for speeds higher than the critical speed. The effect of dry friction forces, which cause whirl at the rotor's natural frequency, was studied by Taylor (1924), Billet (1965), Black (1968) and Ehrich (1969).

The response and stability of shafts with dissimilar stiffness in two perpendicular lateral directions were studied by Stodola (1927) and Soderberg (1931) who identified the unstable region and the subcritical speed due to the rotor's weight in horizontal shafts. The problem is discussed in detail in the Dimentberg (1959) and Tondl (1965) monographs. Electromagnetic effects were introduced by Freise and Jordan (1962).

Instability problems due to the flow though seals in turbomachinery appeared in the 1940's, termed "steam whirl". Thomas (1956) explained the mechanism of instability and developed a stability criterion on the basis of the leakage excitation alone and general damping, which was subsequently applied by Alford (1965) and for this reason it is misnamed by some authors as "Alford's effect". Dimarogonas (1970) introduced circumferential flow effects and bearing linear and nonlinear properties and modified accordingly the Thomas Stability Criterion. Kollman (1962) and Ehrich (1967) studied the effect of trapped fluids inside rotors which, under certain conditions, can initiate violent whirl.

Dynamic response of cracked rotors was identified by Dimarogonas (1970, 1976) who used the elastic hinge modeling to observe that the response is similar with the rotors with dissimilar moment of inertia. Pafelias (1974) reported on an extensive numerical and experimental investigation. Gash (1976), Henry and Okah-Avae (1976) and Mayes and Davis (1976) presented applications to turbomachinery rotors and Dimarogonas and Paipetis (1983) observed the coupling of bending, longitudinal and torsional vibration of cracked rotors.

Vibration of shafts and beams of multistage rotor shapes was studied first by Frahm (1902), in particular, torsional vibration of ship main shafts. Frahm's method was developed in tabular form by Holzer (1907), Guembel (1912) and Tolle (1921). For lateral vibration of bars, the HGT method was developed by van den Dungen (1928) for lumped mass beams and by Hohenemser and Prager (1933) for continuous but discretized beams (Dimarogonas & Haddad 1992) though it is generally misnamed as *Myklestad-Prohl method* (Myklestad 1944, Prohl 1945). This points at an interesting aspect of the development of rotor dynamics: In the 1920's and 30's electric power in USA was in great demand due to the rapidly developing industry and the turbine manufacturers hired engineers from Europe, notable among them Timoshenko, den Hartog and Myklestad who were all hired by Westinghouse. N.O. Myklestad, a Norwegian, attended University in Copenhagen in the late 30's and most probably was familiar with van den Dungen's work that was the standard vibration reference in Europe at that time (Hohenemser, personal communication). Prohl, who most probably was unaware of van den Dungen's work, stated in his paper that the method was outlined in graphical form for him by H. Poritsky, who had a similar background as Myklestad. Prohl's method is essentially identical with the van den Dungen's method but it was the first application of machine computations in rotor dynamics and had very substantial impact on the turbomachinery design in USA. The van den Dungen method was further developed in matrix form by W. Thomson (1950) and was called *transfer matrix method*. This method was applied by Landzberg (1960) for a wide range of dynamic response and stability of rotors. The finite element method was applied to rotor dynamics by Dimarogonas (1970, 1975), Ruhl and Booker (1971) and Nelson and McVaughn (1976). Meirovitch (1974, 1976) has developed the basis for the critical speed computation is discretized rotors.

ROTOR DYNAMICS TEXTS AND MONOGRAPHS.

The first systematic treatise on vibration was part of Lord Rayleigh's *Theory of Sound* or parts of Mechanics textbooks, such as the classic book of August Foeppl. The first engineering vibration textbook was written in 1910 by Hort, an engineer at Siemens-Schuckert in Berlin. This book had a profound effect on vibration teaching since it established vibration as a separate field of mechanics and a course in Mechanical and Electrical Engineering. Moreover, it became a standard reference, though very seldom referenced, for all subsequent vibration textbooks and its outline became the standard for vibration textbooks up to this day.

Hort's book was followed by several vibration textbooks in the 1920's. Short thereafter Timoshenko (1928) and den Hartog (1932) both students of Pandtl at a time, wrote their landmark vibration textbooks. These textbooks were not particularly original but they had a profound influence in both teaching and research on the subject in USA.

Extensive discussions on rotor dynamics appear in Stodola's "Steam and Gas Turbines" (1902 - 1927) and Biezeno and Gammel's "Technische Dynamik" (1953). The first rotor dynamics monograph was written by Dimentberg (1959) who presented an extensive discussion of most rotor dynamic problems. He was followed by Tondl (1965) who focused mainly on his pioneer work on internal damping.

An extensive literature on rotor dynamics was presented in the Loewy and Piarulli monograph (1971). Gash and Pfuetzner (1975) wrote the first rotor dynamics textbook. Kelson, Cymanskii and Yakovlev (1982) wrote a monograph (in Russian) on the dynamics of rotor-bearing systems. Dimarogonas and Paipetis (1983) presented in their monograph analytical methods for most current rotor dynamic problems. Rao (1983) wrote an extensive monograph with most of the classical topics on rotor dynamics. The state of the art is given in the most recent monographs. Vance (1988) focused on turbomachinery problems, Goodwin (1989) presented an extensive treatment of both classical and contemporary subjects on rotor and bearing dynamics, such as identification and diagnostics. Lalanne and Ferraris (1990) focused on industrial and computer applications. An extensive analytical treatment is presented in a recent monograph (in chinese) by Zhang (1990).

References

Akimoff B (1918) Balancing Apparatus. Trans ASME 39:779.

Aristoteles (384-322 BC) On Acoustics, Oxford Edition. Clarendon Press, Oxford.

Alford JS (1965) Protecting Turbomachinery from Self-Excited Rotor Whirl, J. Eng. for Power, 333-344.

Beck T (1900) Beitraege zur Geschichte des Maschinenbaues. Julius Springer, Berlin.

Bernoulli D (1750) Reflections et Eclaircissemens sur les Nouvelles Vibrations des Cordes Exposes dans les Me'moires de l' Academie de 1747 et 1748, Royal Academie of Berlin, p. 147.

Bernoulli D (1751) De vibrationibus...laminarum elasticarum, Commentarii Academiae Scientarum Imperialis Petropolitanae, vol. 13.

Biezeno CB, Grammel R (1953) Technische Dynamik. Springer Verlag, Berlin.

Billet RA (1965) Shaft Whirl Induced by Dry Friction, The Engineer, 220,5277:713:714.

Bishop RED, Gladwell GML (1959) The Vibration and Balancing of a Flexible Rotor, J Mech Eng Sci, 1:66-67.

Black HF (1968) Interaction of a Whirling Rotor with a Vibrating Stator across a Clearance Annulus, J. Mech. Eng. Sci., 10,1.

Boethius (AD 480-524) In: Lindsay RB (1972) Acoustics: Historical and Philosophical Development. Stroudsburg, Pa: Dowden, Hutchinson & Ross. Concerning the Principles of Music.

Boyer CB (1968) A History of Mathematics. Princeton University Press, Princeton N.J.

Chladni EFF (1802) Die Akoustik, Leipzig.

D' Alembert J (1759) Rechersches sur la nature et la propagation du son. Miscellanea Taurinensia, v. 1.

Den Hartog JP (1952) 4th ed. Mechanical Vibration. Mc Graw-Hill, New York.

Dimarogonas AD (1970) Analysis of Steam Whirl. General Electric Technical Information Series, DF70LS48, Schenectady, NY.

Dimarogonas AD (1970) Dynamic Response of Cracked Rotors. General Electric Technical Information Series, DF70LS86, Schenectady, NY.

Dimarogonas AD (1973) Newkirk Effect: Thermally Induced Dynamic Instability of High Speed Rotors, ASME Gas Turbine Conference, paper 73-GT-26, Washington, D.C.

Dimarogonas AD (1975) A General Method for Stability Analysis of Rotating Shafts, Ingenieur Archiv, 44,9-20.

Dimarogonas AD (1976) Vibration Engineering. West Publ, St. Paul.

Dimarogonas AD, Haddad SD (1992) Vibration for Engineers. Prentice Hall, Englewood Cliffs, NJ.

Dimarogonas AD (1978) Lectures in History of Technology (in Greek), 2 vol. Patras University Press.

Dimarogonas AD, Paipetis SA (1983) Analytical Methods in Rotor Dynamics. Elsevier-Applied Science Publishers, London.

Dimarogonas AD (1990) The Origins of Vibration Theory, Journal of Sound and Vibration, 140,2, 181-189.

Dimentberg F (1959) Flexural Vibrations of Rotating Shafts. English Translation (1961) Butterworths, London.

Dungen MF-H van den (1926) Cours de technique des vibrations. Editions de la Revue de l' Ecole Polytechnique.

Dungen MF-H van den (1928) Les Problemes Genereaux de la Technique des

Vibrations. Mem. Sci. Phys. L' Academie des Sciences. Gauthier-Villars, Paris.

Dunkerley S, Reynolds O (1883) On the Whirling and Vibration of Shafts. Phil. Trans. A, p. 279-359.

Eason AB (1923) The Prevention of Vibration and Noise. Oxford Technical Publications, London.

Ehrich FF (1967) The influence of Trapped Fluids on High Speed Rotor Vibration, ASME Jr Eng Ind, 89:806-812.

Ehrich FF (1969) The Dynamic Stability of Rotor/Stator Radial Rubs in Rotating Machinery. J. Eng. for Industry, 1025-1028.

Euler L (1744) Aditamentum, 'De curvis elasticis', in Methodus inveniendi lineas curvas maximi minimive proprietate gaudentes, Lausanne.

Euler L (1767) Novi Commentarii Academiae Scientarum Imperialis Petropolitanae, 10:243.

Foeppl A (1895) Das Problem der DeLaval'schen Turbinenvelle. Der Civilingenieur, 61:333-342.

Foeppl A (1899) Forlesungen ueber der Technischen Mechanik. Verlag von Julius Springer, Berlin.

Foeppl A (1904, 1933) Vorlesungen ueber Technische Mechanik, Band 4: Dynamik. Teubner, Leipzig.

Foeppl O (1923) Grundzuege der Technischen Schwingungslehre. Verlag von Julius Springer, Berlin.

Frahm, H (1902) VDI Zeitschrift, 797.

Freise W, Jordan H (1962) Einseitige magnetische Zugkraefte in Drehstrommaschinen, ETZ-A83:299-303

Galileo Galilei, 1638) Discorsi e demonstrationi matematiche intorno a due nuove scienze attenenti alla mechanica e ai movimenti locali, Translated by Crew H de, Salvio A (1939), Evanston and Chicago.

Gash, R (1976) Dynamic Behavior of a Simple Rotor with a Cross-Sectional Crack, IME Conf. Vibrations in Rotating Machinery, paper C178/76.

Gash R, Pfuetzner H (1975) Rotordynamik. Springer Verlag, Berlin.

Geiger J (1927) Mechanische Schwingungen und ihre Messe. Verlag Julius Springer, Berlin.

Goodman TP (1964) A Least Square Method for Computing Balance Corrections, ASME Jr Eng Ind 86B:273:279.

Goodwin MJ (1989) Dynamics of Rotor-Bearing Systems. Unwin Hyman, London.

Greenhill AG (1883) On the Strength of Shafting when Exposed to Torsion and to End Thrust, Proc. Inst. Mech. Eng April:182-225.

Guembel E (1912) Zeitschrift VDI, 56:1025

Gunther EJ (1966) Dynamic Stability of Rotor-Gearing Systems, NASA SP 113.

Hagg AC Sankey, GO (1958) Oil Film Properties for Unbalance Vibration Calculations. Trans. ASME, J. Appl. Mech 25:141-143.

Hartenberg RS, Denavit J (1964) Kinematic Synthesis of Linkages. New York:Mc Graw-Hill Book Co.

Henry T.A, Okah-Avae BE (1976) Vibrations of Cracked Shafts, IME Conf. Vibrations in Rotating Machinery, paper C162/76.

Hohenemser K, Praeger W (1933) Dynamic der Stabwerke. Verlag Julius Springer, Berlin.

Holzer H (1907) Schifbau, 8:823, 866, 904.

Hort W (1910, 1922) Technische Schwingungslehre. Verlag von Julius Springer, Berlin.

Housner GW, Hudson DE (1950) Applied Mechanics:Dynamics. New York:D. Van Norstand Co.

Hunt FV (1978) Origins in Acoustics. Yale University Press, New Haven & London.

Hurwitz A (1895) Ueber die Bedingungen, unter welchen eine Gleichung nur Wurzeln mit negativen reelen Teilen besitzt, Mathematische Annalen, 46:273 -284.

Jeffcott HH (1919) The Lateral Vibration of Whirling Shafts in the Neighborhood of a Whirling Speed: The Effect of Want of Balance, Phil. Mag. March:304-314.

Kelson AS, Cymanskii HP, Yakovlev BH (1982) Dynamics of Rotor-Bearing Systems. Nauka, Moskow.

Kimball AT (1924) Internal Friction Theory of Shaft Whirling, General Electric Review, 27, 4:244-251.

Kimball AT (1925) Internal Friction as a Cause of Shaft Whirling. Philosophical Magazine, Series 6, 49:724-727.

Klebsch A (1862) Theorie der Elasticitaet fester Koerper, Leipzig.

Klotter K (1960) Technische Schwingungslehre. Springer Verlag, Berlin.

Kollmann FG (1962) Experimentelle und Theoretische Untersuchungen Ueber die Critiscen Drehzahlen Fluessigkeitsgefuelter Hohlkoerper, Fors auf dem Geb des Ing, B,28:115-123, :147-153.

Kraemer, E (1984) Maschinendynamik. Springer Verlag, Berlin.

Kroon RP (1940) Erratic Vibration, a Case for the Specialist. Power, 66:212-214.

Lagrange J-L (1759) Recherches sur la Nature et la Propagation du Son, Miscellanea Taurinensia, v. I.

Lalane M, Ferraris G (1990) Rotordynamics Prediction in Engineering. John Wiley and Sons, New York.

Landzberg AH (1960) Stability of a Turbine-Generator Rotor. Types of steam and bearing excitations. Trans. ASME, Jr. Appl. Mech Ser. E, 27:410-416.

Lazan BJ (1968) Damping of Materials and Members in Structural Mechanics. Pergamon Press, Oxford.

Lewis F (1932) Vibrations during acceleration through the critical speed, Transactions ASME, 54:23.

Liddel WG, Scott R (1879) Greek-English Lexicon. Clarendon Press, Oxford.

Lindsey RB (1966) The story of acoustics. Jr of Acoust Soc of Am, 39(4):629-644.

Loewy R, Piarulli V (1969) Dynamics of Rotating Shafts, Shock and Vibration Monographs, 4.

Love AEH (1927) A Treatise on the Mathematical Theory of Elasticity. Dover Publications, New York.

Lund JW et al (1965) Design Handbook for Fluid Film Type Bearings, Rotor-Bearing Dynamics Design Technology, III. Mechanical Technology Inc, Technical Report AFAPL-TR-65-45 to Wright-Paterson Air Force Base, Ohio.

Mayes IW, Davis WGR (1976) The Vibrational Behaviour of a Cracked Shaft Containing a Transverse Crack, IME Conf. Vibrations in Rotating Machinery, paper C168/76.

Meirovitch L (1974) A New Method of Solution of the Eigenvalue Problem for Gyroscopic Systems, AIAA Jr, 12,10:1337-42.

Meirovitch L (1976) Modal Analysis for the Response of Linear Gyroscopic Systems, Jr Appl Mech 42,2:446-450.

Myklestadt NO (1944) A New Method of Calculating Natural Modes Of Uncoupled Bending Vibration of Airplane Wings and other Types of Beams. Jr of Aeron Sci:153-162.

Navier Claude-L-M-H (1821) Me'moir Academie des Sciences, Paris.

Needham J (1962) Science and Civilization in China, v. 4,I. Cambridge: At the University Press.

Nelson HD, McVaughn JM (1976) The Dynamics of Rotor-Bearing Systems Using Finite Elements, ASME J. Eng. Ind, 98.

Newkirk BL (1924) Shaft Whipping. General Electric Review, vol 27, 3:169-178.

Newkirk BL (1925) Shaft Whipping due to Oil Action in Journal Bearings. General Electric Review, 25, 8:559-568

Newkirk BL, Taylor HD (1925) Shaft Whipping due to Oil Action in Journal Bearings. General Electric Review, 25, 8:559-568.

Newkirk BL (1926) Shaft Rubbing, Mechanical Engineering, 48:830-834.

Pafelias TA (1974) Dynamic Behavior of a Cracked Rotor, General Electric Technical Information Series, DF74LS79, Schenectady, NY.

Pestel EC, Leckie FA (1963) Matrix Methods in Elastomechanics. Mc Graw-Hill Book Co, New York.

Poisson D (1829) Me'moire sur l' equilibre et le mouvement des corps elastiques, Paris, Mem. de l' Academie des Sciences, 8.

Pontryagin LS (1962) Ordinary Differential Equations. Addison-Wesley, Reading.

Prohl M (1945) A General Method for Calculating Critical Speeds of Flexible Rotors, Journal of Applied Mechanics, 12:142.

Rao JS (1983) Rotordynamics.J. Wiley Eastern, New Delhi.

Rayleigh JWS (1894) Theory of Sound. Dover Publ. ((1946), New York.

Rankine WJM (1869) On the Centrifugal Force of Rotating Shaft. Engineer (London) 27:249.

Rieger NF (1965) Rotor-Bearing Dynamics Design Technology, Technical Report AFAPL-TR-65-45, Part I, Wright-Paterson AF Base, Ohio.

Ritz W (1909) Crelle's Journal, 85.

Robertson D (1933) Whirling of a Journal in a Sleeve Bearing, Phil Mag, Ser. 7, 15:113-130.

Ruhl RL, Booker JF (1971) A Finite Element Model for Distributed Parameter Turborotor Systems, ASME paper 71-Vibr-56.

Sandars NK (1968) Prehistoric Art in Europe. Harmondsworth, London.

Schneider E (1928) Mathematische Schwingungslehre, Julius Springer, Berlin.

Seireg A (1969) Mechanical Systems Analysis. Int Textbook Co, Scrandon.

Skudrzuk E (1954) Die Grundlagen der Akoustik. Springer Verlag, Wien.

Soderberg R (1931) On the Subcritical Speeds of the Rotating Shaft, ASME paper APM-54-4.

Sternlicht B (1959) Elastic and Damping Properties of Cylindrical Journal Bearings. Transactions ASME, Series D, 81:101-108

Stodola AB (1916) Neuere Beobachtungen ueber die kritischen Umlaufzahlen von Wellen. Schweiz. Bauzeitung, vol 68:210-214.

Stodola A (1918) Neue kritische Drehzahlen als Folge der Kreiselwirkung der Laeufraeder. Z zur Gesamten Turbinennwesen, 15:269-275

Stodola A (1904, 24) Dampf- und Gasturbinen. Verlag von Julius Springer, Berlin.

Stodola A (1925) Kritische Wellenstoerung infolge der Nachgiebkeit des Oelposters im Lager, Schweizerische Bauzeitung, 85:265.

Stodola A (1927) Steam and Gas Turbines. Mc Graw-Hill Book Co, New York.

Strutt RJ 3rd Baron Rayleigh (1877, 1878) The Theory of Sound. London. Dover Edition, New York (1945).

Szabo' I (1979) Geschichte der Mechanischen Principien. Birkhaeuser Verlag, Basel, Boston, Stuttgart.

Spectral Dynamics Corp (1990) Vibration Handbook.

Theon of Smyrna (2nd century AD) On Mathematical Matters Useful in Reading Platon, French Ed by Jean Dupuis (1892) Paris: Librairie Hachette et Cie.

Taylor HD (1924) Rubbing Shafts above and Below Resonant Speed, General Electric Technical Information Series, 16709.

Thomson W (1950) Matrix Solution for the Vibration of Non-uniform Beams, Journal of Applied Mechanics:337-339.

Thomas H-J (1956) Instabile Eigenschwingungen von Turbinenlaeufern, angefaecht durch Spaltroemungen. AEG-Sonderdruck 1150. Also, Bull. d' AIM, 71, (1958),1039-64.

Timoshenko S (1928) Vibration Problems in Engineering. London.

Tolle M (1921) Regelung der Kraftmachinen, Berlin.

Tondl A (1965) Some Problems of Rotor Dynamics. Chapman & Hall, London.

Timoshenko SP (1916) Philosophical Magazine, 41:744 and 43:125.

Timoshenko SP (1953) History of Strength of Materials. Mc Graw-Hill Book Co, New York.

Vance JM (1988) Rotordynamics of Turbomachinery. J. Wiley & Sons, Inc, New York.

Vitruvius (1rst century BC) De Architectura. In: Lindsey, R.B. (1972) Acoustics:Historical and Philosophical Development. Dowden, Hutchinson & Ross, Stroudsburg, Pa.

Zhang W. (1990) Theoretical Foundations of Rotor Dynamics, Sc Press, Beijing.

SESSION 1   BALANCING

Theoretical and Practical Aspects in the Field Balancing of Large Turbogenerator Rotors

C. Frigeri   M. Tonsi   G.A. Zanetta

ENEL, Automatica Research Centre, Cologno Monzese, MI Italy

ABSTRACT

The paper reviews two experiences of computer aided field balancing of large turbogenerator rotors.

In the first part, the principles of a multi-plane computer program, based on the influence coefficient method, are shortly discussed. Correction weights are found by means of a weighted least squares technique; an iterative procedure is also included, owing to the need to control the mass quantity and to accept different residual vibration levels at various locations and speeds. In addition, an alternative single plane graphic method and traditional graphic displays allow more insight about the problem.

In the second part, the on site balancing of two 320 MW units is illustrated. In both cases, theoretical critical speeds and mode shapes of the shaft line helped for the best choice of the balancing planes and computed influence coefficients were succesfully used to balance the two generators. A few practical aspects, determinant for the success of a balancing test on turbogenerator rotors, are focused; some remarks about the experimental behaviour of the two generators are finally given.

NOTATION

$\{V_r\}$ Residual Vibration Vector	$|V_{ri}|$ Amplitude of the ith Residual Vibration
$\{V_r\}^*$ Conjugate transposed of $\{Vr\}$	$|B_j|$ Amplitude of the jth Balancing Mass
$\{V_o\}$ Initial Vibration Vector	R_i Admissible value of the ith Residual Vibration
$\{B\}$ Balancing Mass Vector	M_j Admissible value of the jth Balancing Mass
$\{B\}^*$ Conjugate transposed of $\{B\}$	p_i ith Vibration Weighting Factor
$[C]$ Influence Coefficient Matrix	q_j jth Mass Weighting Factor
$[C]^*$ Conjugate transposed of $[C]$	N Number of Vibrations
$[P]$ Diagonal Matrix of Vibr. Weighting Factors	M Number of Balancing Planes
$[Q]$ Diagonal Matrix of Mass Weighting Factors	Ω Relaxation Factor

1.INTRODUCTION

The in-field balancing of large turbogenerator (TG) rotors has been effectively dealt with since long by means of the influence coefficient method (ICM), in conjunction with some kind of optimization technique. In general, these methods can take into account a larger number of vibrations than balancing planes, thus including vibrations at different locations, speeds and power levels. Moreover, when dealing with this kind of problems, it is requested to attach more or less importance to each vibration and, possibly, to keep the balancing masses under control. One such algorith m was also developped by ENEL and, by now, it has been in use for some years.

Traditional graphic displays and an additional graphic method for the mass computation with a single balancing plane were included in the program, in order to retain good physical insight about a balancing problem.

Should the true influence coefficients (IC) be available for a given TG shaft line, a balancing test would be a little more than a mere exercise. In fact, it is well known to each practitioner how this is still scarcely verified: systematic measurements have been performed since a comparatively short time and previous balancing tests were neither suitably stored nor clearly documented; trial runs made on purpose are too expensive; under many circumstances non repeatable or non linear behaviour prevents from obtaining consistent results; similar TG units of approximately the same design often show appreciable differences in the structural characteristics. These are a few limiting factors of the potentiality of the ICM.

Therefore, cases arise where one has to decide not only amplitude and phase of the first trial mass, but the balancing planes too. Two examples will be discussed, where the choice of these quantities was effectively assisted by a computer simulation of the shaft line dynamics. Emphasis is given to some practical aspects that allowed to maximize the results and the information obtained from those balancing tests, while reducing the downtime costs.

2. THE MULTIPLANE BALANCING COMPUTER CODE

2.1. Balancing Mass Computation

The balancing program can take into account a practically unlimited number of vibrations and balancing planes, running in the present versions on a IBM compatible PC or on a HP1000 series A700 computer.

The algorithm for the computation of balancing weights is based on the ICM, assuming the mechanical system to be linear and repeatable. The residual vibrations for a set of balancing masses are given by:

$$\{V_r\} = \{V_o\} + [C]\{B\} \tag{2.1.1}$$

If the number M of balancing planes equals the number N of vibrations, [C] is a square matrix, there is a unique solution for $\{B\}$ and the residual vibrations can be reduced to zero.

In general N is greater than M and the system is overdetermined: it is necessary to resort to some optimization scheme. Moreover, further requirements have to be satisfied.

The amplitude of residual vibrations should be less than some given value, different for each vibration:

$$|V_{ri}| \leq R_i \quad , \quad 1 \leq i \leq N \tag{2.1.2}$$

The amplitude of balancing weights should not exceed a definite quantity in order not to overstress the shaft; computed masses could reach useless great values when using planes with proportional IC. The constraints are expressed as:

$$|B_j| \leq M_j \quad , \quad 1 \leq j \leq M \tag{2.1.3}$$

If both $\{V_r\}$ and $\{B\}$ are introduced in the objective function to be minimized, a minimun for this function in the presence of the non linear constraints (2.1.2) and (2.1.3) can be found if at least one point of the domain of $\{B\}$ is known that satisfy the inequalities; but, should we know this point, we had solved the engineering problem.

The solution adopted consists in assigning to each vibration and mass a starting weighting factor and applying an iterative procedure, where the weighting factors (WF) are adjusted at every step until the constraints are satisfied, if possible.

The solution at each step is then found by a weighted least square method, the objective function being:

$$\varepsilon^2 = \sum_{i=1}^{N} (p_i V_{ri})^2 + \sum_{j=1}^{M} (q_j B_j)^2 = \{V_r\}^* [P] [P] \{V_r\} + \{B\}^* [Q] [Q] \{B\} \tag{2.1.4}$$

Substituting (2.1.1) for $\{V_r\}$ and differentiating the scalar ε^2 with respect to the unknowns $\{B\}$, one finally obtains:

$$\{B\} = ([C]^* [P]^2 [C] + [Q]^2)^{-1} (- [C]^* [P]^2 \{V_o\}) \tag{2.1.5}$$

So far, the method is quite similar to that described in [1].

The starting WF can be chosen by the operator or computed by the code, looking for a better conditioning of the system matrix; help is given for the choice of [Q]. If $\{B\}$ does not satisfy the constraints, the iterative procedure is started. At the first iteration the new WF will be given by:

$$p_i{}^{(1)} = \frac{|V_{ri}|^{(0)}}{R_i} \Omega\, p_i{}^{(0)} + (1 - \Omega)\, p_i{}^{(0)} \tag{2.1.6}$$

From the second iteration step on, the WF are given by a linear combinations of those of the two previous steps:

$$p_i{}^{(n)} = p_i{}^{(n-1)} + \frac{p_i{}^{(n-1)} - p_i{}^{(n-2)}}{|V_{ri}{}^{(n-1)}| - |V_{ri}{}^{(n-2)}|} \left(R_i - |V_{ri}{}^{(n-1)}| \right) \tag{2.1.7}$$

If the difference between the residual vibrations of the last two steps is less than an assigned threshold, the WF remain unchanged; if a WF becomes negative, it is substitued by the ratio of the previous value and the corresponding vibration admissible value; if the starting value was zero it remains zero; in any case, the WF cannot exceed a fixed value. The same applies to the mass WF.

The method is essentially that described in [2]. Although the convergence could not be demonstrated, a number of simple numerical tests gave satisfactory results. When the convergence is not reached, it may be that a domain of {B} satisfying the constraints does not exist or is too small: either the constraints are exceedingly strict or a wrong plane is being used. In any case, at each iteration step the results in terms of balancing masses and residual vibrations are shown and the operator is requested whether or not to continue, thus enabling to evaluate a number of different solutions in negligeable time.

The best way to use the algorithm is to start with no WF on the masses and with unity WF on all the vibrations. The results will suggest how to start another iteration, if any.

2.2. Graphical Resources

It was deemed essential to have at any moment a full capability of listing or displaying on a Nyquist plane every IC and vibration of interest (initial, final, residual calculated or differences between each other). In certain cases this allows to judge quickly about the best choice of the balancing planes or the quality of the results.

It is also possible to perform vectorial operations and to display vibration orbits with phase information (Fig. 1), which assists to start a balancing without any prior knowledge of IC.

Another effective way of evaluating the fitness of a single plane is the so called Circle Method. When IC, initial vibrations and the corresponding admissible residuals are known, all the possible values of the balancing mass for each vibration are contained in a circle in the Nyquist plane; if a common area among the circles exists all the mass values in the area satisfy the problem. The value of the balancing mass can then be chosen by a cursor on a graphic display (Fig. 2) or computed as the barycenter of the common area or, also, as the point that leaves the least squares for the residual vibrations.

2.3. Additional Options

Other options of the code are vectorial operations on the masses to help their positioning on the rotor, computation of residual vibrations for an assigned mass and a sensitivity analysis of the residual vibrations with respect to the phase angle of a balancing mass.

This last result is obtained varying of given quantities the phase angle of one mass at a time around a mean position and computing the residual vibrations: in case they show strong variations a good result is unlikely.

3. EXPERIENCES OF FIELD BALANCING 320 MW TURBOGENERATOR ROTORS

A recent computer aided balancing experience on two 320 MW TG twin units, in operation at two different power plants, is illustrated.

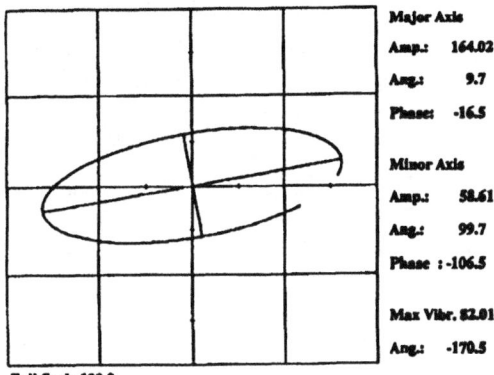

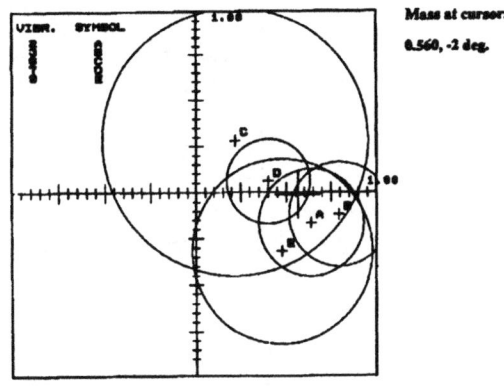

Fig. 1 - Example of Orbit Display

Fig. 2 - Circle Method Display

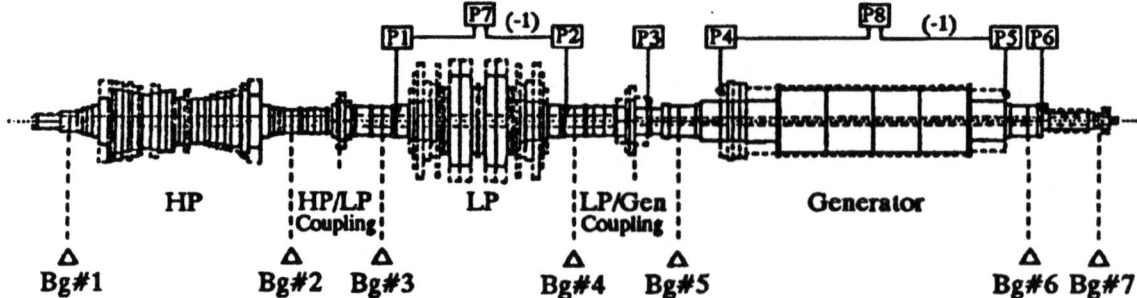

Fig. 3 - Shaft Line Idealization.

The shaft line of the two units consists of the HP-MP turbine, the LP turbine and the generator, each one supported by two bearings, plus one bearing at the exciter end (Fig. 3).

Experimental data were collected by means of multichannel digital acquisition systems installed on mobile labs, during a measurement campaign after a major overhaul, which is now a standard at ENEL for units of that size.

Both TG sets showed high vibrations on one generator bearing, strongly correlated to the excitation current (Fig. 4). Nevertheless, the dynamic behaviour of the two turboset was quite distinct (Fig. 5). On unit A, high vibrations concerned the bearing 6 towards the exciter side and the LP bearings too, due to the excitation of a second mode, with critical speed just above the nominal speed. On unit B, high vibrations interested the LP side bearing 5, while LP vibrations were acceptable at any load.

A large number of TG of that type has been in operation at ENEL power stations for many years now. In spite of that, experimental IC were available for the LP planes only. In the past, measurements were done on the problem, they were not standardized and the tests were poorly documented. In addition, field balancing of generator rotors are unusual, because access to the balancing planes is

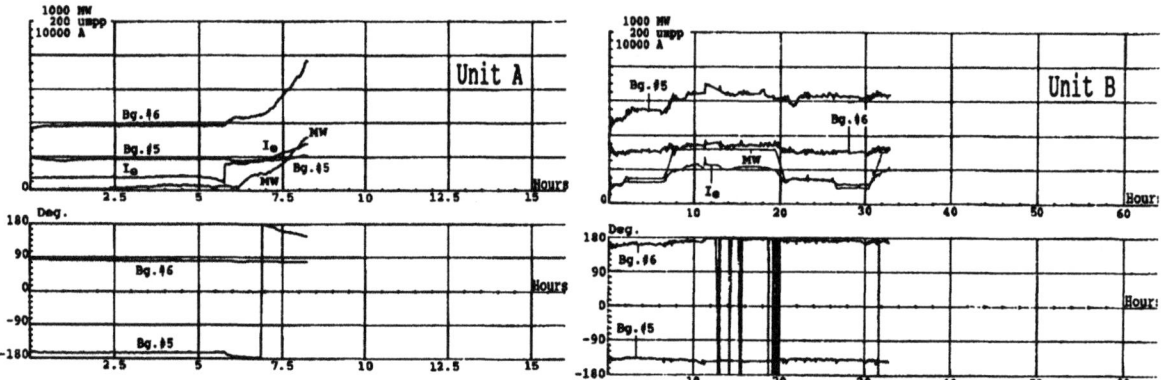

Fig. 4- Vibrations of Unit A and Unit B Generators in Operation at Variable Load before Balancing

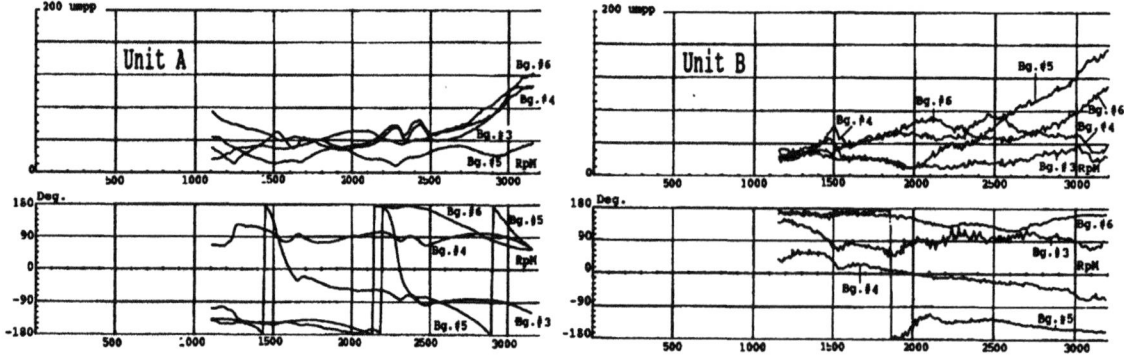

Fig. 5 - Synchronous Component of the Absolute Shaft Vibrations at the LP and Generator Bearings during a Run-down before Balancing on Unit A and Unit B.

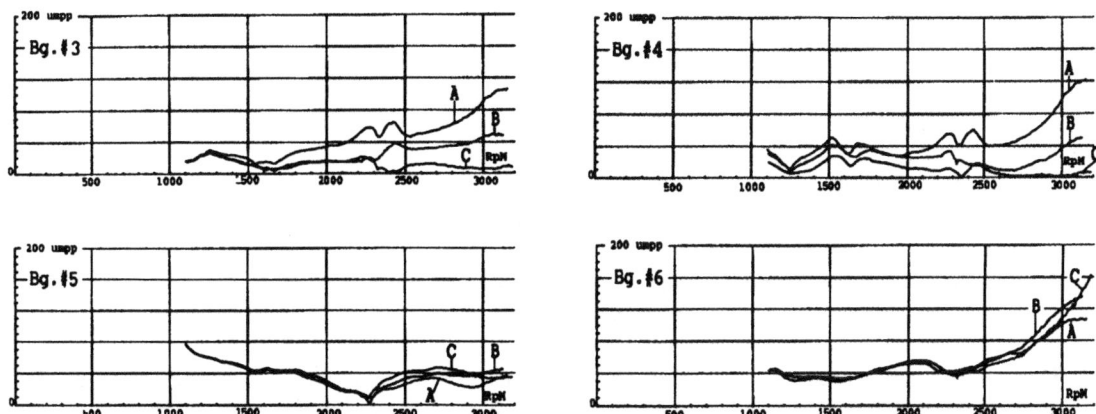

Fig. 6 - Unit A. Amplitudes of Vertical Shaft Motion at the LP and Generator Bearings at Transient Speed: before Balancing (A); after the First Trial on the Plane P7 (B); after the Second Trial on the Plane P7 (C).

difficult and takes long time; finally, often the hurry prevents from reaching repeatable conditions.

The unit B generator underwent in the past an unsuccessful attempt of field balancing. The relevant IC had to be discarded, for lack of coherence and uncertainties on mass positions.

Now, the field balancing began on the LP turbine of unit A, also with the goal to get greater margins for the next generator balancing. Since the LP second mode of vibration was clearly prevalent at the rated speed, it was decided to put two masses in counterphase in the planes P1 and P2 (Fig. 3). This combination was considered as a single plane (plane P7) in mass computations.

Experimental IC and vertical vibrations, with maximum load and rotor current, at bearings 2, 3, 4 were used in computations; against a request of 0.400 Kg masses, cautiously, only 0.250 Kg were actually placed. At the first shot a 43% reduction of LP bearing vibrations was obtained; subsequently the masses were increased up to the requested value and rotated of 30 deg.: LP vibrations were in practice zeroed, with just minor changes at the other bearings (Fig. 6). The three sets of influence coefficients obtained from these two runs showed a very good agreement for turbine bearings; inconsistent results were obtained for the generator bearings.

Up to then, the generator was thought to be dominated at the nominal speed by its second mode of vibration, that could explain neither of the deformation shapes measured on the two units. A critical speed analysis was undertaken in order to understand the dynamic behaviour of the shaft line and to define the most effective balancing planes.

The model of the shaft line (Fig. 3) was built-up by means of the code [3]. Good agreement was obtained between theoretical and experimental critical speeds in the working frequency range, varying the support stiffness; the corresponding mode shapes were defined (Fig. 7).

The investigation pointed out the presence of a critical speed well over 3000 RpM, mainly connected to the deformation of the LP-generator coupling and heavily involving an in-phase displacement at the generator bearings (Fig. 7). A sensitivity analysis indicated that this critical speed in any case ranged between 4000 RpM and 5000 RpM. The combination of this mode with the second mode of vibration could effectively explain the actual deflection shape of the two TG shaft lines. Further critical speeds involving the generator bearings were by all means far apart (above 6000 RpM).

The forced response analysis produced numerical IC for all the six possible balancing planes (Fig. 3). The agreement with experimental IC was very good for the masses in counterphase on the LP planes

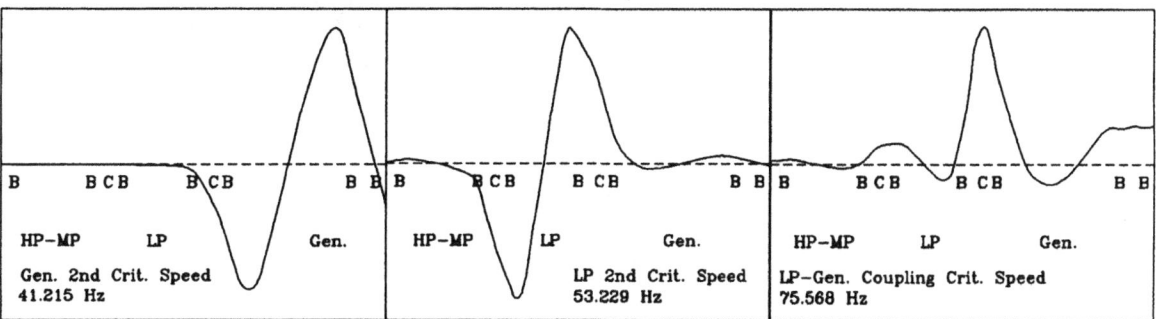

Fig. 7 - Dominant Mode Shapes of the Shaft Line at 3000 RpM for the LP and Generator Bearings.

Bearing No.		1	2	3	4	5	6	7
Plane P7 (Unit A)	Amp.	/	+7	+7	+7	-31	/	/
	Phase	/	-16	+11	+11	+22	/	/
Plane P3 (Unit B)	Amp.	/	/	+13	+11	-21	/	/
	Phase	/	/	+31	+6	-7	/	/
Plane P4 (Unit A)	Amp.	/	/	+75	0	/	-2	-58
	Phase	/	/	-19	-30	/	+18	+18

Table 1 - Per Cent Amplitude and Absolute Phase Deviations between Theoretical and Experimental IC for the Vertical Absolute Shaft Motion.

(Tab. 1); moreover, the experimental frequency responses were reproduced qualitatively well by the model in the 1800 RpM-3200 RpM speed range, for unbalances placed in the possible balancing planes, thus giving enough confidence in using the numerical IC for the trial mass computations.

They were first used to balance the unit B. Suitably weighted vibrations at maximum load and at a reduced load were taken into account for the bearings 2, 3, 4, 5, 6. One plane at a time was first considered then, computed residual vibrations were balanced using a different plane; as a final counter-check the

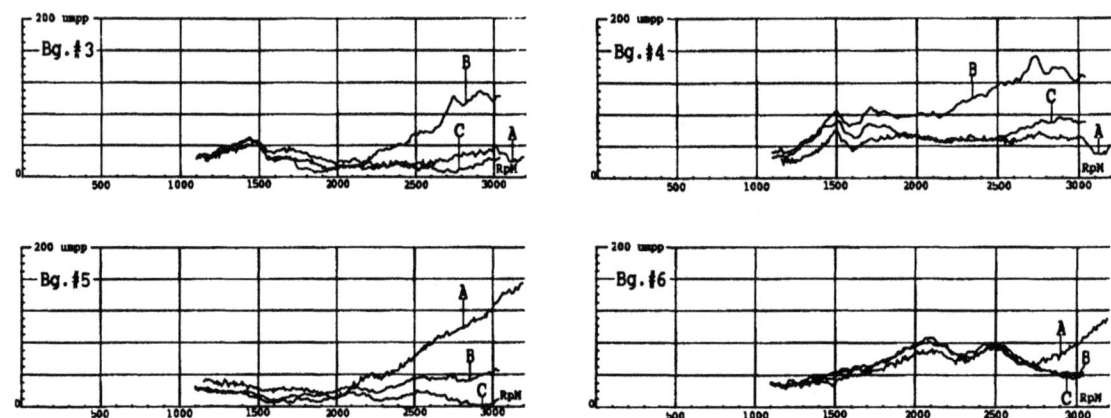

Fig. 8 - Unit B. Amplitudes of the Vertical Shaft Motion at the LP and Generator Bearings: before Balancing (A); after Balancing on Plane P3 (B); after First Balancing on Plane P7 (C).

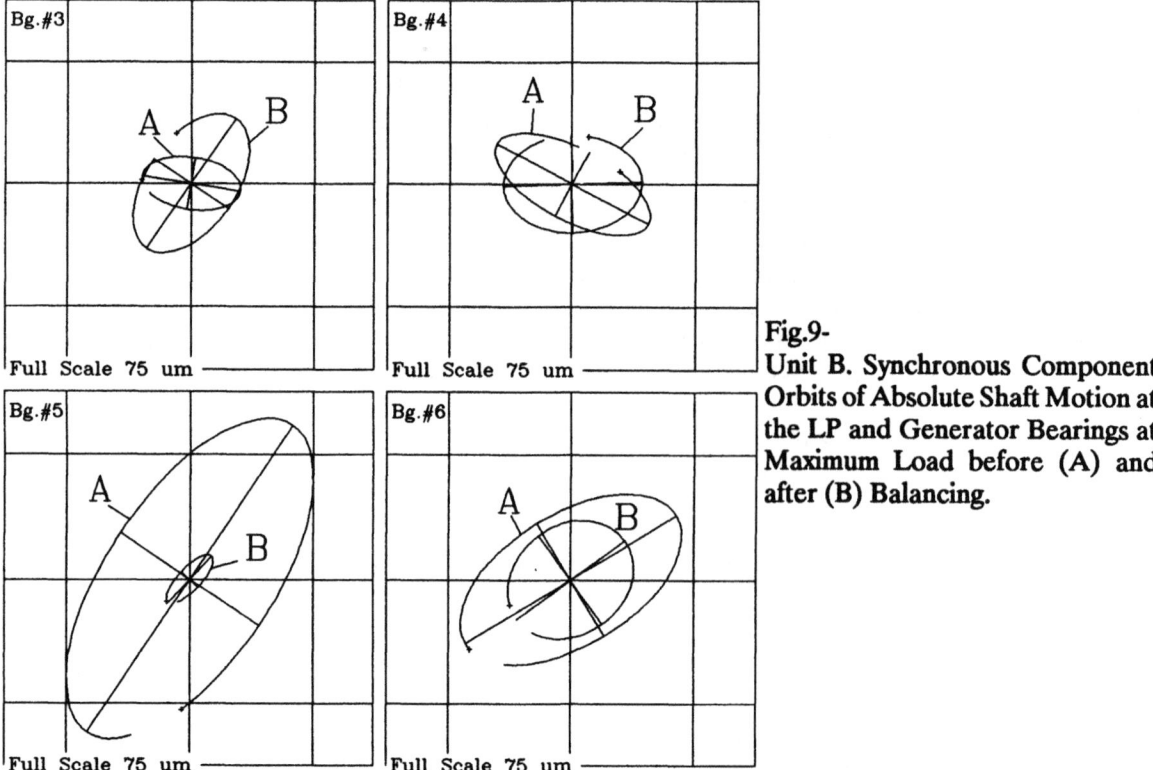

Fig.9-
Unit B. Synchronous Component Orbits of Absolute Shaft Motion at the LP and Generator Bearings at Maximum Load before (A) and after (B) Balancing.

chosen planes were accounted for together. The plane P3 resulted the best one, also for ease of access, although it could significantly increase LP vibrations, that could be reduced by plane P7. A mass of more than 1 Kg was requested at 130 deg. on the coupling (plane 3) and masses of 0.425 Kg at a 120 deg. angle were called for on plane P7.

The coupling plane alone was first used in order to obtain the corresponding experimental IC; for caution a 0.600 Kg trial mass was actually used. A fairly long stop of the machine for the mass positioning considerably bowed the shaft; after the run-up, vibrations greater than 200 umpp were recorded on the LP turbine bearings; none the less, it was possible to prevent the immediate shut-down and the unit was kept at a reduced speed until lower and stable vibrations were obtained. After stabilization at partial and full load, a first set of experimental IC was calculated and resulted in good agreement with the theoretical ones. Immediate data processing confirmed the initial request of mass. In a second run, the total mass resulted to be 0.936 Kg at 134 deg., very close to the initially requested mass. The expected substantial increase of LP bearing vibrations was controlled by an addition of two 0.445 masses at 120 deg. on plane P7 (Fig. 8). The unit was ready to start before the due time. One last trim balancing was finally done a week later to further improve LP vibrations, by a rotation of 20 deg. of the masses on plane P7. This was due to the fact that the second critical speed of the LP rotor is a little bit under the 3000 RpM on this unit (Fig. 8). Vibration orbits, before and after balancing, accounts for the results obtained (Fig. 9).

Interestingly enough, the vibration dependence on the load was significantly reduced after balancing. On the other hand, this partly spoiled the IC.

The balancing of the unit A was resumed with more confidence on the theoretical IC. The computation procedure of the balancing mass was the same as that of unit B. The best plane turned out

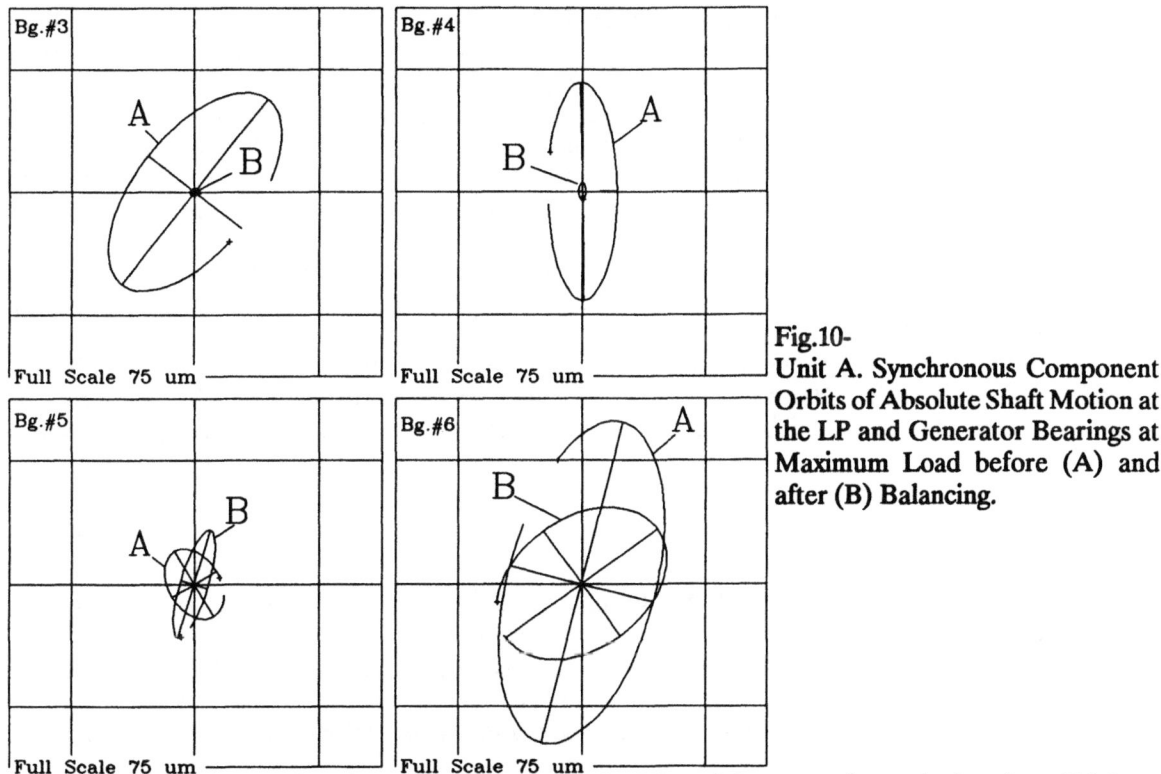

Fig.10-
Unit A. Synchronous Component Orbits of Absolute Shaft Motion at the LP and Generator Bearings at Maximum Load before (A) and after (B) Balancing.

to be the plane P4 internal to the generator towards the LP end, just opposite to the bearing of highest

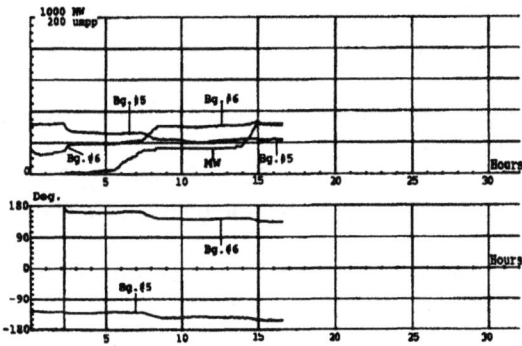

Fig. 11 -
Synchronous Vibrations of Unit A Generator Bearings in Operation at Variable Load after Balancing.

18

vibration. As a rule of thumb the trial mass is put on the same side of the vibrating bearing, as it is also witnessed in the literature [5], where the contrary is defined an interesting paradox. But considering the mode shapes of the two dominant critical speeds for the generator at 3000 RpM and the fact that the nominal speed is between the two critical speed, it is clear that the contributions of the two modes elude on the side of the applied force and add on the opposite side of the generator rotor. The balancing program called for a mass of more than 1 Kg at approximately -20 deg. on plane P4; in addition another mass of 1 kg at 90 deg. on plane P6 could further reduce the in-phase component of the two generator bearings.

No more than two planes at a time were taken into account. This is because a high degree of confidence is required on IC, otherwise possible errors propagate out of control [4].

A mass of 0.802 Kg at -20 deg was added on plane P4 only. Once again, the other plane was not used at first, paying more importance at this stage to the possibility of obtaining experimental IC for each plane. Vibration orbits before and after any balancing are reported (Fig. 10). The other plane has not yet been used at the time of the present writing.

For this unit too, the load influence on generator vibrations was so reduced after balancing (Fig. 11), that the resulting IC were not completely reliable, no matter how much care was taken in trying to achieve repeatable conditions. A comparison with the theoretical IC is given in Tab. 1.

All the balancing tests described above could be carried out during week-end stops of the two units, thus having no down-time costs.

4.CONCLUSIONS

The ICM is well suited for the application to the in-field balancing of TG rotors, but extensive, reliable sets of experimental IC are requested to fully exploit the capabilities of the method. This is not yet the case, since balancing on site is not yet a standardized procedure.

In such conditions, mathematical models of the shaft line can greatly help to chose the most effective balancing planes, by identifying the likely modal contributions to the total deflection, and, possibly, they can provide numerical IC substituting for the experimental ones. In the illustrated examples, one shot balancing was achieved thus keeping down-time costs as low as possible, while obtaining more than satisfactory vibration levels.

The balancing will benefit by a systematic introduction of digital monitoring systems, but the need for obtaining repeatable conditions and for using well definite vibration phase and mass angle conventions must be stressed.

The multiplane balancing code in use at ENEL demonstrated to be very effective and very flexible to assist in field balancing TG shaft lines. The introduction of techniques ([6], [7]) to improve IC processing should be considered, but first this would require to collect a number of comparable sets of such IC.

REFERENCES

[1] Chevalier R.   Balancing of Highly Flexible Shaft Lines on Their Critical Bending Speeds. IFToMM-3rd Int. Conf. on Rotordynamics; Lyon, France; Sept. 10+12.1991; 257+261

[2] Borgese D., Di Pasquantonio F., Bigret R.   Méthode des coefficients d'influence et méthode modale pour la réduction des amplitudes de vibration des machine tournantes (Equilibrage). GAMI; Journées d'étude sur la correction des vibrations des rotors dits flexibles; Paris; June 28+29.1977

[3] DYTS04 USER'S GUIDE - CEGB - RD/L/P - 15/80 - JOB No.VE316

[4] Sanderson A.F.P.   Turbine Generator Trim Balancing Using Optimized Least Squares Methods. IMechE Int. Conf. Vibrations in Rotating Machinery; Edinburgh; Sept. 13+15.1988; C308/88; 491+498

[5] Gunter E.J., Gunter W.E.   Field Balancing 70 MW Gas Turbine- Generators.   IFToMM Int. Conf. on Rotordynamics; Tokyo; Sept. 14+17.1986; 135+143

[6] Larsson L.O.   On the Determination of the Influence Coefficients in Rotor Balancing, Using Linear Regression Analysis.   IMechE Int. Conf. Vibrations in Rotating Machinery; Cambridge; Sept. 15+17.1976; C173/76; 93+97

[7] Lund J.W., Tonneson J.   Analysis and Experiments in Multiplane Balancing of Flexible Rotors. Trans. of ASME Journal of Engineering for Industry; Vol.94, No. 1, p.233; 2.1972

ACTIVE BALANCING OF A HIGH SPEED ROTOR IN MAGNETIC BEARINGS

C.R. Knospe, R.R. Humphris, E.H. Maslen, P.E. Allaire

UNIVERSITY OF VIRGINIA

ABSTRACT

This paper describes the application of an open loop control technique for the attenuation of unbalance response in a high speed rotor. The test rig is a mock up of a dual wheel compressor with an operating speed range of 30,000 to 70,000 rpm. The rotor has a third critical speed (first bending mode) at 24,050 rpm which is lightly damped. Excessive rotor synchronous response occurs during run—up at this speed and intermittently causes loss of support. An open loop control is used to introduce a rotating magnetic force to balance the rotor and reduce rotor synchronous response at this speed. With this control, the third critical speed is essentially canceled. Also demonstrated is the use of the open loop control to attenuate housing structure vibrations when the rotor is operating at one of the housing's natural frequencies, 42,300 rpm.

INTRODUCTION

Active magnetic bearings provide a number of advantages over conventional bearings for a variety of applications. These include elimination of the lubrication system, friction free operation, decreased power consumption, and operation at temperature extremes. While these advantages have primarily motivated research and development, active magnetic bearings also permit much greater flexibility in tailoring the dynamics of the rotor. Through active control, unusual relationships can be established between bearing journal motion and bearing force. In this paper we describe the application of one example of this flexibility to a high speed model compressor.

The University of Virginia began research on active magnetic bearings for a high speed compressor in 1988 in conjunction with an industrial company. The goal of this work was to greatly reduce the power consumed in the bearings. To this end, the machine was retrofitted with permanent magnet biased active magnetic bearings driven by a switching power amplifier [1–3]. The retrofitting placed severe limitation upon the design in terms of bearing size and location. Nevertheless, the project was an outstanding success with the power consumed in the bearings being greatly reduced, — from 3000 Watts for the conventional process fluid lubricated bearings to 207 Watts total power (losses in bearing coils, switching amplifier, controls, and power supply) [3]. While the original power conservation goal was met, the rotor was not operable above 23,000 rpm, near the third critical speed, due to excessive synchronous vibration at this speed. The unbalance response caused large bearing displacements resulting intermittently in loss of support. This paper describes an open loop control balancing method which allows the rotor to pass through and *operate on* the third critical speed with greatly reduced unbalance response.

ROTOR AND FORCED RESPONSE

The test rig, shown in Figure 1, is a model of a dual wheel compressor. The model constructed at the University is driven by an air turbine off laboratory compressed air. This method of powering the rotor does not permit rapid acceleration of the rotor.

The rotor shown in Figure 2 has a disk at each end to simulate axial–inlet, radial–outlet compressor wheels, two radial magnetic bearing journals, a magnetic bearing thrust disk, and a disk to simulate a midspan motor rotor. The rotor's mass is 3.7 lbs (1.7 Kg) and has a bearing span of 7.4 inches (18.8 cm). The rotor was suspended with adjustable PID control. The stiffness of the radial bearings during our tests was 4000 lbf/in.

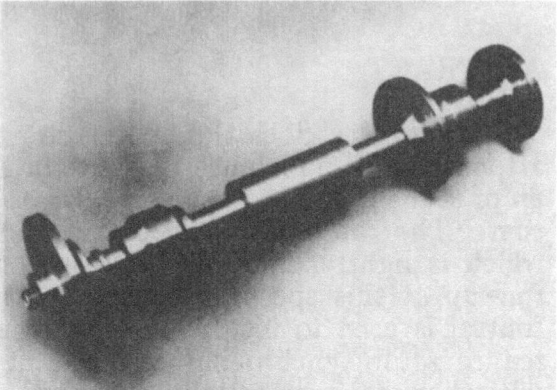

Figure 1: Model compressor retrofitted with magnetic bearings

Figure 2: Rotor for model compressor

The designed operating speed range for this machine is 30,000 to 70,000 rpm, almost between the third and fourth critical speeds of the rotor. A critical speed map for the rotor is shown in Figure 3. Notice from this map that the third and fourth critical speed lines are flat over most of the range of bearing stiffness. This indicates that added bearing damping will have little affect on these modes' amplification factors.

The radial magnetic bearings used are permanent magnet biased. The load capacity of each radial bearing was 18 lbf (N). Forces of this level can only be obtained by magnetically saturating the stator poles. This load capacity was limited by the available space in which to place the magnetic bearings in the model compressor. Use of a higher saturation flux density material for the stator and rotor laminations would have permitted approximately twice the load capacity.

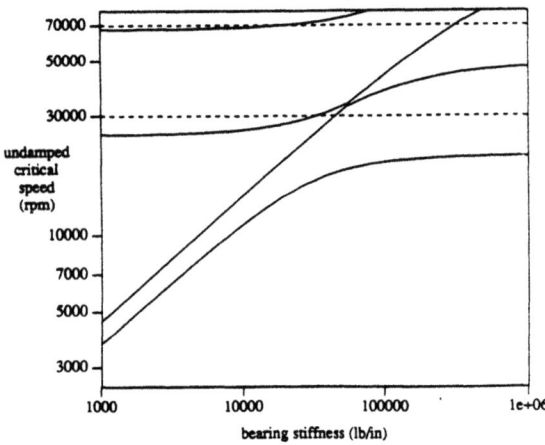

Figure 3: Undamped critical speed map

The vibration amplitudes for this rig were quite small until the third critical speed where the turbine–end bearing exhibits large vibration amplitudes. This large vibration was associated with bearing flux saturation and loss of support. Attempts at balancing proved ineffective at alleviating this problem possibly due to the effect of the rotor's interference fits during operation. The response to unbalance on this machine is still highly variable and has made testing difficult.

OPEN LOOP CONTROL

Many researchers have investigated the use of magnetic bearings to reduce synchronous response. Two methods of synchronous vibration reduction have been advocated: feedback control and open loop control. These two methods, while fundamentally different, are often confused; both have been referred to as "automatic balancing".

Feedback control reduces the synchronous response by altering the eigenstructure of the rotor system. The simplest and most common form is the insertion of a notch filter at the running speed in the feedback path used to achieve stable rotor suspension [4]. This results in reduced effective stiffness and damping at the operating speed and thus in reduced transmitted synchronous vibration. Another method of feedback control is disturbance accommodation/frequency weighting estimator—based full—state feedback [5,6]. These methods have many disadvantages which are detailed in [7,8]. In brief, they yield sub—optimal performance, compromise robustness, and may have poor transient/asynchronous response.

Since the rotor's unbalance response is well correlated, we may cancel it using an open loop control approach. Fundamentally, feedback control is only needed to stabilize a system or tailor its response to unknown/uncorrelated inputs. From a control theory perspective, the synchronous response of a rotor is not a *feedback property*. Feedback properties are properties of a system which may only be changed through feedback, such as stability and transient response. Thus, using an open loop control method, we may cancel the unbalance response without altering the stability or the transient response. This is in marked contrast to the feedback methods discussed earlier which only achieve vibration attenuation through altering the system's dynamics (natural frequencies and mode shapes). The open loop control method is therefore analogous to conventional rotor balancing which involves the addition of canceling forces without changing the bearings' properties (stiffness and damping) or the system's dynamics (natural frequencies and mode shapes). (Note: balancing can affect stability for nonlinear fluid film bearings.)

The open loop vibration control method is represented in Figure 4. Synchronous open loop control signals are superimposed on the stabilizing feedback control to drive the magnetic bearing coils. Thus, synchronous magnetic forces are exerted on the rotor. The amplitudes and phases of these forces are adjusted to attenuate the vibration. With proper adjustment, the vibration may be canceled at any N points on the shaft for an N bearing rotor or may be reduced (but not canceled) at a larger number of points [9]. Alternatively, the vibration can be eliminated throughout the housing if the magnetic bearings are the only transmission path of synchronous vibration from the rotor (i.e., no conventional bearings, seals, or significant rotor—casing coupling through the process fluid). If not, the vibration may still be attenuated throughout the structure. Several researchers have investigated open loop control methods both theoretically and experimentally [7,10,11,12].

It should be noted that the applied correction forces of open loop control are not inertial in contrast to those of conventional balancing. The condition under which the bearing journals have no synchronous vibration does not correspond to the condition under which no synchronous vibration is transmitted through the bearings. This is because the exertion of magnetic correction forces upon the shaft results in equal and opposite magnetic forces upon the bearing stators. Thus, a synchronous force is transmitted to the housing or structure even though the bearing deflection is zero. However, a condition of no transmitted synchronous force can also be obtained through open loop control. This condition can be obtained at any operating speed since the open loop control does not alter the system stability. In this condition, the rotor spins about its

22

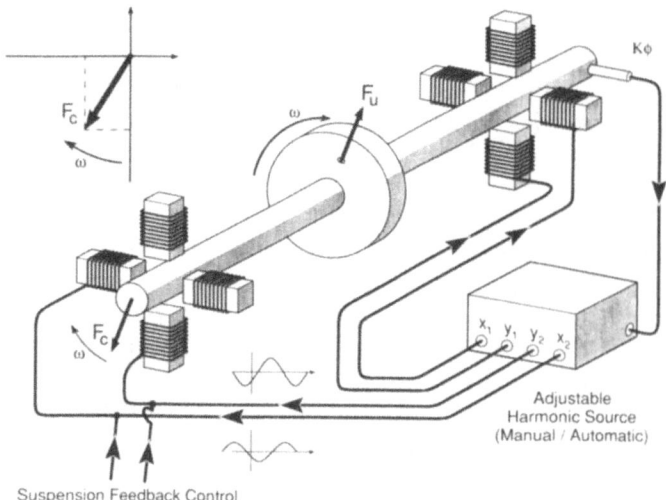

Suspension Feedback Control

Figure 4: The open loop control method

inertial axis. This is what notch filter controllers attempt and, in turn, sacrifice robustness to achieve.

A rotor supported in magnetic bearings with its housing is described by the dynamical equations

$$[M]\{\ddot{x}\} + [C]\{\dot{x}\} + [K]\{x\} = \{f_u\} + [B][\{f_f\} + \{f_o\}]$$ (1)

where $\{x\}$ is the displacement vector, $[M], [C]$, and $[K]$ are the n x n mass, damping, and stiffness matrices respectively, $\{f_u\}$, $\{f_f\}$, and $\{f_o\}$ are the n x 1 unbalance force vector, and m x 1 feedback control force vector, and the m x 1 open loop control force vector and $[B]$ is the n x m bearing selection matrix. Using the Laplace transform, Eqn. (1) becomes

$$\left[[M]s^2 + [C]s + [K] + [B]\,[G(s)]\right]\{x(s)\} = \{f_u(s)\} + [B]\,\{f_o(s)\}$$ (2)

where $[G(s)]$ is the transfer function matrix of the feedback controller. Since the unbalance force vector and open loop control force vector are harmonic at the operating speed ω, Eqn. (2) can be evaluated for the forced response

$$\{X\} = \left[-[M]\omega^2 + j[C]\omega + [K] + [B][G(j\omega)]\right]^{-1}\left[\{F_u\} + [B]\{F_o\}\right]$$ (3)

or

$$\{X\} = \{X_u\} + [T]\,\{F_o\}$$ (4)

where $\{X_u\}$ is the n x 1 force response without open loop control and $[T]$ is the n x m influence coefficient matrix relating the m synchronous open loop control force magnitudes to the n synchronous displacement magnitudes. If the vibration at p locations are chosen ($p \geq m$) to be suppressed with the open loop control, then these p rows of Eqn. (4) can be chosen to yield

$$\{X\}_p = \{X_u\}_p + [T]_p\{F_o\}$$ (5)

The open loop control forces $\{F_o\}$ may be chosen to reduce the vibrations $\{X_p\}$. An adaptive controller might minimize the weighted sum of the squares of the vibration, yielding the open loop control law

$$\{F_o\} = -\left[[T]_p^T\,W[T]_p\right]^{-1}[T]_p^T W\{X_u\}_p$$ (6)

where W is a diagonal weighting matrix. Other cost functions could also be chosen to determine an open loop control law — the quadratic sum has the advantage of yielding a simple analytic solution. In order to employ Eqn. (6), an estimate of the matrix $[T]_p$ and vector $\{X_u\}_p$ must be obtained. For some systems this may be done off–line. Burrows and Sahinkaya [10] employ a probing scheme on–line with least square estimation. Methods of estimation may also be employed on–line without probing via simultaneous identification and control.

EXPERIMENTAL RESULTS

The open loop control was employed on the high speed magnetic bearing rig using a simple setup. A key phasor was used to synthesize a harmonic signal synchronous with rotor speed for each bearing axis. These open loop control signals are added as a perturbation to the feedback control signals for each magnetic bearing axis. The magnitude and phase of the open loop signals are individually adjusted, or "tuned" at a given speed to minimize the unbalance response of interest. For the four axes of the rig, there were a total of eight open loop adjustments to be made at any given speed.

In the first test, the rotor was brought up to its third critical speed, 24050 rpm, and the open loop control was adjusted to minimize the midspan synchronous response and reduce the bearing orbits. As shown in Figures 5 and 6, the orbits at inboard, outboard, and midspan have been greatly reduced using the magnetic balancing. The midspan synchronous vibration went from greater than 9 mils (230 μm) to less than 1 mil (25 μm). The control currents in the four magnetic bearing coils with and without the open loop control are given in Table 1. Note that the currents required for balancing the rotor are not large — the rated current for these coils is 2–5 amps.

With the open loop control still adjusted for midspan minimum vibration at the third critical speed, the rotor was again run up through the third critical. The open loop control was turned "on" at approximately 21,000 rpm and "off" at 27,000 rpm. The synchronous response of the rotor at the inboard bearing, outboard bearing, and midspan during this run up are shown in Figure 7. Note that the third critical speed has essentially been canceled. Despite the fact that the third critical speed is lightly damped and has a high amplification factor, the open loop control set for 24050 rpm works very effectively over a wide range of frequencies around the critical speed.

To demonstrate the effectiveness of the open loop control method in minimizing the vibration transmitted to the housing and foundation, the rotor was brought up to 42300 rpm, a natural frequency of the housing structure. The open loop control was then adjusted to eliminate the synchronous control current

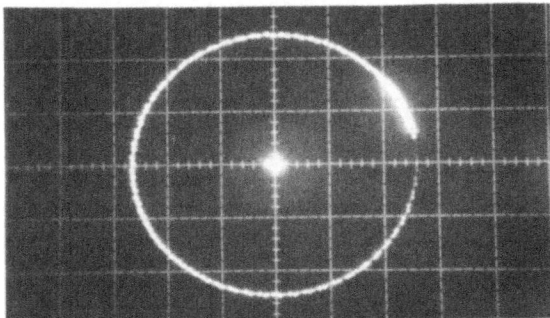

Figure 5: Photograph of the mid–span shaft orbit at the third critical speed, 2405 rpm, with and without open loop control

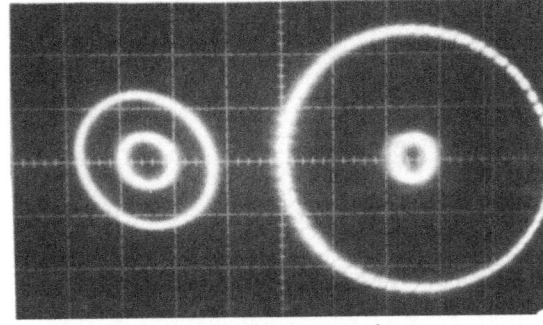

Figure 6: Photograph of the orbits at inboard and outboard bearing at the third critical speed, 24050 rpm, with and without open loop control

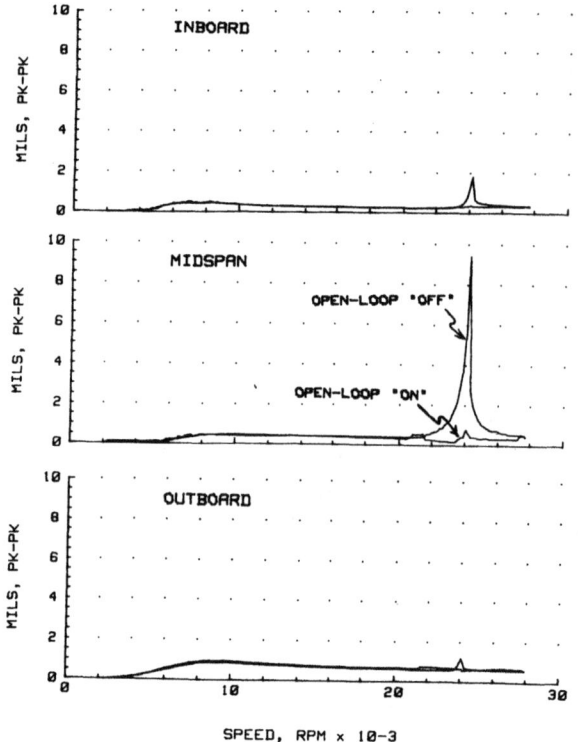

Figure 7: Vibration amplitudes
vs. speed with and without open
loop control; control adjusted
to minimize midspan vibration at
the third critical speed, 24050 rpm

into the bearings. This causes the bearings to exert no force upon the rotor in response to the shaft orbits. Since no synchronous force is exerted upon the bearings, no synchronous force is transmitted through them to the housing. Under this setting, the rotor spins about its inertial axis as in the notch filter feedback method. However, the open loop control does not degrade the rotor's stability margin as does the notch filter method. Figure 8 shows the frequency spectrum of vibration with and without open loop control from accelerometers mounted at several locations on the housing and foundation. Approximately 75% reduction in synchronous vibration was obtained throughout the structure. Note that accelerometer measurements were not necessary for the adjustment of the open loop control; only the availability of the total synchronous current in the bearing coils is

necessary. Because of the conventional bearings in the air turbine used to drive the magnetically suspended rotor, even with no synchronous control current, some synchronous vibration is transmitted into the housing. Using measurements from an accelerometer on the housing near the outboard bearing, the open loop control was re–adjusted to minimize the synchronous vibration at this location. This resulted in nearly complete suppression of the housing response at this natural frequency as shown in Figure 9.

TABLE 1

Currents Required for Magnetic Bearings for a Compressor Model at Third Critical Speed of 24,050 RPM

Axis	Open Loop "OFF" (Amp)		Open Loop "ON" (Amp)	
	dc	rms	dc	rms
Inboard Vertical	−0.11	0.50	−0.10	0.54
Inboard Horizontal	0.19	0.55	0.19	0.59
Outboard Vertical	−0.37	0.33	−0.37	0.22
Outboard Horizontal	−0.08	0.32	−0.08	0.19

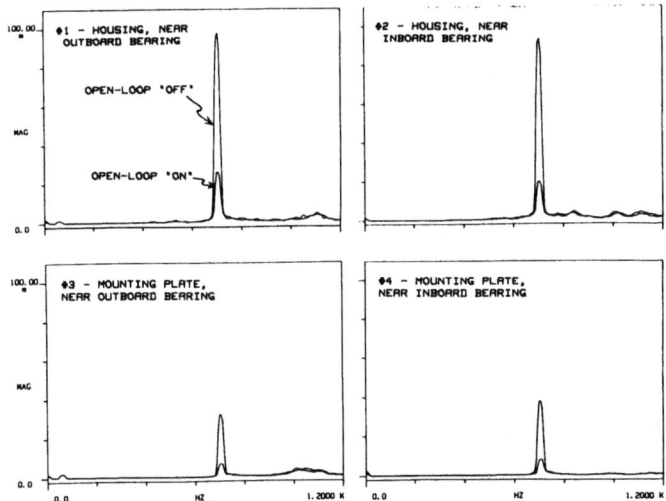

Figure 8: Frequency spectra showing vibration amplitudes at several locations on the housing and foundation. With and without open loop control adjusted for minimum synchronous currents. Rotor operating at 42300 rpm.

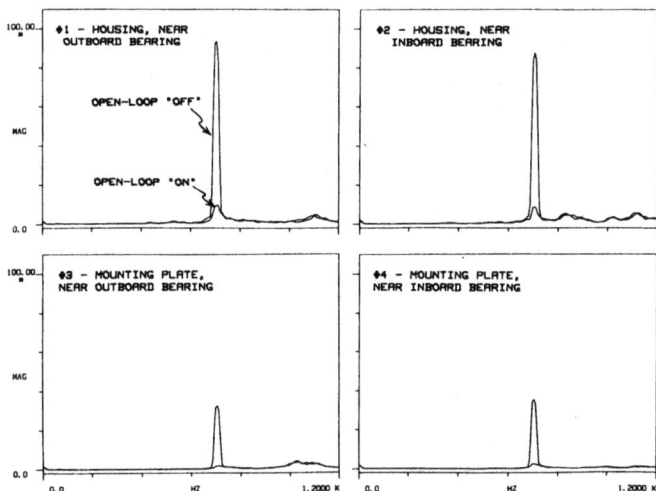

Figure 9. Frequency spectra showing vibration amplitudes at several locations on the housing and foundation. With and without open loop control adjusted for minimum synchronous vibration on the housing near the outboard bearing.

CONCLUSION

Open loop control was shown to allow the operation of a high speed model compressor at its third critical speed, 24050 rpm, with greatly reduced shaft vibration. While the open loop control method does not degrade rotor stability, it does permit the virtual elimination of the large vibration associated with lightly damped critical speeds.

It was demonstrated that there was a wide range of speeds around the critical speed in which the attenuation of vibration is quite effective.

The method was also shown in another test to be successful in greatly reducing the synchronous vibration transmitted to the housing of the rig. It was demonstrated that this could be accomplished in two fashions: tuning to eliminate the synchronous control currents or tuning to reduce the vibration measured at a point on the housing.

Open loop control methods for unbalance response attenuation are gaining rapid acceptance in the magnetic bearing and rotordynamic communities.

Many interesting areas for investigation remain, adaptive control strategies and simple hardware implementations, for example. But, an equally important area for these communities to consider are the specifications used for acceptable rotor performance. Do specification on the amplification factors for critical speeds make sense when the rotor's vibration at these speeds is virtually eliminated? How should rotor stability be measured if not with the amplification factor? Questions such as these will become increasingly important because of the increased dynamic flexibility of magnetic bearings.

ACKNOWLEDGEMENTS

This research was funded in part by the Commonwealth of Virginia's Center for Innovative Technology and the Army Research Office.

REFERENCES

[1]   Maslen, E.H., Allaire, P.E., Scott, M.A., and Hermann, P. "Magnetic Bearing Design for a High Speed Rotor," Proceeding of 1st International Symposium

[2]   Keith, F.J., Maslen, E.H., Humphris, R.R., and Williams, R.D. "Switching Amplifier Design for Magnetic Bearings," Proceedings of the 2nd International Symposium on Magnetic Bearings July 12–14, 1990, Tokyo, Japan, pp. 211–218.

[3]   Sortore, C.K. "Design of Permanent Magnet Biased Magnetic Bearings for a High Speed Rotor, "Master's Thesis, University of Virginia, January 1990.

[4]   Haberman, H. and Brunet, M. "The Active Magnetic Bearing Enables Optimum Damping of Flexible Rotors," ASME Paper 84–GT–117, 1984.

[5]   Maslen, E.H. "Magnetic Bearing Synthesis for Rotating Machinery," Ph.D. Dissertation, University of Virginia, August 1990.

[6]   Reinig, K. and Desrochers, A. "Disturbance Accommodating Controllers for Rotating Mechanical Shaft," ASME Journal of Dynamic Systems, Measurement and Control, Vol. 108, p. 24–31, March 1986.

[7]   Knospe, C.R., Humphris, R.R., and Sundaram, S. "Flexible Rotor Balancing using Magnetic Bearings," Proceedings of the Conference on Recent Advances in Active Control of Sound and Vibration, Blacksburg, VA. April 15–17, 1991, pp. 420–429.

[8]   Knospe, C.R. "Stability and Performance of Notch Filter Control for Unbalance Response," International Symposium on Magnetic Suspension Technology, NASA Langley Research Center, August 19–23, 1991.

[9]   Tessarzik, J.M., Badgley, R.H., and Anderson, W.J., "Flexible Rotor Balancing by the Exact Point–Speed Influence Coefficient Method," Journal of Engineering for Industry, ASME Transactions, Series B, Vol. 94, No. 1, February 1972, p. 148.

[10]   Burrows, C. and Sahinkaya, M., "Vibration Control of Multi–Mode Rotor–Bearing Systems," Proceedings of the Royal Society–London, Vol. 386, 1983, pp. 77–94.

[11]   Kanemitsu, Y., Ohsawa, M., and Watanabe, K., "Real Time Balancing of a Flexible Rotor Supported by Magnetic Bearings," 2nd International Symposium on Magnetic Bearings, Tokyo, Japan, July 12–14, 1990, pp. 265–272.

[12]   Chen, H.M. and Ku, R.C. "Forced Responses of Submerged Rotors in Magnetic Bearings," ROMAG '91 Magnetic Bearing and Dry Gas Seal Conference, March 13–15, 1991.

SESSION 2   TORSION

A New Generation of Torsional Stress Analyzers for Turbine—Generators

Hans L. Berger, Stefan T. Kulig

Siemens A G, Mulheim a.d. Ruhr, Germany.

ABSTRACT

Torsional stress analyzers (TSA) are used in power plants to monitor turboset shaft lines with respect to unallowable torsional stresses. This paper describes a new generation of these devices measuring torques at the shaft line directly at one or several points.

The principle design of the new TSA is explained, the basics of the mathematical procedure to calculate stresses and fatigue are laid down.

The advantages of this new development compared to those types already on the market are pointed out.

1.   INTRODUCTION

Today's large turbosets produce an electrical output of up to 1400 MW. Of principle importance with respect to turbosets is the necessity of assuring on-load operation of the units after failures in the electrical system. I.e., system failures should not result in unit trips; if this cannot be avoided, however, the generator must be re-connected to the grid as soon as possible.

In case of an electrical failure in the system, e.g. a short-circuit, very high transient currents will flow in the generator, causing appropriate transient electrical torques. Thus, the shaft line is induced to torsional vibrations causing increased stresses in the bearing journals and at the couplings. In the past, this has led to several severe shaft failures [1]. Consequently, the effects of system faults on the turboset have been intensively investigated in the last decade.

The resulting new findings on failure effects have led to modifications of the appropriate design standards and evaluation criteria. The most important results are:

—   Interaction of system and turboset may result in loads representing a higher danger for the shaft line than those characteristic for the three-phase terminal fault. These loads are faulty synchronization, high-speed reclosing and out-of-step operation.

—   It is neither possible nor reasonable to dimension the shafts in such a way that the above failures can be endured just as often as they come up.

With nearly all severe failures, part of the shaft line life is consumed due to mechanical overstressing. This especially applies to older units that were already designed some time ago when the new findings had not been available.

The natural consequence has been the development of a monitoring device to record and evaluate the torsional vibrations occuring with generator faults. The first fully applicable device was already developed in the 70'ies by SIEMENS/KWU and installed in 1977 when commissioning the Swiss nuclear power plant Gösgen. Subsequently, also other manufacturers developed and used similar devices [2—4]. These devices were named torsional stress analyzers or just TSA-devices.

In parallel to the progressing development in the area of probe and computer technology a completely new device was developed in the late 80'ies that is going to be described in the following.

2.   BASIC REQUIREMENTS ON THE TSA

Electrical system faults normally occur all of a sudden. They are transferred to the generator by the unit transformer stopping or accelerating the rotor accordingly. The resulting torsional vibrations of the shaft line are slightly damped, decaying after approx. 30—60 s only. In case of most failures, we are not dealing with individual faults but several surges in short intervals, exciting the vibrating shaft line again. I.e., there are superpositions resulting with respect to the relevant phase position in new excitations or reductions of the appropriate vibrations.

Fig. 1 shows the example of torque oscillations at the coupling between generator and the last turbine section after a three-phase terminal fault. In case (a), fault reclosing takes place at the vibration maximum. This leads to a partial elimination of the two vibration modes, i.e. shaft load is smaller than before fault clearing. In case (b), however, it has been assumed that the fault is cleared in the first stage of line protection at a minimum of the torsional vibration mode. Due to the phase similarity of the two vibrations involved, vibration rises severely after switching, resulting in very high shaft stresses.

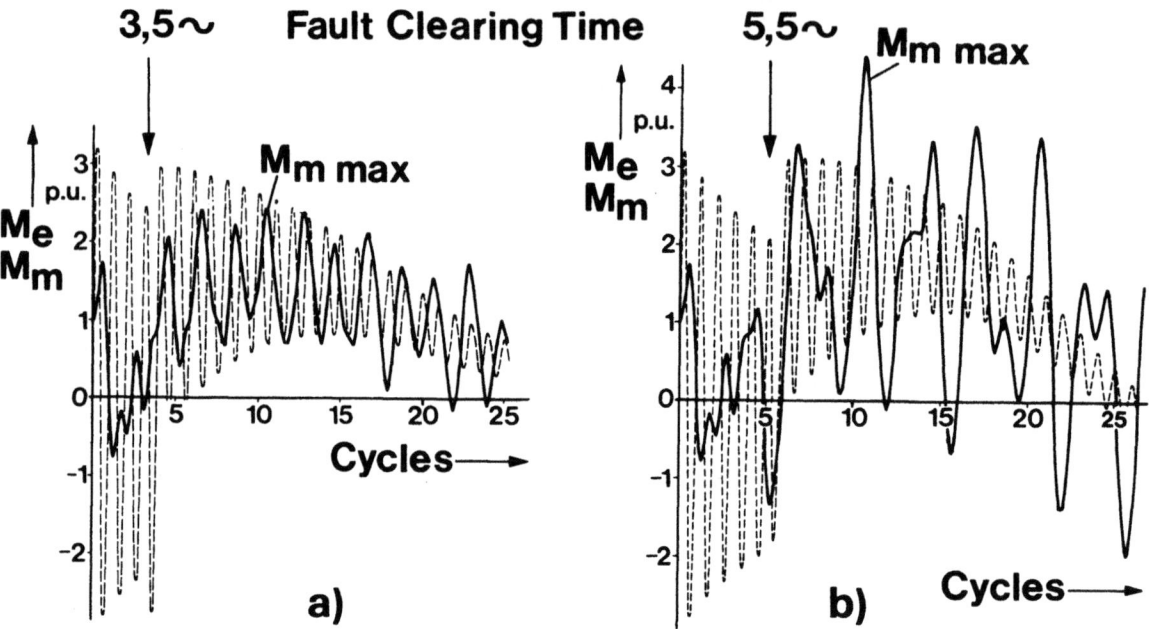

Fig. 1   Electrical and mechanical torques during and after clearing
a three-phase fault near the power plant

In case of a sequence of switching operations as with three-phase reclosing, this effect can be superimposed several times. Fig. 2 gives an example of torsional torques after three-phase system high-speed reclosure. It is standard practice of system protection procedures that following fault occurance the faulty line is switched off at both ends after approx. 80—100 ms and connected in again after approx. 1 s. This reflects the assumption that the fault has been cleared on its own in this period. If this is the case, normal operation proceeds. If not, connecting in results in repeated faults, this time already with mechanical preload, leading finally to fault clearing after 80—100 ms. In the example at hand, the torsional torques increase to 8-fold rated torque caused by repeated switching operations at unfavourable switching point. It is evident that similar failures may result in totally different unit loads.

Accordingly, the degree of load is highly dependent on the relevant phase position. To assess the effects of a failure, the twist of a shaft line before the failure must be definetely known. This will be only possible via on-line measurements at the shaft. Failure analysis only on the basis of the electrical load is not possible, as the relevant torsional vibrations cannot be accurately derived.

3.   TSA—DEVICES OF THE FIRST GENERATION

Such devices were offered by various manufacturers [2—4]. They generally worked on the principle of the shaft line being simulated as a multimass system by a computer (mostly an analog computer) with the torques being characteristic for a fault determined as operands from this model. Input data for the model have mostly been the measured values of the electrical

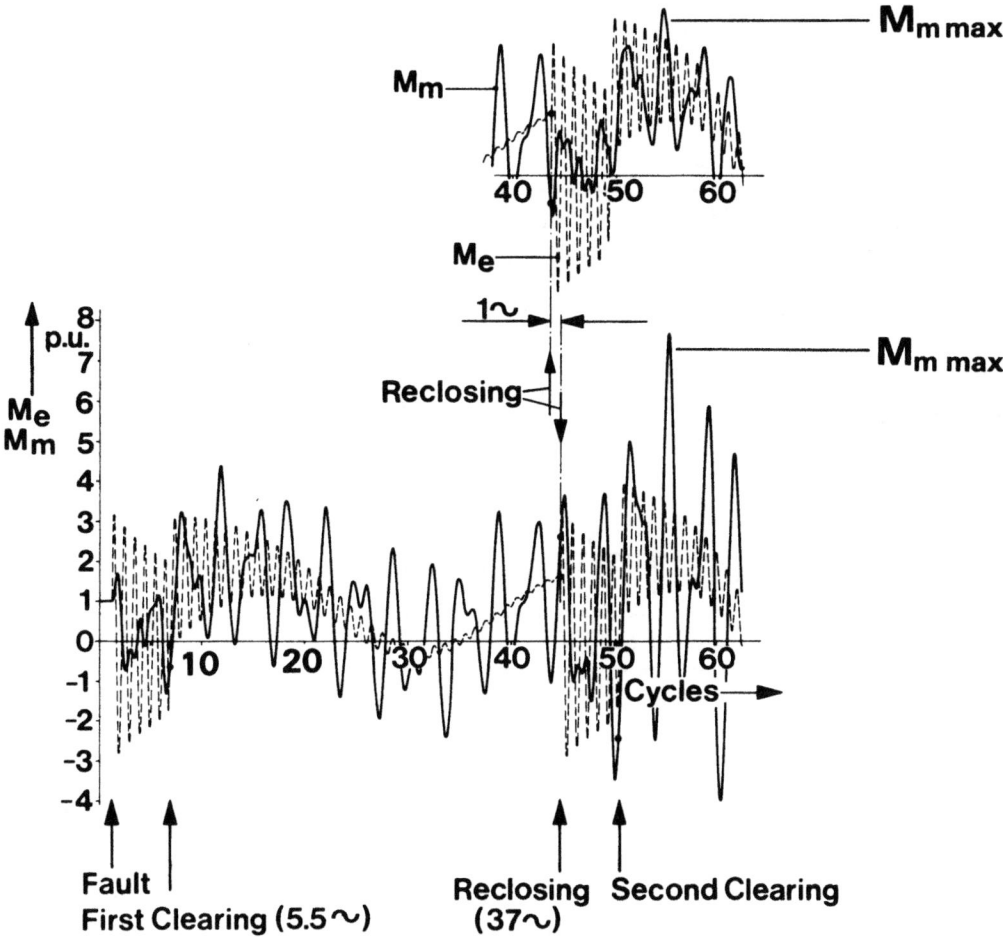

Fig. 2   Reclosing after clearing a three-phase fault — unsuccessful reclosing —

generator load as well as the steam conditions as unit for the current turbine load. The phase specific correction of the devices took place by measuring the angular velocity of the shaft using one to three planes according to the appropriate type and manufacturer. In Fig. 3 the principle mode of operation of a TSA-device acc. to [2] is explained by means of a block diagram. With this model the angular velocity $\dot{\varphi}$ (t) of the shaft is measured via a digitron counting wheel on the HP turbine. It is fed together with the generator and turbine load data to the model of the shaft line stored in the computer. Determination of $\dot{\varphi}$ (t) is required to correct the phase angle of the model according to the angle of the actual shaft line. In addition, mechanical damping of the shaft line is considered via $\dot{\varphi}$. At the analog computer output the time functions of the appropriate mechanical torsional coupling torques are gained in real time. They are used as input data for a series-connected digital processor performing the evaluation of the coupling stresses and the determination of shaft fatigue. This is in accordance with the Rainflow Method [5]. The results are printed and in parallel a signal is transmitted to the control panel. TSA-devices of this type are operated in a number of power plants [6].

4.   TSA—DEVICES OF THE NEW GENERATION

With respect to the principle design the new TSA-device of SIEMENS company decisively differs from the units having been installed, yet. At one or two points of the shaft line torque proportional signals are directly measured and fed to a personal computer for further processing. This has the great advantage that extensive measurements of angular velocity, steam parameters and electrical torque can be dispensed with as well as the series-connected analog computer to calculate torsional torque.

Regarding this concept, it has been a benefit that some years ago a proximity torque transducer emerged on the market that also under power plant conditions has been fully applicable [7]. The relevant measuring principle is based on the magnetostrictive effect on the shaft surface. With any shaft transmitting a torque there are tensile and compressive stresses

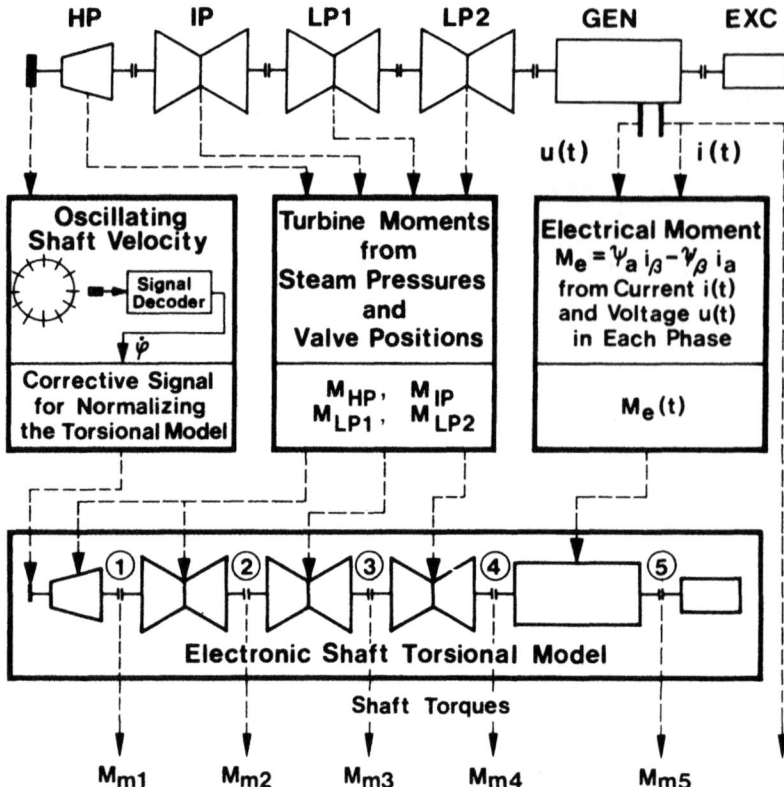

Fig. 3 First generation of torsional stress analyzer (TSA)

on the surface. They are vertical to each other forming an angle of ± 45° with the appropriate axis vector. The maximum mechanical stresses are coming up immediately on the shaft surface with the shear stress component being directly proportional to the torsional torque and inversely proportional to the 3rd power of the shaft radius. The changes of the mechanical stresses in the shaft surface are again proportional to the permeability changes and these are measured by the transducer.

Fig. 4 shows the principle design of a TSA-device of the 2nd generation; in this case with only one transducer in the shaft area LP2/Gen. The measuring signals are processed in a signal processor and subsequently passed to a PC. There the calculations of the actual coupling stresses and shaft fatigue are performed in accordance to the mentioned Rainflow Method. The fatigue values are determined and summed up for the total operating time of the unit. Also in this case the results are printed while in parallel, a signal is sent to the control panel.

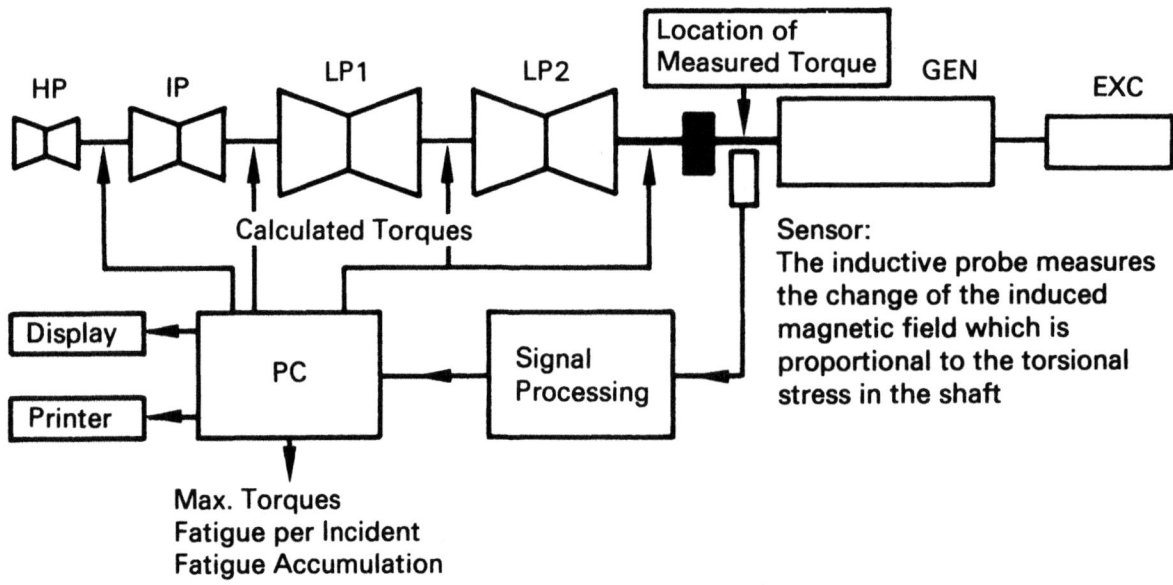

Fig. 4 New generation of the torsional stress analyzer (TSA)

Under normal conditions, the installation of one transducer on the shaft line will suffice; it is located in that shaft area exhibiting most clearly the most important natural modes, i.e. those of the lowest natural frequencies. With respect to those shaft lines in case of which such areas cannot be established or due to pre-calculations resonance monitoring of shaft sections is requested, an additional 2nd transducer is used.

Fig. 5 a illustrates the transducer on the shaft and in front of it the signal processing unit including its components amplifier, phase selective rectifier, phase shifter and oscillator. This picture has been taken in the test stage. The complete set up of the prototype with programming keyboard, PC, operator's console and monitor is shown in Fig. 5b. Normally, the unit is located in the auxiliary control room.

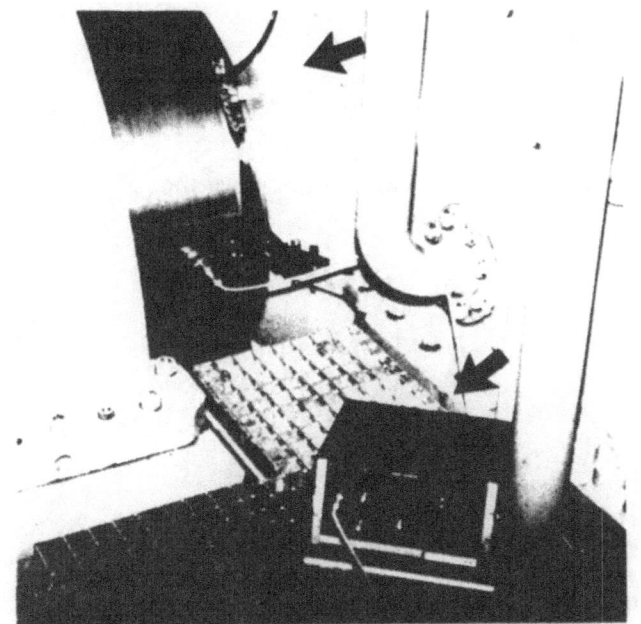

Fig. 5a Transducer and signal processing unit   Fig. 5b Complete set up

5. COMPUTATION OF THE TORSIONAL TORQUES FROM THE MEASURING SIGNAL

There are two ways of determining the coupling stresses and shaft fatigue:

— The torsional torque is directly measured at the point of the shaft line with the highest stresses to be expected. Further processing takes place also for this measuring plane only; the remaining shaft areas of interest are to be analyzed on the basis of experimental values.

— The torsional torque is measured at a point of the shaft line being location of the most important natural modes. Starting with this measuring signal by means of a mathematical procedure the torsional torques of all the appropriate shaft sections are calculated. This mathematical procedure was developed by our company in cooperation with Kaiserslautern University, department "Technomathematik" and converted to a computation programme [8]. In the following the basics of this procedure.

The basis is the modal description of the shaft line:

$$(x_T) \cdot (J) \cdot (x) \cdot (q) + (x_T) \cdot (k) \cdot (x) \cdot (q) = (x_T) \cdot (M) \tag{1}$$

This equation results from the known reel torsional vibration equation by transforming with the eigenvector matrices (x) or (x_T). Where:

(J) = Matrix of mass moments of inertia
(k) = Matrix of torsion spring coefficients
(M) = Matrix of external moments
(q) = $(x_T) \cdot (\varphi)$ = Matrix of modal twist

Before designing the TSA-device for a specific shaft line, the modal parameters of the matrices, incl. natural frequencies and eigenvectors are to be determined with a digital computer. If these are known, the time function of the torsional torque $M_{ti}(t)$ picked up by the transducer may be described as sum of the modal moments. The following applies:

$$M_{ti}(t) = k_{ij}(\varphi_i(t) - \varphi_j(t)) \tag{2}$$

The angle of twist φ_i may be given with

$$\varphi_i(t) = \underbrace{(x_{i1}, x_{i2}, \ldots x_{in})}_{\bar{x}_i^T} \cdot \begin{pmatrix} q_1(t) \\ q_2(t) \\ \cdot \\ \cdot \\ \cdot \\ q_n(t) \end{pmatrix} \tag{3}$$

The same applies to angle φ_j. If these angles are added to equation (2), one of the following two equations will result:

$$M_{ti}(t) = k_{ij}(x_{i1} - x_{j1}, x_{i2} - x_{j2}, \ldots x_{in} - x_{jn}) \cdot \begin{pmatrix} q_1(t) \\ \cdot \\ \cdot \\ \cdot \\ q_n(t) \end{pmatrix} \tag{4}$$

or

$$M_{ti}(t) = k_{ij} \sum_{\nu=1}^{n} (x_{i\nu} - x_{j\nu}) \cdot q_\nu(t) \tag{5}$$

The modal coordinate $q_\nu(t)$ describes the movement of a one lump model being thus representable as

$$q_\nu(t) = Q_\nu(t) \cdot \sin(\omega_\nu t + \varphi_\nu) \tag{6}$$

where ω_ν is the natural angular frequency of the appropriate natural mode. Entering this equation in the torque equation (5) results in:

$$M_{ti}(t) = k_{ij} \sum_{\nu=1}^{n} (x_{i\nu} - x_{j\nu}) \cdot Q_\nu \sin(\omega_\nu t + \varphi_\nu) + M_{eli} \tag{7}$$

where M_{eli} is that portion of the electrical moment being effective at point i. The theoretical treatment of electrical failures has proven that electrical torques can be very well approximated in short time periods ≤ 100 ms by equations in the form of:

$$M_{eli} \approx M_0 + M_1 \sin(2 \cdot \pi \cdot f_N \cdot t + \beta_1) + M_2 \cdot \sin(2 \cdot \pi \cdot 2 \cdot f_N \cdot t + \beta_2) \tag{8}$$

The equation (7) supplemented by (8) describes adequately the time function of the torsional torque at point i. On the right side there are the unknown quantities Q_ν and φ_ν, i.e., amplitude and phase angle of modal vibration as well as the coefficients of the electrical torque M_0, M_1 and M_2 and the phase angles β_1 and β_2. If one restricts oneself in case of a turboset to e.g. $n = 5$ natural modes, $5 \cdot 2 + 5 = 15$ unknown quantities are gained. As the measured time function of the torsional torque at point i is available the unknown quantities can be calculated accordingly. It is assumed that the time function equation (7) corresponds to each point of the measuring curve

$$M_{ti}(t_0) = k_{ij} \sum_{v=1}^{n} (x_{iv} - x_{jv}) \cdot Q_v \sin(\omega_v \cdot t_0 + \varphi_v) + \ldots$$

$$+ M_0 + M_1 \cdot \sin(2 \cdot \pi \cdot f_N \cdot t_0 + \beta_1) + M_2 \sin(2 \cdot \pi \cdot 2 f_N \cdot t_0 + \beta_2)$$

$$\tag{9}$$

$$M_{ti}(t_m) = k_{ij} \sum_{v=1}^{n} (x_{iv} - x_{jv}) \cdot Q_v \sin(\omega_v \cdot t_m + \varphi_v) + \ldots$$

$$+ M_0 + M_1 \cdot \sin(2 \cdot \pi \cdot f_N \cdot t_m + \beta_1) + M_2 \sin(2 \cdot \pi \cdot 2 f_N \cdot t_m + \beta_2)$$

Theoretically it would suffice to set up the equation for 15 moments of time only. To increase the appropriate quality, however, corrections are performed. In this case equation system (9) is set up as an over-defined system while the failures on the right side, e.g. in the natural values, neglection of higher natural frequencies and measuring inaccuracies, are eliminated by the least-square-fit procedure. Though in the existing equations damping has been deliberately neglected, agreement between measurement and calculation is extremely good in the total time range over several seconds, due to the fact that this total range has been divided in a number of small windows with 70 measuring points each and the parameter calculation has been repeated for each of these windows. Damping is thus automatically taken into consideration.

Based on the coefficients determined, it is possible to determine the time function of the mechanical torsional torque for each shaft section and analyze subsequently, even if there are no measurements available for this point.

This is the procedure in its principles. With respect to practical considerations, however, a number of problems had to be solved, the most important ones shall be briefly outlined in the following:

— If in the period considered torque jumps due to another line fault occur, this has to be reflected by application of the wavelet transformation.
— The natural modes being only to some extent present in the measuring signal have to be especially conditioned to avoid the formation of incorrect weighting factors.
— The calculation algorithm has to be organized in such a way that the quantities calculated are available as soon as possible after fault occurrence.

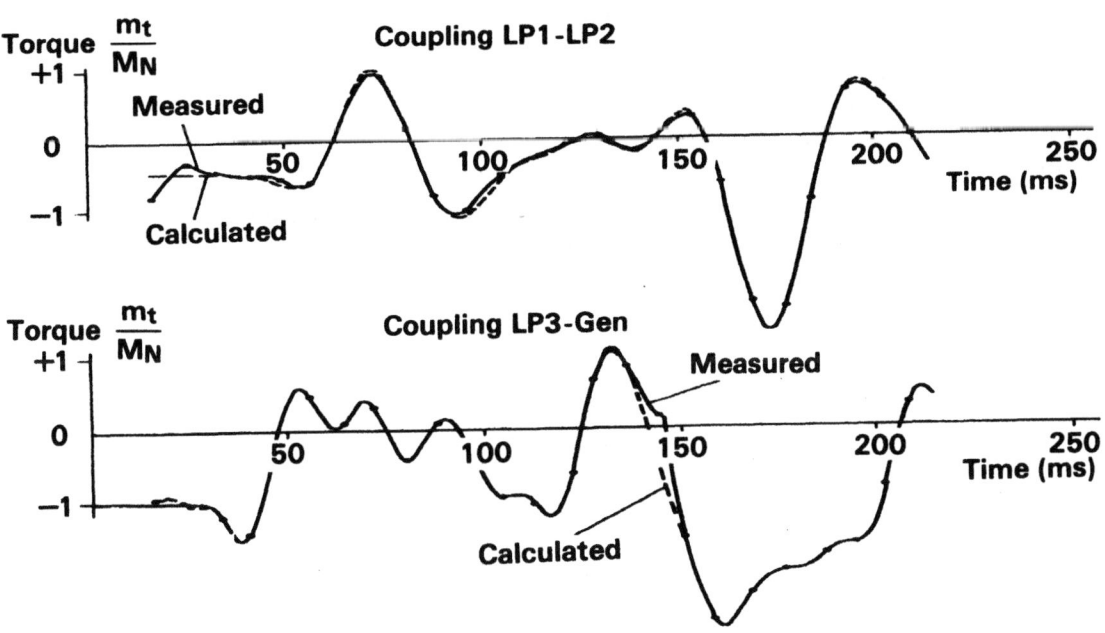

Fig. 6 Comparison of calculated and measured torques after clearing of three-phase fault

These problems have been taken into consideration when realizing the TSA-devices introduced above, assuring such that the device is in a position to assess the effects of system faults on a turboset completely and accurately. As an example, Fig. 6 shows the comparison of two time functions of the calculated and measured torsional torques at the LP1/LP2 and LP3/Generator coupling following a clearing of three-phase fault. They are in good agreement with respect to the evaluation of the failure.

6. Prospects

In the last 10 years, intensive research has taken place with respect to both, improving material knowledge and extracting further information on failures. The newly developed TSA-unit allows a more accurate calculation and evaluation of electrical system fault effects on power plant turbosets as with exactly calculable critical torques it immediately utilizes at least one directly measured torsional torque in the shaft line. Main areas of application for these new devices should therefore be older units as well as large, highly utilized turbosets.

7. References

[1]   D.N. Walker, C.E.J. Bowler, R.L. Jackson:
      Results of Subsynchronous Resonance Test at Mohave.
      Paper T75 176-3, presented at the 1975 Winter Meeting of the IEEE Power Engineering Society

[2]   H. Fick, J. Stein:
      The Torsional Stress Analyzer for Continously Monitoring Turbine-Generators.
      IEEE Transactions, Vol. PAS-99, No. 2, 1980

[3]   E.E. Gibbs, D.N. Walker:
      Torsional Vibration Monitoring.
      Paper presented at the Pacific Coast Electrical Association Engineering and Operating Conference, 1980

[4]   J.D. Hurley, W.H. South:
      Torsional Monitor Equipment for Turbine-Generator Units.
      Paper presented at the American Power Conference, Chicago, 1979

[5]   N.E. Dowling:
      Fatigue Failure Predictions for Complicated Stress-Strain Histories.
      J. of Mats. (JMLSA) No. 1, Bd. 7, 1972

[6]   D. Lambrecht, T.S. Kulig, W. Berthold, J. van Horn, H. Fick:
      Evaluation of the Torsional Impact of Accumulated Failure Combinations.
      Paper 11-06 presented at Cigre-Conference, Paris, 1984

[7]   H. Winterhoff, E.H. Heidler:
      Berührungslose Drehmomentmessung.
      AEG-Kanis Firmenschrift

[8]   B. Claus:
      Torsionsschwingungen von Turbosätzen und die Wavelet-Transformation.
      Diplomarbeit, FB Mathematik, Universität Kaiserslautern, 1989

Coupled axial-torsional vibration in turbine generator rotors

R. A. Driver and T. S. Wilkinson*

NEI Parsons, Heaton Works, Newcastle-upon-Tyne, U.K.

ABSTRACT

Vibration of long turbine blades may provide a coupling mechanism between the axial and torsional modes of vibration of bladed rotors. A mathematical model is presented, together with results for a large Canadian turbine generator.

NOTATION

ω	Angular frequency of vibration (rad/s)
k	Stiffness of thrust bearing (N/m)
λ	Damping coefficient of thrust bearing (Ns/m)
γ	Axial displacement of rotor (m)
p	Axial force in rotor (N)
θ	Torsional displacement of rotor (rad)
t	Torque in rotor (Nm)
L	Length of rotor section (m)
M_S	Mass per unit length of rotor section (kg/m)
A_S	Axial force per unit axial strain of rotor section (N)
I_S	Moment of inertia per unit length of rotor section (kgm)
T_S	Torque per unit torsional strain of rotor section (Nm^2)
M_B	Mass of blade (kg)
N_B	Number of blades in blade row
Z_M	Distance from shaft axis of centre of mass of blade (m)
Z_R	Distance from rotor axis of blade root (m)
Ω_i	Angular frequency of blade eigenmode i (rad/s)
Φ_i	Angle to rotor axis of motion of blade c. of m. in eigenmode i (rad)
K_i	Ratio of axial root b.m. to tangential root s.f. in eigenmode i (m)
T	Forcing torque acting on generator rotor (Nm)

INTRODUCTION

In a turbine generator rotor system, electrical load unbalance gives rise to a generator air gap torque at twice the electrical supply frequency. Therefore, modes of vibration of the rotor system which can be excited by this forcing - notably torsional modes close to this frequency - must be well damped.

Large turbine blades rigidly supported at their roots can resonate at frequencies well below twice the supply frequency, so the dynamic behaviour of these blades must be included in a rotor vibration model if it is to give realistic results up to this frequency. This has already been done considering only torsional vibrations (1). In general however, a row of vibrating blades exerts both an axial force and a torque on the rotor, so the blade vibration interacts with both axial and torsional modes of rotor vibration.

In this paper, a mathematical model for the axial-torsional vibration of a bladed rotor system is presented, in which the two modes are coupled by the vibration of the blades. The effects of this coupling on the damping of torsional rotor vibrations is also analysed.

BACKGROUND

In 1985, a large 60Hz turbine generator, manufactured by a competitor of the authors' company, failed catastrophically in Taiwan. The failure has been attributed to fatigue cracking of turbine blading, caused by torsional vibration of the rotor system (2). Such vibration may be excited by unbalanced stator currents, and a significant response may obtain at 120Hz if there is a natural frequency of torsional oscillation close to that frequency. Nevertheless, the stresses induced in the long LP turbine blading will almost certainly remain insignificant if that blading is not receptive near 120Hz. However, the constraints on designers are such that four pole (half speed) machines of this size tend to have blading with at least one natural frequency close to 120Hz. In this event the blading will be responsive in any of the modes of torsional vibration of the complete rotor system near 120Hz and, although the mass of a blade row is small compared to that of a turbine rotor, the blade flexibility will affect the natural frequencies of the rotor system.

It was considered prudent therefore, to consider whether unbalanced electrical loading might excite dangerous vibrations in the machines designed by the authors' company for the Bruce power station in Ontario, Canada which, at least superficially, are similar to the Maanshan, Taiwan machine.

The vibration characteristics of the NEI Parsons machines were checked. The last row blades have a second modal frequency near 125Hz, and the penultimate row blades have a first modal frequency near 127Hz. In addition, the complete rotor system has a torsional resonance just below 120Hz. These calculated natural frequencies are close to 120Hz, and their accuracy cannot be guaranteed. One might reasonably ask, therefore, what would have happened if there had been coincidence?

In an attempt to answer this question, it was suggested that as energy from the torsional rotor vibration would be transferred to axial vibration through the coupling effect of the vibration of the large blades, the axial shaft vibration would then be

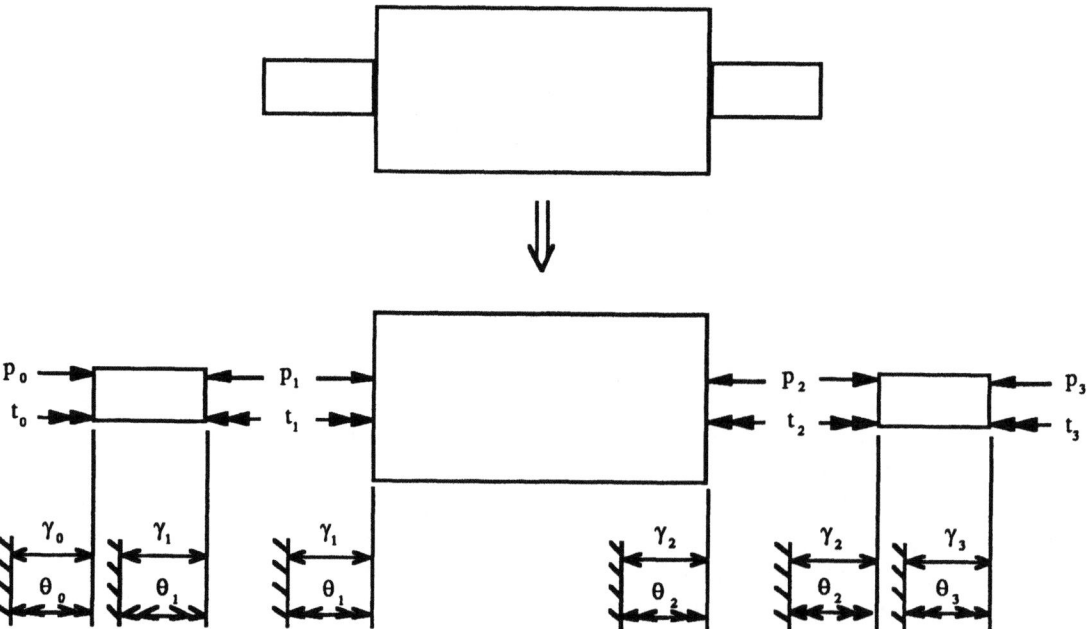

Fig.1. Illustration of dividing a rotor into elements, showing the boundary conditions at the interfaces between elements.

damped by viscous effects in the thrust bearing oil film. This suggestion led to the development of the mathematical model presented here for the coupled axial-torsional vibration of bladed rotors (3).

THE TRANSFER MATRIX MODEL

The rotor system is thought of as a number of elements connected in series, as shown in Fig.1. An element may be a section of shaft, a thrust bearing, a blade row or any other feature which affects the transmission of axial or torsional stress waves along the rotor system. The axial and torsional displacements of the left hand end of each element are the same as the displacements of the right hand end of the previous element. Similarly, the axial force and torque acting on the left hand end of each element are equal and opposite to the force and torque acting on the right hand end of the previous element.

Assuming each element is a linear system, it can be represented by a four-by-four transfer matrix which multiplies the displacement-force-displacement-torque vector at its left hand end to give the displacement-force-displacement-torque vector at its right hand end, as shown in Eq.(1).

$$\begin{bmatrix} \gamma \\ p \\ \theta \\ t \end{bmatrix}_i = [F_i] \begin{bmatrix} \gamma \\ p \\ \theta \\ t \end{bmatrix}_{i-1} \tag{1}$$

Substituting the result of each equation into the equation for the next element gives Eq.(2).

$$\begin{bmatrix} \gamma \\ p \\ \theta \\ t \end{bmatrix}_N = [F] \begin{bmatrix} \gamma \\ p \\ \theta \\ t \end{bmatrix}_0 \tag{2}$$

where

$$[F] = [F_N][F_{N-1}]\cdots[F_2][F_1]$$

Thus the entire rotor system can be represented by a four-by-four transfer matrix which is simply the product of the transfer matrices of all the elements which make up the system.

The Condition for Resonance

No axial forces or torques act on the free ends of the rotor system, so the second and fourth lines of Eq.(2) give Eq.(3).

$$\begin{bmatrix} f_{(2,1)} & f_{(2,3)} \\ f_{(4,1)} & f_{(4,3)} \end{bmatrix} \begin{bmatrix} \gamma \\ \theta \end{bmatrix}_0 = \begin{bmatrix} 0 \\ 0 \end{bmatrix} \tag{3}$$

$$\therefore \quad \frac{\theta_0}{\gamma_0} = -\frac{f_{(2,1)}}{f_{(2,3)}} = -\frac{f_{(4,1)}}{f_{(4,3)}} \tag{4}$$

However, when the rotor system resonates the displacements at the free ends will be non-zero. So, the condition for resonance is that the determinant of the sub-matrix in Eq.(3) is zero. When this condition is satisfied, the ratio of the axial and torsional displacements at the left hand end of the rotor system is given by Eq.(4). Eq.(1) can then be used to calculate the displacements and forces at each boundary, and hence the mode shape.

Derivation of the Transfer Matrix for a Rotor Section

There is no coupling between axial and torsional vibration in a normal rotor section, so the two modes can be considered separately. The transfer equation for each mode is derived by setting up the wave equation for vibration in that mode, solving the equation using the boundary conditions at the left hand end of the section, and using the result to find the boundary conditions at the right hand end. The result of this is the transfer matrix shown in Eq.(5).

$$
[F_i(\omega)] = \begin{bmatrix} \cos\alpha L & -\dfrac{\sin\alpha L}{A_s\alpha} & 0 & 0 \\ A_s\alpha\sin\alpha L & \cos\alpha L & 0 & 0 \\ 0 & 0 & \cos\beta L & -\dfrac{\sin\beta L}{T_s\beta} \\ 0 & 0 & T_s\beta\sin\beta L & \cos\beta L \end{bmatrix}
\tag{5}
$$

where

$$
\alpha = \omega\sqrt{\dfrac{M_s}{A_s}} \qquad \beta = \omega\sqrt{\dfrac{I_s}{T_s}}
$$

Derivation of the Transfer Matrix for a Thrust Bearing

The thrust bearing is represented by a linear spring of stiffness k which decreases the axial force in the rotor in proportion to the axial displacement at the position of the thrust bearing. The bearing has no effect on the transmission of torsional stresses. Therefore, the transfer matrix for the thrust bearing is simply as shown in Eq.(6).

$$
[F_i(k)] = \begin{bmatrix} 1 & 0 & 0 & 0 \\ -k & 1 & 0 & 0 \\ 0 & 0 & 1 & 0 \\ 0 & 0 & 0 & 1 \end{bmatrix}
\tag{6}
$$

Derivation of the Transfer Matrix for a Blade Row

Coupling between the axial and torsional modes of shaft vibration occurs at the large, low pressure turbine blade rows because the blades move in a direction which has both axial and torsional components. The transfer matrix shown in Eq.(7) is derived assuming that the blades are constrained to vibrate in a combination of their first and second eigenmodes, which were computed using the method described by Montoya (4). This is a reasonable assumption as the resonant frequencies corresponding to the higher modes are considerably higher than the frequencies under consideration for shaft vibration.

$$
[F_i(\omega)] = \begin{bmatrix} 1 & 0 & 0 & 0 \\ M_B N_B \omega^2 g_{11} & 1 & -Z_M M_B N_B \omega^2 g_{12} & 0 \\ 0 & 0 & 1 & 0 \\ M_B N_B \omega^2 g_{21} & 0 & -Z_M M_B N_B \omega^2 g_{22} & 1 \end{bmatrix}
\tag{7}
$$

where

$$[G] = \begin{bmatrix} \cos\Phi_1 & \cos\Phi_2 \\ (K_1 - Z_R)\sin\Phi_1 & (K_2 - Z_R)\sin\Phi_2 \end{bmatrix} \begin{bmatrix} \dfrac{\Omega_1^2}{\Omega_1^2 - \omega^2} & 0 \\ 0 & \dfrac{\Omega_2^2}{\Omega_2^2 - \omega^2} \end{bmatrix} \begin{bmatrix} \cos\Phi_1 & \cos\Phi_2 \\ \sin\Phi_1 & \sin\Phi_2 \end{bmatrix}^{-1}$$

IMPLEMENTATION OF THE MODEL

An interactive program is used to assemble and edit a file containing all the data needed for the model of the rotor system. The data does not include the stiffness of the thrust bearing, as this can vary with the electrical load on the generator. From the previous section, it is clear that the transfer matrix of the rotor system is a function of both frequency and thrust bearing stiffness. To define the ranges of these variables in which to search for shaft resonances, the user also generates a small file which defines a grid of points in the stiffness-frequency plane. A log scale is used for the stiffness axis of the grid.

The program which finds resonances of the rotor system reads the data from both of these files. This program includes a subroutine which uses the data from the first input file to calculate the determinant of the matrix in Eq.(3), given values of frequency and thrust bearing stiffness. To find resonances, the main program calls this subroutine with all the pairs of values of frequency and thrust bearing stiffness in the grid defined by the second input file. The values of the determinant returned by the subroutine are stored in a two-dimensional array, so that values calculated for adjacent points are stored in adjacent array elements. The program then searches through the array, looking for determinants of opposite sign stored in adjacent elements. When it finds such a pair, it uses a convergence method to locate the point between the corresponding grid points at which the determinant is zero - a resonance point. When a resonance point is found, the main program calls a second subroutine which calculates the mode shape at that resonance. It then writes the position of the resonance point (ie. the values of frequency and thrust bearing stiffness), and the mode shape data to an output file.

This output file is read by an interactive program which enables the user to generate plots showing the positions of all the located resonance points in the stiffness-frequency plane, or the mode shape at any resonance.

RESULTS AND DISCUSSION

Fig.2 shows the positions of the resonance points located for the turbine-generator rotor systems installed at Bruce power station. The number next to each point is simply for the purpose of identification, and the horizontal dashed lines represent the modal frequencies of the blades. The resonance points occur in lines which are either horizontal or which rise in frequency with increasing thrust bearing stiffness. It appears that the horizontal lines represent modes of vibration which are independent of thrust bearing stiffness - ie. predominantly torsional modes, whereas the rising lines represent modes with a greater axial component. Inspection of the mode shapes confirms this.

Fig.2 reveals a drawback of this method of finding modes of resonance. Some lines of resonance points which are next to each other disappear halfway across the graph, but they always do so in pairs. This is not because the modes disappear at high values of thrust bearing stiffness, but because the program cannot detect two (or four or six etc.) changes in the sign of the determinant between adjacent points on the grid. This problem can be alleviated by using a finer grid, but there is always a possibility that the program will miss some modes of resonance.

Recent calculations by Trevor Evans, a student on placement at NEI Parsons,

40

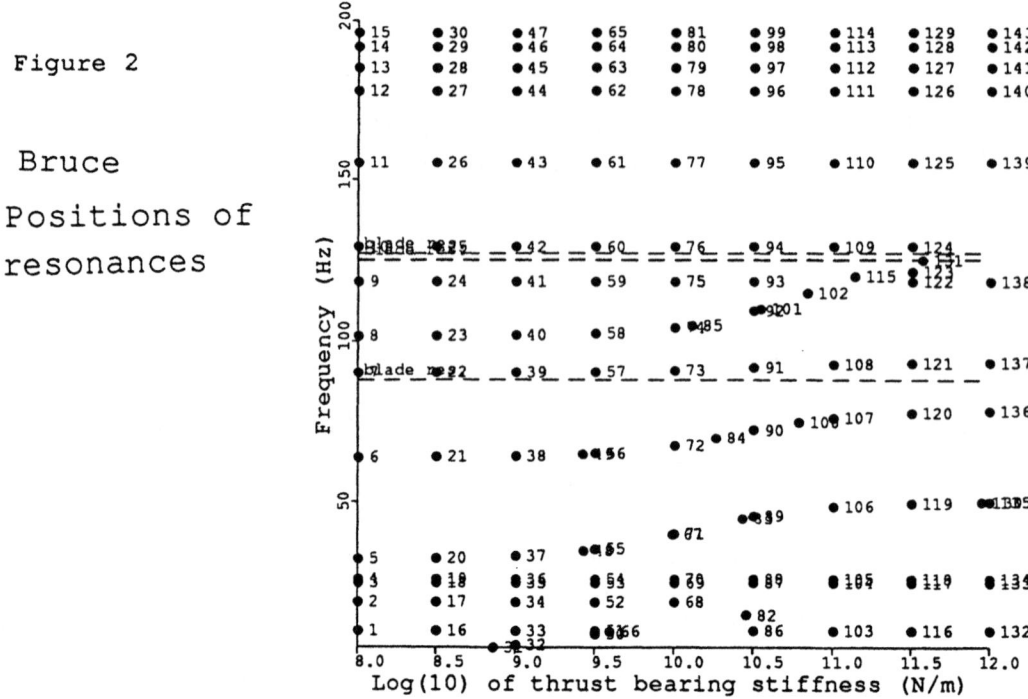

Figure 2

Bruce
Positions of
resonances

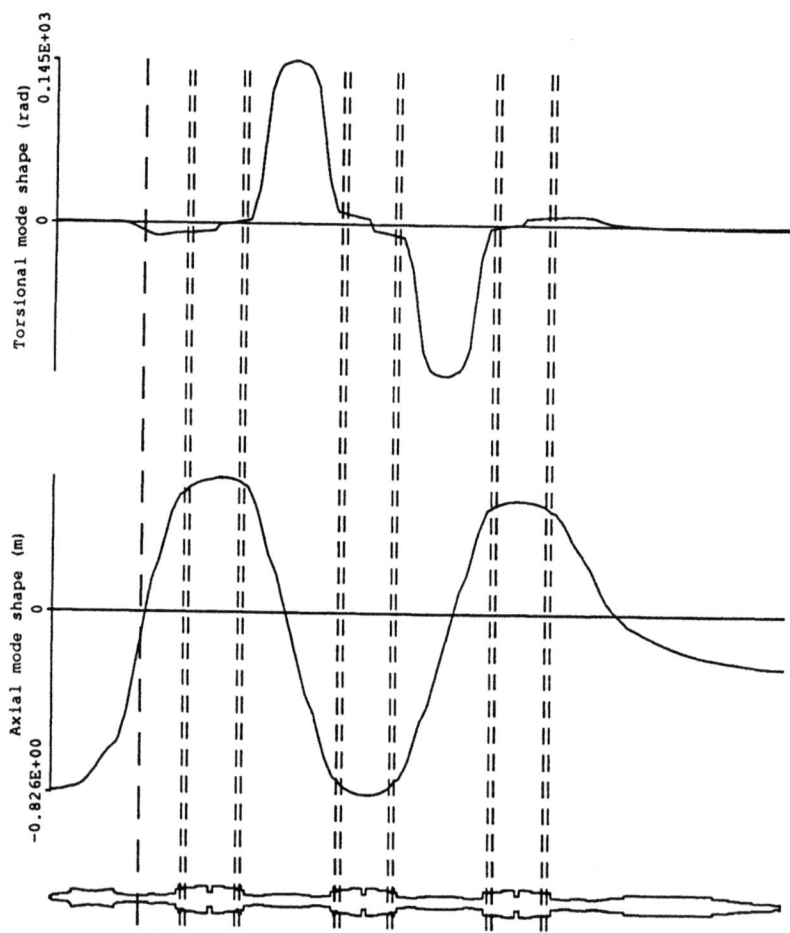

Thrust bearing stiffness: 0.100E+10   N/m
Mode frequency: 118.480500 Hz

Resonance No.41

Figure 3

show the dynamic stiffness of the thrust bearing oil film on the Bruce machines to be approximately 1.6GN/m under full load conditions, and the damping coefficient to be 380GNs/m (5). The forcing torque on the Bruce machines is at 120Hz. Fig.3 shows the mode shape calculated for resonance point 41 (see Fig.2) which occurs close to these conditions, at 1.0GN/m and 118.5Hz. Therefore, forced vibration of the rotor system will occur mainly in this mode. The amplitudes shown in Fig.3 are not absolute - they are only given to show the relative magnitudes of the axial and the torsional components. The coarse dashed line indicates the position of the thrust bearing, and the fine dashed lines indicate the positions of the blade rows.

No experimental work has been done to confirm these results. However, some measurements of amplitudes of axial displacement at the ends of turbine-generator shafts are planned.

Estimation of the Amplitude of Forced Vibration

The power fed into the vibration by the forcing torque is proportional to the amplitude of the mean torsional displacement of the generator rotor, as shown in Eq.(8).

$$P_N = \tfrac{1}{2}T\omega\theta_{GEN} \tag{8}$$

The power dissipated by viscous effects in the thrust bearing oil film is proportional to the square of the amplitude of the axial displacement at the thrust bearing, as shown in Eq.(9).

$$P_{OUT} = \tfrac{1}{2}\lambda(\omega\gamma_{TB})^2 \tag{9}$$

For the vibration to be maintained at a steady amplitude, the power dissipated must be equal to the power fed in. This leads to Eq.(10).

$$\frac{\theta_{GEN}}{T} = \frac{1}{\omega\lambda}\left(\frac{\theta_{GEN}}{\gamma_{TB}}\right)^2 \tag{10}$$

The torsional receptance at the generator rotor varies with the square of the ratio of the mean torsional displacement of the generator rotor to the axial displacement at the thrust bearing. As the forced vibration of the Bruce rotor systems will occur mainly in the the mode shown in Fig.3, this ratio - and hence the receptance - can be calculated from the mode shape data for this mode. The result is a receptance of 0.07e-12rad/Nm. This is extremely low, due to the proximity of the blade frequencies at 125Hz and 127Hz which gives rise to a high degree of coupling with the axial mode and hence a strong damping effect. For comparison, the next torsional shaft resonance is resonance point number 43 (see Fig.2). The frequency of this resonance is 155Hz - well away from the blade frequencies - so the mode is weakly damped. The receptance calculated from the mode shape data for this resonance is 0.14e-6rad/Nm. Therefore the receptance to a forcing torque near 155Hz would be two million times greater than the receptance to a forcing torque near 120Hz.

CONCLUSIONS

In turbine generator rotor systems, axial vibration can be excited by torsional vibration and vice versa, the coupling effect being greater close to the modal frequencies of the blades. This effect can be important when a torsional mode of resonance occurs close to a forcing frequency, enabling energy to be dissipated by viscous effects in the thrust bearing oil film. From this, it can be seen that modal frequencies of blades being close to forcing frequencies brings advantages as well as disadvantages.

ACKNOWLEDGEMENTS

We would like to thank the directors of NEI Parsons for granting us permission to publish this work. Our thanks also goes to all our colleagues who have helped us to write the computer programs and assemble the data necessary to complete the work.

REFERENCES

(1)   Okabe A, Otawara Y, Kaneko R, Matsushita O and Namura K (1991) An equivalent reduced modelling method and its application to shaft-blade coupled torsional vibration analysis of turbine-generator set. IMechE Journal of Power and Energy 1991 Vol 205 No A3 pp 173-181.

(2)   Stress Technology Incorporated (1988) Steam turbine off-frequency operation. Report No 711 G 624

(3)   Driver RA (1991) Analysis of axial and axial-torsional shaft resonance, with results for Bruce, Drax and Pickering shafts. NEI Parsons report No RDL 91-73.

(4)   Montoya J (1966) Coupled bending and torsional vibrations in a twisted rotating blade. The Brown Boveri Review Vol 53 No 3 March 1966 pp 216-230.

(5)   Evans TP (1991) Stiffness and damping of thrust bearings. A masters project report submitted in partial fulfilment of the requirements for the award of the degree of MSc in Computer Integrated Engineering of the Loughborough University of Technology, September 1991.

Nonstationary Torsional Vibrations during Clutch Operation

S.Yanabe, Professor, Nagaoka University of Technology.
A.Okada, President, Nippon Vulkan Co.,LTD.
N.Yamagishi, Manager, Niigata Converter Co.,LTD.

ABSTRACT

This paper deals with nonstationary torsional vibrations of ship power transmission system during clutch operation. Main concern is to clarify the effects of various parameters, such as a friction coefficient of clutch plates and characteristics of clutch oil pressure, on the clutch engaging process and the maximum torsional angle of the shaft coupling.

INTRODUCTION

A friction clutch is one of the essential machine elements in power transmission systems. The clutch transmits a torque by means of a friction force between clutch plates. Because of this, it is often pointed out that the clutch causes various vibrations of the clutch itself or the power transmission system. Especially in automobile, it is well known that a clutch squeal noise or an abnormal vibration in start often occurs in some car. These phenomena are very vibrational and it seems to originate from the nature of automobile transmission sytem. While, the power transmission system in ships, which is composed of an engine, shaft coupling, clutch, gears and propeller, has a little different nature.

In this study, considering the data measured during various clutch operations in a real fishing ship, we identified a dynamic model with torsional four degree of freedom for the powere transmission system of the ship. Next, we carried out various calculations based on this model and investigated effects of some parameters like clutch friction, clutch oil pressure, coupling damping and rigidity, and engine output torque on the clutch engaging process.

MEASURED RESULTS AND POWER TRANSMISSION SYSTEM

Measured Results

Variations of several parameters like rotational speed of engine and propeller shaft, relative torsional angle of coupling, clutch oil pressure and propeller shaft torque were measured during various clutch operations. Figure 1 shows one example of

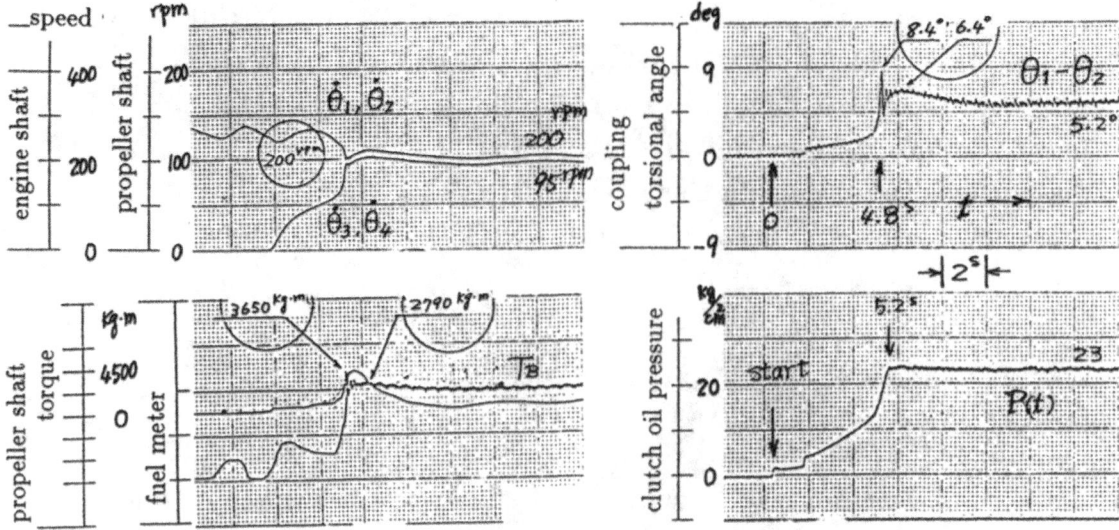

Fig.1 Example of measured results under clutch engaging process of a ship

measured data. It was obtained under the conditions that the ship stopped and the clutch was shifted from neutral to forward. The figure shows that it takes 4.8 s to complete clutch engagement, and that the maximum coupling torsional angle reaches 8.6 °.

Power Transmission System

Power transmission systems of ships can be modelled as shown in Fig.2(a). System dimensions of the tested ship are shown in Table 1. Figure 3 (μ -V curve) shows frictional characteristics of the clutch plates used in the tested ship. In the following calculations the old μ - V curve is used. Figure 4 shows time varying patterns of the clutch oil pressure P(t) and the curve A is the measured one shown in Fig.1.

Table 1 System dimensions

Moment of inertia	I_1	I_2	I_3	I_4	I_5	I_6
(kg.m^2)	316	31.3	0	1.57	2.76	111
Torsional spring		K_i			K_o	
coeff. (kN.m/rad)		110			8460	
Damping coeff.		C_i				
(N.m.s/rad)		100				
Gear ratio	$i_i =$	r_3/r_2		$i_o =$	r_5/r_4	
		1.0			2.1	

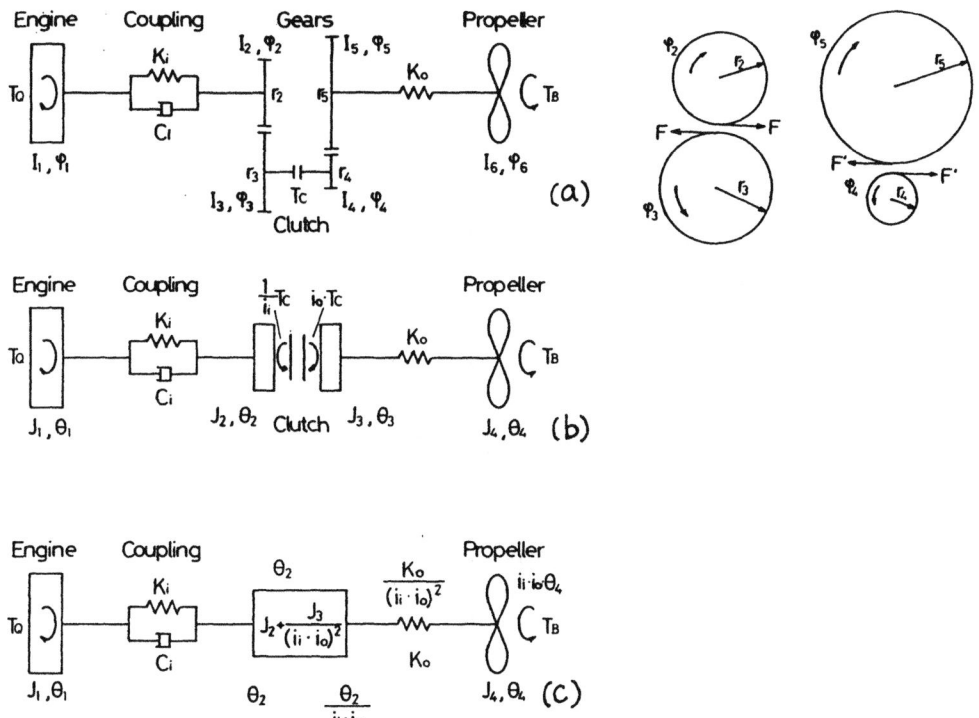

Fig.2 Analytical model of power transmission system of a ship

Figure 5 shows approximate characteristic curves of the engine output torque T_Q. Referring to the rated point (1000 PS, 355 rpm, 100 %), the tested T_Q characteristic (50 %, 260 rpm) is decided. In the same figure, the propeller load torque T_B is also shown, which is assumed to be proportional to the squar of the propeller speed.

DYNAMIC MODEL OF POWER TRANSMISSION SYSTEM

The equations of motion of the power transmission system shown in Fig.2(a) can be written as follows. φ_i denotes a rotational angle of i-th moment of inertia.

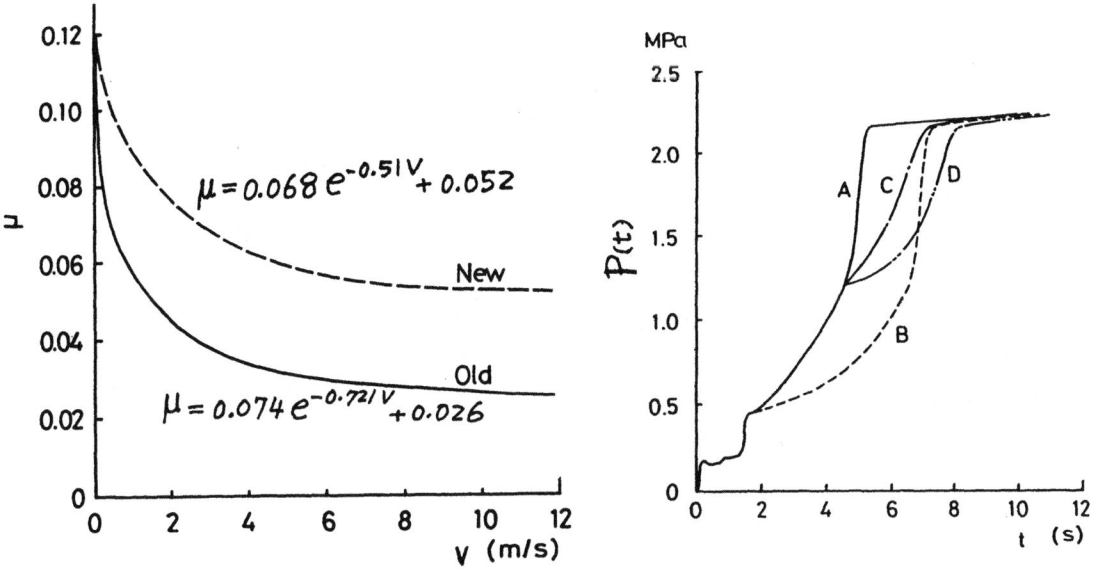

Fig.3 μ - V curve of clutch plate

Fig.4 Clutch oil pressure

46

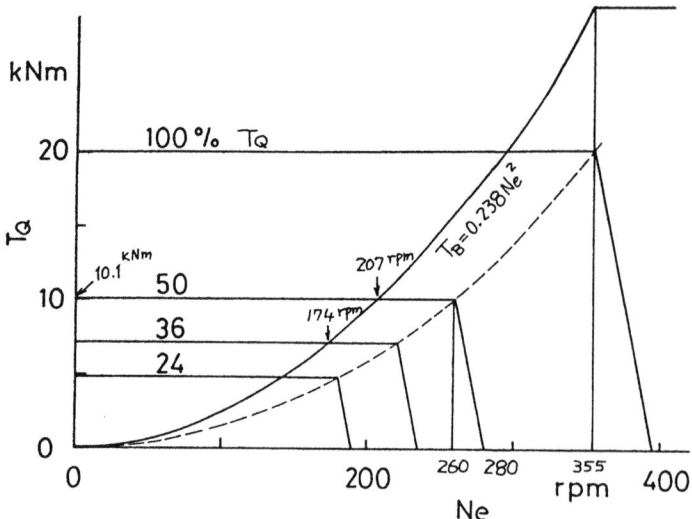

Fig.5 Assumed engine shaft output torque

$$I_1\ddot{\varphi}_1 = -K_i(\varphi_1 - \varphi_2) - C_i(\dot{\varphi}_1 - \dot{\varphi}_2) + T_Q(\dot{\varphi}_1)$$
$$I_2\ddot{\varphi}_2 = -K_i(\varphi_2 - \varphi_1) - C_i(\dot{\varphi}_2 - \dot{\varphi}_1) - Fr_2$$
$$I_3\ddot{\varphi}_3 = Fr_3 - T_C(t)$$
$$I_4\ddot{\varphi}_4 = T_C(t) - F'r_4 \tag{1}$$
$$I_5\ddot{\varphi}_5 = -K_o(\varphi_5 - \varphi_6) + F'r_5$$
$$I_6\ddot{\varphi}_6 = -K_o(\varphi_6 - \varphi_5) - T_B(\dot{\varphi}_6)$$

$$r_2\varphi_2 = r_3\varphi_3, \quad r_4\varphi_4 = r_5\varphi_5, \quad i_i = r_3/r_2, \quad i_o = r_5/r_4$$

where $T_Q(\dot{\varphi}_1)$ and $T_B(\dot{\varphi}_6)$ are the engine output torque and the propeller load torque respectively. $T_C(t)$ is the clutch transmission torque and is calculated by Eq.(2). μ, z, R, A and F denote friction coefficient of clutch plates, the number of plate sets, mean radius of clutch plate, piston area and clutch preload spring force respectively.

$$T_C(t) = \mu z R\{P(t) * A - F\}, \quad \mu = \mu(V), \quad V = (\dot{\theta}_3 - \dot{\theta}_4) * R \tag{2}$$

Eliminating F and F', Eq.(1) can be rewritten as Eq.(5) by using Eqs.(3) and (4).

$$J_1 = I_1, \quad J_2 = I_2 + I_3/i_i^2, \quad J_3 = I_5 + I_4 * i_o^2, \quad J_4 = I_6 \tag{3}$$

$$\theta_1 = \varphi_1, \quad \theta_2 = \varphi_2, \quad \theta_3 = \varphi_5, \quad \theta_4 = \varphi_6 \tag{4}$$

$$J_1\ddot{\theta}_1 = -K_i(\theta_1 - \theta_2)C_i(\dot{\theta}_1 - \dot{\theta}_2) + T_Q(\dot{\theta}_1)$$
$$J_2\ddot{\theta}_2 = -K_i(\theta_2 - \theta_1) - C_i(\dot{\theta}_2 - \dot{\theta}_1) - \{T_C(t)/i_i\} \tag{5}$$
$$J_3\ddot{\theta}_3 = -K_o(\theta_3 - \theta_4) + T_C(t)i_o$$
$$J_4\ddot{\theta}_4 = -K_o(\theta_4 - \theta_3) - T_B(\dot{\theta}_4)$$

The dynamic model of Eq.(5) is shown in Fig.2(b). Equation(5) holds good only for the case where the driving and driven side have different speeds (slip exists). When the slip disappears (Eq.(6)), the clutch engagiment has finished and both sides of the clutch parts rotate together.

$$\theta_3 = \theta_2/(i_i i_o) \tag{6}$$

Substituting Eq.(6) into Eq.(5) and eliminating T_C, Eq.(5) can be written as

$$J_1\ddot{\theta}_1 = -K_i(\theta_1 - \theta_2) - C_i(\dot{\theta}_1 - \dot{\theta}_2) + T_Q(\dot{\theta}_1)$$
$$\{J_2 + J_3/(i_i i_o)^2\}\ddot{\theta}_2 = -K_i(\theta_2 - \theta_1) - C_i(\dot{\theta}_2 - \dot{\theta}_1) \qquad (7)$$
$$-\{K_o/(i_i i_o)^2\}(\theta_2 - i_i i_o \theta_4)$$
$$J_4\ddot{\theta}_4 = -K_o\{\theta_4 - \theta_2/(i_i i_o)\} - T_B(\dot{\theta}_4)$$

These are the equations of motion after clutch engagement and the dynamic model is shown in Fig.2(c).

Nonstationary clutch operation processes can be calculated as follows.

(1) System parameters shown in Table 1, μ, $P(t)$, T_Q, T_B shown in Figs.(3),(4),(5) and initial speed of engine shaft are inputted as data.

(2) Eq.(5) is numerically solved by the Runge-Kutta method.

(3) After Eq.(6) is satisfied, Eq.(7) is solved continuously in the same manner as the above. A time interval of numerical integration was decided to be $\Delta t \fallingdotseq 0.002$ s.

PREDICTED RESULTS

Predicted Results and Outline of Clutch Engaging Process

Figure 6 shows a predicted result corresponding to the measured case shown in Fig.1. Comparing both figures, we can find that the variation process of the coupling torsional angle $(\theta_1 - \theta_2)$ and the speed of engine and propeller shaft qualitatively resemble each other. Further, the predicted value of the clutch engaging time t_f = 4.8 s or the maximum coupling torsional angle $(\theta_1 - \theta_2)_{max} = 8.7$ ° is in good

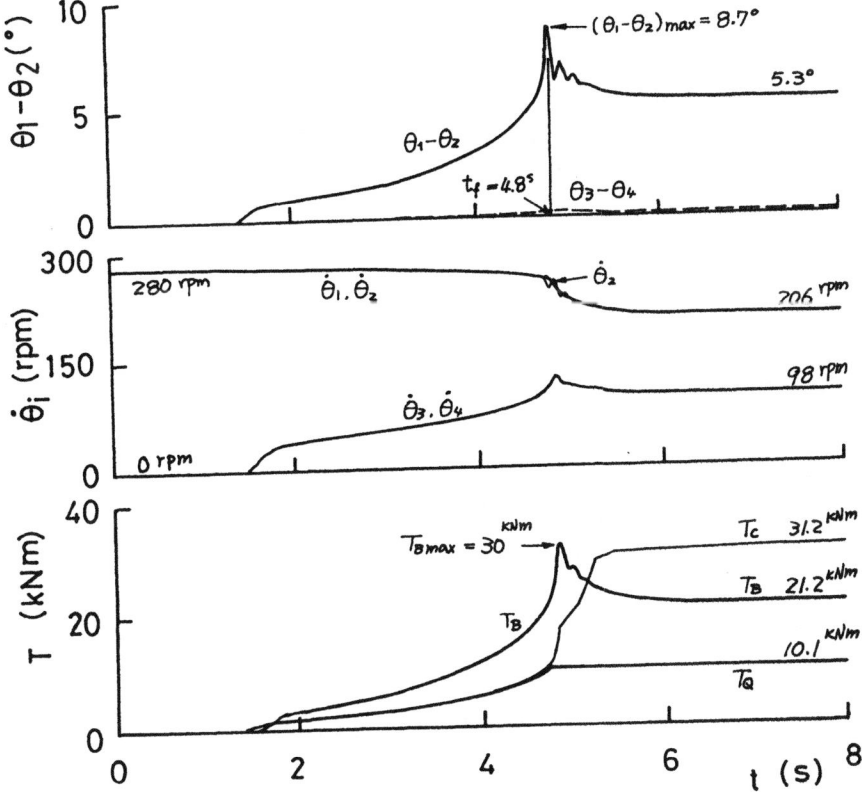

Fig.6 Predicted result corresponding to the measured example

agreement with the measured one. From these, it is thought that the dynamic model shown in Figs.2(b),(c), the developed calculation program and some assumptions for several characteristic curves have enough validity to analyze the clutch engaging process.

Here we outline the clutch engaging process. It should be noted that T_C, which is given as a time function, leads the process. As T_C increases, the rotation of J_2 is disturbed and $(\theta_1 - \theta_2)$ increases. Then the engine speed gradually decreases. When T_C exceeds T_Q, a stable torque balance point disappears. This causes the sudden drop in $\dot{\theta}_2$ and the rapid increase in $(\theta_1 - \theta_2)$. While $\dot{\theta}_3$ gradually increases with T_C and the clutch finishes the engagement when θ_3 reaches the relation of Eq.(6). This corresponds to the time when $(\theta_1 - \theta_2)_{max}$ occurs. Once the clutch engages, the slip between the driving and the driven shaft will not occur because T_C continues to increase rapidly. After this, each speed of engine and propeller shaft approaches the speed where T_Q balances with T_B, having nothing to do with T_C.

Effects of μ -V curve on the clutch engaging process

Among the calculation data used in Fig.6 (this is called standard data), μ -V curve was changed from old clutch plate to new one(see Fig.3). The predicted result for this case is plotted in Fig.7 by dotted lines. Solid lines denote the result of Fig.6. This figure shows that both t_f and $(\theta_1 - \theta_2)_{max}$ become larger when the μ -V curve becomes worse.

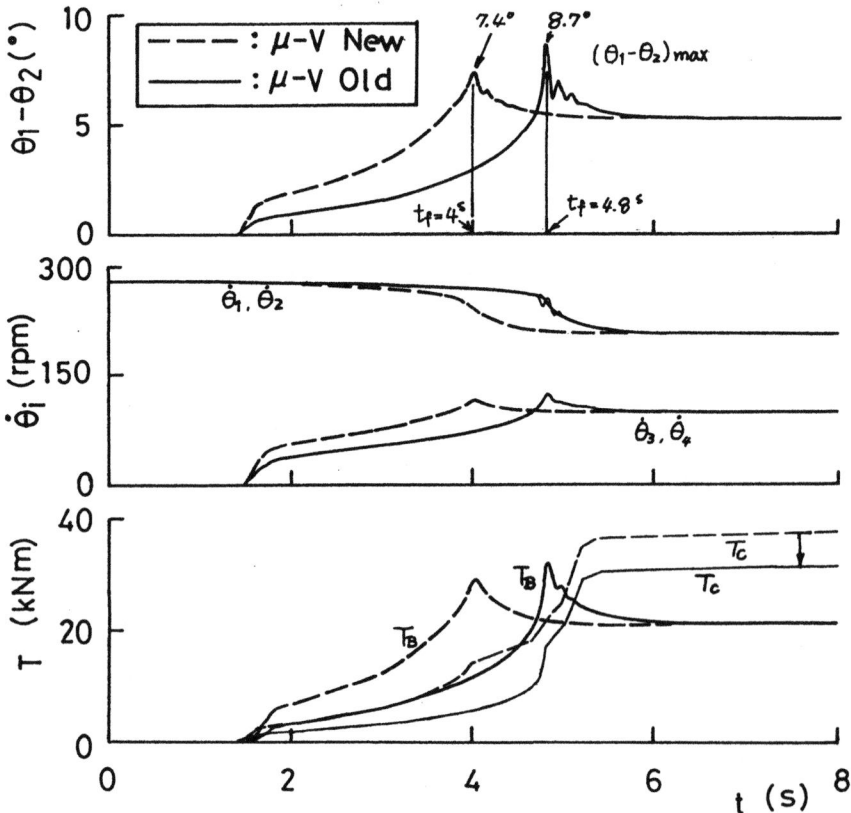

Fig.7 Effects of μ - V curve on the clutch engaging process

Effects of P(t)

Among the standard data, the time varying function of clutch oil pressure P(t) was changed to three cases B,C,D as shown in Fig.4. The results are plotted in Fig.8. Curve A is the result of Fig.6. It is found that t_f largely changes according to the time increasing rate of P(t) at lower pressure level and that $(\theta_1 - \theta_2)_{max}$ seems to be slightly affected by the increasing rate of P(t) at just before clutch engagement.

Effects of Coupling Damping and Rigidity C_i , K_i

Effects of coupling damping are shown in Fig.9. For the case of $C_i = 100C$, the vibrational component is not observed and $(\theta_1 - \theta_2)_{max}$ decreases more than 20 % (from 8.7 to 6.9 °). For the case of 10C, which is not shown, $(\theta_1 - \theta_2)_{max}$ decreased 7 % (8.1 °).
Effects of coupling rigidity are shown in Fig.10. Statically the coupling torsional angle is inversely proportional to the coupling rigidity. The figure shows that $(\theta_1 - \theta_2)$ together with its maximum value decreases to half level and the frequency of vibrational component becomes higher (10 Hz) when $K_i = 2K$.

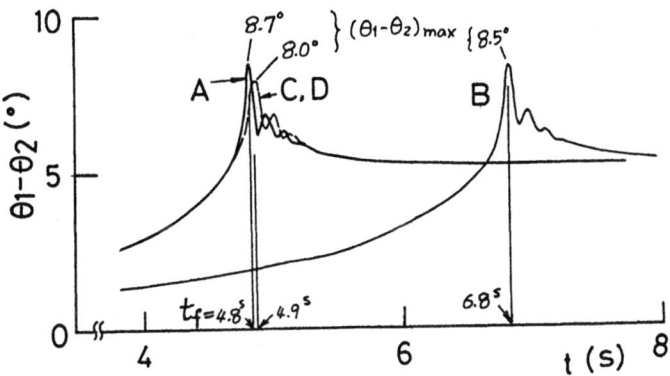

Fig.8 Effects of oil pressure rise pattern on the variation of coupling torsional angle

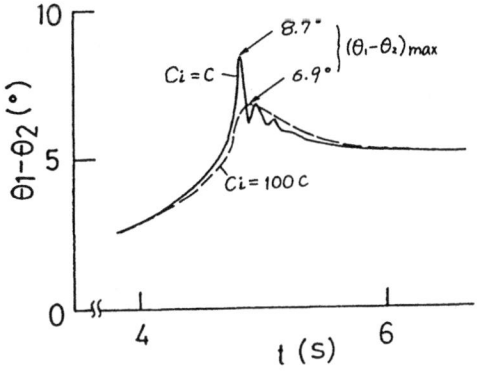

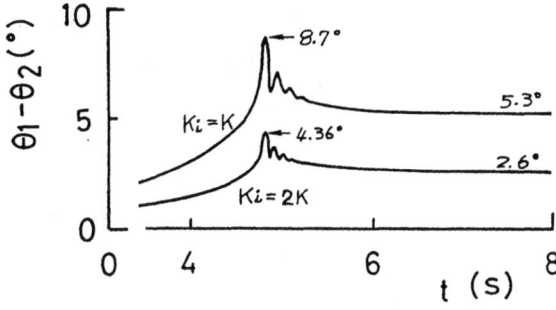

Fig.9 Effects of coupling damping
on the coupling torsional angle

Fig.10 Effects of coupling stiffness
on the coupling torsional angle

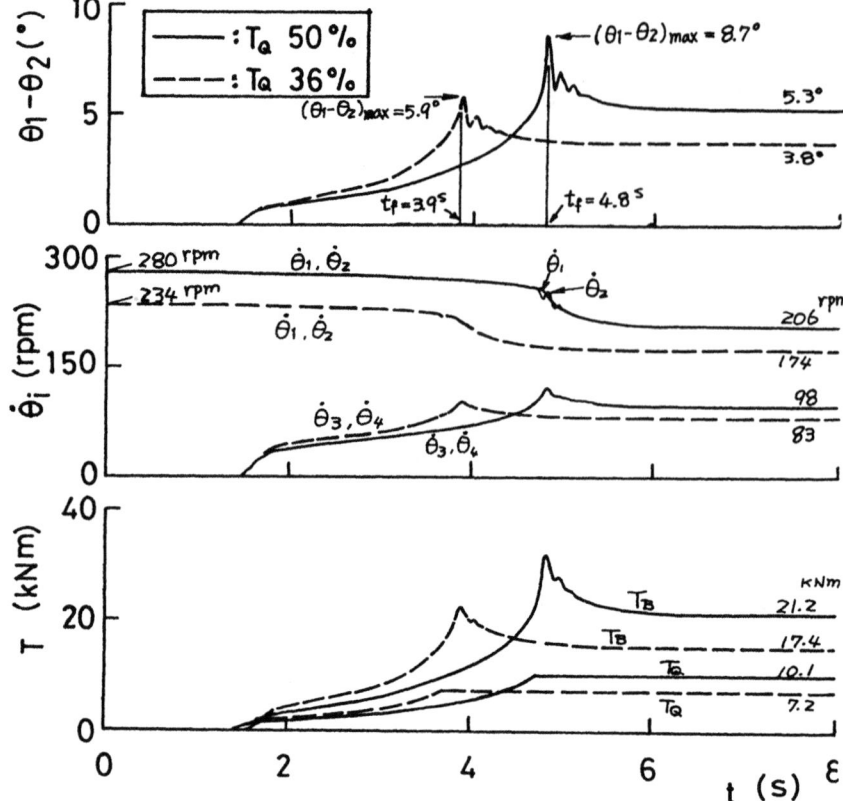

Fig.11 Effects of engine shaft torque on the clutch engaging process

Effects of Engine Output Torque T_Q

Among the standard data, T_Q was changed from 50 % to 36 % in Fig.5. At the same time, the initial speed of engine was set to be the zero output torque point. The predicted result is shown in Fig.11. It is found that both t_f and $(\theta_1 - \theta_2)_{max}$ become smaller as T_Q decreases. After clutch engagement, the engine speed approaches 174 rpm where T_Q balances with T_B.

CONCLUDING REMARKS

Effects of various parameters on the clutch engaging process especially on the maximum torsional angle of the coupling were qualitatively clarified. Quantitative estimation of the maximum value should be investigated in future.

References

1. Kitago K, Tanaka M. Study on dynamic characteristics of friction clutch. Trans. Japan Soc. of Mech. Eng. 1972, 38-311, p.1905 (in Japanese).
2. Matsushita O. Vibrations during friction clutch engaging process. Trans. Japan Soc. of Mech. Eng. 1974, 40-329, p.146 (in Japanese).

SESSION 3   CRACKED STRUCTURES

A FINITE ELEMENT MODEL FOR TRANSVERSE CRACKS IN ROTORS INCLUDING SOME FRACTURE MECHANICS ASPECTS

D.P.W.Horrigan, J.F.Williams, Z.A.Parszewski.
Dept. of Mechanical Engineering,
University of Melbourne.

ABSTRACT

The model presented is a rotating finite element which has a crack at its mid plane. It has been formulated using the F.E. package, PAFEC. The stiffness matrix for this element is a function of the orientation of the bending moment vector at the crack plane and crack depth. The stress intensity factors for this geometry have been computed by the authors allowing the incorporation of a crack growth law.

INTRODUCTION

During the past 15-20 years, from the numerous investigations into the dynamics of cracked rotors, it has become evident that in their modelling there are two major problems to be addressed, namely the effect which a crack has on the stiffness of a shaft, and the mechanism by which the crack opens and closes during rotation and its resultant effect on the stiffness characteristics of the shaft.

Previous authors have attempted to overcome these problems by making simplifying assumptions and consequently the results obtained have been rather varied and approximate. The usual assumptions imply that a linear stress distribution at the crack still exists, for example, by using a step change in the second moment of area to model the weakening of the shaft. If a finite length of a shaft is considered, it demonstrates almost linear characteristics if the orientation of the bending moment vector at the crack plane is constant relative to the crack front. For small perturbations, the force - displacement relations behave almost linearly. Thus, the element displays restricted linear characteristics.

Accordingly, the aim of this study is to develop an accurate cracked beam element for vibration analysis by using data computed on the Finite Element package PAFEC to calculate a series of response surfaces.

REVIEW OF PREVIOUS WORK

Early papers on this topic are those due to Gasch [4] and Henry and Okah-Avae [5]. In order to simulate the effect of a crack in a shaft both authors define a 'crack parameter' to alter the stiffness of the rotor. In addition, the breathing action of the crack is determined to be open or closed depending on the sign of the deflection at the crack plane. From their work, Henry and Okah-Avae [5] conclude that even for relatively large cracks, the change in stiffness of the shaft is quite small. From their analogue computer study, they find that the presence of the crack only affects the magnitude of the system response while running below the main critical speed and only if the dominant mode shape is such that it causes some bending moment at the crack. Their work confirms the predictions of Tondl [10] of instability between the first and second critical speeds which is reduced with increased amounts of damping.

Mayes and Davies [6] investigate the stability and dynamics of cracked rotors by obtaining a Green's Function for the shaft and describe the crack by use of the energy release rate for a circumferential notch. Their results are mainly concerned with the existence of instability and its dependence on the phase of the out of balance due to the crack. From their Green's function approach, they derive expressions for the nth natural frequency and use these to find the approximate location of the crack.

Nelson and Nataraj [7] develop a cracked element by assuming a step change in the second moment of area and applying a variational technique. As with Henry and Okah-Avae [5] they find that the first modes are not especially excited by the presence of the crack but that subcritical resonances occur at odd integer fractions of the shaft speed. Each harmonic of the response is a function of the rotor spin speed causing the harmonics to change whirl direction as the rotor speed changes.

Dimaragonas et. al. [1,2] investigated the stability of cracked rotors using a model which estimates the additional compliance of the crack using Paris' equation. They have computed the effects of disk weight, gravity, damping and unbalance on the stability of the system. In addition to this, using the same model, Dimaragonas and Papadopoulos [3] have looked at the coupling of longitudinal and lateral vibrations caused by the presence of a crack. They conclude that the most pronounced effect appears in the vibration spectrum when both lateral and longitudinal vibrations coexist and can therefore be used as an unambiguous method of crack detection.

MATHEMATICAL MODEL

The model presented is based upon data calculated using the Finite Element package PAFEC. By calculating the ratio of cracked to uncracked stiffness for each coefficient of the stiffness matrix at different crack depths and various angles of bending moment, a theoretical element may be modified to incorporate the effect of a crack. A small section of circular shaft was modelled as shown in Figure (1).

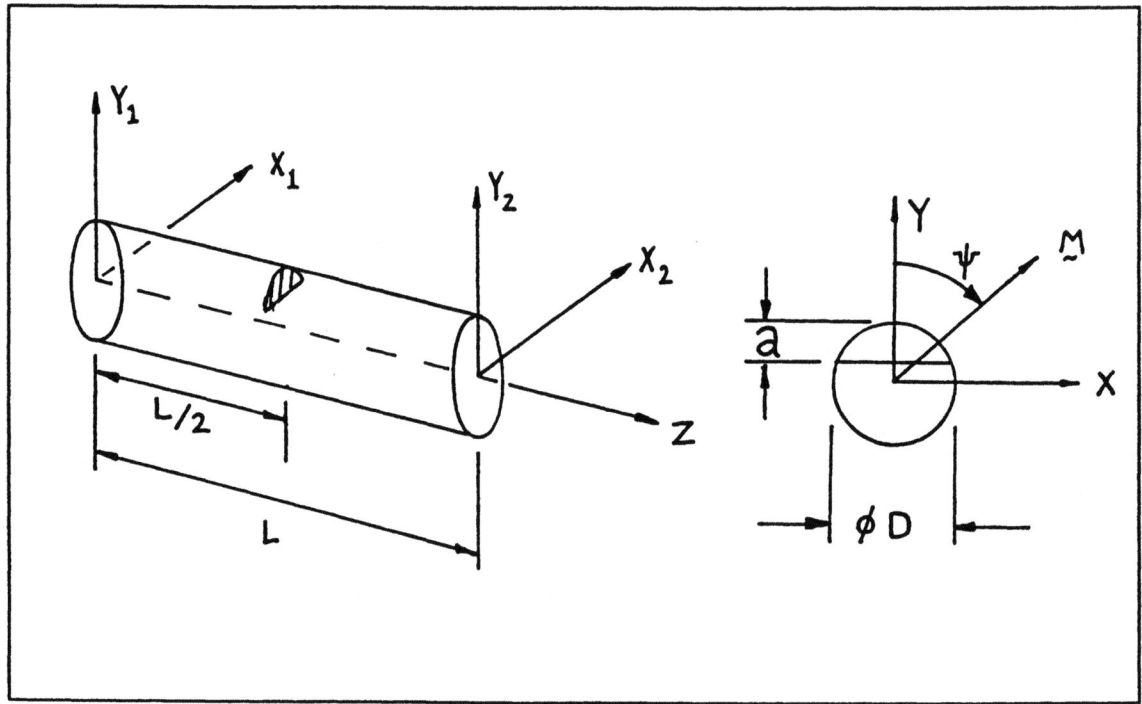

Figure 1 - Element, Axes and Cross Section.

Four loads may be applied at each end, comprising two forces and two moments. The assumed load - displacement relationship is written:

$$P_i = A_i + B_{ij} u_j + C_{ijk} u_j u_k + D_{ijkl} u_j u_k u_l + \ldots \qquad (1)$$

This may be continued into cubic terms and beyond. Thus the column vector of loads {P} may be related to the vector of displacements {u} and their multiples by a rectangular matrix [S]. This may be expressed as:

$$\{P\} = [S] \{u\} \qquad (2)$$

The matrix [S] is rectangular and the column vector {u} has been extended accordingly. Since [S] is to be constant, the columns {P} and {u} may be augmented to matrices by taking numerous sets of loads and the corresponding displacements:

$$[\{P\}^1 \mid \{P\}^2 \mid \{P\}^3 \mid \ldots \mid \{P\}^m] = [S] [\{u\}^1 \mid \{u\}^2 \mid \{u\}^3 \mid \ldots \mid \{u\}^m] \qquad (3)$$

To solve for the as yet unknown matrix [S], least squares regression analysis shows that [S] may be expressed as:

$$[S] = [[u]^T [u]]^{-1} [u]^T [P] \qquad (4)$$

The matrices [u] and [P] are the collection of data columns shown in Equation (3). The computed matrix [S] is in general a non-linear model of the force-displacement characteristics for the cracked element. An example of a quadratic model is given in Appendix C.

Application to Produce a Linearised Finite Element

The aim of this exercise is to produce a finite element which can be used readily in future computation. Since an element of this type displays the near linear characteristics mentioned previously, the model is constrained to the first two terms of equation (1). These are A_i, a set of constants and the B_{ij}, the linear terms. Both of these depend upon the crack depth and the angle ψ of orientation of the bending moment vector.

The general stiffness, damping and mass characteristics of a rotating beam element have been formulated, for example that derived by Parszewski et.al. [8]. The stiffness of the element includes bending, shear and rotary inertia effects and the overall stiffness is the sum of the bending, shear and rotary inertia terms:

$$[K] = [K_B] + [K_S] + [K_I] \qquad (5)$$

Stiffness matrices calculated using Equation (4) will contain the bending ($[K_B]$) and shear ($[K_S]$) components and since the "test specimen" is static, it is assumed that the effect of the crack is contained within these two terms and the effects on the damping and mass matrices are neglected. Evaluations of [K] have been made at intervals of $\Delta\psi=15°$ from the crack being fully closed ($\psi=-90°$) to fully open ($\psi=90°$) and have been normalised against an uncracked version computed with the same element geometries to counter small variations in the system stiffness. The results obtained are thus presented in this normalised form.

RESULTS AND DISCUSSION

The stiffness properties for an element constrained as a cantilever have been computed using the method outlined. The resulting matrix of coefficients is defined in terms of the axes in Figure (1) and the general form of the matrix is as shown in Equation (7) of Appendix A. For comparison, using the same dimensions and material properties, the uncracked matrix from theory as well as the uncracked and cracked matrices from computation are presented in Appendix B.

The model considered here is constrained to constant and linear terms only. Since a zero force - zero displacement condition is applied, the value of the constants A_i should be zero. This is in fact the case with typical values of the magnitude of these coefficients being less than $10^{-2}N$. These values are simply a result of numerical round-off during computation and possibly small errors in the results. However, these coefficients are insignificant and are subsequently neglected.

For the coefficients b_{ij}, 95% confidence intervals have been computed from the regression statistics. These show that for all of the major coefficients, typical estimates of the error in the coefficients are under 0.5%. In addition to this, R squared values for all of the regressions are very close to unity. This indicates that the each response is virtually completely contained in the linear term.

The most significant changes with ψ and "a" occur in the coefficient b_{22}, plotted in Figure (2). This coefficient relates to the bending moment M_X, which acts to close the crack fully when $\psi = -90°$ and open it at $\psi = 90°$. As the crack closes, the value of b_{22} will approach that of an uncracked element but as it opens it assumes a lower value which is dependent on crack depth. Interestingly, the family of curves in Figure (2) has a minimum at around $\psi=30°$. This is possibly due to a combination of the effect of the crack closing and asymmetry of the cross section due to the crack. This behaviour is also seen in b_{44}, which relates to the bending moment M_Y. The crack has less effect in this direction and consequently the resulting curves do not display as great a variation in stiffness but the general trend is the same.

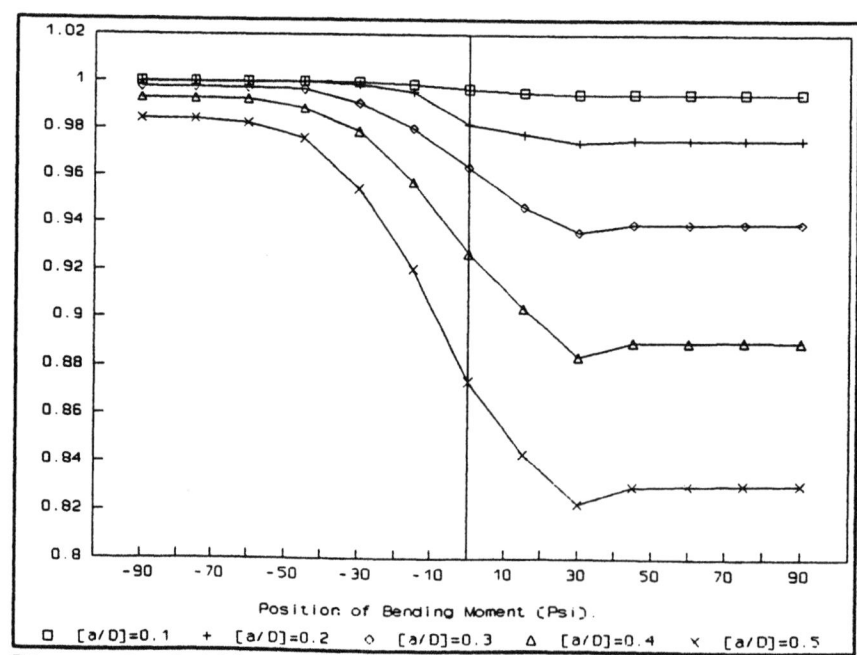

Figure 2: Stiffness Ratio for Coefficient b_{22} v ψ

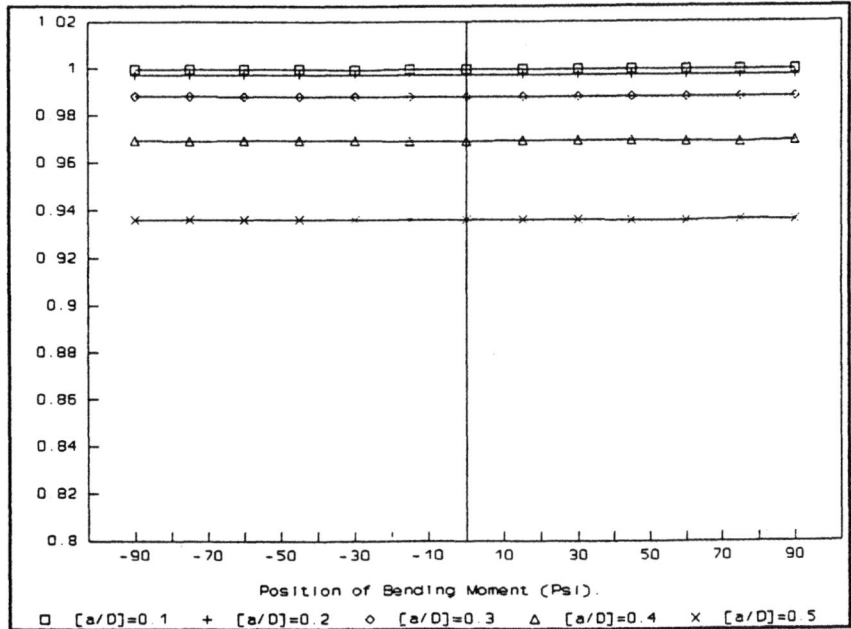

Figure 3: Stiffness Ratio for Coefficient b_{11} v ψ

The remaining coefficients do not change greatly with ψ, as for example b_{11} which is plotted in Figure (3). In general, coefficients which relate to displacements are not as sensitive to the orientation of the bending moment to the crack front and show relatively small variation with crack depth. It is interesting to note that while the changes in stiffness are not the same for each coefficient the proportional change in each remains approximately constant for all the coefficients regardless of ψ. That is, the proportional change from one crack depth to the next to the change from no crack to 50% crack depth is approximately constant.

The theoretical uncracked element derived by Parszewski et. al. [8] in Equation (8) predicts that the only terms which are non zero are those on

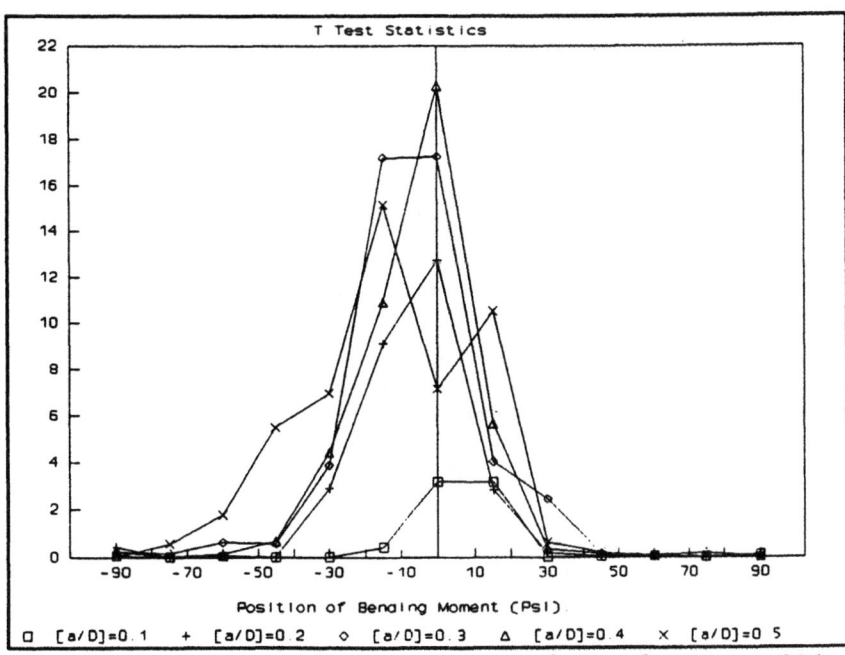

Figure 4: Test Statistics. Critical T value is 2.306.

the major and minor diagonals of the matrix. This is clearly reflected in the calculated uncracked matrix in Equation (9) which is very close to that given in Equation (8) apart from the small residual values. However, with a crack of 20%, as in Equation (10), there appears to be some growth in these off-diagonal terms, an effect which tends to increase with crack depth.

The off-diagonal coefficients are of the same order of magnitude as the errors for the larger coefficients. However, they will cause changes to the vibration spectrum which may be unique to a crack. As a result, it is necessary to assess whether the appearance and growth of these terms is due to random error or whether they are statistically significant. Each of these was tested at a 95% level of confidence against a null hypothesis that the mean value of the coefficient is zero. Figure (4) shows a plot of the T test statistics for b_{24} against ψ for a range of crack depths. In this particular case, the value of the test statistic rises with crack depth except for deeper cracks where the values seem to level out. This suggests that this coefficient becomes almost certainly non zero since the test statistic becomes large relative to the cut off value.

The presence of non zero off-diagonal terms mean that loads applied in one plane cause displacements in a perpendicular plane. The effect of this is to couple the vibrations in the two planes X-Z and Y-Z. Consequently, the vibrations in these two planes will be coupled by both internal and bearing damping as well as by the crack.

For this crack geometry the Mode I, II and III stress intensity factors due to a bending moment have been computed using PAFEC. The results obtained show that most of the crack front the value of K_I is very similar to the value for a rectangular beam as given by Tada et. al. [9] and is in fact at least an order of magnitude larger than Modes II and III. Thus, the growth rate may be estimated using K_I with a rate law of the form:

$$\frac{da}{dN} = C(\Delta K)^m \qquad where \quad \Delta K = \lambda \Delta \sigma \sqrt{\pi a} \qquad (6)$$

where λ is a shape factor and $\Delta \sigma$ the bending stress range.

The crack growth rate is integrated over time to update the depth of the crack "a", which allows the estimation of the stiffness coefficients by interpolation between crack depths. Inspection of Figures 2 and 3 shows that for a fixed value of ψ the coefficients change in a similar proportion.

CONCLUSION

The method outlined in this work has been used to produce a more accurate rotating finite element with a crack at its mid plane. This element is suitable for incorporation into finite element computations for analysing the dynamic effects which cracks have on rotors. The use of fracture mechanics allows the inclusion of a crack growth law for further investigations into life expectancy and prolongment.

The results obtained showed that the coefficient b_{22}, which is related to the opening and closing bending mode, undergoes the highest changes and rates of change during rotation. The moment and rotation associated with this should be the best indicators of the presence of a crack. The large changes in this coefficient suggest that system identification techniques may be able to identify the presence and location of a crack.

Finally, the emergence of off diagonal terms in this model suggests that the presence of a crack in a rotor will produce out of plane effects which may give some hint of the presence of a crack.

REFERENCES

[1]   Dimaragonas, A.D., Papadopoulous,C.A. - 1983
      "Vibration of Cracked Shafts in Bending."
      J. Sound and Vibration, Vol. 91 No. 4   pp 583-593.

[2]   Dimaragonas, A.D., Paipetis, S. - 1983
      "Analytical Methods in Rotor Dynamics."
      Applied Science Publishers, London.

[3]   Dimaragonas, A.D., Papadopoulous,C.A. - 1987
      Coupled Longitudinal and Bending Vibrations of a .."
      J. Sound and Vibration, Vol. 117, No. 1   pp 81-93.

[4]   Gasch, R - 1976
      "The Dynamic Behaviour of a Simple Rotor with a .."
      I.Mech.E., Paper C178/76.

[5]   Henry,T., Okah-Avae, B.E. - 1976
      "Vibrations of Cracked Shafts."
      I.Mech.E. Paper C162/76.

[6]   Mayes,I.W., Davies, W.G.R. - 1976
      "The Vibrational Behaviour of a Rotating Shaft System .."
      I.Mech.E. Paper C168/76

[7]   Nelson, H.D., Nataraj, C. - 1986
      "The Dynamics of a Rotor System with a Cracked Shaft."
      Trans. A.S.M.E. Vol. 108, p189.

[8]   Parszewski, Z.A., Kirby, E.D., Krynicki, K. - 1987
      "F.E.M. in Turbogenerator Dynamics - Rotor Modelling."
      Proc. I.C.F.E.M. 5, pp 192-198.

[9]   Tada, H., Paris, P.C., Irwin, G. - 1974
      "Stress Analysis of Cracks Handbook."
      Del Research Corp., Hellertown, Pennsylvania.

[10] Tondl, A. - 1965
      "Some Problems of Rotor Dynamics."
      Chapman - Hall, London.

[11] Walpole, R.E., Myers, R.H.,
      "Probability and Statistics For Engineers and Scientists."
      Collier MacMillan, 3rd Edition.

APPENDIX A

The general form used for the 4*4 cantilever matrix is:-

$$\begin{Bmatrix} F_x \\ M_x \\ F_Y \\ M_Y \end{Bmatrix} = \begin{bmatrix} b_{11} & b_{12} & b_{13} & b_{14} \\ b_{21} & b_{22} & b_{23} & b_{24} \\ b_{31} & b_{32} & b_{33} & b_{34} \\ b_{41} & b_{42} & b_{43} & b_{44} \end{bmatrix} \begin{Bmatrix} U_x \\ \theta_x \\ V_Y \\ \phi_Y \end{Bmatrix} \qquad (7)$$

where the coordinates are as defined in Figure (1).

APPENDIX B

For the element with dimensions as defined in Figure (1) and material properties given below, the following is a comparison of the stiffness matrices in bending and shear. The material and section properties are:

Young's Modulus: $2.05*10^{11}$ Nm^{-2}

Poisson's Ratio: 0.3

Length to diameter ratio: 2.5

Using the element formulation of Parszewski et. al. [8], the combined bending and shear stiffness matrix for the uncracked section is:

$$[K] = [K_B + K_S] = \begin{bmatrix} 1.429*10^8 & 0.0 & 0.0 & -4.465*10^6 \\ 0.0 & 202434.3 & 4.465*10^6 & 0.0 \\ 0.0 & 4.465*10^6 & 202434.3 & 0.0 \\ -4.465*10^6 & 0.0 & 0.0 & 202434.3 \end{bmatrix} Nm^{-1} \quad (11)$$

In comparison, the method outlined predicts a stiffness matrix of:

$$[K] = [K_B + K_S] = \begin{bmatrix} 1.413*10^8 & -6.2 & -268.6 & -4.383*10^6 \\ -16.2 & 196553.0 & 4.292*10^6 & 0.4 \\ -276.1 & 4.371*10^6 & 1.409*10^8 & 5.9 \\ -4.307*10^6 & 0.2 & 8.9 & 196893.2 \end{bmatrix} Nm^{-1} \quad (9)$$

For an element with a 20% crack and the orientation of the bending moment vector at $\psi=60°$, the predicted matrix is:

$$[K] = [K_B + K_S] = \begin{bmatrix} 1.4083*10^8 & 108.0 & 3134.4 & -4.3686*10^6 \\ 78.4 & 191674.3 & 4.3008*10^6 & -1.8 \\ 3225.4 & 4.3729*10^6 & 1.4088*10^8 & -87.7 \\ -4.2936*10^6 & -4.7 & -140.5 & 196072.4 \end{bmatrix} Nm^{-1} \quad (10)$$

APPENDIX C

Using all the previous load and displacement data computed, the following are the coefficients A,B and C as in equation (1) for the quadratic model of the element with a 50% crack.

$$A^T = [14.64 \quad -828.4 \quad 59.89 \quad -2.542]$$

$$B = \begin{bmatrix} 1.320*10^8 & -3220.0 & -1.191*10^5 & -4.094*10^6 \\ -1.645*10^6 & 2.494*10^4 & 8.051*10^5 & 6.184*10^4 \\ -1.804*10^7 & 4.518*10^6 & 1.453*10^8 & 5.396*10^4 \\ -9.442*10^7 & 1.737*10^4 & 8.130*10^5 & 2.933*10^6 \end{bmatrix}$$

$$C = \begin{bmatrix} 2.32*10^{10} & -3.45*10^8 & -1.11*10^{10} & -1.40*10^9 & 2.41*10^7 & 1.61*10^9 & 1.12*10^7 & 2.67*10^{10} & 3.63*10^9 & 2.06*10^8 \\ -2.56*10^{10} & -1.25*10^{10} & -4.03*10^{11} & 5.82*10^8 & 3.20*10^8 & 1.88*10^{10} & 3.86*10^8 & 2.78*10^{11} & 1.24*10^{10} & 1.57*10^7 \\ 2.00*10^{12} & -5.31*10^{11} & -1.70*10^{11} & -1.25*10^{11} & -8.99*10^7 & -5.80*10^9 & 1.65*10^{10} & -9.34*10^{10} & 5.30*10^{11} & 1.94*10^9 \\ 9.42*10^{12} & -2.51*10^{12} & -8.07*10^{13} & -5.85*10^{11} & -3.48*10^8 & -2.25*10^{10} & 7.81*10^{10} & -3.65*10^{11} & 2.51*10^{12} & 9.07*10^9 \end{bmatrix}$$

Thus the model matrix [S] is the combination of A,B and C:

$$[S] = [A \mid B \mid C]$$

Some similarity can be seen in the entries for B as for the above calculated linearised B matrix. The large value in A corresponding to M_x reflects the trouble that the quadratic function has in trying to cope with the almost bilinear nature of the crack in the opening direction.

THE EFFECTS OF A TRANSVERSE CRACK ON THE DYNAMICS OF A CIRCULAR SHAFT

P. S. LEUNG
Department of Mechanical Engineering and Manufacturing Systems
Newcastle Polytechnic, U. K.

ABSTRACT

The effect of an "open" transverse crack on the dynamics of a uniform circular shaft is examined in this study. An empirical equation was obtained to describe the dynamic properties of the shaft in terms of the crack size, crack location, crack orientation and slenderness of the shaft. The application of this equation for crack detection in a rotating machine is then explored.

1 INTRODUCTION

Fatigue initiated cracks have been found in a number of rotating machines. With increased slenderness of modern machine structure and extended period of operation, cracks are more likely to occur in machines in the future. It is therefore increasingly important to study the effect of a crack on a rotating machine in order to detect the fault as early as possible. However, many publications [ie, 1 to 5] in this subject tend to be associated with sophisticated theoretical analyses of the problem. There is a lack of fundamental experimental investigation to examine the effect of a crack on a uniform shaft.

In this study, the stiffness of a uniform circular shaft with a transverse crack is measured with different crack depths, crack location, crack orientation and slenderness ratio of the shaft. These experimental results were used to formulate an empirical equation to describe the stiffness of the shaft. The application of this empirical equation to detect the existence of cracks is then explored.

Attempts have also been made to observe the variation of natural frequencies of the shaft with the presence of a crack. The result will also be discussed.

2 DETERMINATION OF STIFFNESS EQUATION

The purpose of this part of the work is to observe the stiffness of a shaft and relate it to the presence of a transverse crack. The circular shaft is chosen to be studied here because it is the most commonly used geometry.

The stiffness of the shaft on pin-pin supports is measured, with and without a crack. The crack was made by a saw cut across the shaft in this study. This type of saw cut would simulate a crack which would remain "open" during all rotational motion of the shaft. The stiffness of the shaft was measured with different crack depths, crack orientation, crack locations and shaft slenderness ratio. The characteristics of the shaft was observed by varying one parameter at a time whilst keeping the other conditions constant. An empirical equation was then formulated to describe the stiffness of the cracked shaft, taking into account all the above parameters. In order that the result can be applied to other shafts with similar geometry, the empirical equation was made non-dimensional.

The arrangement of the shaft test rig is shown in Fig 1. If Ko is the stiffness of the shaft without a crack, and Kc is the stiffness of the same shaft with the presence of a crack, the following relationships were observed.

(2a) Effect of Crack Depth Ratio (C/D)

The stiffness of a cracked shaft was measured with a range of crack depth, C, to shaft diameter, D, ratio (C/D), from a value of 0 to 0.6. All other parameters were kept constant during the test. It was found that the general pattern of the stiffness changes could be described by the following equation,

$$\left(\frac{Kc}{Ko}\right) \propto \left(1 - \frac{C}{D}\right)^{n1} \tag{1}$$

The value of n1 was found constant during a set of test, but the value changed when other parameters were changed. Therefore, n1 is a function of other parameters which will be discussed later.

(2b) Effect of Shaft Slenderness Ratio (L/D)

The effect of the shaft slenderness ratio (L/D), ie, the ratio between the shaft length and the shaft diameter, was also investigated. It was found that the presence of a transverse crack could have a significant effect on the stiffness of a short shaft whilst the same crack would have a much lesser effect on the stiffness of a longer shaft. The result is reasonable because the effect of local reduction in shaft stiffness due to the crack would become small when the rest of the

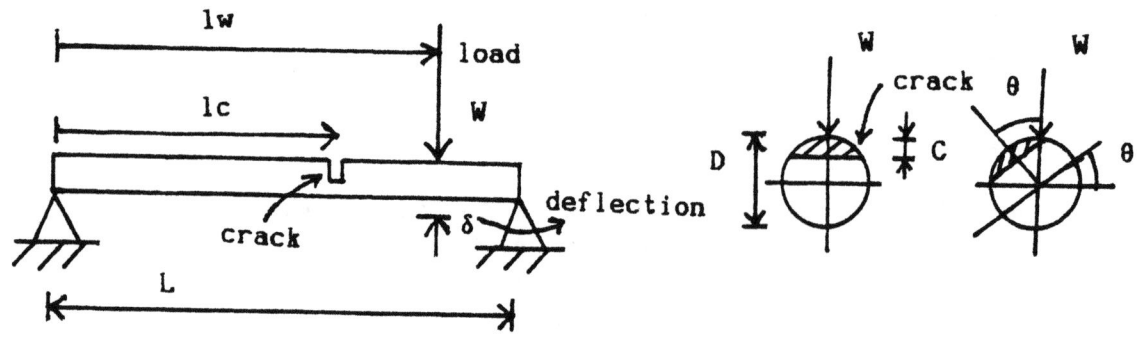

Fig 1 Arrangement of measurements

system is flexible. It was found that a crack of any sizes would have a negligible effect on the shaft stiffness if the (L/D) ratio exceeded a value of 40. The experiment showed that the general pattern of shaft stiffness with different (L/D) ratios could be represented by the following equation,

$$\left(\frac{Kc}{Ko}\right) \propto \left(n2\right)^{\left(\frac{n3}{(L/D)^{n4}}\right)} \tag{2}$$

where n2,n3,n4 are functions of other parameters relating to the crack problem.

(2c) Effect of Crack Orientation (θ)

The stiffness of the shaft was found to be dependent on the orientation of the crack (Fig 1). From the experiment, it was found that the shaft stiffness may be approximated by a sinusoidal function as follows,

$$\left(\frac{K(\theta)}{K(\theta=0)}\right) \propto \left(1 + n5\ SIN^2\ \theta\right) \tag{3}$$

Equation (3) compares the shaft stiffness when the crack has an angle θ with the reference axis where θ is equal to zero. Generally, the value of n5 is a function of the geometry of the shaft cross sectional area, and hence also a function of the (C/D) ratio. In this particular study, it was found reasonably accurate to represent n5 by the function, n5 = 0.2112 x (C/D).

(2d) Effect of Crack Location (lc/L)

The location of the crack also has an effect on the shaft stiffness. In the tests, the position of the applied load was fixed at a point. The shaft stiffness at the loading point was measured when the position of the crack was moved along the shaft (Fig 1). It was found that the influence of the crack on the stiffness was negligible at the edges of the shaft. However, the effect of the crack gradually increased as its position moved towards the middle of the shaft. The shaft would be most flexible when the crack is located at the middle. The general relationship between the shaft stiffness and the crack location may be approximated by the following equation,

$$\left(\frac{Kc}{Ko}\right) \propto \left(\frac{1 - n6}{0.25}\right)\left(\frac{lc}{L} - \frac{1}{2}\right)^2 + n6 \tag{4}$$

where n6 is a function of other parameters.

(2e) Effect of Load Position (lw/L)

It was found that the position of the applied load also has an effect on the stiffness of the shaft. If the position of the crack is fixed at a position and the applied load is moved along the shaft, the effect of the crack was found to be at its maximum when the load was applied at the middle of the shaft. The variation of the shaft stiffness with the load position (lw/L) could be described by an equation similar to equation (4), as follows,

$$\left(\frac{Kc}{Ko}\right) \propto \left(\frac{1 - n7}{0.25}\right) \left(\frac{lw}{L} - \frac{1}{2}\right)^2 + n7 \qquad (5)$$

where n7 is a function of other parameters.

(2f) The Combined Stiffness Equation

The effect of individual parameters, ie, crack depth, crack position .. etc, on the shaft stiffness have been investigated in sections (2a) to (2e). When one parameter was investigated, all other parameters were kept constant. It is the purpose of this section to combine these individual characteristics into one equation which could describe the stiffness of a cracked shaft taking into account all the parameters studied here.

As found in sections (2a) to (2e), there are interaction between the crack parameters to determine the values of the empirical coefficients n1 to n7 in equations (1) to (5). Therefore, these individual equations (1 to 5) were combined together with care, so that the final combined equation would be able to preserve the original characteristics of the individual equations without contradicting each other. A number of curve fitting techniques have been tested and used to achieve the best optimum result.

A possible equation to describe the stiffness of a cracked shaft is proposed as follows,

$$\left(\frac{Kc}{Ko}\right) = \left[4 \ (1 - m1) \ \left(\frac{lc}{L} - \frac{1}{2}\right)^2 + m1 \right] \ (1 + m5 \ \text{SIN} \ \theta)^2 \qquad (6)$$

where $\quad m1 = 4 \ (1 - m2) \ \left(\frac{lw}{L} - \frac{1}{2}\right)^2 + m2 \qquad (6a)$

$$m2 = \left(1 - \frac{C}{D}\right)^{\left(\frac{m3}{(L/D)^{m4}}\right)} \qquad (6b)$$

$$m3 = 1150 \tag{6c}$$

$$m4 = 2 \tag{6d}$$

$$m5 = 0.2112 \ (C/D) \tag{6e}$$

m3, m4 and m5 are empirical values obtained from the experiments.

It was found that equation (6) was able to give an approximate value of the stiffness of an cracked shaft with the idealised "crack" made by a saw cut. The application of the empirical equation for the detection of cracks was investigated and discussed in the following sections.

3 DETECTION OF CRACKS BY DYNAMIC TESTINGS

Equation (6) suggests that the shaft stiffness is a function of the crack parameters, such as the crack depth, crack location, and shaft slenderness ratio etc. The change in stiffness would cause a change in the dynamic behaviour of the shaft. If the relationship between the dynamic behaviour and the shaft stiffness is known, it may be possible to detect the presence of a crack, with the aid of dynamic testing techniques. It is now relatively straight forward to measure the basic dynamic behaviour of a rotating machine. In many cases, such as turbogenerators, monitoring equipments are installed in the machine as standard facilities.

There are two common properties to be measured in determining the dynamic behaviour of a rotating machine.

(1) the natural frequencies
(2) the amplitude of oscillation

The feasibility of using these two parameters in detecting the existence of crack is discussed as follows,

(3a) Natural Frequencies

It was found in this study that the frequencies of the shaft did not change significantly with the presence of a crack, even when the stiffness of the shaft had been reduced significantly. This is due to the fact that, in general terms, the natural frequencies of a shaft is proportional to the square root of the stiffness, ie,

$$\omega n \ \alpha \ \sqrt{K} \tag{7}$$

For example, a 10% reduction in stiffness only causes about 5% reduction in natural frequency. Therefore, a deep crack could have existed before a noticeable change in frequency is observed. It was rather difficult to detect a crack by measuring the natural frequencies.

(3b) Amplitude of Oscillation

Although a rotating machine could be balanced to reduce the vibration
to a satisfactory level, there are always some residual unbalance in
the shaft. These residual unbalance would force the shaft to oscillate
with a small amplitude. A simple relationship exists between the
unbalance force F, shaft stiffness at the same location K, and the
corresponding amplitude of oscillation X, as follows,

$$X = \frac{F}{K} \tag{8}$$

For a 10% reduction of shaft stiffness, there would be about 11%
increase in amplitude of oscillation. Therefore, it would be much
easier to detect the existence of a crack by measuring the amplitude of
oscillation rather than by measuring the natural frequencies.

It is of course possible to detect the presence of a crack by measuring
the shaft stiffness directly. However, it is usually necessary to shut
down the plant before the shaft stiffness can be measured. It is
beneficial to be able to detect the crack whilst the machine is still
running; and this can be achieved by measuring the amplitude of forced
oscillation as suggested in equation (8). The principle and procedures
of using response testing for crack detection is described in the
following section.

4 RESPONSE TESTING OF SHAFT

Multi-plane balancing is commonly used to balance a modern rotating
machine [6]. The size of the residual unbalance, F, (Fig 2) and the
corresponding phase angle ϕ can be identified by standard procedures
[6]. The position of the measuring probes are usually known to the
engineers. For an uncracked shaft, the relationship between the
amplitude, the unbalance force and the stiffness in the same plane
would be as follows,

$$Xoi = \frac{Fi}{Koi} \tag{9}$$

where i indicates the position of the measuring plane.

For a shaft with a transverse crack, equation (9) could be modified as
follows,

$$Xci = \frac{Fi}{Kci} \tag{10}$$

Combining equations (9) and (10),

$$\left(\frac{Xoi}{Xci}\right) = \left(\frac{Kci}{Koi}\right) = f\left(\frac{C}{D}, \theta, \frac{lc}{L}, \frac{lw}{L}, \frac{L}{D}\right) \tag{11}$$

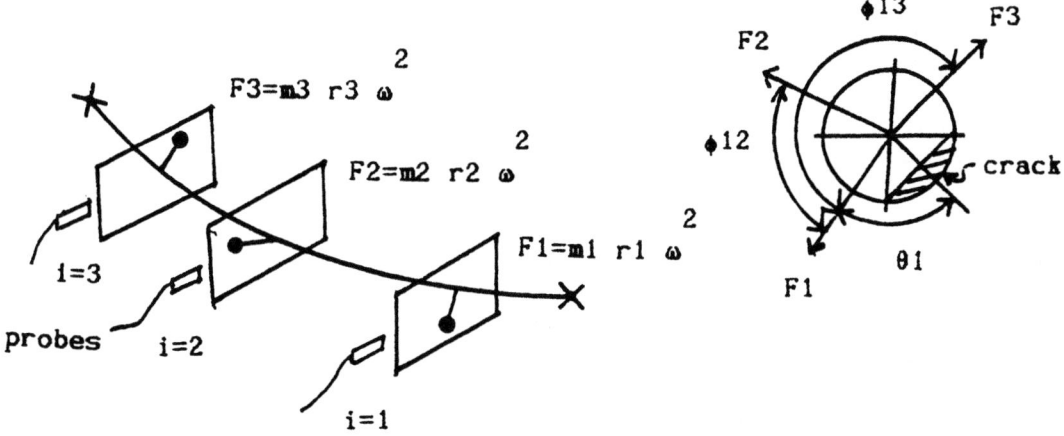

Fig 2   Balancing planes

Equation (11) indicates that the ratio of amplitudes between the uncracked shaft and the cracked shaft is equal to the non-dimensional stiffness equation obtained in equation (6). Therefore, the amplitude ratio is a function of the crack parameters such as (C/D), θ, and (lc/L). The shaft slenderness ratio (L/D) of the machine is usually available to the engineers investigating the problem, and the value of (lw/L) is given by the position of the measuring probes. Therefore, the only unknowns in equation (11) are (C/D), θ, and (lc/L). If three arbitrary balancing planes were used in the test, three independent equations would be obtained.

$$\left(\frac{Xo1}{Xc1}\right) = f_1 \left(\frac{C}{D}, \ \theta 1, \ \frac{lc}{L}, \ \frac{lw1}{L}, \ \frac{L}{D}\right) \qquad (12a)$$

$$\left(\frac{Xo2}{Xc2}\right) = f_2 \left(\frac{C}{D}, \ \theta 2, \ \frac{lc}{L}, \ \frac{lw2}{L}, \ \frac{L}{D}\right) \qquad (12b)$$

$$\left(\frac{Xo3}{Xc3}\right) = f_3 \left(\frac{C}{D}, \ \theta 3, \ \frac{lc}{L}, \ \frac{lw3}{L}, \ \frac{L}{D}\right) \qquad (12c)$$

If θ1 is the phase lag of the unbalance force to the crack in plane 1 (Fig 2), the corresponding phase angles in other planes can be obtained by the following equation,

$$\theta i \ = \ \theta 1 + \phi 1i \qquad (13)$$

Therefore, equations (12a) to (12c) can be rearranged to form a set of non-linear simultaneous equations, g's, as follows,

$$g_1\left(\frac{C}{D}, \frac{lc}{L}, \theta 1\right) = f_1 - \left(\frac{Xo1}{Xc1}\right) = 0 \qquad\qquad (14a)$$

$$g_2\left(\frac{C}{D}, \frac{lc}{L}, \theta 1\right) = f_2 - \left(\frac{Xo2}{Xc2}\right) = 0 \qquad\qquad (14b)$$

$$g_3\left(\frac{C}{D}, \frac{lc}{L}, \theta 1\right) = f_3 - \left(\frac{Xo3}{Xc3}\right) = 0 \qquad\qquad (14c)$$

Equation (14) can be solved by the standard Newton's iteration method [7], and gives a value of (C/D), (lc/L) and θ1. If the variation of amplitude is caused by the existence of a crack, the solution would converge and gives a sensible result.

Therefore, it is possible to estimate the size, the location and the orientation of the crack, by measuring the vibration amplitudes of the shaft at three different locations.

5 DISCUSSION OF RESULTS

The response testing method described in section 4 has been used to detect the presence of transverse crack in a uniform shaft. Reasonable results (within 10% accuracy) were obtained under the idealised conditions of laboratory testings. It shows a positive direction of the proposed method.

During the test, it was observed that the proposed method worked best when the crack was situated at the middle of the shaft. There were some difficulties in identifying the crack when it was located near the supporting edges of the shaft.

The proposed method was suitable for shafts with a slenderness ratio (L/D) below a value of 40. It was very difficult to detect a crack when the (L/D) ratio was higher than 40. Since the (L/D) ratio of many machines would be lower than 40, this limitation would not normally present a problem.

The above observations provide two possible explanations why, sometimes, deep cracks were not detected until failure.

It should, however, be pointed out that the amplitude of oscillation in a complex machine may change due to other reasons. The application of the proposed method on complex machinery is yet to be tested. However, if the dynamic behaviour of the machine is observed over a period of time, and the measurements and calculations always provide a consistent answer that a crack exists at a particular location and that the crack is deepening, it should be a fair indication of crack existence.

The advantages of the proposed method are that it does not require machine shut down for measurement. The requirement for measuring devices is not critical; and in many cases, the measuring devices already exist in the machine. The proposed method is also able to estimate the location and the size of the crack.

It is recognized that the empirical equation developed in this work (equation 6) may be rearranged and expressed in other forms, by using different curve fitting techniques. This work, nevertheless, demonstrates the point that the stiffness of a cracked shaft could be represented by a simple algebraic equation. This equation can be used to detect the presence of a crack by means of response testings.

This method is relatively simple and provides an alternative solution to the problem of crack detection in rotating machines.

6 CONCLUSIONS

The dynamic behaviour of a circular shaft with an "open" transverse crack was investigated. An empirical equation was obtained to describe the stiffness of the cracked shaft in terms of the crack size, crack location, crack orientation and slenderness ratio of the shaft. The application of this empirical equation for crack detection was then investigated. It was possible to incorporate the equation into the response testing of a rotating machine. The presence of a crack, its size and location could then be estimated, based on the measured dynamic response of the machine. The method is relatively simple and does not require machine shut down.

REFERENCES

1. Mayes I.W. & Davies W.G.R.
   The vibrational behaviour of a rotating shaft system containing a transverse crack.
   I.Mech.E. - Vibrations in rotating machinery 1976

2. Dimarogonas A.D. & Papadopoulos C.A.
   Vibration of cracked shafts in bending
   Journal of sound & vibration   1983   vol 91   pp.538-593

3. Grabowski B.
   The vibrational behaviour of a turbine rotor containing a crack
   ASME conference 1979,   No 79-DET-67

4. Gasch R.
   Dynamic behaviour of a simple rotor with a cross sectional crack
   I.Mech.E. - Vibrations in rotating machinery 1976

5. Henry T.A. & Okah-Avae B.E.
   Vibration in cracked shafts
   I.Mech.E. - Vibrations in rotating machinery 1976

6. Rieger N.F.
   Flexible rotor-bearing system dynamics III - Unbalance response and balancing of flexible rotors in bearings
   ASME publication   1973

7. Shoup T.E.
   Applied numerical methods for the micro-computer
   Prentice-Hall 1984

Vibrations of Cracked Rotating Blades

J. Wauer

*Institut für Technische Mechanik, Universität Karlsruhe,
Kaiserstraße 12, W-7500 Karlsruhe 1, Germany*

ABSTRACT

A rotating blade with a single transverse crack is considered. The proposed model is a torsionally rigid, non-pretwisted Bernoulli-Euler beam of extremely different bending stiffnesses. Two fields are connected by a local spring element characterizing the reduced stiffness of the crack region. The governing nonlinear boundary value problem is solved in two steps: first, the stationary pre-deformation due to the centrifugal forces is calculated and subsequently, the linearized equations of motion describing the superimposed small vibrations are analysed. Quantitative results concerning the stationary deformation and the natural frequencies of the coupled, free extensional-flexural vibrations are presented if crack depth and location are varied.

INTRODUCTION

The dynamics of cracked rotating shafts and blades is an actual and important problem of vibrational research. While for shafts it has been studied widely in the past (see Wauer, 1990), for blades one is presently in the beginning (see Chen and Chen, 1988 and 1990, Datta and Ganguli, 1990, for instance). Since in this case the structural component rotates about a transverse axis, there is a stationary deformation due to the centrifugal force creating a stiffening effect for the superimposed vibrations. Neglecting this influence leads to incorrect results for higher angular speeds as stated in a new fundamental work on modelling and generation of equations of motion by Wauer (1991). On the other hand, due to the above mentioned pre-deformation, a non-breathing, permanently open crack during motion can be assumed.

The objective of the present paper is to supplement the formulated equations of motion by quantitative results. To limit the numerical expense, a torsionally rigid, non-pretwisted Bernoulli-Euler beam of extremely different bending stiffnesses with a single transverse crack is considered. First, the governing equations of motion, namely the originally nonlinear boundary value problem, the stationary equations of the steady-state response in the centrifugal field and the equations of motion for superimposed small vibrations, are reduced to the described simplified problem. Subsequently, solutions for the stationary pre-deformation are computed and exemplified by diagrams in which the extensional and transverse displacements of the blade tip versus angular speed varying crack depth and location are depicted. Finally, an approximate solution procedure calculating the free and forced vibrations is proposed and a first result concerning the natural frequencies as a function of rotational speed for the same crack parameters as used in discussing the steady-state response are presented.

FORMULATION

Consider a cantilevered uniform (except for the crack) Euler-Bernoulli beam (naturally non-twisted) of length L with an axi-symmetric, non-circular cross-section, mass per unit length μ, extensional rigidity EA (Young's modulus E, cross-sectional area A) and governing bending stiffness EI (sectional moment of inertia I). It is fixed in a rigid rotor hub of radius R rotating with a constant angular speed Ω about the Z-axis of an inertial X, Y, Z reference frame (see Fig 1). Additionally, two body-fixed coordinate systems are introduced: a first x, y, z frame is chosen

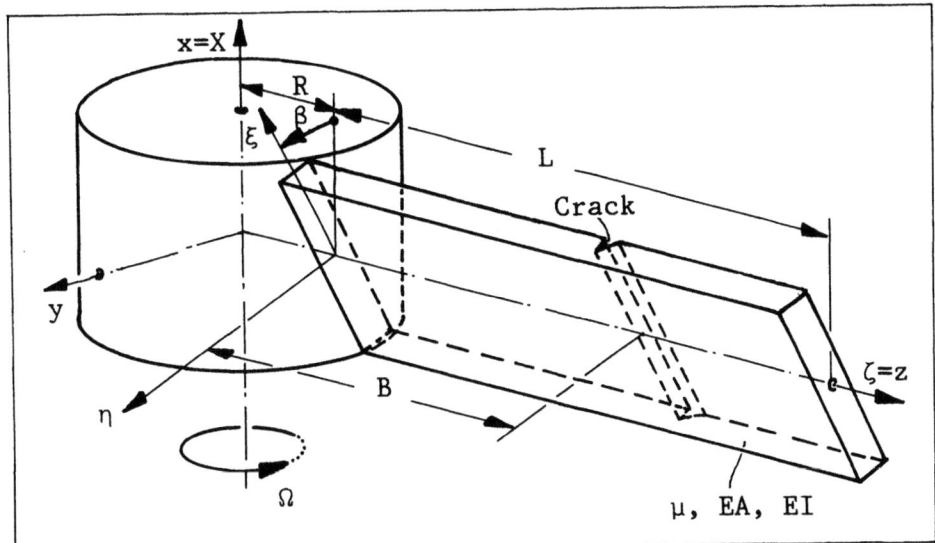

Figure 1: Model of the cracked rotating blade

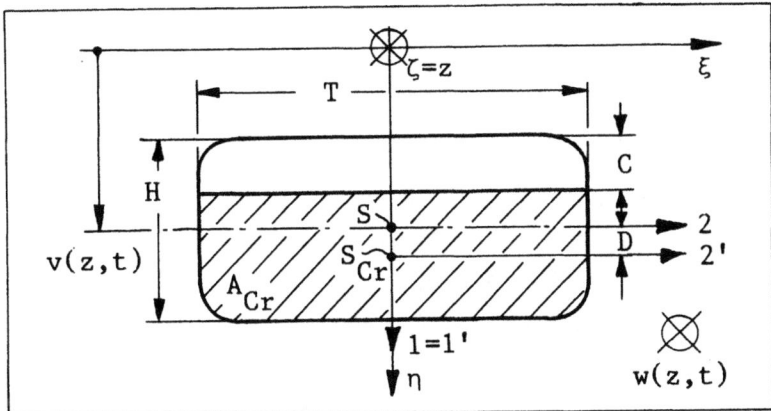

Figure 2: Geometry of cross-section with crack

in such a form that the x-axis coincides with the X-axis and the z-axis is in the longitudinal direction of the blade; a second ξ, η, ζ reference frame has its origin in the blade base, the ξ- and the η-axes are in the principal directions $1, 2$ of the cross-section of the beam (with EI as the smaller of the two bending stiffnesses) and the ζ-axis coincides with the z-axis. The constant angle of attack (between the ξ- and x-axis) is denoted β. A transverse surface crack is located at the position B $(0 < B < L)$ and the cross-section there is depicted in Fig 2. The crack is assumed to have a straight edge and to be oriented such that its edge is parallel to the 2–axis, which is the worst orientation. Then, the 1–axis is symmetry–axis also for the cracked region and this has simplifying consequences for the coupling of the equations of motion. The depth of the stationary crack is denoted C. The distance between the centres of the cracked and non–cracked cross-sectional area is D. An external excitation is assumed in the form of a space– and time–dependent load per unit length $p(z, t)$ in the direction of the rotational $x(= X)$-axis. Damping and weight influences are neglected. The space– and time–dependent displacements are $v(z, t)$ in the weak transverse direction and $w(z, t)$ in the axial direction, both measured in the rotating $\xi, \eta, \zeta(= z)$ reference frame. Subscripts t and z will denote partial derivatives with respect to time and space. The model is such that the entire blade is uniform except for the region of the crack. The discontinuity in stiffness — when the crack is open — is represented by massless distributed longitudinal springs (stiffness property c_L) connecting the faces B_-, B_+ of the beam components for that part A_{Cr} of the cross-section in the damaged region which is uncracked (in Fig 2 it is hatched).

The governing nonlinear boundary problem (including bending vibrations about the strong transverse direction and also torsional deformations) was derived by Wauer (1991). Reducing it to the simplified case considered here yields (for an open crack) the field equations

$$-\mu v_{tt} + 2\mu\Omega\cos\beta w_t + \mu\Omega^2\cos^2\beta v + EA(w_z v_z)_z - EI(v_{zz} + w_z v_{zz} - v_z w_{zz})_{zz}$$
$$-EI(v_{zz} w_{zz})_z = -p(z,t)\sin\beta,$$
$$-\mu w_{tt} - 2\mu\Omega v_t\cos\beta + \mu\Omega^2(R+z+w) + EA[w_z + (v_z^2 + 3w_z^2)/2]_z$$
$$+EI[(v_z v_{zz})_{zz} + (v_{zz}^2)_z/2] = 0,$$
$$0 < z < B_- \quad\text{and}\quad B_+ < z < L, \tag{1}$$

the boundary conditions at the outside boundaries

$$v = w = v_z = w_z = 0 \quad\text{at}\quad z = 0,$$
$$-EI(v_{zz} + w_z v_{zz} - v_z w_{zz})_z + EAv_z w_z - EIv_{zz}w_{zz} =$$
$$EI(v_{zz} + w_z v_{zz} - v_z w_{zz}) =$$
$$EA[w_z + (v_z^2 + 3w_z^2)/2] + EI[(v_z v_{zz})_z + v_{zz}^2/2] =$$
$$EIv_z v_{zz} = 0 \quad\text{at}\quad z = L, \tag{2}$$

and — using the abbreviation $\Delta[\ldots] = [\ldots](B_+,t) - [\ldots](B_-,t)$ and the definitions $k_A := c_L A_{Cr}$, $k_2 := c_L I_{Cr}$ (sectional moment of inertia I_{Cr} of A_{Cr} with respect to its governing principal axis $2'$) — the transition conditions at the crack location

$$\Delta[v] = 0,$$
$$EI(v_{zz} - v_z w_{zz} + w_z v_{zz})|_{z=B_-} = (k_2 + D^2 k_A)\{\Delta[v_z] - \Delta[w_z v_z] - \Delta[v_z]w_z(B_-)\}$$
$$-Dk_A\{\Delta[w] - \Delta[w]w_z(B_-)\},$$
$$[EA(w_z + v_z^2/2 + 3w_z^2/2) + EI(v_z v_{zz})_z + EIv_{zz}^2/2]|_{z=B_-} = k_A\Delta[w] - Dk_A(\Delta[v_z] - \Delta[v_z w_z]),$$
$$EIv_z v_{zz}|_{z=B_-} = -Dk_A\Delta[w]v_z(B_-) + (k_2 + D^2 k_A)\Delta[v_z]v_z(B_-),$$
$$EI\Delta[(v_{zz} - v_z w_{zz} + w_z v_{zz})_z + v_{zz}w_{zz}] - EA\Delta[w_z v_z] = 0,$$
$$EI\Delta[v_{zz} - v_z w_{zz} + w_z v_{zz}] = Dk_A\Delta[w]\Delta[w_z],$$
$$EA\Delta[w_z + v_z^2/2 + 3w_z^2/2] + EI\Delta[(v_z v_{zz})_z + v_{zz}^2/2] = 0,$$
$$EI\Delta[v_z v_{zz}] = Dk_A\Delta[w]\Delta[v_z] - (k_2 + D^2 k_A)\Delta^2[v_z]. \tag{3}$$

In comparison to Wauer (1991), there are some slight changes in the transition conditions caused by a more accurate evaluation of the corresponding crack potential. Additionally, it is noted that also within the scope of a nonlinear formulation perhaps stiffness properties k_{11}, k_{22} and k_{12} (instead of $k_2 + D^2 k_A, k_A$ and Dk_A, respectively,) would be more consistent, if they would be obtained by using principles of fracture mechanics (see Papadopoulos and Dimarogonas, 1988, for instance) and not by such a "geometric" method utilized here.

In order to determine the stationary pre–deformation due to the centrifugal force (without external exitation), the fact is exploited that all variables will attain time–independent values

$$v(z,t) = v_0(z), \quad w(z,t) = w_0(z). \tag{4}$$

The resulting boundary value problem in a linear approximation — sufficient in the most applications — can also be reduced from the relations derived by Wauer (1991):

$$\mu\Omega^2\cos^2\beta v_0 - EIv_{0,zzzz} = 0,$$
$$\mu\Omega^2(R+z+w_0) + EAw_{0,zz} = 0,$$
$$0 < z < B_- \quad\text{and}\quad B_+ < z < L, \tag{5}$$

$$v_0 = v_{0,z} = w_0 = 0 \quad\text{at}\quad z = 0,$$
$$v_{0,zz} = v_{0,zzz} = w_{0,z} = 0 \quad\text{at}\quad z = L, \tag{6}$$

$$v_0(B_-) = v_0(B_+),$$
$$EIv_{0,zz}(B_-) = (k_2 + D^2 k_A)[v_{0,z}(B_+) - v_{0,z}(B_-)] - Dk_A[w_0(B_+) - w_0(B_-)],$$
$$EAw_{0,z}(B_-) = k_A[w_0(B_+) - w_0(B_-)] - Dk_A[v_{0,z}(B_+) - v_{0,z}(B_-)],$$
$$v_{0,zzz}(B_-) = v_{0,zzz}(B_+),$$
$$v_{0,zz}(B_-) = v_{0,zz}(B_+),$$
$$w_{0,z}(B_-) = w_{0,z}(B_+). \tag{7}$$

Discussing the small (forced) vibrations of the pre-loaded blade, the external excitation $p(z,t)$ is included and solutions in the form

$$v(z,t) = v_0(z) + V(z,t), \quad w(z,t) = w_0(z) + W(z,t) \tag{8}$$

are assumed. The linearized equations of motion can also be found in the paper by Wauer (1991) if they are simplified to flexural–extensional vibrations considered here:

$$-\mu V_{tt} + 2\mu\Omega\cos\beta W_t + \mu\Omega^2\cos^2\beta V + EA(w_{0,z}V_z + v_{0,z}W_z)_z - EI(V_{zz} - v_{0,z}W_{zz} + w_{0,zz}V_z$$
$$+w_{0,z}V_{zz} + v_{0,zz}W_z)_{zz} - EI(v_{0,zz}W_{zz} - w_{0,zz}V_{zz})_z = -p(z,t)\sin\beta,$$
$$-\mu w_{tt} - 2\mu\Omega\cos\beta V_t + \mu\Omega^2 W + EA(W_z + v_{0,z}V_z + 3w_{0,z}W_z)_z$$
$$+EI[(v_{0,z}V_{zz} + v_{0,zz}V_z)_{zz} + (v_{0,zz}V_{zz})_z] = 0,$$
$$0 < z < B_- \quad \text{and} \quad B_+ < z < L, \tag{9}$$

$$V = V_z = W = W_z \;\; = \;\; 0 \quad \text{at} \quad z = 0,$$
$$-EI(V_{zz} + w_{0,z}V_{zz} + v_{0,zz}W_z - v_{0,z}W_{zz} - w_{0,zz}V_z)_z$$
$$+EA(v_{0,z}W_z + w_{0,z}V_z) - EI(v_{0,zz}W_{zz} - w_{0,zz}V_{zz}) \;\; = \;\; 0,$$
$$EI(V_{zz} + w_{0,z}V_{zz} + v_{0,zz}W_z - v_{0,z}W_{zz} - w_{0,zz}V_z) \;\; = \;\; 0,$$
$$EA(W_z + v_{0,z}V_z + 3w_{0,z}W_z) + EI[(v_{0,z}V_{zz} + v_{0,zz}V_z)_z + v_{0,zz}V_{zz}] \;\; = \;\; 0,$$
$$EI(v_{0,z}V_{zz} + v_{0,zz}V_z) \;\; = \;\; 0 \quad \text{at} \quad z = L, \tag{10}$$

$$\Delta[V] = 0,$$
$$EI(V_{zz} - v_{0,z}W_{zz} - w_{0,zz}V_z + w_{0,z}V_{zz} + v_{0,zz}W_z)|_{z=B_-} = (k_2 + D^2 k_A)(\Delta[V_z]$$
$$-\Delta[w_{0,z}V_z + v_{0,z}W_z] - \Delta[v_{0,z}]W_z(B_-) - w_{0,z}(B_-)\Delta[V_z])$$
$$-Dk_A(\Delta[W] - \Delta[w_0]W_z(B_-) + w_{0,z}(B_-)\Delta[W]),$$
$$\{EA(W_z + v_{0,z}V_z + 3w_{0,z}W_z) + EI[(v_{0,z}V_{zz} + v_{0,zz}V_z)_z + v_{0,zz}V_{zz}]\}|_{z=B_-}$$
$$= k_A\Delta[W] - Dk_A(\Delta[V_z] - \Delta[v_{0,z}W_z + w_{0,z}V_z]),$$
$$EI(v_{0,z}V_{zz} + v_{0,zz}V_z)|_{z=B_-} = -Dk_A(\Delta[w_0]V_z(B_-) + v_{0,z}(B_-)\Delta[W])$$
$$+(k_2 + D^2 k_A)(\Delta[v_{0,z}V_z(B_-) + v_{0,z}(B_-)\Delta[V_z]),$$
$$EI\Delta[(V_{zz} - v_{0,z}W_{zz} - w_{0,zz}V_z + w_{0,z}V_{zz} + v_{0,zz}W_z)_z + v_{0,zz}W_{zz} + w_{0,zz}V_{zz}] = 0,$$
$$EI\Delta[V_{zz} - v_{0,z}W_{zz} - w_{0,zz}V_z + w_{0,z}V_{zz} + v_{0,zz}W_z] = Dk_A(\Delta[w_0]\Delta[W_z] + \Delta[w_{0,z}]\Delta[W]),$$
$$EA\Delta[W_z + v_{0,z}V_z + 3w_{0,z}W_z] + EI\Delta[(v_{0,z}V_{zz} + v_{0,zz}V_z)_z + v_{0,zz}V_{zz}] = 0,$$
$$EI\Delta[v_{0,z}V_{zz} + v_{0,zz}V_z] = Dk_A(\Delta[w_0]\Delta[V_z] + \Delta[v_{0,z}\Delta[W])$$
$$-2(k_2 + D^2 k_A)\Delta[v_{0,z}]\Delta[V_z]. \tag{11}$$

Remember that due to the stationary pre–deformation a permanently open crack during the superimposed vibrations is assumed.

STEADY–STATE RESPONSE

Because of the unsymmetry of the one–sided edge crack, the radially directed centrifugal force do not only cause a finite extensional deformation $w_0(z)$ but also a non–vanishing transverse displacement $v_0(z)$ and even in the linear formulation presented here, they have to be calculated from a coupled boundary value problem in $v_0(z)$ and $w_0(z)$, see Eqs (5) to (7). Since uniform sections of the blade are assumed, an exact solution is possible. For this purpose, a frequency of reference $\Omega_0 := \sqrt{EA/(\mu L^2)}$ and different non–dimensional parameters

$$\alpha = \frac{\Omega}{\Omega_0}, \quad \kappa^2 = \frac{I}{AL^2}, \quad r = \frac{R}{L}, \quad \gamma = \frac{B}{L} \tag{12}$$

are introduced. Then — using the abbreviation $p^4 = \frac{\alpha^2\cos^2\beta}{\kappa^2}$ — , the general solutions of the differential equations (5) are

$$v_0(z) \;\; = \;\; C_{1i}\sinh p\frac{z}{L} + C_{2i}\cosh p\frac{z}{L} + C_{3i}\sin p\frac{z}{L} + C_{4i}\cos p\frac{z}{L},$$

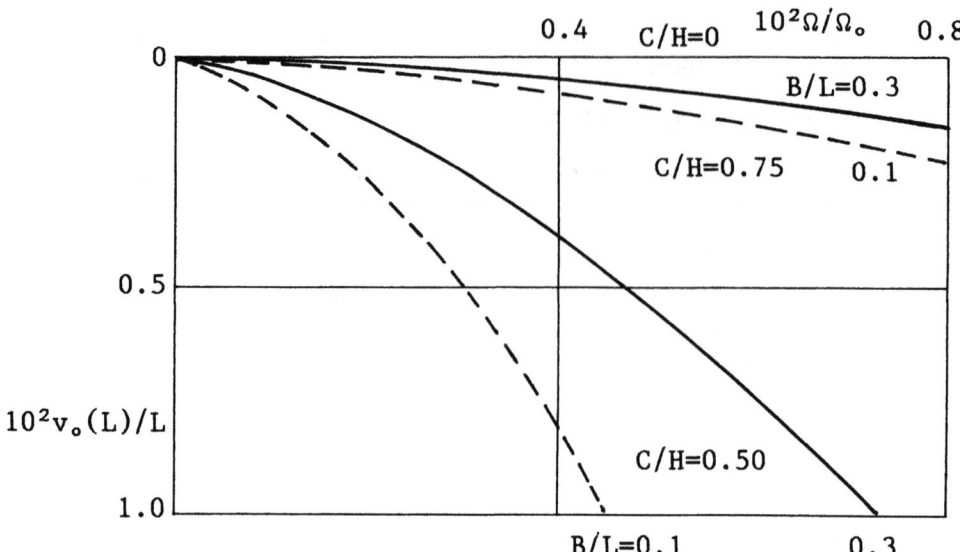

Figure 3: Flexural displacement of the blade tip versus rotational speed

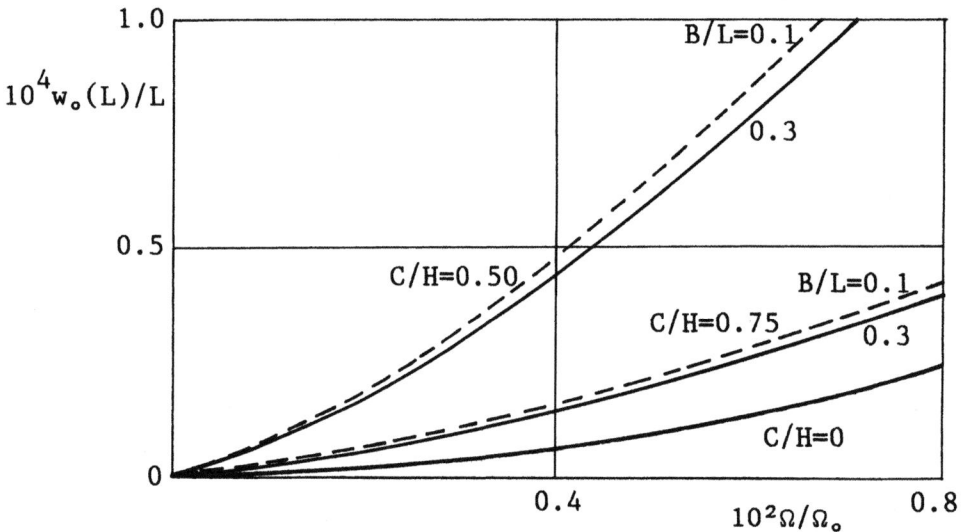

Figure 4: Extensional displacement of the blade tip as function of rotational speed

$$w_0(z) = C_{5i} \cos \alpha \frac{z}{L} + C_{6i} \sin \alpha \frac{z}{L} - \left(\frac{z}{L} + r \right),$$

$$i = 1 \quad : \quad 0 < \frac{z}{L} < \gamma_-, \qquad i = 2 \quad : \quad \gamma_+ < \frac{z}{L} < 1, \tag{13}$$

where the $2 \times 6 = 12$ integration constants C_{1i} to C_{6i}, $i = 1, 2$ have to be computed applying the boundary and transition conditions (6) and (7), respectively. To perform this procedure analytically — in the present case, a system of 12 inhomogeneous, linear algebraic equations must be solved — , a computer–aided manipulation program package can be used. Doing so, the dependence of the stationary deformations on the different parameters can be seen explicitely. But the analytical results fill some pages so that within this chapter only numerical results are presented. Therein, the "blade" is assumed to be a beam of length $L = 500$mm with a rectangular cross–section of width $T = 60$mm and height $H = 10$mm and the following other properties:

$$\beta = 45^0, \quad R = 250\text{mm}, \quad c_L = 6 \cdot 10^5 \frac{A_{cr}}{A} \text{Nmm}^{-1} \tag{14}$$

The flexural and extensional displacements $v_0(L)$ and $w_0(L)$, respectively, are plotted in Fig 3 and 4 versus rotational speed Ω for different crack locations B and crack depths D. The speed scale reaches approximately to 40% of the lowest natural bending frequency for a uniform non–rotating beam without crack. The crack is arranged on that side of the surface shown in Fig 1 and 2. As expected, both displacements are greater than zero and increase with increasing rotational speed and also with an increasing crack depth, where the influence on the bending deformation is much stronger than for the axial one. Furthermore, a crack located close to the base of the blade has a much more significant effect on the displacements than a damage in a greater distance from it. Another important result — not illustrated here — is that due to the centrifugal force which is a tension, a gaping stationary crack appears, i.e., indeed, a finite $\Delta w_0 > 0$ at the crack location results. It also increases if roational speed and crack depth increase and the crack moves towards the blade base.

From these results, it appears that the stationary behaviour does not change substantially due to small cracks, but to find the combined influence of damage and rotational motion on the vibrations of the blade, the correction has to be taken into consideration.

BLADE OSCILLATIONS

Since the governing boundary value problem (9) to (11) in $V(z,t)$ and $W(z,t)$ has space–dependent coefficients [owing to the stationary deformations $v_0(z)$ and $w_0(z)$], an analytical solution procedure as used by Papadopoulos and Dimarogonas (1988), for instance, is not practicable. For that reason, approximate solution techniques have to be used. For cracked shafts — as proposed by Wauer (1990b) and tested by Collins et al. (1991) — , a concept has been proved effective that the entire shaft is assumed to be uniform and the discontinuity in stiffness at the (permanently open) crack is represented by a self–equilibrating generalized force vector. Subsequently, Galerkin's method can be applied to obtain a set of ordinary differential equations to be solved numerically.

But for cracked blades this procedure is also not applicable, because the transition conditions (11) are much more complicated due to the appearing space–dependent stationary deformation. Here, a variational problem corresponding to the governing boundary value problem (9) to (11) is chosen as the starting point. The advantage is that then all dynamic boundary and transition conditions in (10) and (11) are included implicitly and kinematically admissable functions are sufficient as trial functions. Assume solutions in the form of a Ritz–series

$$
\left(\begin{array}{c} V(z,t) \\ W(z,t) \end{array} \right) = \sum_{k=1}^{N} \left(\begin{array}{c} \Phi_k(z) \\ \Psi_k(z) \end{array} \right) T_k(t),
\tag{15}
$$

where $[\Phi_k(z), \Psi_k(z)]^T$, $k = 1, 2, \ldots, N$ are appropriate shape functions in vector form and $T_k(t)$, $k = 1, 2, \ldots, N$ are unknown time functions to be determined in such a way that the functional

$$
\begin{aligned}
\mathcal{F} = & \int_{t_0}^{t_1} \Big\{ \frac{\mu}{2} \int_{(z)} [V_t^2 + W_t^2 - 2\Omega(V_t W - W_t V)\cos\beta + \Omega^2(W^2 + V^2\cos^2\beta)]dz \\
& - \frac{EA}{2} \int_{(z)} [W_z^2 + w_{0,z}(V_z^2 + W_z^2) + 2(v_{0,z}V_z + w_{0,z}W_z)W_z]dz \\
& - \frac{EI}{2} \int_{(z)} [V_{zz}^2 - 2(v_{0,z}V_{zz}W_{zz} + v_{0,zz}v_z W_{zz} + w_{0,zz}V_z V_{zz}) + w_{0,z}V_{zz}^2 + 2v_{0,zz}W_z V_{zz}]dz \\
& - \frac{k_A}{2}\Delta^2[W] + Dk_A\Delta[W]\Delta[V_z] \\
& - Dk_A(\Delta[w_0]\Delta[W_z V_z] + \Delta[W]\Delta[w_{0,z}V_z] + \Delta[W]\Delta[v_{0,z}W_z]) \\
& - \frac{k_2 + D^2 k_A}{2}(\Delta^2[V_z] - 2\Delta[v_{0,z}]\Delta[V_z W_z] - 2\Delta[V_z]\Delta[v_{0,z}W_z] - 2\Delta[V_z]\Delta[w_{0,z}V_z]) \Big\}dt, \\
& 0 < z < B_- \quad \text{and} \quad B_+ < z < L
\end{aligned}
\tag{16}
$$

becomes stationary. This leads to the inhomogeneous system of ordinary differential equations of second order

$$
\sum_{k=1}^{N} \left(m_{jk}\frac{d^2 T_k}{dt^2} + g_{jk}\frac{dT_k}{dt} + k_{jk}T_k \right) = q_k(t), \quad k = 1, 2, \ldots, N
\tag{17}
$$

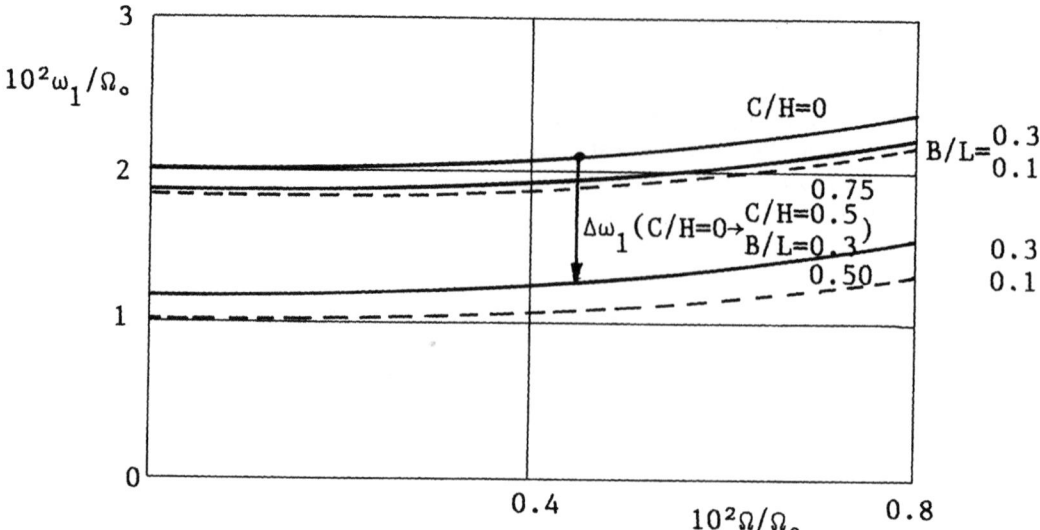

Figure 5: Lowest natural frequency versus rotational speed varying crack depth and location

where

$$m_{jk} = \mu \int_0^L (\Phi_j \Phi_k + \Psi_j \Psi_k) dz,$$

$$g_{jk} = 2\mu\Omega \int_0^L (-\Psi_j \Phi_k + \Phi_j \Psi_k) dz,$$

$$k_{jk} = \mu\Omega^2 \int_0^L (\Phi_j \Phi_k \cos^2\beta + \Psi_j \Psi_k) dz - EA \int_0^L [\Psi_{j,z}\Psi_{k,z} + w_{0,z}(\Phi_{j,z}\Phi_{k,z} + \Psi_{j,z}\Psi_{k,z})$$

$$+ (v_{0,z}\Phi_{j,z} + w_{0,z}\Psi_{j,z})\Psi_{k,z})]dz$$

$$- EI \int_0^L \Big[\Phi_{j,zz}\Phi_{k,zz} - (v_{0,z}\Phi_{j,zz} + v_{0,zz}\Phi_{j,z})\Psi_{k,zz} - w_{0,zz}\Phi_{j,z}\Phi_{k,zz}$$

$$+ w_{0,z}\Phi_{j,zz}\Phi_{k,zz} + v_{0,zz}\Psi_{j,z}\Phi_{k,zz}\Big]dz - k_A \Delta[\Psi_j]\Delta[\Psi_k]$$

$$+ Dk_A\{\Delta[\Psi_j]\Delta[\Phi_{k,z}] - \Delta[w_0]\Delta[\Psi_{j,z}\Phi_{k,z}]$$

$$- \Delta[\Psi_j](\Delta[w_{0,z}\Phi_{k,z}] + v_{0,z}\Psi_{k,z})\} - (k_2 + D^2 k_A)\{\Delta[\Phi_{j,z}\Phi_{k,z}]\Delta[\Phi_{k,z}]$$

$$- \Delta[v_{0,z}]\Delta[\Phi_{j,z}\Psi_{k,z}] - \Delta[\Phi_{j,z}](\Delta[v_{0,z}\Psi_{k,z}] + \Delta[w_{0,z}\Phi_{k,z}])\},$$

$$q_k(t) = \int_0^L \Phi_k p(z,t) \sin\beta \, dz. \tag{18}$$

For a first result, a 1–term approximation is applied. The lowest natural frequency ω_1, corresponding to bending, as a function of rotational speed if crack depth and location are varied, are calculated. An appropriate trial vector function $[\Phi_1(z), \Psi_1(z)]^T$ is assumed in form of the solution of the boundary problem

$$EI\Phi_{1,zzzz} = q_0, \quad EA\Psi_{1,zz} = 0, \quad 0 < z < B_- \text{ and } B_+ < z < L,$$
$$\text{boundary and transition conditions (6), (7) in } \Phi_1, \Psi_1, \tag{19}$$

where q_0 is a non–vanishing constant load per unit length. The function satisfies all geometric boundary conditions in (10) and is a fairly good approximation for the lowest (bending) mode of the cracked blade. Using the same data as assumed determining the stationary pre–deformations $v_0(z), w_0(z)$ in the last chapter, i.e., also the functions $v_0(z), w_0(z)$ themselves obtained there, this is illustrated in Fig 5. As expected, the frequency drop $\Delta\omega_1$ increases if the crack depth and the distance between the crack position and the free end of the blade increase. The frequency ω_1 itself increases with increasing angular speed Ω and is lower than for a uniform non–cracked blade.

CONCLUSIONS

The dynamics of a cracked rotating blade model with a transverse stationary crack was examined. If a blade rotates about a transverse axis, the stiffening effect due to the centrifugal force has to be taken into account. Therefore, in a first step the originating stationary pre–deformation was computed and subsequently, the superimposed vibrations were studied.

For the steady–state response, the deformations leading to a gaping crack were determined by an analytical solution. Since a one–sided edge crack was assumed, the pre–deformation is not only an extensional but also a flexural one. The deformations increase as the crack moves from the free end of the blade towards the fixed end. They also increase if crack depth and rotational speed increase.

For the superimposed vibrations, an approximate variational method was applied. Owing to the gaping crack during stationary motion, a permanently open crack during vibration was assumed so that a completely linear boundary value problem (but with space–dependent coefficients) had to be solved. A modal truncation was carried out and for unforced vibrations, the fundamental frequency of the coupled extensional–flexural oscillations (essentially, it is associated with bending) was determined. As expected, due to the stiffening effect, this frequency increases if the rotational speed increases. Since a damage reduces the effective stiffness of the considered system, the frequencies of the cracked blade are lower than for the undamaged blade. The frequeny change increases if both the crack depth increases and the crack moves from the free end of the blade towards the fixed end.

For forced vibrations, no quantitative results have been reached yet; such investigations will be kept in reserve for a future paper.

From the results of this study it appears that only when a crack causes a substantial reduction of stiffness (as assumed in this study) it has a significant effect on the frequencies of free vibrations. For small cracks — also in the case of rotating blades — , the influence is not very considerable. However, to obtain the dominant increase of the (bending) frequencies due to the stationary angular speed, an originally nonlinear formulation of the governing equations of motion in connection with a step–by–step solution is absolutely necessary.

REFERENCES

Chen L-W, Chen C-L (1988) Vibrations and stability of cracked thick rotating blades. Computers & Structures 28: 67-74.

Chen L-W, Chen J-L (1990) Non–conservative stability of a cracked thick rotating blade. Computers & Structures 35: 653-660.

Collins KR et al (1991) Detection of cracks in rotating Timoshenko shafts using axial impulses. J Vibration & Acoustics 113: 74-78.

Datta PK, Ganguli R (1990) Vibration characteristics of a rotating blade with localized damage including the effects of shear deformation and rotatory inertia. Computers & Structures 36: 1129-1133.

Papadopoulos CA, Dimarogonas AD (1988) Coupled longitudinal and bending vibrations of a cracked shaft. J Vibration Acoustics Stress & Reliability in Design 110: 1-8.

Wauer J (1990a) On the dynamics of cracked rotors: a literature survey. Applied Mechanics Reviews 43: 13-17.

Wauer J (1990b) Modelling and formulation of equations of motion for cracked rotating shafts. Int J Solids Structures 26: 901-914.

Wauer J (1991) On the dynamics of cracked rotating blades. Applied Mechanics Reviews 44: S273-278.

SESSION 4   CONDITION MONITORING I

DEVELOPMENT OF COMPUTERIZED ON LINE CONDITION MONITORING OF ROTATING
MACHINERY - A CUSTOM BASED APPROACH

K.Ramakrishna, A.Rajamani   and   S.Narayanan,
Vibration Laboratory,   Dept. of Applied Mechanics,
BHEL(R & D),Vikasnagar,   Indian Institute of Technology,
Hyderabad-500 593,India.   Madras-600 036, India.

ABSTRACT

This paper describes development of a custom based software for on
line condition monitoring of rotating machinery using parameters such
as shaft centre locus, unbalance vector etc., for preventive
maintenance and diagnostics. The software also provides for display
of cascade plots, Bode plots, orbit plots, etc. The application of
the same for few cases are discussed.

INTRODUCTION

Online condition monitoring of rotating machinery especially the high
speed/critical machinery is well accepted as an important activity
for trouble shooting and preventive maintenance. Data storage,
retrieval, faster computation and diagnosis necessitates the use of a
computer based online condition monitoring system. This paper
describes the methodology and utility of custom based software for
shaft centreline monitoring, vibration vector monitoring, trend
displays etc. The software makes corrections in the angular
deviations arising out of constraints in mounting of probes at sites.
Provision for cascade plot, order plot, vibration amplitude spectrum
plot with any other parameter (load/temperature/pressure,flow etc.),
orbit plot, Bode plot, Nyquist plot etc., is made to help in fault
diagnosis. A procedure for phase evaluation and monitoring from two
digitized time records is dealt with. Application of the software to
a few rotating machinery is given. A computer interfaced to the
Digital spectrum analyzer is used for transfer of various types of
data such as time record/spectrum/orbit etc.

SHAFT CENTRE MONITORING

Overall machine bearing vibration monitoring is not sufficient to
evaluate the machine performance. Shaft centre locus monitoring gives
good information on the actual rotor condition in a journal bearing.
Changes in shaft centreline position can reveal hidden and potentially
catastrophic problems that include bearing wear, bearing degradation

caused by electrostatic discharge, external and internal preloads, loose bearing fits, bearing instability, shaft misalignment etc. [Ref. 1,2]. The computation of shaft centre is based on the DC gap voltage measured using two mutually perpendicular proximity probes. The probe outputs are digitized in an A/D system and transferred to the computer. The digitized output is calibrated and displayed on the monitor in a circle as a locus of shaft centre, with bearing centre as origin and radius equal to bearing radial clearance (Fig. 1) For the given radial bearing clearance the computed values of eccentricity ratio and attitude angle at a given speed are continuously displayed along with the shaft centre locus. The eccentricity ratio provides significant insight into the stability characteristics of the journal bearing system. For example, a value less than 0.7 indicates a higher probability that instability mechanism will take place. A value in the range of 20 to 50 degrees for attitude angle is exhibited usually by a sleeve bearing (less for tilted pad bearing)[Ref 1]. The software also corrects for deviations in the orientation of probes from vertical and horizontal directions [Ref. 3].

If DY and DX are the angular deviations (negative sign if deviation is outwards of the quadrant) in the vertical and horizontal respectively, the actual displacements of proximity probes in terms of observed displacements are computed by the following expressions:

$$X_{act} = \frac{X_{obs}Cos(DY) - Y_{obs}Sin(DX)}{Cos(DX+DY)} \quad (1)$$

$$Y_{act} = \frac{Y_{obs}Cos(DX) - X_{obs}Sin(DY)}{Cos(DX+DY)} \quad (2)$$

Fig. 1 also indicates change in the shaft centre for a typical turbine generator rotor under excitation and varying load conditions.

Shaft centreline curve (catenary) for a multisupported rotor system can be obtained using the probes located at the same level (of water column) at all the bearings. The catenary can be monitored especially before or after shutdown/overhaul/bearing replacements for comparison with previous data to avoid unnecessary operation problems usually encountered after overhauls.

VIBRATION VECTOR MONITORING

This part of the software utilizes the filtered dynamic signal of the proximity probes and the optical probe output which gives the speed pulse. The phase and amplitude of the running speed component (unbalance vector) are evaluated from the time records. While the amplitude is the averaged maximum level of the time record, phase is evaluated from the time lag between the speed pulse and the heavy spot obtained from the time record of shaft displacement. The concept of phase difference evaluation between time records is illustrated in Fig.2. Vibration and speed pulse time records are transferred from a digital analyzer to a computer, with speed pulse signal being used for external triggering. The vibration signal time record is searched for two successive zero crossings (using sign change detection algorithm) to get T_1 and T_2. The phase lag is evaluated in degrees by the following expression:

$$\text{Phase difference} = (\ (T_1 + T_2) * 180 / 2)\ /\ (T_2 - T_1) \qquad (3)$$

The amplitude and phase thus obtained are displayed on the monitor in calibrated units as vibration vector (Fig 3). It gives status of unbalance at any instant. Thus using the influence coefficient data, the correction mass and phase can be instantly obtained at any moment whenever the unbalance vector exceeds the severity limit. Provision is made to get the locus of the vector during any period. This locus is in fact a Nyquist plot taken during coastdown or coastup. The amplitude and phase plotted with respect to speed gives the Bode plot.

Cascade plots can be obtained by the plotting of spectrums transferred from a spectrum analyzer as a waterfall display with respect to any operating parameter like temperature, pressure, flow, speed, time, location etc., as against ready made software which offers restricted flexibility. Fig 4 gives a cascade plot generated for a gas turbine coast down using the software. Fig 5 is a typical display of orbits for a hydro turbine shaft for different operating conditions. It gives an indication of shaft vibration due to change in parameters (in this case load).

BAR CHART DISPLAYS

For machines having a large transmission system (Fig 6) containing shafts rotating at different speeds, the severity and trend at various locations can be easily depicted by generating bar chart displays as shown for a boiler feed pump in Fig 7. It clearly indicates that the hydraulic coupling zone is the critical vibration zone. Fig 8 shows another type of display generated to show the dominance of different spectral components contributed by shafts which rotate at different speeds at each location, in the form of stacked bar chart. Predominance of third harmonic of motor speed in most of the locations again indicates the turbo coupling zone as the source of vibration.

DIAGNOSTIC CHARTS

Charts based on the information available in the literature and field experiences can be presented in the form of bar charts for identifying the causes of vibration for various symptoms [Ref. 4]. Guides for severity assessment based on available standards, computational programs for balancing, critical speeds, etc. are also incorporated in the software.

CONCLUSION

A custom based software is developed for shaft centre monitoring, unbalance vector monitoring, cascade plot generation, trend plots, diagnostic charts etc. Utility of the software to a few site problems is illustrated.

ACKNOWLEDGEMENT

The authors wish to acknowledge the BHEL Management for permission to present this paper in the conference.

REFERENCES

1. BILL PRYOR (1984) Determining shaft centreline position in four easy steps, Orbit, March: 6-7.
2. (1989) Shaft centreline radial position, Orbit, April: 24-26.
3. DONALD E. BENTLY (1989) Compensation of rotor dynamic motion data where probes are not installed at 90 degrees and how to rotate coordinate system, Orbit, April: 3-5.
4. CHARLES JACKSON (1979) A Practical Vibration primer, Gulf publishing company, Houston, Texas.

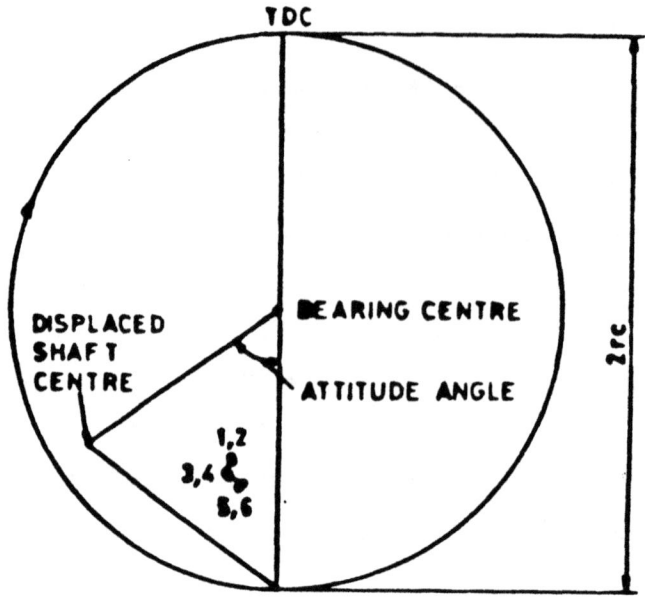

ECCENTRICITY RATIO = 0·6
ATTITUDE ANGLE = 23°
rc — RADIAL BEARING
       CLEARANCE

FIG 1. BEARING CENTRELINE LOCUS FOR A TURBO-
GENERATOR ROTOR
(1 NO LOAD 6 FULL LOAD (210 MW), OTHER POINTS IN ORDER
EXC., 60MW, 120MW, 180MW)

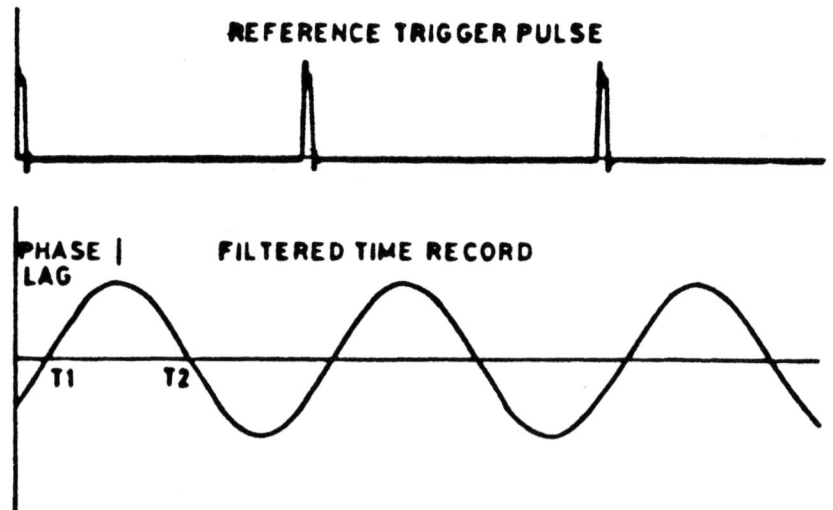

PHASE LAG = ((T1+T2) x 180/2)/(T2 - T1) DEGREES

FIG 2. PHASE DIFFERENCE EVALUATION FROM TWO TIME
RECORDS

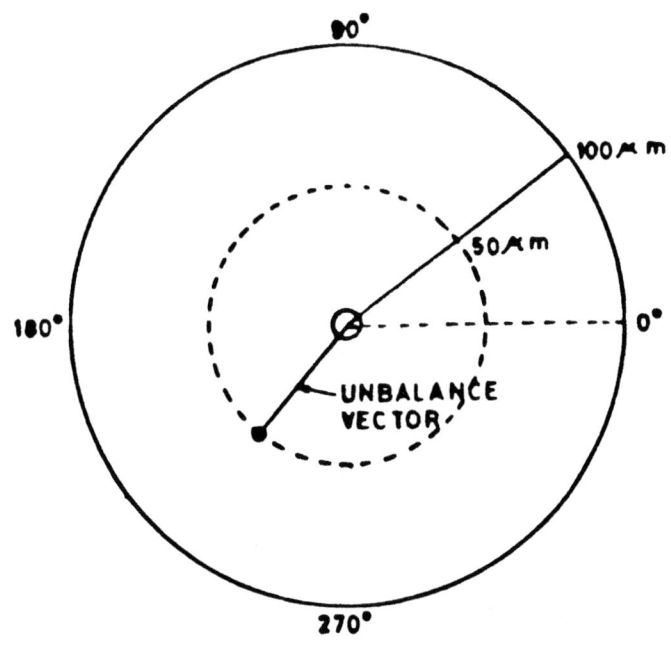

FIG 3. UNBALANCE VECTOR MONITORING

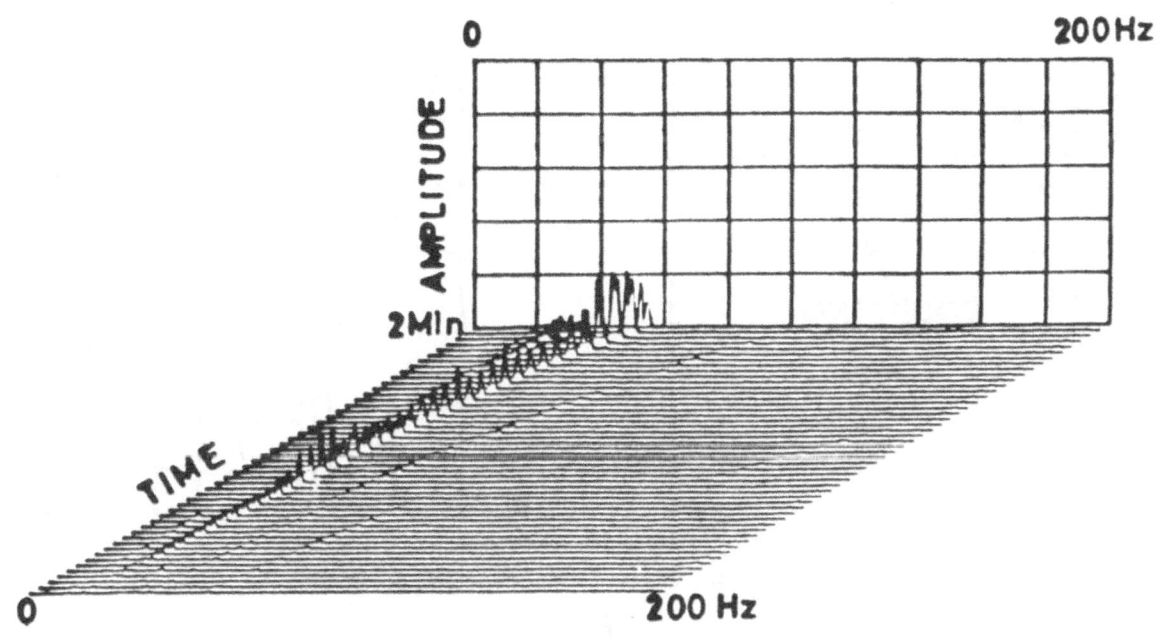

FIG 4. CASCADE PLOT FOR A GAS TURBINE

82

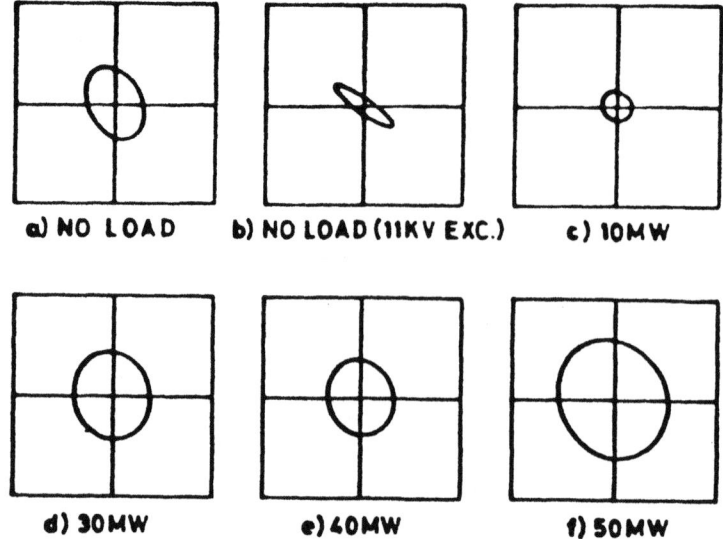

a) NO LOAD b) NO LOAD (11KV EXC.) c) 10MW

d) 30MW e) 40MW f) 50MW

FIG 5. SHAFT CENTRE ORBITS FOR A HYDRO TURBINE ROTOR

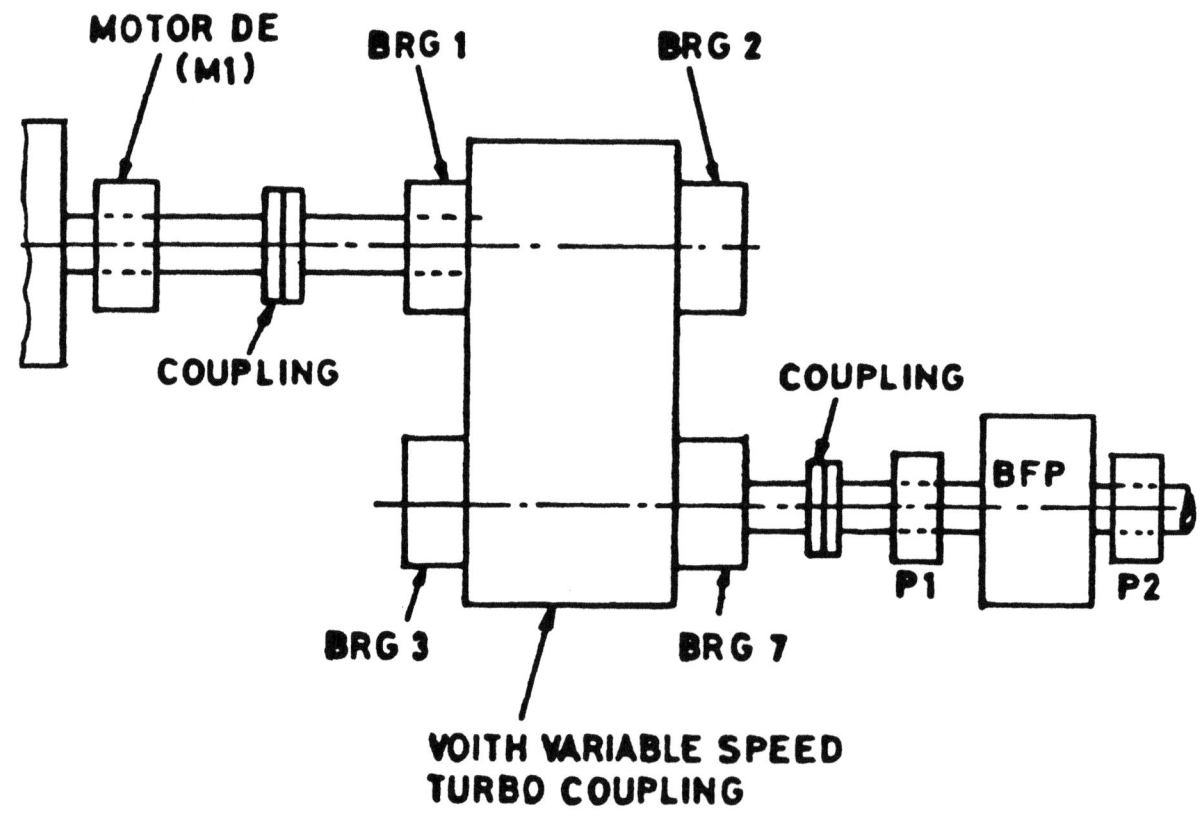

MOTOR DE (M1)   BRG 1   BRG 2

COUPLING

COUPLING

BFP

P1   P2

BRG 3   BRG 7

VOITH VARIABLE SPEED TURBO COUPLING

FIG 6. SCHEMATIC OF A BOILER FEED PUMP SYSTEM

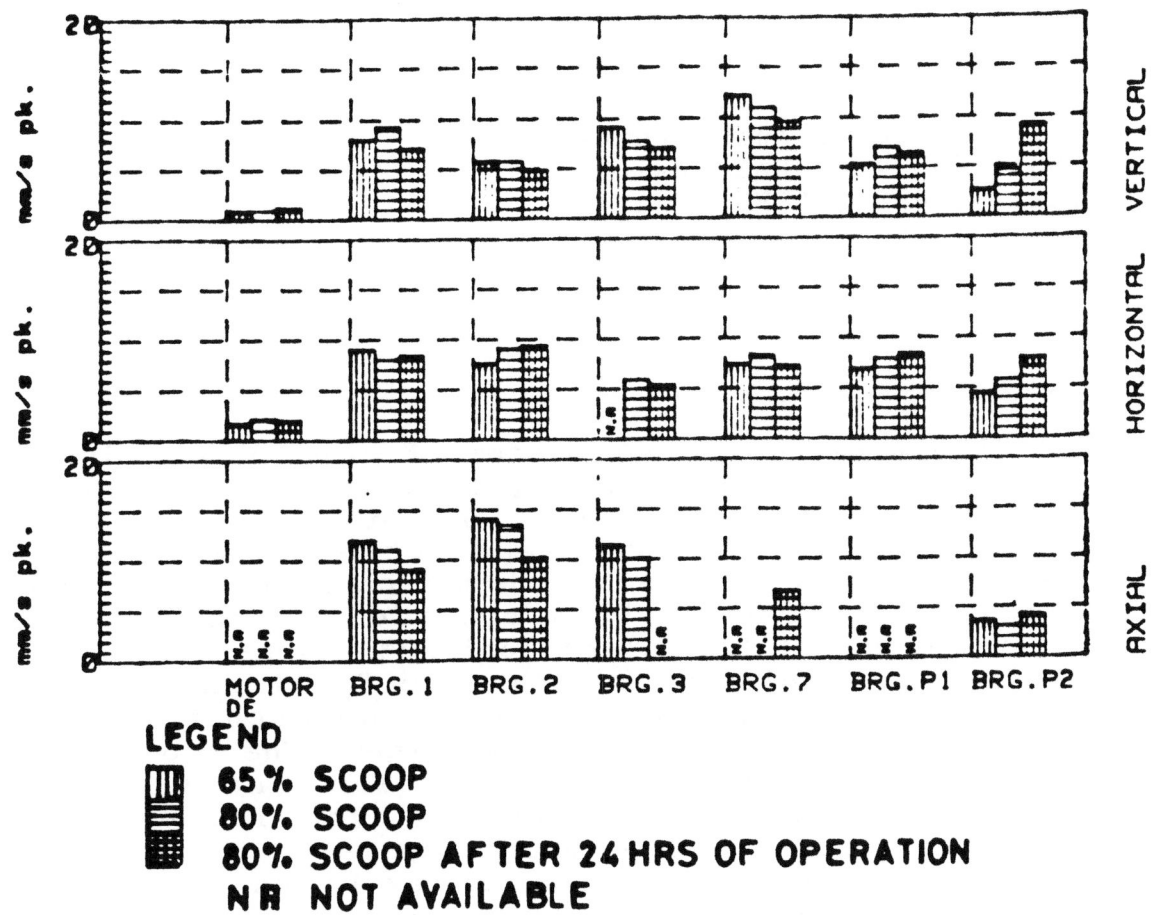

LEGEND

|||| 65% SCOOP
|||| 80% SCOOP
|||| 80% SCOOP AFTER 24 HRS OF OPERATION
N A NOT AVAILABLE

FIG 7. BAR CHART OF OVERALL VIBRATION LEVELS FOR B.F.P.

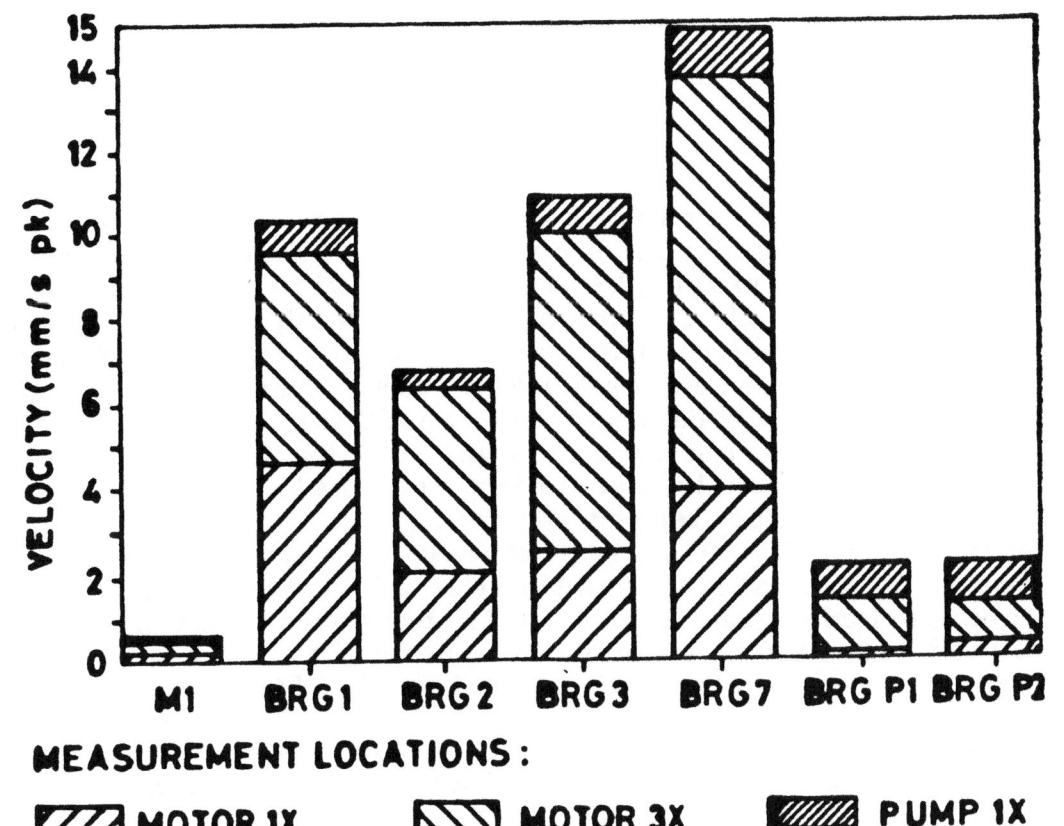

MEASUREMENT LOCATIONS:

|////| MOTOR 1X |\\\\| MOTOR 3X |/////| PUMP 1X

FIG 8. VIBRATION SPECTRAL COMPONENTS FOR B F P

An Expert System of Crack Monitoring and Diagnosing for Rotating Machines*

M Zhao   Z H Luo

Vibration, Shock & Noise Institute
Shanghai Jiao Tong University
Shanghai, 200030, P.R.China

ABSTRACT

In this paper an expert system of crack monitoring and diagnosing for rotating machines, with simulated experience of the experts, is proposed. The difference in the vibrational behaviors between a crack and other defects which may appear to give same information as that of a crack is discussed. A flow chart is provided to show how the expert system works. The system contains three data moduli: the first one furnishes the standard fault information, the second stores the condition of the machine and the third records the faults of the machines. The whole system is divided into two parts: Block A is for tiny crack while block B is for more serious happenings. The diagnosis process is divided into three main steps:
1. Prechecking the total vibration level on all measured points of a machine.
2. Processing the signals from transducers into proper form.
3. Diagnosing whether there is a crack on the shaft by means of the direct method of pattern identification in fuzzy mathematics. The system possesses a self-learning function.

INTRODUCTION

Once a crack or other faults occur on the rotating machines, the production line will be forced to stop and what is worse, catastrophe may ensue with breaking of the shaft.(Jack 1977) It is therefore the wish of all operators to be able to predict the crack, if any, ahead of time. A perfect expert system can help the operators to judge whether there is a crack on the shaft of a machine or not and whether the machine can run safely or not from the information obtained by measuring and analyzing vibration of a machine as an experienced expert.

From the early 80's, a number of enterprises started to monitor the vibration condition of important rotating machines in order to change their periodical maintenance to a planned way with economical benefit. Afterwards the progress of research in diagnosing faults and condition

---

* The project supported by National Natural Science Foundation of China

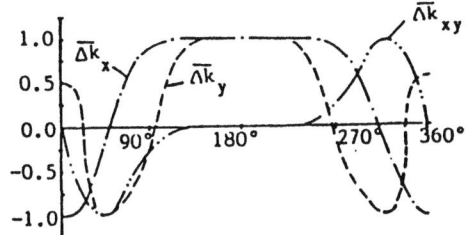

Fig. 1 The Curves of $\overline{\Delta k}_x$, $\overline{\Delta k}_y$ and $\overline{\Delta k}_{xy}$ vs. ωt

monitoring gives further benefit. At present, the widespread interest is on the expert system. However, so far, to the authors' knowledge, most expert systems for diagnosing faults in rotating machine remain yet to undergo the crucial test, (Shi et al. 1991, Guan et al. 1991) and some of the so-called diagnosis softwares as commodities have actually only the function of condition monitoring.

It is believed that a successful expert system should own the self-learning function, i.e. it is able to learn the diagnostic experience accumulated by itself during operation and correct erroneous diagnosis caused by the particular machine or structure.

this paper proceeds along this line.

THE VIBRATIONAL BEHAVIOR OF A CRACKED SHAFTING

A crack may be detected by ultrasonic wave, X-ray, magnetic-particle, color permeation, electric potential difference, acoustic-emission method, etc. But for rotating machinery the vibration measurement method is more favorable, because it requires only simpler instrument and can be processed while the machine is operating. In this instance the information used for monitoring and diagnosis comes from vibration signals of rotating machines. Therefore it is particularly important to understand the vibration behavior of cracked shaft.

Once a crack appears on a shaft, the bent stiffnesses change with time due to the open-close effect. For a well balanced horizontal shaft under its first critical speed, the unbalanced force acting on it is generally smaller than the gravity. For this reason, whether a crack is open or close is decided mainly by the gravity. The stiffnesses of a cracked shaft k_x, k_y and k_{xy} can be expressed as follows: (Zhao & Luo 1990)

$$k_x = k_0 - 1/2 \Delta k_x (1 - \overline{\Delta k}_x) \tag{1}$$

$$k_y = k_0 - 1/2 \Delta k_y (1 - \overline{\Delta k}_y) \tag{2}$$

$$k_{xy} = 1/2 \Delta k_{xy} (1 - \overline{\Delta k}_{xy}) \tag{3}$$

where k_0 is the stiffness of the uncracked shaft; Δk_x, Δk_y and Δk_{xy} are the differences between maximum and minimum values of the stiffnesses of a cracked shaft respectively; and $\overline{\Delta k}_x$, $\overline{\Delta k}_y$ and $\overline{\Delta k}_{xy}$ are the normalized stiffness increments corresponding to stiffnesses k_x, k_y and k_{xy} respectively. Fig. 1 shows the curves $\overline{\Delta k}_x$, $\overline{\Delta k}_y$ and $\overline{\Delta k}_{xy}$ vs.

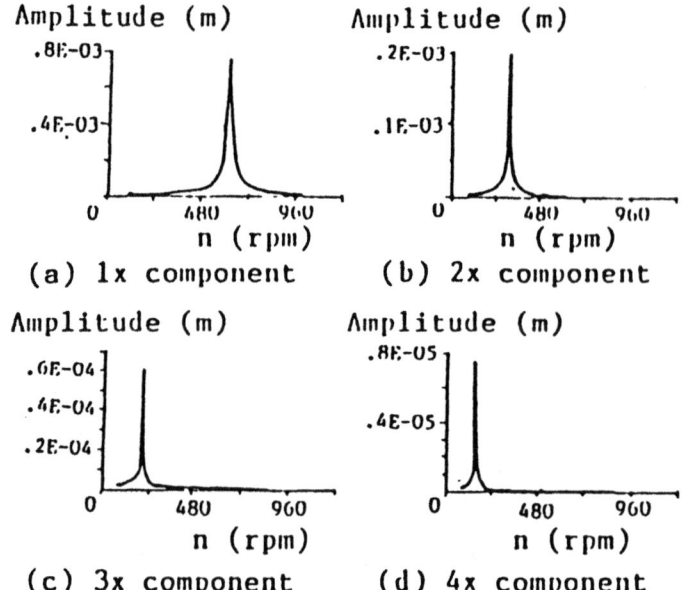

(a) 1x component   (b) 2x component

(c) 3x component   (d) 4x component

Fig.2 The amplitude vs. speed in x-direction (a/D=0.25)

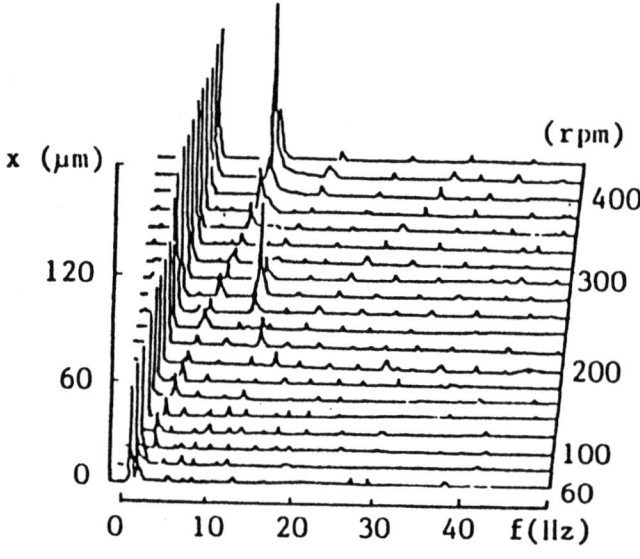

Fig. 3   The cascade while the cracked shafting
is accelerating

ωt, in which ω is the angular velocity of a shaft, and the ratio of crack depth to shaft diameter a/D is equal to 0.25.

With the above model the vibrational behavior of a cracked shafting can be summarized as follows:(Zhao & Luo 1990)

1. Sub-harmonic resonance appears for cracked shaft, i.e. when the rotor runs through 1/n critical speed (n is equal to 2, 3, 4, ...), the amplitude of the n times harmonic component reaches maximum value (see Fig. 2). This characteristic can be reflected clearly from the cascade in Fig. 3.

2. The first critical speeds of a cracked shafting in x-and y-directions are different. Both are slightly lower than that of uncracked shafting.

3.When uncracked shafting operates under first critical speed the orbit of axle center is a circle. But when a crack

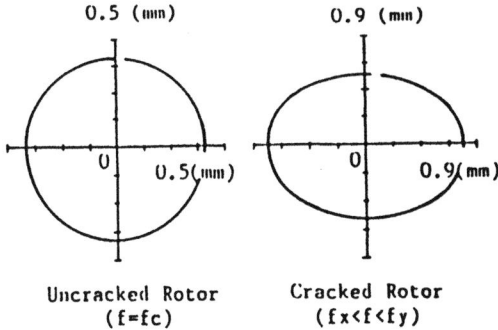

Uncracked Rotor
(f=fc)

Cracked Rotor
(fx<f<fy)

Fig. 4   The Orbit of shaft
        Center

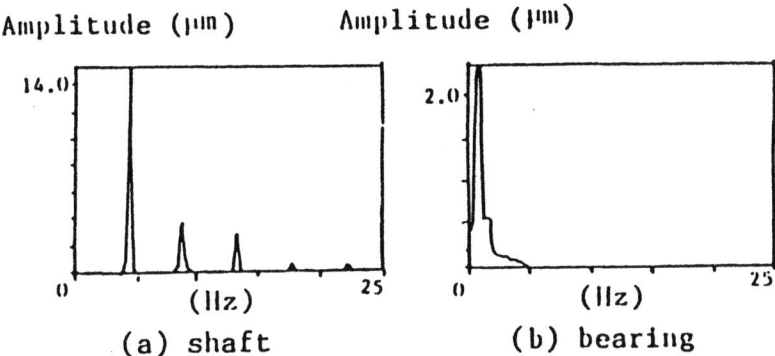

(a) shaft

(b) bearing

Fig. 5 Frequency spectrum

appears the orbit changes from circle to ellipse (see Fig. 4).

4. From running test of a cracked shafting it is obvious that a crack excites 2x, 3x and 4x components of the shaft displacement, but not of the bearing. (see Fig. 5)

5. When the rotor operates above the first critical speed, the deeper the crack, the more rapid increase the 2x component of the steady response. The rate of increase for the 2x component is much more than that of 1x component. (see table 1)

6. For vertically shaft, a crack can excite 3x and 5x components.

Table 1   The Effect of Crack Depth on Response

order	direction	response (μm) crack depth a/d	0.0	0.1	0.2	0.3
1	x		51.28	56.00	75.86	108.70
1	y		51.28	53.72	62.76	73.97
2	x		0.0	0.1511	0.7855	1.83
2	y		0.0	0.1530	0.7218	1.43

THE CHARACTERISTICS OF OTHER FAULTS

The above-mentioned vibrational behaviors of a cracked shafting form the basis for crack monitoring, yet they are not sufficient alone to serve for the purpose, because other defects may give similar signals. Only if the vibrational difference between the crack and other defects is distinguished clearly, then the crack monitor can be carried out.

Hence an examination of other faults is in order.

Unbalance

The vibrational behavior of an unbalanced rotor has been discussed in the rotor-dynamics or vibration text. The unbalance can excite a 1x component of vibration (displacement, velocity or acceleration). If the 1x component of the response of a rotor is much higher than 2x, 3x and 4x components, it may attribute to the rotor unbalance.

Misalignment

If a misaligned coupling is used for connecting two shafts or the center line of bearings is not co-linear, generally, the 2x component of the response of the rotor system will be prominent. But in some case such as the misaligned flange coupling (Lu et al. 1987), the vibration behavior of the rotor system exhibits (b + 1) times components (where the b is the number of the binding bolts for the coupling), because the axis of misalignment makes the stiffness of the flange and the bolts change with time (the period is 2 /b). Thus it is difficult to distinguish between a crack and the misalignment by the vibrational behavior. Here other auxiliary measurements can be taken up, e.g. the bearing oil film pressure or the bearing load or the bearing temperature. If the vibrational signal of a rotor system consists of 2x, 3x and 4x components of running frequency, at the same time, the bearing temperature rises, or the bearing oil film pressure changes, the misalignment can be associated with this vibrational signal.

Friction

Sometimes rubbing friction exists between the rotor and the shell or the stator for rotating machinery if the rotor is installed incorrectly or the rotor is bent by heat. Though the friction may cause the same vibration components as that of a cracked rotor, it can easily be distinguished from a crack by means of the vibration signals in time domain or the orbit of the shaft center, because the friction flattens the waveform or truncates the signals and makes un-smooth orbit (see Fig. 6).

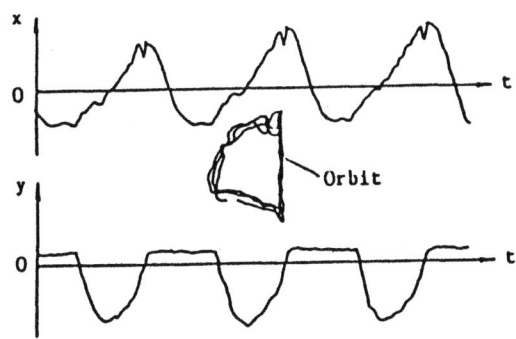

Fig. 6 Waveform and orbit due to friction (Mitchell , 1981)

Looseness

The looseness between a rotor system and its foundation can cause a string of harmonic components of running frequency at abnormally high amplitudes for both the shaft and bearing. Therefore if some very large amplitudes appear in he signals, the vibration of both the shaft and the bearing should be measured in order to check whether there is a looseness between a rotor system and its foundation.

STRUCTURE OF EXPERT SYSTEM FOR CRACK MONITORING AND DIAGNOSIS

A crack monitoring and diagnosis system, with self-learning function, must consist of data collection, signal analysis, crack diagnosis and self-learning of symptoms. In order to complete the first two purposes, the following are necessary, for example, displacement transducers and acceleration transducers, temperature sensors, with corresponding meters, multi-channel recorder and data analysis and storage system with frequency spectrum analysis, the orbit of shaft center, trend analysis etc.

In order to carry out other two purposes, a "knowledge storage" is built up with not only the information for a crack but also for other faults in rotating machine. This forms the basis for determining a crack. At the same time a "data storage" should be set up with the operating record of the machine. Of course, in the beginning, this data storage is empty. After the machine has been installed or repaired the information from normal operation can be recorded as a "health card". By continuous recording of operation the trend of vibration amplitude vs. time can be obtained and any abnormal occurring with the symptom of crack can be detected(Imam et al., 1989). Also, another "data storage" should be set up, with the date, diagnostic result, actual fault and the vibration information for the fault diagnosis together with information on check-up and overhaul, as a basis for self-learning. Fig. 7 shows the flow diagram for the expert system.

In the flow chart, block A completes the function of detecting a tiny crack. Once something abnormal (small magnitude) is discovered, the machine may not be required to shut down until the date for scheduled repair with particular inspection. Block B accomplishes the process of diagnosing a crack or other defects of a rotating machine. In other words, the whole system is divided into two parts: Block A is for minor dis-order or tiny crack while Block B

90

is for more serious happenings which will call for immediate attention.

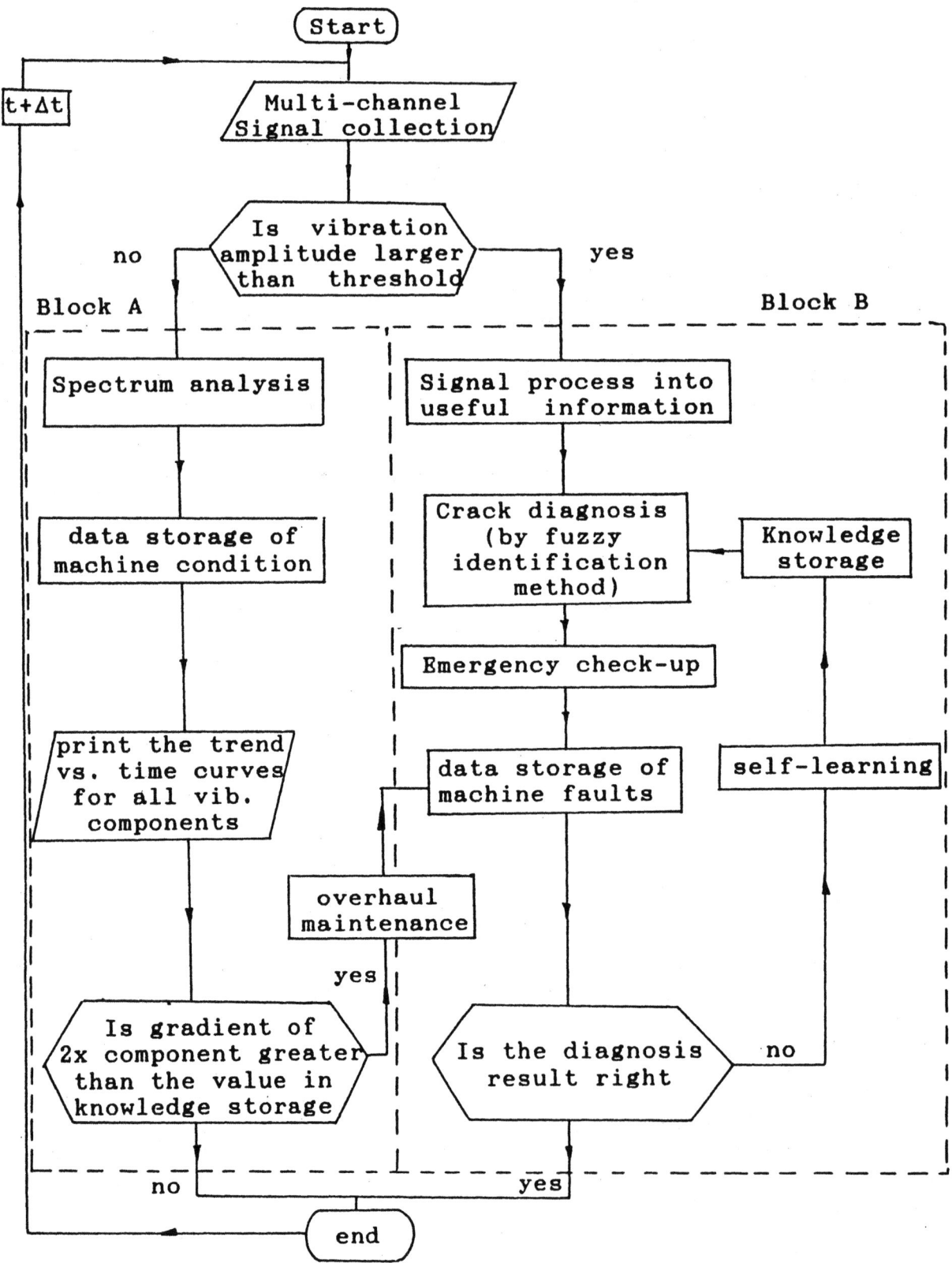

Fig. 7 Flow Chart of the Expert System

REALIZATION OF SOFTWARE AND EXPANDING OF EXPERT SYSTEM

For designing the software of the expert system the turbo-prolog language is proposed so that the standard fault information in knowledge storage can be added and deleted automatically. In this way a new machine can utilize the experience from similar machines or experts in attendance to make diagnosis more accurate. Expert experience can also be kept into the knowledge data storage by person to computer dialogue. By expanding the contents of three data storages and accumulating the skillful experience of engineering personnels the expert system can be made to diagnose all kinds of faults.

REFERENCES

Zhao M & Luo ZH. (1991), Crack Monitoring for a rotating Shaft by Vibration Measurement, Proc. 11th ICPR, Hefei, China

Shi TL et al. (1991), Diagnostic Expert System for Turbo-alternator DEST, Proc. 3rd Trouble Diagnosis Conf. (in chinese), Tianjin, China

Guan HL et al.(1991), A Study on Expert System Of Fault Diagnosis for Turbo-compression Sets ibid.

Zhao M & Luo ZH. (1990), The Vibration Behavior of Rotor Shafting with a Transverse Crack, Proc. ICHDBRSD

Zhao M & Luo ZH (1990), A Method of Fault Diagnosis For Shafting by Vibration Spectrum, Proc. 15th ICMES New Castle, UK.

Imam I et al. (1989) Development of an On-line Rotor Crack Detection and Monitoring System, J. of Vib., Acoustics, Stress and Reliability in Des. 111:241-250

Lu Z et al.(1987), Vibration Behavior excited by the Misalignment of the rotor and test, Proc. of 3rd Conf. on vibration theory with its application, Harbin (in Chinese)

Xu M (1986), Prospects for the Fault Diagnosis of Mechanical Equipments and the Process Technique of Vibration Signal "Dynamic Analysis and Test Technique" No. 2 (In Chinese)

Thomas DL (1984), Vibration Monitoring Strategy for large Turbo-Generators, Proc. 3rd Intl. Conf. of Vib. in Rot. Mach. pp. 91-98

Shrokea Munhoku(1984), Lecture Notes on Mechanical Vibration, Zhenzhou Machinery Institute 1984 (In Chinese)

Armor AF(1983), On-line Monitoring of Turbine-generator Shaft Cracking, ASME paper 83-JPG-Pwr-7

Ziebarth H (1981), Early Detection of Cross-Sectional Rotor Cracks by Turbine Shaft Vibration Monitoring Techniques, ASME paper 81-JPGC-Pwr-26

Mitchell JS (1981), An Introduction to Machinery Analysis and Monitoring, Penn Well Publishing Company

Jack AR et al. (1977), Cracking in 500MW L.P. Rotor Shafts, I.Mech. E. Conf., The Influence of Environnent on Fatigue

ALIGNMENT IDENTIFICATION OF MULTI–BEARING ROTOR SYSTEMS

J. M. Krodkiewski, J. Ding, Z. A. Parszewski

Department of Mechanical Engineering, University of Melbourne, Parkville, Victoria 3052, Australia

ABSTRACT

A general approach has been developed for on–site identification of the actual configuration (alignment) of multi–bearing rotor systems (e.g. a turbogenerator unit). Identification procedure is based on the monitored trajectories of the relative motion of rotor journals with respect to the bearings and a nonlinear mathematical model of the system considered. The mathematical model includes the dynamic properties of rotors, their foundation and supporting structures as well as the nonlinear properties of the oil bearings. Time integration of the nonlinear equations of motion along the measured trajectories yields the desired configuration parameters (relative transverse positions of the bearing centers). The derived procedure of the identification was verified by means of a computer simulation as well as an experimental investigation on a four–bearing rotor test installation. Results may be used for diagnosis of vibrations of rotating machinery as well as their vibration response correction or optimization.

INTRODUCTION

Recent studies [1 – 5] have shown that the bearing alignment changes do have significant influence on rotor dynamic behavior during operation in the case of multi–supported rotor bearing system. Slight changes in a bearing alignment can cause, for example, a significant change of both synchronous and nonsynchronous vibrations as well as the changes of stability threshold. Studies and investigation have been reported specifically on the effect of the alignment changes on the rotor–bearing interaction. Hori et.al. [2] studied the effect of bearing alignment on the stability boundary. Hashemi [3] demonstrated the influence of some of the alignment changes caused by the dissimilarities of oil film thickness on the rotor bearing interaction of steam turbine generators.

In an approach to system dynamics developed by Parszewski and Krodkiewski [1], the concept of the system analysis in terms of the configuration parameters was presented, which allows for better understanding of machine behavior at operating conditions as well as proper handling of static indeterminacy in multi–supported rotor bearing system. Some aspects of rotor dynamics in terms of system configuration parameters have been carried out [4, 5] concerning stability threshold and post stability behavior as well as dynamically optimal configuration for a multi–bearing rotor system.

Because of the thermal expansion, large deformation of the supporting structures or loosening of connections, the system may shift from its initial alignment configuration. It has been shown that the dynamic characteristics of a multi–bearing system will change with the variation of its configuration, and this is due to the nonlinearity of the bearing dynamic forces.

DEVELOPMENT OF THE IDENTIFICATION APPROACH

The equations of motion for a multi–bearing rotor system can be constructed, to include the influence of its configuration on the system dynamic response, as follows,

$$M\ddot{r} + Kr = H(q, \dot{q}) + P + F(t). \tag{1}$$

As illustrated in Figure 1, here r is a vector describing displacements of rotor's nodal points in both xoz and yoz planes. The rotor displacements are referenced with respect to an inertial coordinate system. P is a vector containing static load components and $F(t)$ contains dynamic force components (e.g. excitation). $H(q, \dot{q})$ is composed of the hydrodynamic force components which are functions of the relative journal displacements q

and the relative journal velocities $\dot{q}$ with respect to the bearings. M and K are the mass matrix and the stiffness matrix.

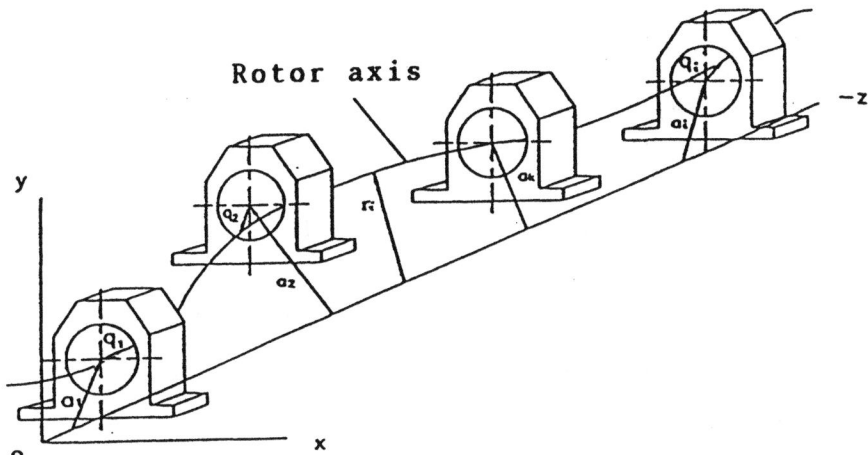

Figure 1. Rotor bearing coordinate system

The discrete model of the rotor has N nodes, some of which are at the N_b journal bearings incorporated in the system. If all the rotor nodal points connected with the bearings are numbered from 1 to N_b, vector r can be written as,

$$
r = \begin{Bmatrix} r_1 \\ r_2 \\ \vdots \\ r_{N_b} \\ r_{N_b+1} \\ \vdots \\ r_N \end{Bmatrix} = \begin{Bmatrix} q_1 \\ q_2 \\ \vdots \\ q_{N_b} \\ r_{N_b+1} \\ \vdots \\ r_N \end{Bmatrix} + \begin{Bmatrix} a_1 \\ a_2 \\ \vdots \\ a_{N_b} \\ 0 \\ \vdots \\ 0 \end{Bmatrix} = \begin{Bmatrix} q \\ r' \end{Bmatrix} + \begin{Bmatrix} a \\ 0 \end{Bmatrix}, \tag{2}
$$

where

$$
r' = (r_{N_b+1}, \ r_{N_b+2}, \cdots, \ r_N)^T, \tag{3}
$$

$$
a = (a_1, a_2, \cdots, a_{N_b})^T. \tag{4}
$$

The vector, a, describes bearing transverse positions (i.e. the alignment) with respect to the stationary system of reference. If equation (1) is partitioned in the same manner as described above, it can be rewritten as,

$$
\left[\begin{array}{c|c} M_{11} & M_{12} \\ \hline M_{21} & M_{22} \end{array} \right] \begin{Bmatrix} \ddot{q} \\ \ddot{r}' \end{Bmatrix} + \left[\begin{array}{c|c} K_{11} & K_{12} \\ \hline K_{21} & K_{22} \end{array} \right] \begin{Bmatrix} q \\ r' \end{Bmatrix} + \left[\begin{array}{c|c} K_{11} & K_{12} \\ \hline K_{21} & K_{22} \end{array} \right] \begin{Bmatrix} a \\ 0 \end{Bmatrix} = \begin{Bmatrix} H(q,\dot{q}) \\ 0 \end{Bmatrix} + \begin{Bmatrix} P_1 \\ P_2 \end{Bmatrix} + \begin{Bmatrix} F_1(t) \\ F_2(t) \end{Bmatrix}, \tag{5}
$$

or

$$
M_{11}\ddot{q} + M_{12}\ddot{r}' + K_{11}q + K_{12}r' + K_{11}a = H(q,\dot{q}) + P_1 + F_1(t), \tag{6}
$$

$$
M_{21}\ddot{q} + M_{22}\ddot{r}' + K_{21}q + K_{22}r' + K_{21}a = P_2 + F_2(t). \tag{7}
$$

A solution which represents the steady state motion, if periodic, can be predicted in terms of Fourier series as follows,

$$q=q_0+\sum_n(q_{an}\cos n\omega t+q_{bn}\sin n\omega t),\ r'=r_0+\sum_n(r_{an}\cos n\omega t+r_{bn}\sin n\omega t), \tag{8}$$

where $\omega=\dfrac{2\pi}{T}$ is the base frequency.

Upon introducing equation (8) into equation (6) and integrating both sides with respect to time t over the interval $[0,\ T]$, the following equation is obtained,

$$K_{11}q_0+K_{12}r_0+K_{11}a=\frac{1}{T}\int_0^T H(q,\dot{q})\mathrm{d}t+P_1\ . \tag{9}$$

Similarly, introduction of equation (8) into equation (7) and integration yields,

$$K_{21}q_0+K_{22}r_0+K_{21}a=P_2. \tag{10}$$

Equation (10) can be easily solved for vector r_0

$$r_0=K_{22}^{-1}(P_2-K_{21}q_0-K_{21}a). \tag{11}$$

Introduction of (11) into (9) produces

$$T(\bar{K}q_0+\bar{K}a)=T(P_1-K_{12}K_{22}^{-1}P_2)+\int_0^T H(q,\dot{q})\mathrm{d}t, \tag{12}$$

where

$$\bar{K}=K_{11}-K_{12}K_{22}^{-1}K_{21}. \tag{13}$$

Upon denoting the integral of the hydrodynamic force on the right hand side of equation (12) as

$$h_0=\frac{1}{T}\int_0^T H(q,\dot{q})\mathrm{d}t, \tag{14}$$

and solving it for vector a, the following equation is obtained,

$$a=\bar{K}^{-1}(h_0+P_1-K_{12}K_{22}^{-1}P_2-\bar{K}q_0). \tag{15}$$

Equation (15) gives the relationship between the configuration vector a and the relative motion q of the journals $(1, 2, \cdots, N_b)$ with respect to the bearing housings. The relative motion q can be measured as a function of time. Introduction of the experimental data into expression (14) yields vector h_0, whereas vector q_0 according to (8) may be obtained from the following equation,

$$q_0=\frac{1}{T}\int_0^T q(t)\mathrm{d}t. \tag{16}$$

NUMERICAL VERIFICATION

In order to verify the developed procedure for an on–site identification of a multi–bearing rotor system configuration, a computer system for the nonlinear simulation of the rotor response $q(t)$ has been developed. This nonlinear simulation system was applied to a four–bearing test installation (Figure 2). The installation is composed of: a rotor and four journal bearings with adjustable pedestals specially designed for the configuration changes, a rigid foundation, a driving system and a hydraulic system.

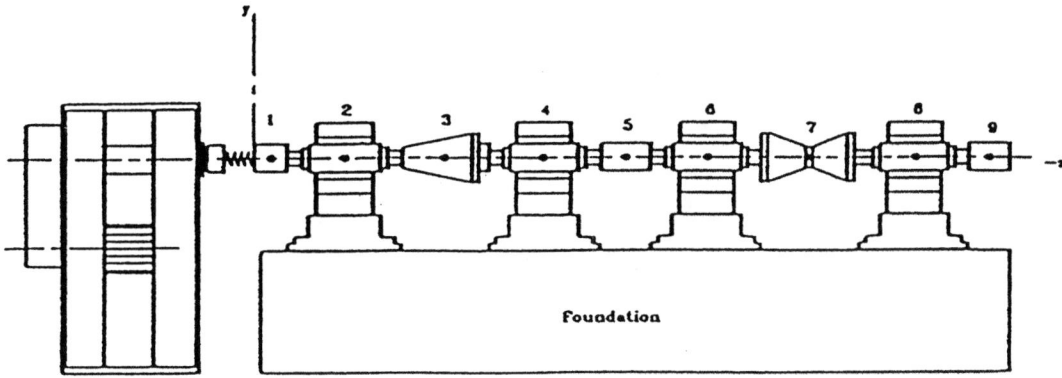

Figure 2. The four—bearing rotor test installation.

The rotor was modeled with nine master nodes condensed from 55 sub—elements. Some information about the rotor is given in Table 1.

Table 1. Physical properties of the test rotor

Rotor length:	1.918 m
Total mass:	37.07 kg
Mass center:	0.969 m
Young's modulus:	2.06×10^{11} N/m²
Mass density:	7800 kg/m³
Poisson's ratio:	0.3

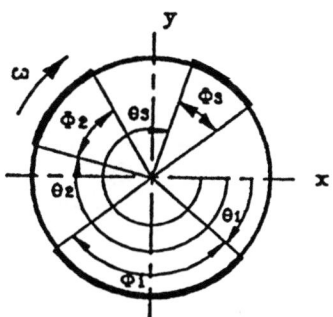

Figure 3. Three—lobe bearing geometry. $\theta_1 = 7.7°$, $\phi_1 = 150°$, $\theta_2 = 192.5°$, $\phi_2 = 65.5°$, $\theta_3 = 288.5°$, $\phi_3 = 47$, R=25 mm, L/D=0.83

Four identical three—lobe journal bearings were used to support the rotor at nodes 2, 4, 6, and 8. The geometry of the bearing is shown in Figure 3. The radial clearance is 1.2 %.

For an assumed configuration, the computer simulation system was applied for obtaining relative journal's trajectories, time history of the journal's displacements and hydrodynamic forces. Figure 4 (a) illustrates transient and steady state journal's motion from one of the computer simulations of the test installation. These numerical data were used to produce time history diagrams of hydrodynamic forces $H(q, \dot{q})$ (Figure 4 (b)). Integration of these forces, according to equation (14), yields vector h_0. Vector q_0 was obtained by integration of the steady state motion (Figure 4 (a)) with accordance to equation (16). Both q_0 and h_0 were introduced into equation (15) to calculate the configuration, vector a.

Four cases presented in this paper are shown as examples of the identification procedure. Operating speed was 3000 RPM. The assumed configuration and identified result are compared in Table 2.

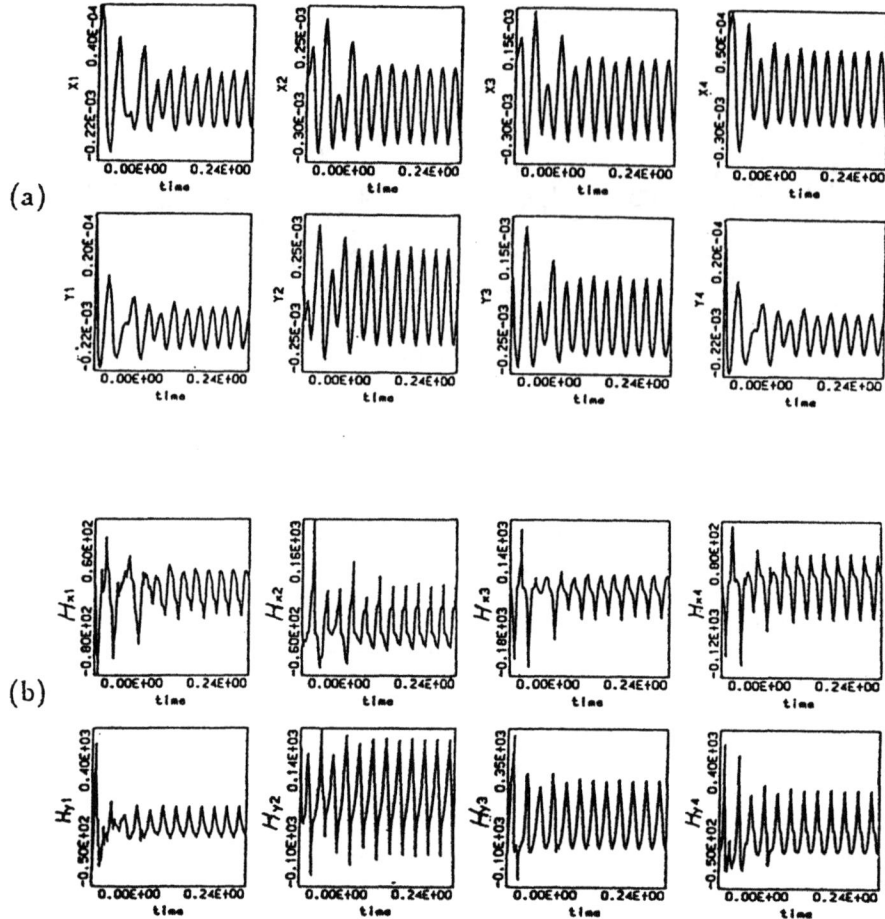

Figure 4. Nonlinear simulation of the test rotor response. (a) Journal displacements. (b) Hydrodynamic forces.

Table 2. The identified results compared with the assumed bearing transverse positions for the test rig

Case	Assumed config x (mm)	y (mm)	Identified config x (mm)	y (mm)	Bearing Number
1	0	0	0	0	B1
	−0.004	−0.15	−0.0041	−0.158	B2
	−0.004	−0.15	−0.0045	−0.155	B3
	0	0	0	0	B4
2	0	0	0	0	B1
	0	−0.108	−0.0001	−0.109	B2
	0	−0.150	−0.0007	−0.148	B3
	0	0	0	0	B4
3	0	0	0	0	B1
	0	−0.150	−0.0007	−0.157	B2
	0	−0.108	−0.0008	−0.108	B3
	0	0	0	0	B4
4	0	0	0	0	B1
	0	−0.108	−0.0009	−0.109	B2
	0	−0.108	−0.0012	−0.109	B3
	0	0	0	0	B4

The differences between the assumed and identified configurations are all less than 5 percent and this shows the accuracy of the numerical computation.

EXPERIMENTAL VERIFICATION

Experiments have been carried out on the above mentioned four—bearing rotor test installation. The supporting structure was specially designed to allow the bearing pedestals to be shifted horizontally and vertically with high accuracy as shown in Figure 5.

The measurement instrumentation is shown in Figure 6. The journal—to—bearing displacement was measured using a BENTLY NEVADA 7200 series proximity transducer system. A calibration unit was used to transform the signal of journal surface displacement to journal—to—bearing center displacement. Each bearing was mounted with four eddy current transducers. Each two transducers were used to measure displacements of both ends of the journal in one direction. This gives the average displacements of the journal central point. A DAS20 A—D/D—A conversion board was installed in an IBM AT/PC computer for acquiring data simultaneously from eight channels (two for each bearing).

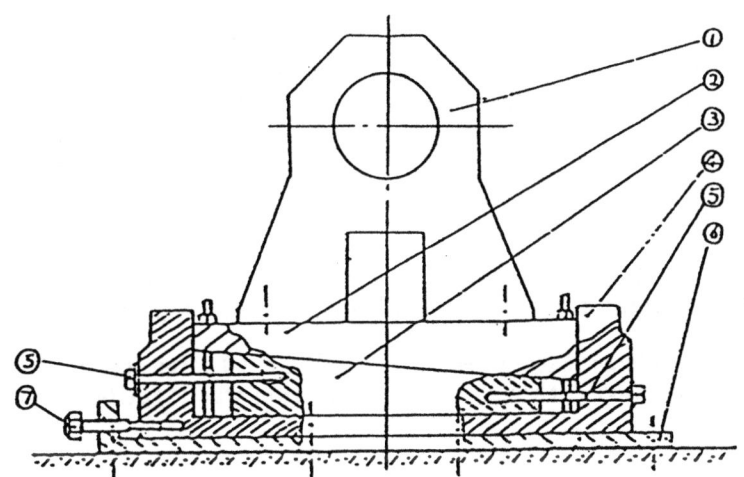

Figure 5. Bearing pedestal shifting mechanism. 1. bearing; 2. vertical shift upper wedge; 3. vertical shift lower wedge; 4. horizontal shift block; 5. vertical shift adjustment bolts; 6. base plate; 7. horizontal shift adjustment differential bolt.

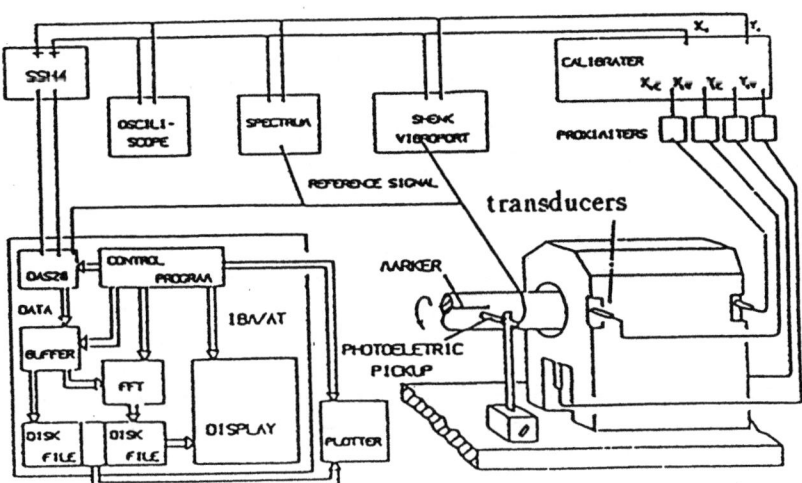

Figure 6. Measurement instrumentation.

First, all the bearing pedestals are adjusted in such a way that all four bearing centers are on one straight line. Bearings 1 and 4 provides a reference line and the following shifts of the inner bearings gives various configurations of the system. The initial system configuration was set such that the second bearing was shifted by −0.95 mm (downward) and the third bearing by −0.5 mm. All experiments were done at the same rotor speed of 3000 RPM. Next, the second bearing had been subject to a series of shifts upward from the initial configuration. The shift range is from −0.95 mm to 0.4 mm with an increment of 0.15 mm. Despite of the fact that the shifts of the bearings are large (from −0.95 mm to 0.4 mm), the relative journal–to–bearing static displacements were within the range of 84% of the radial clearance. For each alignment, the relative journal–to–bearing displacements in all bearings were collected from the test installation as shown in Figure 7. These data are used to produce dynamic bearing forces via the bearing model. Both the journal orbits and the bearing dynamic forces are integrated to allow the computation of configuration vector a from equation (15).

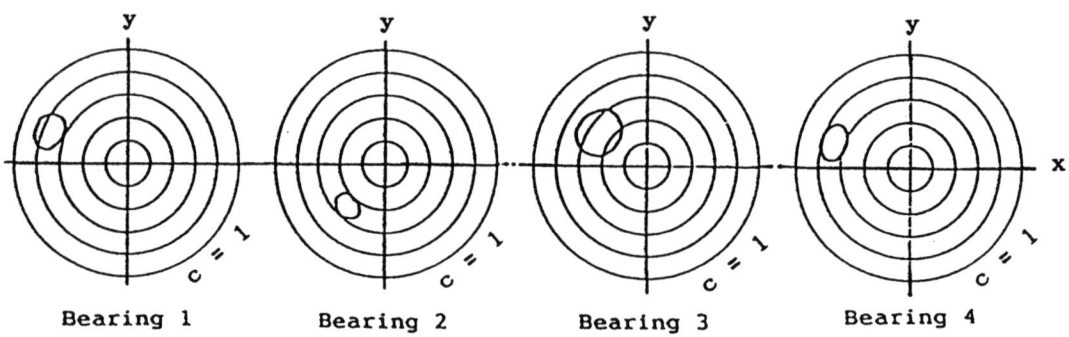

Figure 7. The measured orbits of the journals; c − nondimensional radial clearance; The imposed configuration (mm) : $x_2=0$, $y_2=0.25$, $x_3=0$, $y_3=−0.5$.

Figure 8 shows the identification results for all the 10 alignments corresponding to the shifts of the second bearing. Although the configuration was identified not as close to

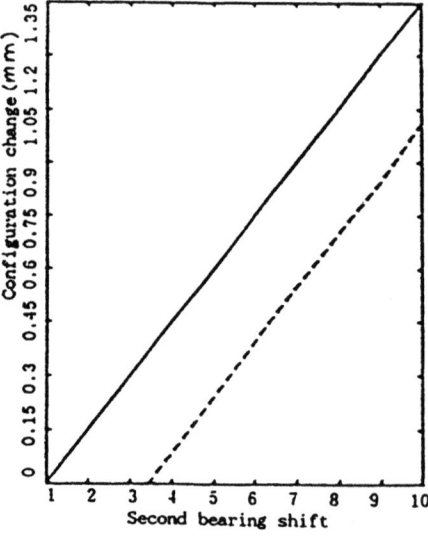

Figure 8. Actual and identified configuration versus 10 upward shifts of bearing 2.
————————————, actual; − − − − − − − , identified.

the assumed one as it was in the numerical verification, the change in the measured configuration are almost the same as change in the identified one (Figure 8). The difference between the actual and identified configurations of the test rig can be due to the incompleteness of journal bearing modelling. The rubber sealing may play a significant role but it is omitted in mathematical model. The omission of the angular displacements of the journal in the adopted hydrodynamic theory plays less significant role since the ratio of the journal length to the distance between the bearings was 0.08 and the ratio of length to diameter of the bearings was 0.83.

CONCLUSIONS

An approach for the identification of the alignment of multi–bearing rotor systems has been developed for on–site application providing the basis for the optimization of the system vibration response. This identification procedure is of practical significance. It makes use of the relative motions of the journals with respect to the bearings. These can be easily monitored on the rotor bearing system. These journal trajectories can be integrated to find the time–dependent bearing dynamic forces. Both the journal's motion and the hydrodynamic forces are used for the calculation of the system configuration parameters. The numerical verification carried out on a four–bearing rotor test installation showed high accuracy of this identification procedure. The experimental verification on the same installation showed satisfactory agreement between the actual and identified configuration changes.

REFERENCES

1. Parszewski, Z. A. and Krodkiewski, M. J., "Machine Dynamics in Terms of the System Configuration Parameters", IFToMM JSME. Inter. Conf. of Rotordynamics, Tokyo, 1986.
2. Hori, Y. and Uematsu, R., "Influence of Misalignment of Support Journal Bearings on Stability of a Multi–Rotor System", *Tribology International*, Vol.13, No.5, 1980, pp.249.
3. Hashemi, Y., "Alignment Changes and their Effects on the Operation and Integrity of Large Turbine Generators: Experience in the CEGB South Eastern Region", Steam and Gas Turbine Foundation and Shaft Alignment, I. Mech. Eng. Conf. Publ. 1983, pp. 19.
4. Parszewski, Z. A., Krynicki, K. and Kirby, E., "Effect of Bearing Alignment on Stability Threshold and Post–Stability Behavior of Rotor Bearing Systems", Conf. C366, I.Mech. E, 1988.
5. Parszewski, Z. A. and Li, D. X., "Dynamically Optimum Configuration for a Multi–Bearing Rotor System—Theory and Experiments", *International Symposium for Dynamics and Design*, 1989, pp.279–284.
6. Lund, J. W., "The Stability of an Elastic Rotor in Journal Bearings with Flexible, Damped Supports", *Journal of Applied Mechanics*, Tran. ASME, Ser. E Vol.87, No.4, 1965, pp.99.
7. Morton, P. G., "Analysis of Rotors Supported upon Many Bearings", *Journal* of Mech. *Eng. Science*, Vol.14, No.1, 1972, pp.25.
8. Kirk, R. G. and Gunter, E. J., "Nonlinear Transient Analysis of Multi–Mass Flexible Rotors—Theory and Application", NASA CR–2300, 1973.
9. Lund, J. W., "Modal Response of a Flexible Rotor in Fluid–Film Bearings", *Journal of Engineering for Industry*, Transactions of the ASME, 1974, pp.525.

Smart Rotors

A. Dimarogonas, A. Kollias

School of Engineering and Applied Science, Washington University in St Louis

ABSTRACT

The phenomenon of electro-rheology relates to changes in rheology of certain dispersions upon application of electrical fields and appears as an increase to the resistance of flow, and in some cases conversion from a fluid to solid behavior, under an increasing electrical field. Most ER fluids consist of a dispersion of fine particles in a liquid medium with the addition of a surfactant agent, for example silica particles in transformer oil with water added as surfactant. The shear stress consists of the Newtonian resistance plus a constant shear which depends on the applied electric field.

Through the controlled bearing properties, active control of the rotor dynamic behavior can be achieved. Thus, for example, transition through the critical speed can be virtually eliminated through active control of the bearings.

STATE OF THE ART

"Smart" materials and structures is a term used to characterize materials or structures which have sensing, actuating or control capabilities either by themselves or through imbedded sensors and thus they are capable of self-adapting to changes in the external conditions to maintain a specified goal. "Smart" materials, in particular, have properties which can be used for sensing, actuating or control of the structural properties. Some of these materials are (Roberts, Barker & Jager 1988) optical fibers (Claus, Jackson & May 1985), piezoelectric crystals or ceramics (Bailey & Hubbard 1985), shape memory alloys (Rogers & Robertshaw 1988), and electrorheological fluids (Gandhi & Thomson 1988).

The phenomenon of electro-rheology (ER) was first reported by Winslow in 1947 and occasionally bears his name. It relates to changes in rheology of dispersions upon application of electrical fields. It appears as an increase to the resistance of flow, and in some cases conversion from a fluid to solid behavior, under an increasing electrical field (Block and Kelly 1988).

Most ER fluids consist of a dispersion of fine particles in a liquid medium with the addition of a surfactant agent, for example silica particles in transformer oil with water added as surfactant. The shear stress consists of the Newtonian resistance $\mu \partial \gamma / \partial t$ plus a constant shear τ_y which depends on the square of the applied electric field (V/mm),

$$\tau_y = A(V/h)^2 \qquad (1)$$

where μ is the fluid viscosity, V the applied voltage over a gap h, A is a constant which depends on the ER fluid and is constant over a range of frequencies from DC to 100 Hz (Klass and Martinek 1967).

The influence of temperature is very complex, since it affects the fluid dielectric properties and the surfactant action. For example, a fourfold increase of the shear force was observed with a temperature increase from 25 to 65° C (Klass and Martinek 1967 a,b).

ER effects are strong for fields of the order of kV/mm. It appears that there is a low threshold field of the order of x100 V/mm (Winslow 1949) and a high value of the field, of the order of x10 kV/mm, which causes dielectric breakdown. Particle concentrations of the order of 10% gave substantial ER effects while maintaining fluidity without electric field (Block and Kelly 1988).

Winslow (1947, 1949) observed that the particles within the ER fluid fibrilate on the application of an electric field. Such fibrils have been considered fundamental to the ER effects, forming mechanical links which lead to a static shear stress to break them (Arguelles et al 1974; Shul'man et al 1977).

Coulombic interactions between polarized particles have been studied with a statistical mechanics approach (Brooks et al 1986) giving good correlation with experimental results at very high electric fields.

Several investigators have studied the effect of water surfactant, rendering it fundamental to the ER effects. However, the advent of the anhydrous EH fluids tends to suggest that the water involving mechanisms are not essential in the ER effect.

A fluid which can be converted essentially to a solid on the application of a voltage replacing thus mechanical links, may have a great number of engineering applications. This was recognized since Winslow's original work (1947, 1949).

ER clutches or brakes have been proposed by Gorodkin et al (1979), Spronston et al (1983), Bullough (1988). Hydraulic valves without moving parts can have a very fast response (Winslow 1947; Brooks 1982). Active control of vibration dampers can be achieved through ER fluids (Kerr 1981, Strandrud 1966, Bullough and Foxon 1978). Wide band high power vibrators can be achieved through a continuous mechanical motion and controlled force transmission through an ER link (Bullow and Foxon 1978). Robotic controlled systems having very fast response can be based on ER links (Bullough 1988).

SMART BEARINGS

ER fluids can be used to control actively the static and dynamic response and thus the stability threshold of slider bearings over a wide range of frequencies and operating conditions.

Utilization of ER fluids has been recently proposed (Dimarogonas & Kollias 1989) to yield an actively controlled slider bearing. It was demonstrated that such bearing can maintain stability within a wide range of speeds by active control.

ER fluids, as discussed above, respond to shear rate with a shear stress

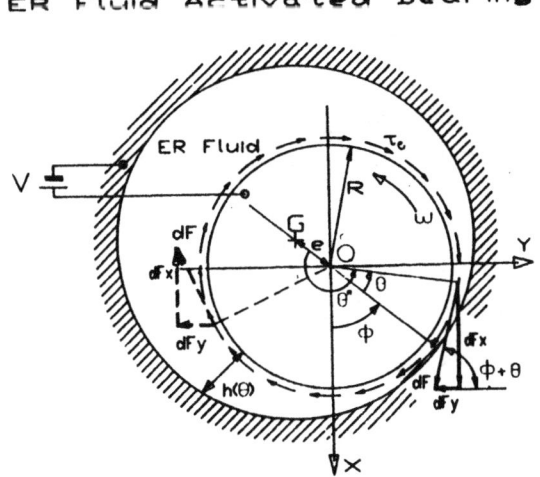

Fig. 1. ER Bearing Geometry.

component τ_y which is independent of the shear rate but depends on the applied voltage and the oil-film thickness. It is apparent from Figure 1 that in the neighborhood of the minimum oil film thickness the shear force on the journal is much higher than anywhere else (shear stress $\tau_y = A[V/h]^2$). This generates a force which forces the journal to a smaller attitude angle and higher eccentricity. Both of these factors combine to give higher threshold of instability.

Most of the existing experimental results apply to shear rates up to 7000 sec^{-1}. There are bearing applications where the shear rate is several orders of magnitude higher. For a journal with angular velocity ω, the average shear rate is $d\gamma/dt = U/C = \omega R/C$. Assuming that $\omega = 377$ rad/sec and a typical value for R/C = 500 (Dimarogonas 1988), the shear rate is 188,500 sec^{-1}, much greater than the ones commonly reported in the literature. For this range, Bullough (1989) has reported tests above 100,000 sec^{-1} with similar results. However, Stangroom (1989) reported that for high shear rates the constant shear rate component remains nearly the same $A(V/h)^2$ while the Newtonian viscosity drops with increasing voltage. This reduces the Sommerfeld Number of the bearing and increases the eccentricity and the threshold of instability.

Though it is possible to use control surfaces in the bearing with high clearances and low shear rates, it is interesting to see what happens at high shear rates not only on theoretical but also on practical grounds, because if the ER bearing can operate at high shear rates without control surfaces, the result is a compact design.

It was demonstrated that an ER slider bearing can increase the stability range by active control. Moreover, through the controlled bearing properties, active control of rotor dynamic behavior can be achieved. Thus, for example, transition through the critical speed can be virtually eliminated through active control of the bearing static and dynamic properties.

It was mentioned above that squeeze-film bearings are used with antifriction bearings to provide damping for high-speed rotors, mostly in aircraft applications (Cooper 1963). The relatively small clearances they are allowed to have result in relatively low shear rates and the damping they provide is consequently small. Again, an ER fluid might be suitable for such application, since the apparent viscosity (shear force/shear rate) becomes extremely high for small shear rates.

In addition to the vertical load, ER fluid static shear stresses were integrated around the journal to yield additional static loads, horizontal and vertical. Then, the bearing linearized coefficients and the threshold of instability of the bearing were computed. The threshold of instability can be substantially increased by applying an electric field to the ER lubricant. A 10% increase in the instability threshold was achieved for a maximum electric field of 8 kV/mm with the fluid (2) for a test 170° bearing operating at So=.155, ε=.765, ϕ=35°, while the maximum electric field with fluid (2) was 0.5 kV/mm, way below the critical field which is well above 10 kV/mm (Dimarogonas & Kollias 1989).

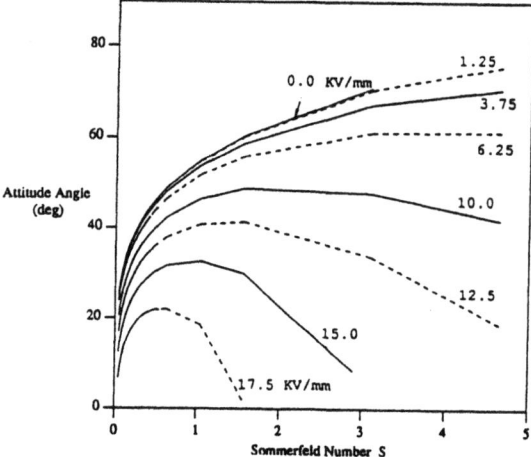

Fig. 2. Eccentricity ratio as function of the electric field

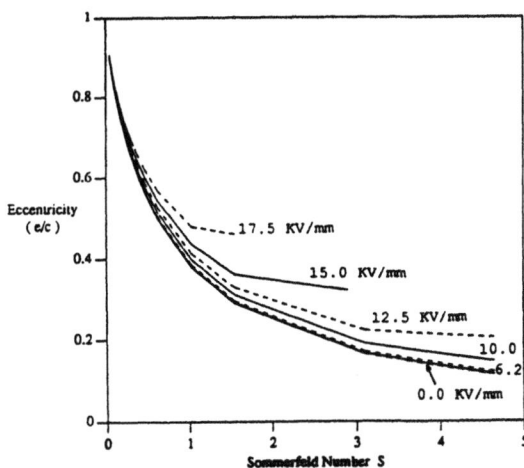

Fig. 3. Attitude angle as function of the electric field

For these results, two rather weak ER fluids were used:
(1), Eige & Peschon (1960), μ=10 poises, $A=7 \times 10^{-9}$ N/V^2.
(2), Bullough (1973), μ=12.5 poises, $A=.28 \times 10^{-10}$ N/V^2.

Bearing eccentricity, attitude angle, threshold of instability, linearized coefficients should be expressed as functions of the Sommerfeld Number So and the ER number Er.

Substantial changes in the bearing parameters can be achieved under elecric field. Figure 2 shows the increase of the eccentricity ratio with the electric field. Figure 3 shows the decrease of the attitude angle with the electric field. Figures 4 and 5 shows the change of the dimensionless linearized bearing coefficients with the electric field.

It must be pointed out that a single voltage was applied to the whole bearing. Higher controllability can be obtained by partitioning the bearing surfaces under different voltages. Moreover, variation of the voltage can be used for real time vibration control with an appropriate contol strategy.

For certain parameter constellations, direct application of ER fluids in lubrication might be inadequate for control or impossible, for example:

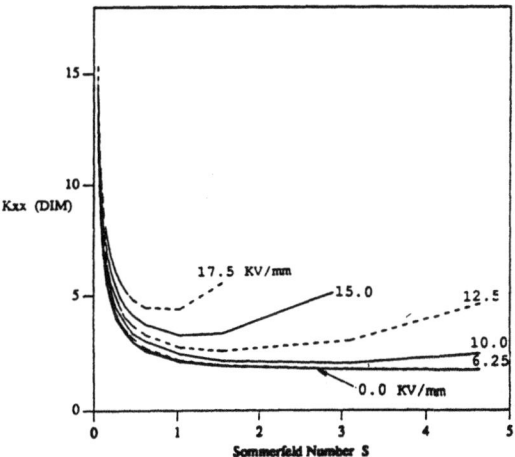

Fig. 4. Bearing stiffness as function of the electric field.

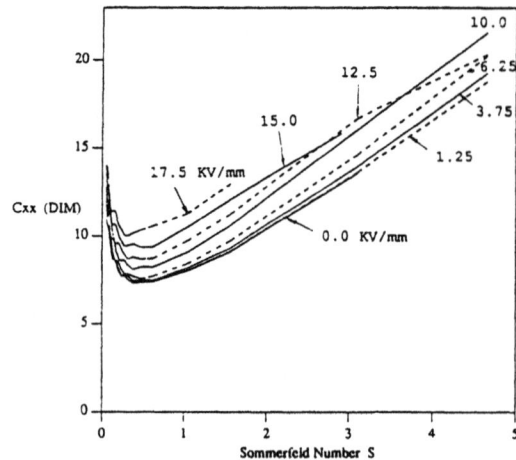

Fig. 5 Bearing Damping as function of the electric field

a. For very small clearances or very high velocities, leading to very small passages for the dispersions or very high shear rate.

b. For Newtonian terms very high as compared with the ER yield stress. This leads to low controllability, defined here as the ratio $\ell = \tau_y/\mu d\gamma/dt$, ER fluid yield stress/Newtonian shear stress. It is apparent that at high $d\gamma/dt$ the controllability is very small.

c. For large extension of the stability range.

In such cases control surfaces might be introduced, axial or radial, with high clearances. The yield stresses remain the same while the shear rate and the Newtonian stresses diminish. For such surfaces, the flow can be analyzed with bearing lubrication methods for e/c very small. It is easier, however, to treat the case as purely shear flow and integrate the shear stresses on the journal directly to yield the bearing additional forces.

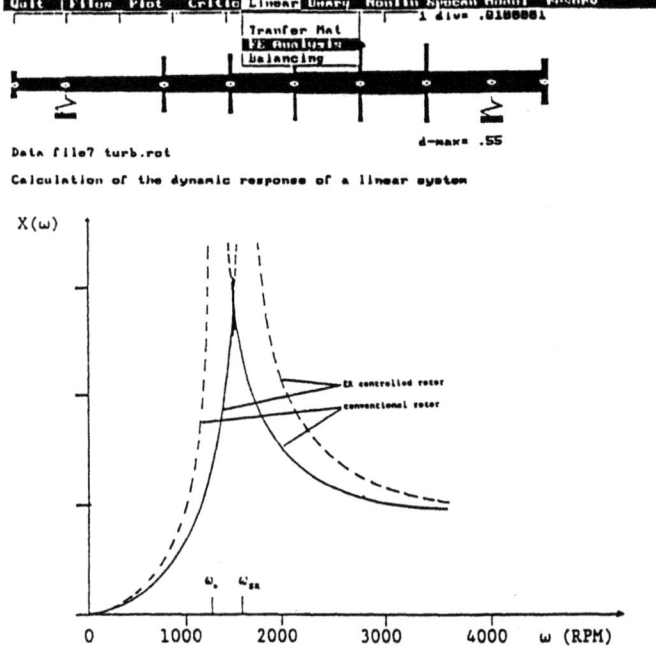

Fig. 6. Critical Speed Vibration Control

SMART ROTORS

Through the controlled bearing properties, active control of the rotor dynamic behavior can be achieved. Thus, for example, transition through the critical speed can be virtually eliminated through active control of the bearings.

The transition control strategy consists of starting the rotor with electrically activated bearings of high stiffness. When the speed approaches the critical speed of the stiff system, the electric field is removed and the critical speeds drops below the operating speed.

The transient effect is insignificant since it has the lower frequency which can be substantially lower than the operating speed at the time of de-activation of the electric field.

References

Arguelles J, Martin HR, Pick R (1974). Rheological Model for Steady Electroviscous Flow between Parallel Plates. Jr. Mech. Eng. Sc. 16(4):232-239.

Bailey T, Hubbard JE (1985). Distributed Piezoelectric - Polymer Active Vibration Control of a Cantilever Beam. Journal of Guidance, Control and Dynamics 8(5):605-611.

Block H, Kelly JP (1988). Electro-Rheology. Jr. of Physics, D21(12):1661-1667.

Boehme G (1987) Non-Newtonian Fluid Mechanics. North Holland, Amsterdam.

Bonnecase RT, Brady JF(1989). Dynamic Simulation of a Suspension Forming an Electrorheological Fluid. 2nd Int. Conf. on Electrorheological Fluids, Raleigh, NC.

Brooks DA (1982). Electrorheological Fluids. Chart. Mech. Eng., vol. 63.

Brooks D et al (1986). Viscoelastic Studies on an Electrorheological Fluid. Colloids Surf 18:293-312.

Bullough WA (1988). Electrorheological Fluids. Engineering, Feb., Tech. File 163:i -iv.

Bullough WA, Stringer J.D. (1973). The Utilisation of the Electroviscous Effect in a Fluid Bearing. 3rd Int. Fl. Power Symp, paper F3, Turin, Italy.

Bullough WA, Foxon MB (1978). A Proportionate Coulomb and Viscously Damped Isolation System. J. Sound & Vib. 56(1):35-44.

Carlson JD, Duclos TG (1989). ER Fluid Clutches and Brakes - Fluid Property and Mechanical Design Considerations. 2nd Int. Conf. on Electrorheological Fluids, Raleigh, NC.

Claus RO, Jackson BS, May RG (1985). NDE of Composites by Optical Time -Domain Reflectometry in Embedded Optical Fibers. IEEE SOUTHEASTCOM 85 Proceedings (Raileigh, NC):241-245.

Conrad H, Chen Y, Sprecher AF. (1989). Electrorheology of Suspensions of Zeolite Particles in Silicon Oil. 2nd Int. Conf. on Electrorheological Fluids, Raleigh, NC.

Cooper S (1963). Preliminary Investigation of Oil Films for the Control of Vibration. IME Lubrication and Wear Convention, Proceedings:305-315.

Dimarogonas AD, Kollias A, Electroreological Fluid Smart Journal Bearings, Society of Tribologists and Lubrication Engineers (to appear).

Dimarogonas AD, Haddad SD (1992). Vibration for Engineers. Prentice Hall, Englewood Cliffs.

Dimarogonas AD, Paipetis SA (1983). Analytical Methods in Rotor Dynamics. Elsevier-Applied Science Publishers, London.

Duclos TG, Coulter JP, Miller, LR. (1988). Applications for Smart Materials in the Field of Vibration Control. Proceedings, ARO Smart Materials, Structures and Mathematical Issues Workshop , Blacksburg, Va.:132-146.

Eige J, Peschon J (1960). Vibration-Shock System. Int. Report, Project 3120, Stanford Univ.

Gast AP, Adriani PM (1989). Microstructural Models of Electrorheological Fluids. 2nd Int. Conf. on Electrorheological Fluids, Raleigh, NC.

106

Ghandi MV, Thomson BS (1988). A New Generation of Ultra Advanced Intelligent Materials Featuring Electrorheological Fluids. Proceedings, ARO Smart Materials, Structures and Mathematical Issues Workshop , Blacksburg, Va.:63-68.

Gorodkin RG, Korobko YV (1979). Fluid Mech.- Soviet Res. vol. 8, 48.

Inoue A (1989). Study of a New Electrorheological Fluid. 2nd Int. Conf. on Electrorheological Fluids, Raleigh, NC.

Kerr J (1981). A Solid Chance to Jam Liquid Flow Lines. The Engineer, July 23:63-64.

Klass DL, Martinek TW (1967). Electroviscous Fluids. I. Rheological Properties. J. Appl. Phys, v. 38, n. 1:67-74.

Klass DL, Martinek TW (1967).Electroviscous Fluids. II. Electrical Properties. J. Appl. Phys, v. 38, n. 1:74-80.

Klingenberg DJ, Zukoski CF (1989). Structure Formation in Electrorheological Fluids. 2nd Int. Conf. on Electrorheological Fluids, Raleigh, NC.

Korobko EV, Sh'ulman ZP, (1989). The Mechanism of Visco-plastic Behavior of Electrorheological Suspensions. 2nd Int. Conf. on Electrorheological Fluids, Raleigh, NC.

Najji B, Bou-Said B, Berthe D, (1989). New Formulation for Lubrication with Non-Newtonian Fluids. ASME Journal of Tribology, 111:29-34.

Opperman G et al, (1989). Applications of Electroviscous Fluids as Movement Sensor Control Devices in Active Vibration Dampers. 2nd Int. Conf. on Electrorheological Fluids, Raleigh, NC.

Papanastasiou TC (1987). Flows of Materials with Yield. Journal of Rheology 31(5):385-404.

Reddi MM, Trumpler PR (1962). Stability of High-Speed Journal Bearings under Steady Load. 1: The Incompressible Film. ASME Journal of Engineering for Industry, ser. B, 84:351-358.

Rogers CA, Robertshaw HH. (1988). Development of a Novel Smart Material. ASME Winter Annual Meeting, Chicago, Ill.

Rogers CA, Barker DK, Jaeger CA. (1988) Introduction to Smart Materials and Structures. Proceedings, ARO Smart Matrials, Structures and Mathematical Issues Workshop , Blacksburg, Va.:17-28.

Stangroom JE (1989). The Bingham Plastic Model of ER Fluids and its Implications. 2nd Int. Conf. on Electrorheological Fluids, Raleigh, NC.

Strandrud HT (1966). Electric-field valves inside cylinder control vibrator. Hydraulics & Pnewmatics, September:139-143.

Shul'man ZP, et al (1986). Structure, Physical Properties and Dynamics of Magnetorheological Suspensions. Int. J. of Multiphase Flow v. 12, n. 6:935-955.

Shul'man ZP, et al (1987). Characteristics of an Electrorheological Damper in a Vibration Insulator. Inz.-Fiz. Zhur, v. 52, n. 2:237-244.

Sproston JL, Stevens NG, Page IM (1983). An Investigation of Torque Transmission using Electrically Stressed Dielectric Fluids. Inst. of Phys. Conf. Ser. (66) 53-58.

Tayal SP, Sinhasan R, Singh D.V. (1982). Analysis of Hydrodynamic Journal Bearings Having Non-Newtonian Lubricants Using the Finite element Method. ASLE Transactions 25 (3) 410-416.

Winslow WM (1947). Methods and means of translating Electrical Impulses into Mechanical Force. US Patent 2,147,850.

Winslow WM (1949). Induced filtration of suspensions.J. Appl. Phys, v. 20:1137-1140.

Winslow WM(1953). Field controlled hydraulic devise. US Patent 2,661,596.

Wong W, Shaw M, (1989). Investigations of the Role of Moisture in Electrorheological Fluids. 2nd Int. Conf. on Electrorheological Fluids, Raleigh, NC.

Fuzzy Input-Neural Net Adoptive Expert Systems for Rotor Diagnosis and Prognosis

Andrew D. Dimarogonas

School of Engineering and Applied Science, Washington University in St Louis

ABSTRACT

Neural nets used in expert system make them capable of learning, self adopting to particular engineering applications and handling fuzzy input information.

Inputs (symptoms) and outputs (causes) are represented by neurons interconnected through synapses, weighted connections among neighboring neurons. Input signals come to the neurons and excite or inhibit the neurons firing activity on the basis of a certain functional relationship. Knowledge is distributed to a large number of weighted synapses which facilitates learning by experience, realized through modification of the synapses weights according to a chosen learning rule.

Available experience for failure diagnosis in turbomachinery was utilized to initially teach the system. Additional diagnoses from the author's experience were taught to the system and additional features and diagnoses defined.

NEURAL NETWORKS IN EXPERT SYSTEMS

Neural networks is a new approach to engineering expert systems which are capable of learning, self adopting to particular engineering applications and handling fuzzy input information.

The properties of neural networks have been studied in relation to the study of the learning process in biological systems (Hebb 1949).

The principal component of a neural network is an artificial neuron. Neurons are interconnected with synapses, weighted connections with neighboring neurons. Input signals come to the neuron from the neurons connected to it by the appropriate synapses and excite or inhibit the neuron's firing activity on the basis of a certain functional relationship. A large class of functions can be represented by appropriate network architecture (Minsky and Papert 1969) which can be learned from examples.

In traditional machine learning, symbolic representations, such as first order predicate calculus, are used to represent knowledge (Michalski, Garbonell and Mitchell 1983). The resulting algorithms are specific to the selected representation and presume an ad-hoc knowledge of the system represented. In the neural network representation, knowledge is distributed to a large number of weighted synapses which facilitates learning by experience, realized through modification of the synapses weights according to a chosen learning rule.

FUZZY SETS.

Traditional classification systems use binary logic. There are two possible outcomes for an event: The two possible states are described by 1 or 0. This assumes a clear distinction between two and only two possible states of an event, key element of the single valued logic in the western culture. However, key elements in the human thinking are not numbers but labels of fuzzy sets (Zadeh 1973), that is, classes of objects in which the transition from non-membership to membership is gradual rather than abrupt. Much of the logic behind human reasoning is not the traditional single valued or multivalued logic but a logic with fuzzy truths, fuzzy connectives and fuzzy rules of inference. This reflects the fact that human expertise, which expert systems are aimed at transferring to machines, is very often domain dependent, incomplete and episodic. Fuzziness, once incompatible with scientific work, is inherent in the human reasoning and the development of the fuzzy sets theory offers the analytical vehicle to relate to traditional scientific fields.

A fuzzy subset A of a universe of discourse U is characterized by a membership function $\mu_A:U\to[0,1]$ which associates with each element of y of U a number $\mu_A(y)$ in the interval [0,1] which represents a grade of membership of y in A. The support of A is the set of points in U at which $\mu_A(y)$ is positive and the crossover point in A is an element of U whose grade of membership in A is 0.5 .

There is a variety of membership functions (Kaufmann 1975) which are selected on the basis of the particular application.

Fuzzy inputs-outputs and fuzzy reasoning together with some extent of learning from examples have been already used in expert systems (Bergadano et al 1987, Kosko 1992). Several fuzzy expert system shells are now in use (Siler et al 1987).

ROTOR FAULT DIAGNOSIS

An important problem in rotating machinery is the diagnosis of incipient failures so that corrective measures could be implemented to avoid failure. This is usually the job of experts who, through extensive experience, can interpret vibration, sound, temperature and other signals in relation to possible mechanisms of failure. It seems that this is a typical place where an expert system can be used.

Inputs

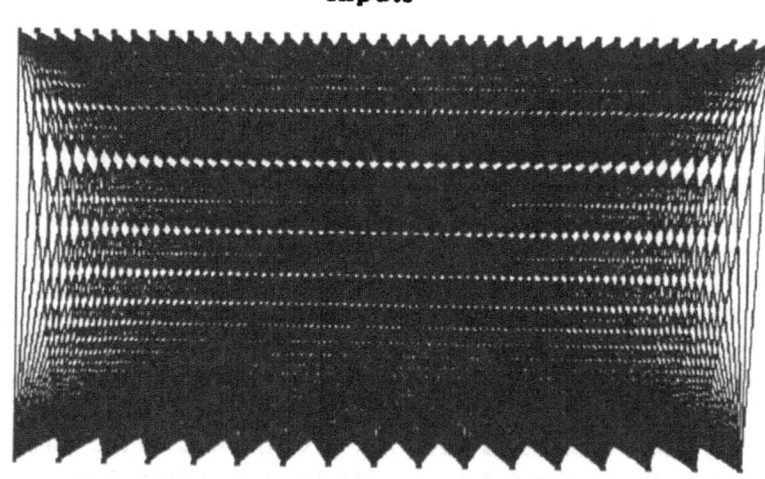

Outputs
Fig. 1. Neural Network Module.

Predicate calculus based expert systems have been already widely used, with frame representation of knowledge, in addition to experience tables and other aids to assist in machinery failure diagnosis (Sohre 1980, Keim & Nordmann 1989). However, there are several shortcomings in this approach:

a. Most of the symptoms are fuzzy and a true-false answer cannot be defined.
b. That the same set of symptoms might lead to multiple or fuzzy diagnoses

cannot be implemented.

c. Complete knowledge should be available in advance. Moreover, the knowledge has to be universal, that is applicable to all cases. This contradicts common experience.

d. Application experience is not utilized and learning acquisition is impossible, tedious or incomplete.

e. It is tedious to estimate the effect of uncertainty of the inputs on diagnosis output.

All these shortcomings are removed with the use of fuzzy logic for definition of input-output and neural network methodology for reasoning and learning from application experience.

A NEURAL NETWORK EXPERT SYSTEM.

Neural network and fuzzy set methodology is used in the development of an expert system shell aimed at developing a wide range of expert systems.

A heteroassociative neural network module (NNM) architecture is selected, figure 1 with separate layers of input and output neurons where the input layer is projecting on the output layers. Moreover, modules can be connected in series and in parallel each receiving input from the output of the preceding modules and sending its output to modules down the hierarchy line, figure 2.

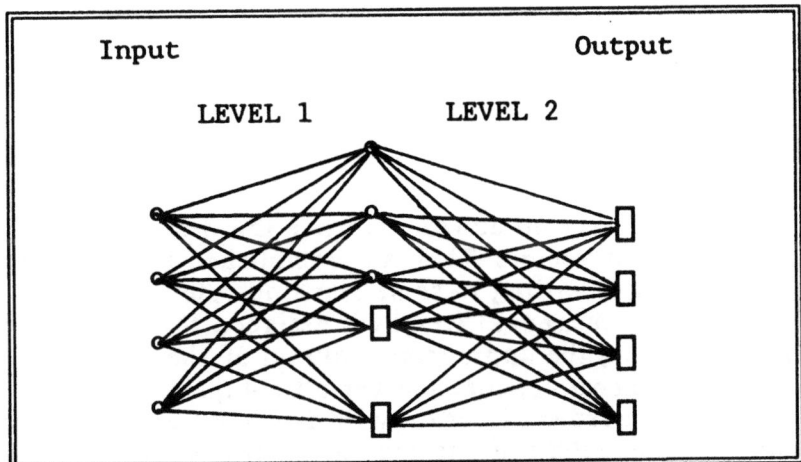

Figure 2. Two-level Neural Network Architecture

The input-output relation in the module is at steady-state

$$a_j = \sum_{i=1}^{m} w_{ij} \phi_i \mu_A(\phi_i) \qquad (1)$$

$$\psi_j = \theta_j(a_j)$$

where,

ϕ_i is the value of the input neuron i, i=1,2,...,m

$\mu_A(\phi_i)$ is a membership function,

w_{ij} is the weight function between input i and output j,

ψ_j is the value of the output neuron j, j=1,2,...n,

$\theta_j(a_j)$ is a threshold function.

Knowledge is represented in this module by:

110

a. The weights w_{ij} of the synapses.
b. The input membership function $\mu_A(\phi_i)$
c. The output threshold values ψ_j
d. The network topology.

This procedure is coded in EXPERTS with the following conventions:

a. For fuzzy input quantities, the input takes values [0,1].
b. For single value input quantities, the input takes values 0 or 1.
b. For inputs which are scalar quantities, the input takes non-negative values from 0 to 1.
c. The inputs neurons are called UNITS and they are grouped in FEATURES. The names UNITS and FEATURES are default names.
   The user can specify names related with the expert system under development, i.e. SYMPTOMS can replace UNITS in a diagnosis system.
d. The output neurons are called RESULTS by default, and are user definable, i.e., DIAGNOSES in a diagnosis system. They can be logic variables, true or false, or fuzzy.
e. Initially the user has to specify at least one feature, one unit in this feature and one output. The built-up of the network is done gradually to EXPERTS during the learning process. More features, units and results can be added also by the learning process.
f. The state vector includes the input vector $\{\phi\}$ and the output vector $\{\psi\}$,

$$\{v\}=\{\ \phi\ |\ \psi\ \} \qquad (3)$$

where curly brackets indicate a transposed column vector.

THE LEARNING PROCESS

The aim of the learning process is to develop and/or improve the desired mapping function between the input and output vectors. This is accomplished by adjusting the weights and the threshold values according to some learning rule.
   If a result is defined for the first time, equal weight is assigned to all the units that are activated for this result. Then, the appropriate column of the weight matrix $[w_{ij}]$ is normalized to a certain value, common for all columns. This correlational rule is expected to be improved later-on with further learning, since one can intuitively conclude that the weight of the several units does not have to be the same.
   Each subsequent teaching session with the same result, and possibly different values of the input vector, is incorporated into the system by way of an error-correcting rule (Hinton, Sejnowski and Ackley 1984; Rumelhart, Hinton and Williams 1986; Sejnowski and Rosenberg 1986).
   Let $\{c_{ij}\}$ the existing value of the column i of the weight matrix and $\{d_{ij}\}$ the same column determined on the basis of the equal weight method but with the present value of the input vector. Then, the column i of the weight matrix is adjusted to the value

$$\{c_{ij}\}^* = \{c_i\} + \lambda\{d_i\} \qquad (4)$$

where λ_i is a correction factor which will determine the weight of the new knowledge on the existing one. Finally, vector $\{c_i\}^*$ is normalized.
   During application of the error-correcting rule, new features and units might be defined during the definition of the input vector for a new case that the system is learning. The system creates the network additions and assigns zero weight to the idle synapses.
   When a new unit is created, a threshold value is assigned. Each time the system

learns a new case, it updates the value of the unit threshold values, if the unit values are less than the threshold ones. This is not formally correct, since threshold values should be different for the different results. It is not used at present in the interest of simplicity and memory saving and because the author feels that for engineering applications the threshold values should not depend strongly on results. This extension, however, would be straight-forward if threshold coupling is essential.

If the same value of the input vector results always to the same value of the output vector, the procedure is equivalent to the first order predicate calculus. The neural network representation allows for a fuzziness, in the form of incomplete or different inputs for a certain output. After a sufficiently large number of learning sessions, each application gives a result vector in the sense of the most frequent result corresponding to the input vector and the ones very close to it.

SENSITIVITY ANALYSIS

Many times, the values of the units, beyond their fuzziness, they are uncertain, in the sense that either they are based on incomplete or uncertain information which might or might not affect substantially the result vector. In the neural network formulation, this sensitivity can be quantified with the Euclidean distance

$$d_j = \{\Sigma[\psi(\phi+\Delta\phi_j)-\psi(\phi)]^2\}^{1/2} \qquad (5)$$

The system computes the Euclidean distance for variations $\Delta\phi_j$ due to the change in value of any unit j.

IMPLEMENTATION.

Available experience for failure diagnosis in turbomachinery (Shore 1980) was utilized to initially teach the system. Additional diagnoses from the author's experience were taught to the system and additional features and diagnoses defined.
Following is a list of features to be monitored:

Feature	Values
PREDOMINANT FREQUENCY, relation to resonant.	Rotor/Stator Resonant 40-50% of Resonant 50-100% of Resonant subsynchronous R / 2 R / 4 R / N
PREDOMINANT FREQUENCY, relation to rotation.	Operation N 2 per rev Multiple of N Very High
ELECTRICAL (LINE), Frequency relation to	Line Frequency L NL Slip Frequency

DIRECTION AND LOCATION of Predominant Frequency	Vertical Horizontal Axial Shaft Bearings Casing Foundation Piping Coupling
The Response to INCREASING SPEED	Unchanged Increases Decreases Peaks Comes suddenly Drops suddenly
The Response to DECREASING SPEED	Unchanged Increases Decreases Comes suddenly Drops suddenly
Vibration Amplitude Relation to the LOAD	No effect In at Full Load Out at Full Load In at Zero Load Out at Zero Load On at Partial Load Out at Partial Load Only at Start-Up With Changing Load With Other Machines
Effect of OIL PRESSURE on Vibration	Lower on increasing Higher on increasing Lower on decreasing Higher on decreasing
Effect of OIL TEMPERA- TURE on Vibration	Lower on increasing Higher on increasing Lower on decreasing Higher on decreasing
The Predominant SOUND during operation	LF Rumble Loud Roar Hum Beat High pitch Loud Scream Squeal Ultrasonic
The HISTORY of the Machine	Initial Start-Up 1rst year 1-10 years >10 years

The Condition of SEALS	Rubbed Leaking
The Condition of SHAFT or ROTOR	Bent Cracked/broken Wear under hubs Components eroded Accumulations
The BEARING Damage.	Thrust Wiped Fatigued Babbitt squeezed
The CASING Damage.	Distorted/cracked Foundation Soleplates Flange leaks
The COUPLING Condition	Misaligned Burned or pitted Bolts loose
The GEAR Condition	Teeth damage Back side
The OTHER PARTS Damage	Sliding surfaces Thermal expansion restrained Fluid marks Salt deposits
PHASE ANGLE of the predominant vibration	Constant Erratic Varying

The diagnoses taught to the system were:

Initial Unbalance	Permanent bow or lost parts
Temporary rotor bow	Temporary casing distortion
Permanent casing distortion	Foundation distortion
Seal rub	Rotor axial rub
Misalignment	Piping forces
Journal and bearing eccentr	Bearing damage
Bearing whirl	Bearing unequal stiffness
Thrust bearing damage	Gear inaccuracy & damage
Coupling inaccuracy & damage	Aerodynamic excitation
Coupling resonance	Overhang resonance
Pressure pulsations	Vibration transmission
Oil seal induced vibration	Cracked Rotor
Electrically excited vibration	Structural resonance of casing
Structural resonance of support	Structural resonance of foundation
Rotor & bearing system resonance	Insuff. tightness in assembly of rotor
Insufficient tightness in assembly of bearing liner	
Insufficient tightness in assembly of bearing case	
Insufficient tightness in assembly of casing & support	

It is expected that as is the system will not lead to unambiguous diagnoses for particular fields of applications. Repeated utilization and learning from application experience can lead to an adoptive orientation to a particular field of application.

References

Baldwin JF (1985) Fuzzy Sets and Expert Systems. Inf Sc, 36:123-156.

Bergandano F, Giordana A, Saitta L (1987). Learning from Examples in Presence of Uncertainty, in Sanchez E, Zadeh LA Approximate Reasoning, Pergamon Press, Oxford.

Bruner JS et al (1956). A Study of Thinking, John Wiley, New York.

Carbonell J and P Langley (1987). Machine learning, in Shapiro EC (Ed.), Encyclopedia of Artificial Intelligence, 1:464-488, John Wiley, New York.

Clayton Labs (1990) EXPERTS: An Adoptive Expert System Shell-User's Manual". St. Louis.

Forsyth R (Ed.) (1984). Expert Systems: Principles and Case Studies, Chapman and Hall, London, U. K.

Hebb DO (1949). The Organization of Behavior, John Wiley, New York.

Hopfield JJ (1982). Neural networks and physical systems with emergent collective computational properties, Proceedings of the National Academy of Science USA, 79:2554-2558.

Kaufmann A (1975). Theory of fuzzy subsets 1, Academic Press, New York.

Keim M, Nordmann R (1989) Application of an Expert System for Diagnosis of Rotordynamic Problems in Turbomachinery, ASME 12th biennial Conf. on Mech. Vib. & Noise: Diagnostics, DE-18-5:85-91

Kohonen T (1984) Self-Organizing and Associative Memory, Springer, Berlin.

Kokar M (1989) Machine learning, in A Kusiak (Ed), Knowledge-Based Systems in Manufacturing, Taylor and Francis, London:45-81.

Kosko B (1992) Neural Networks and Fuzzy Systems, Prentice Hall, Englewood Cliffs.

Levy WB and Desmond NL (1985) The rules of elemental synaptic plasticity, in Anderson JA, Levy WB, and Lehmkuhle S (Eds.), Synaptic Modification, Neuron Selectivity, and Nervous System Organization:105-121, Lawrence Erlbaum, Hillsdale, N. J.

Matheus CJ and Hohensee WE (1987) Learning in artificial neural systems, Computational Intelligence, 3, No. 4:283-294.

Michalski RS (1983) A theory and methodology of inductive learning, in Michalski RS, Garbonell JG, and Mitchell TM (Eds.), Machine Learning: An Artificial Intelligence Approach, 1, Tioga Publishing Co., Palo Alto, Calif.

Michalski RS (1987) Concept learning in S. C. Shapiro (Ed) Encyclopedia of Artificial Intelligence, 1:185-194, John Wiley, New York.

Michalski RS, Garbonell JG, and Mitchell TM (Eds.) (1983) Machine Learning: An Artificial Intelligence Approach, 1, Tioga Publishing Co., Palo Alto, Calif.

Minsky M, Papert S (1969) Perceptrons: An Introduction to Computational Geometry, MIT Press, Cambridge, Mass.

Rosenblatt F (1962) Principles of Neurodynamics, Spartan, Washington, D. C.

Shore JS (1980) Turbomachinery Problems and their Correction. in Sawyer's Turbomachinery Maintainance Handbook, 2, Chapter 7, Connecticut.

Siler W, Buckley J, Tucker D (1987) Functional Requirements for a Fuzzy Expert System Shell, in Sanchez E, Zadeh LA, Approximate Reasoning, Pergamon Press, Oxford.

Wasserman PD, Schwartz T (1987) Neural networks, Part 1, IEEE Expert, 2, No. 4:10-12.

Xu J, Peeken H (1989) Failure Diagnosis Using Fuzzy Logic. ASME 12th biennial Conf. on Mech. Vib. & Noise: Diagnostics, DE-18-5:93-99.

Zadeh LA (1965) Fuzzy Sets, Information and Control, 8:338-353.

Zadeh LA (1973) Outline to a new approach to the analysis of complex systems and decision process, IEEE Trans. Sys, Man and Cybern, SMC-3:28-44.

SESSION 5   STABILITY

PLAIN BEARING STABILISATION USING SWIRL INJECTION

R.D. Brown, M.Sc. (ENG), C.ENG. M.I.MECH.E., Senior Lecturer,
Dept. of Mechanical Engineering,
Heriot-Watt University, Edinburgh, U.K.

ABSTRACT

The design of stable journal bearings for rotating machinery is
conventionally achieved by profiled lobes or tilting pad assemblies.
A different approach is aimed at reducing the de-stabilising
cross-stiffness by flow modification in a plain bearing. A
theoretical analysis is confirmed by experimental work on a small
test rig. Further development work would appear to be justified in
order to obtain the full economic benefit.

INTRODUCTION

A well known problem with plain bearing operation has been a
tendency for unstable behaviour in light load and/or high running
speed conditions. A wide range of anti-whirl bearings has been used
in rotating machinery over the last 50 years. The majority of these
stable bearing designs rely on small deviations from a circular
profile in the bearing bush. Among useful comparisons of
non-circular bearings with plain bearings which have been published
are those of Allaire (1) and Rieger (2).

An early anti-whirl design, the pressure dam bearing, consists
of a rearward facing step in the upper half of the bearing. In
operation this step provides a downward parasitic load which
stabilises the bearing. Other non-circular arrangements include two
lobe designs (lemon-bore and offset-half) and multi-lobe designs
including tilted lobes. The most complex anti-whirl bearings use
pivoted pads which "track" journal motion and hence eliminate the
cross-coupling effects. A more recent stable bearing design makes
use of a groove in the bottom half of the bearing surface to
restrict the pressure field, Morton (3). A common feature of all
these approaches are small but usually precise modifications to the
bearing profile. In the case of tilting pads there is an additional
complexity of the mechanical pivot arrangement.

A somewhat different approach is to accept the plain bearing and
its useful damping and provide a retrograde fluid swirl to reduce
forward cross-coupling and hence improve the stability behaviour.
This approach was originally suggested in connection with
centrifugal pump seals by Black et al (4) in 1981. Ambrosch and
Schwaebel (5) patented stability improvements in compressors and
turbines using flow injection into labyrinth seals. Backward swirl
injection in centrifugal compressors has been successful in
suppressing sub-synchronous whirl, Kirk (6) and some work has been
reported recently in using backward tangential flow to reduce
unbalance response in a rotating shaft, Hart (7).

A theoretical examination of the stability of a short journal bearing demonstrates that reduction of cross-stiffness improves stability. A two-dimensional theory for a fluid annulus indicated that backward tangential flow resulted in a cross-stiffness which was inversely proportional to the clearance and directly dependent on friction. A small scale test using a flexible shaft supported on a hydrodynamic bearing demonstrated that oil whirl could be suppressed with a relatively small flow of lubricant. Experimental variations in radial clearance and surface friction gave results which were in line with theoretical indications. This experimental evidence demonstrates that the concept is worth exploring on a larger scale.

NOMENCLATURE

B_{ij}	damping coefficients	N/(m/sec)
C	bearing radial clearance	m
D	bearing diameter	m
e	eccentricity ratio	
f	friction coefficient	
F_x, F_y	fluid forces	N
h	damper radial clearance, local bearing gap	m
K_{ij}	stiffness coefficients	N/m
L	bearing length	m
M	bearing mass	kg
N	rotational speed	Revs/sec
P	pressure	N/m^2
R	bearing radius	m
$\underline{S}$	Sommerfeld Number	
u	mean circumferential velocity	m/sec
W	bearing load	N
x, y	lateral co-ordinates	m
Z	axial co-ordinate	m
δ	rotor displacement	m
μ	fluid viscosity	$N.sec/m^2$
ρ	fluid density	kg/m^3
τo	wall shear stress	N/m^2
ϕ	attitude angle	RADIANS
ω	angular velocity	RAD/SEC

SHORT BEARING

A useful analytic model for plain bearings is the short bearing simplification of Reynolds equation which is obtained by neglecting the circumferential pressure gradient. The resulting equation can be integrated to give the hydrodynamic bearing forces as functions of eccentricity. Small perturbations in displacement and velocity from an equilibrium position are used to define non-dimensional stiffness and damping coefficients.

Reynolds equation in polar co-ordinates for a short bearing can be written as:-

$$\frac{\partial}{\partial Z}\left(\frac{h^3}{12\mu}\frac{\partial P}{\partial Z}\right) = -\frac{1}{2}\,\omega\,C\,e\,\sin\theta + C\,\dot{e}\,\cos\theta + C\,e\,\dot{\phi}\,\sin\theta \qquad (1)$$

After several integrations the non-dimensional load parameter and the attitude angle can be obtained

$$S\left(\frac{L}{D}\right)^2 = \mu NL\,\frac{D}{W}\left(\frac{R}{C}\right)^2 \cdot \left(\frac{L}{D}\right)^2 = \frac{(1-e^2)^2}{\pi e[16e^2+\pi^2(1-e^2)]^{0.5}} \qquad (2)$$

$$\tan\phi = \frac{\pi\sqrt{1-e^2}}{4.\,e} \qquad (3)$$

The dynamic performance can be obtained by linearised perturbations from the equilibrium position. The resulting four stiffness and four damping coefficients are functions of eccentricity when expressed in non-dimensional form, Smith (8) and Vance (9). In this case the dynamic coefficients are used to investigate the stability of a rigid journal supported on an oil film.

Thus the equations of motion for a rigid mass in free motion are:-

$$\begin{bmatrix} M & 0 \\ 0 & M \end{bmatrix}\begin{Bmatrix} \ddot{x} \\ \ddot{y} \end{Bmatrix} + \begin{bmatrix} Bxx & Bxy \\ Byx & Byy \end{bmatrix}\begin{Bmatrix} \dot{x} \\ \dot{y} \end{Bmatrix} + \begin{bmatrix} Kxx & Kxy \\ Kyx & Kyy \end{bmatrix}\begin{Bmatrix} x \\ y \end{Bmatrix} = 0 \qquad (4)$$

Adopting conventional non-dimensional forms for the coefficients i.e. $K^*_{ij} = \frac{CK_{ij}}{W}$ and $B^*_{ij} = \frac{\omega CB_{ij}}{W}$ then a stability parameter $\omega\sqrt{C/g}$ and a frequency ratio Ω/ω can be shown to be functions of these non-dimensional coefficients.

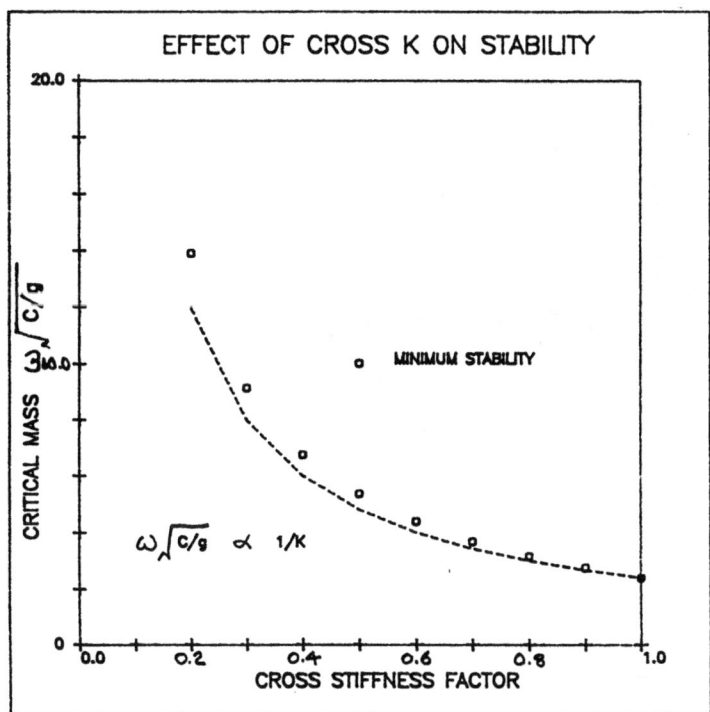

Figure 1. Plain bearing stability - effect of reduction in cross-stiffness.

118

A conventional stability analysis of a rigid mass supported on an oil film plots the stability parameter $\omega\sqrt{C/g}$ against bearing eccentricity. For a short bearing the minimum value is 2.53. A plot of minimum stability against reduction in cross-stiffness Figure 1 demonstrates that the stability parameter is inversely proportional to the cross-stiffness. These results were obtained from a numerical calculation and varying the direct damping terms had a marginal effect. The other coefficients in the system i.e. direct stiffness and cross-damping are conservative terms and hence do not affect the stability.

TWO-DIMENSIONAL MODEL OF AN ANNULAR FLOW DAMPER

From the diagram Figure 2 a circumferential flow is assumed to be maintained in the annular space between an eccentric rotor and a stator. If the mass flow is assumed constant then any change in local clearance modifies the local velocity. Following Hart (7) velocities and shear stresses are integrated round the shaft section leading to a set of restoring forces:-

$$\begin{Bmatrix} Fx \\ Fy \end{Bmatrix} = -\rho\frac{RL}{C}\bar{u}^2\pi \begin{bmatrix} 1 & +f \\ -f & 1 \end{bmatrix} \begin{Bmatrix} x \\ y \end{Bmatrix} \tag{5}$$

The strength of the cross-stiffness terms Kxy, Kyx are clearly seen to be inversely proportional to radial clearance C and proportional to friction factor f.

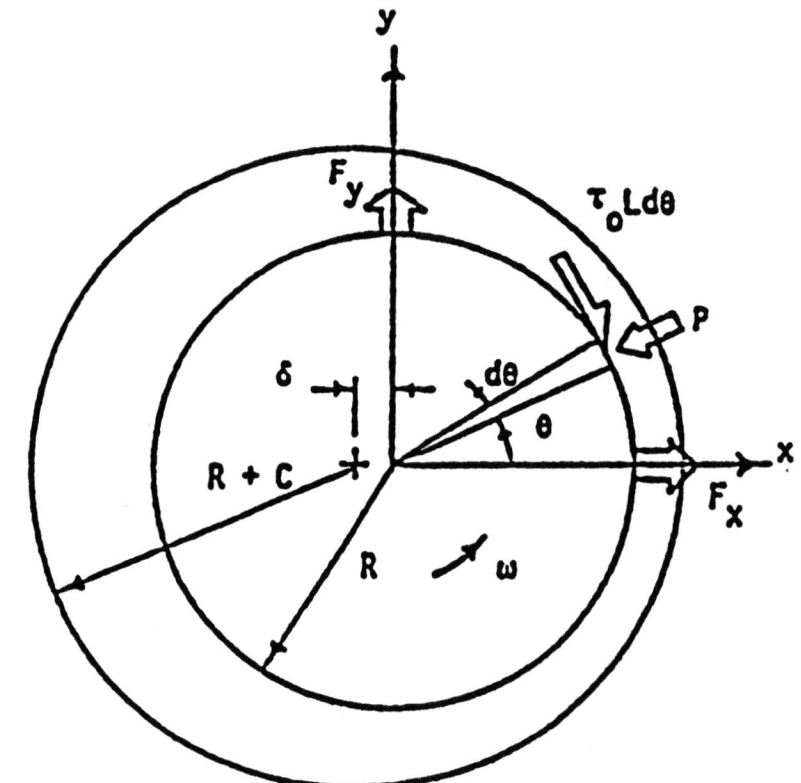

Figure 2.   Geometry of annular gap.

EXPERIMENTAL ARRANGEMENT

It has been shown theoretically and experimentally (7) that reverse swirl can produce a backward cross-stiffness force. The main feature of the test rig was a 50 mm diameter journal with a radial clearance of 0.125 mm. This bearing was mounted at one end of a 10 mm diameter shaft which was supported in a ball race at the other with a bearing span of 411 mm. The rotor was driven by a 0.37 kW motor with a variable speed control. A critical speed analysis, Figure 3, indicated a fundamental critical at approximately 6700 r/min. It is clear that any instability at shaft speeds up to 5000 r/min is likely to be oil whirl rather than a resonant oil whip.

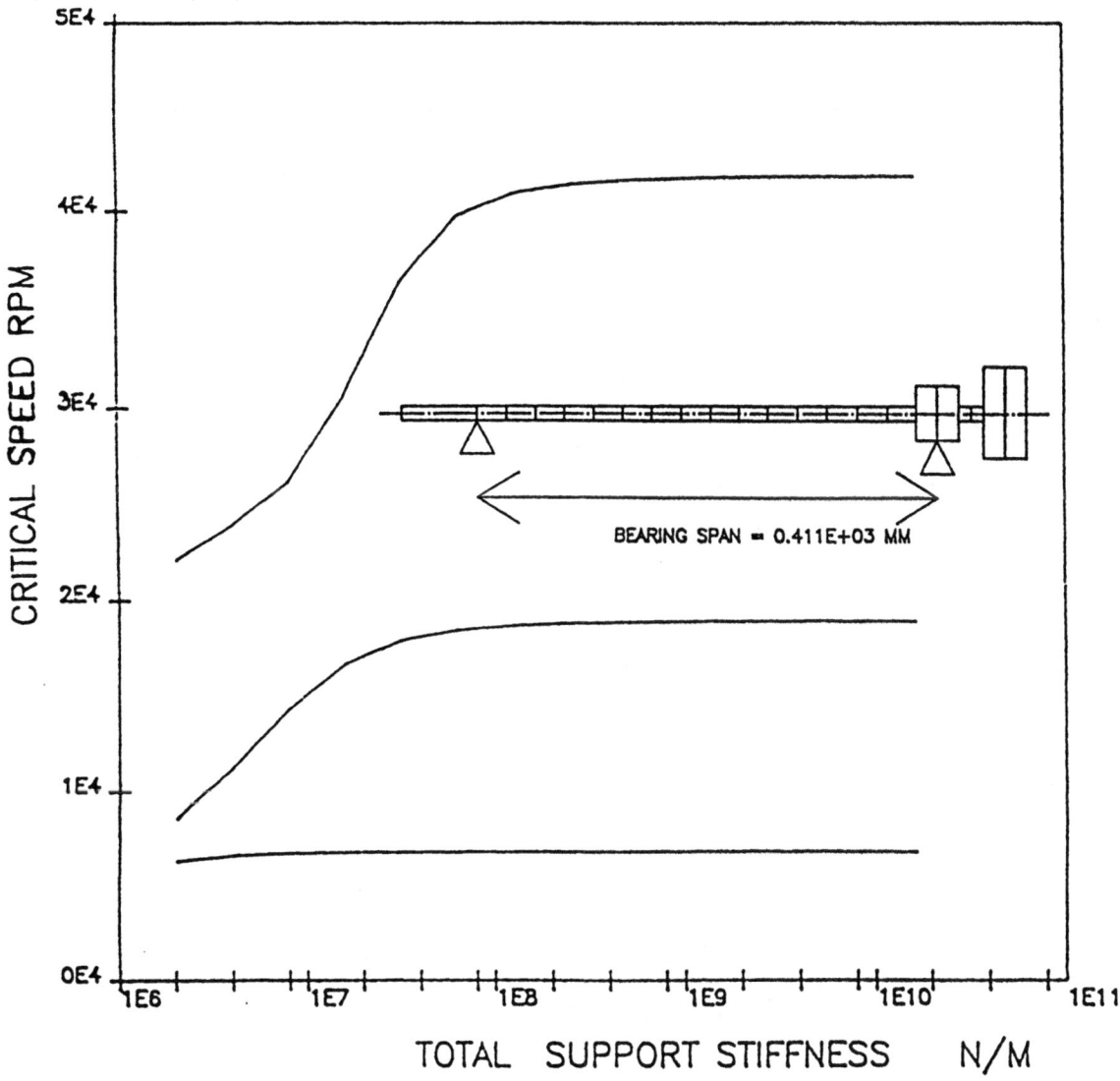

Figure 3.   Flexible rotor and critical speed map.

An additional 75 mm diameter journal was mounted adjacent to the
first with an independent oil supply. The distinguishing feature of
the second journal was that the oil supply entered the 0.5 mm
clearance space through 4 equi-spaced tangential jets. Both the
magnitude of the radial clearance and the surface finish of the
journal were experimental variables. The experimental technique was
to run the shaft at different speeds between 3000 and 5000 r/min
with a bearing pressure of between 0.14 and 0.35 bar. Throughout
this range, unstable behaviour was easily obtained. An example of
the vibration spectra obtained for a specific bearing pressure is
shown in Figure 4. It is quite clear that there is a marked
sub-synchronous frequency at speeds above 3000 r/min. The
corresponding frequency ratio of approximately one half is
additional evidence that oil whirl has been established.

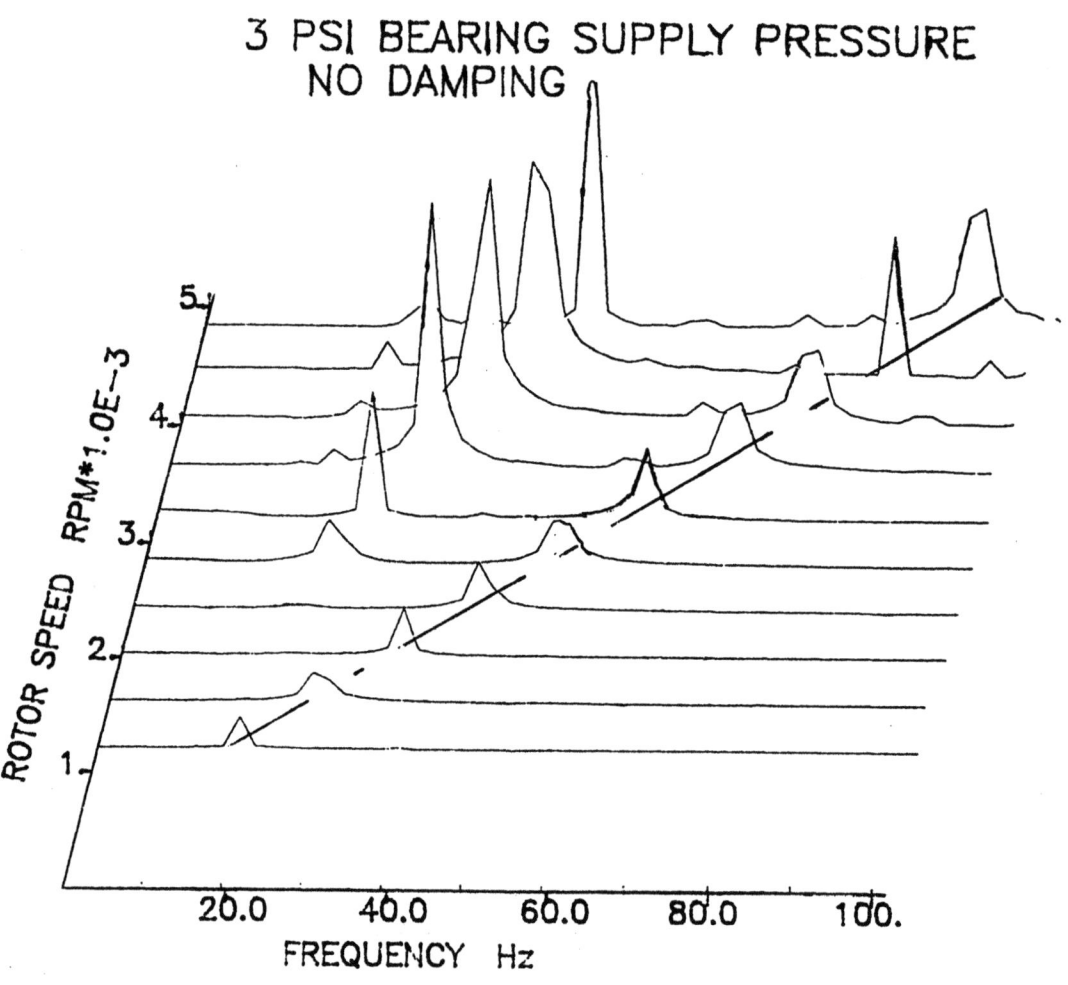

Figure 4.   Oil whirl instability.

As the oil supply was a single source feeding both bearing and damper in parallel it was important to adjust the control valves to maintain the bearing supply pressure when the damper flow was initiated.   When this had been achieved experimental data was taken on the flow and pressure required in a particular damper arrangement to quench the instability.   Figure 5 when compared with Figure 4 clearly demonstrates the complete suppression of oil whirl over a wide speed range.

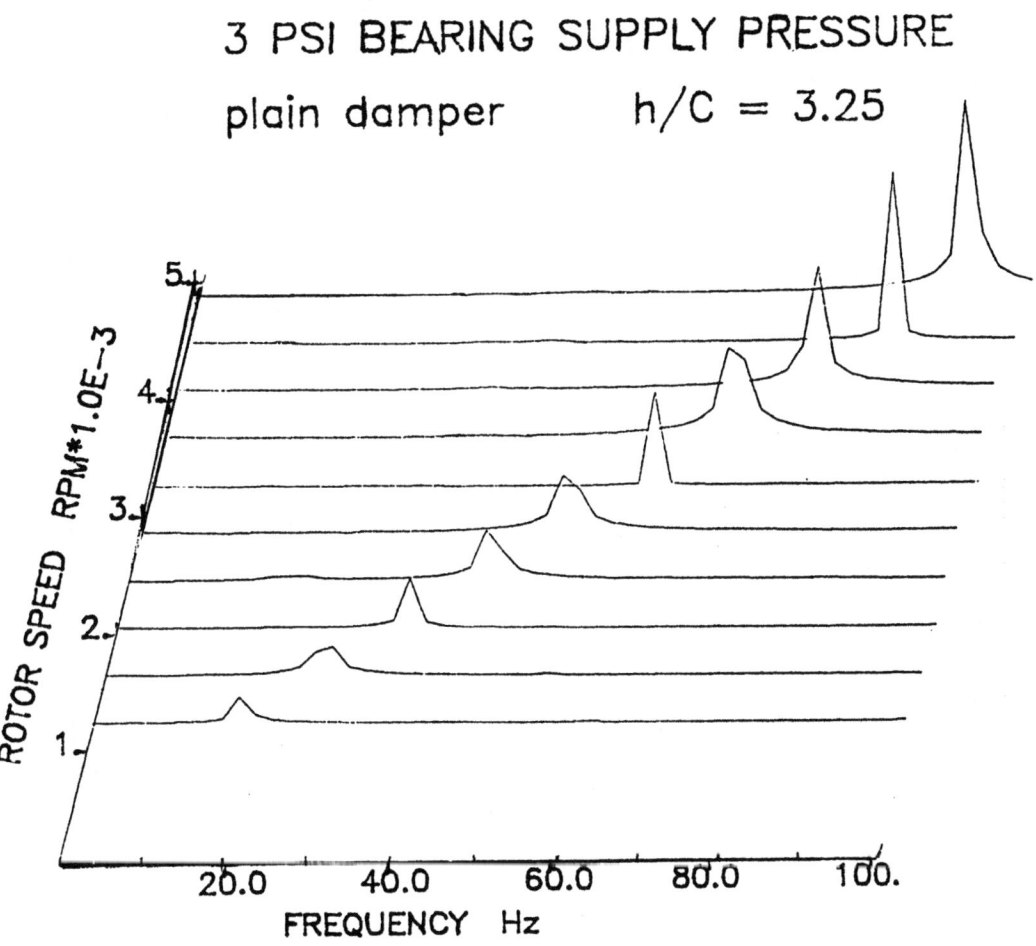

Figure 5.   Suppression of oil whirl.

A summary of all the experimental results is shown in Figure 6. The geometrical variations are clearance ratio h/C and surface roughness variation for the lowest h/C value. The velocity ratio V/ωR required to quench instability is between 2% and 9% of the journal speed. The dominating effect of radial clearance and surface friction correlate well with equation (5). Due to the low velocity ratio the damper was shown to work irrespective of flow direction.

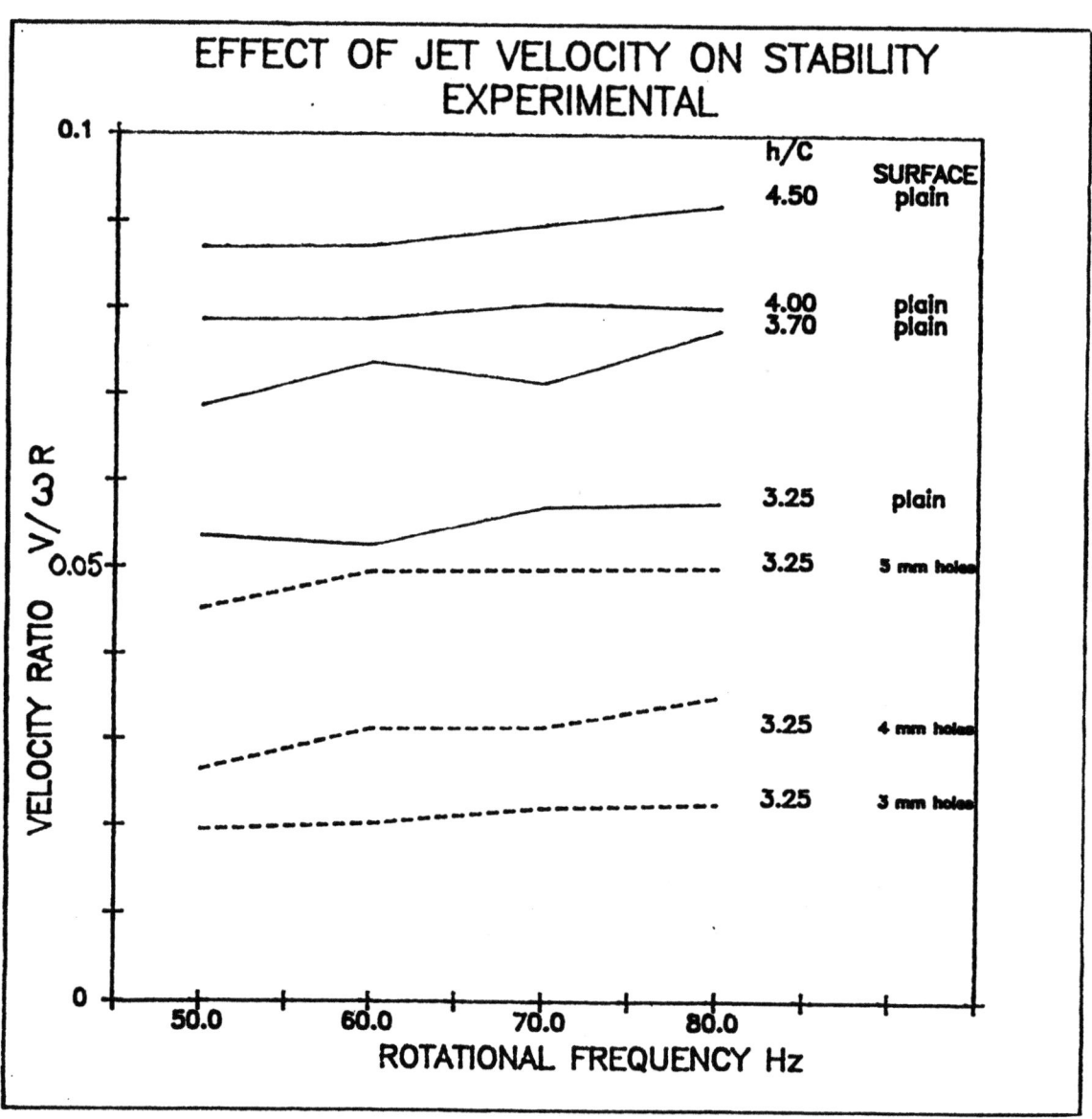

Figure 6. Effect of damper geometry on flow requirements.

CONCLUSIONS

Annular swirl injection has been shown to stabilise a plain bearing, in agreement with a simple theory. The experimental design in these tests allowed for separate control over the conventional bearing supply and the damper supply. However, there is reason to suppose that a single supply source would be adequate, the flow entering the bearing assembly tangentially before leaving with an axial component through the bearing clearance. This would facilitate a simpler design arrangement. Another possibility would be to use the device as a tangential swirl damper which could be located at any convenient plane e.g. an overhung free end.

REFERENCES

1. Allaire P.E., Flack R.D.: Design of Journal Bearings for Rotating Machinery. Proceeding 10th Annual Turbomachinery Conference, Texas A & M University, Dec. 1981 pp 25-45.
2. Rieger N.F.: Chapter 6 Fluid-Film and Rolling-Element Bearings, Vibrations of Rotating Machinery Pt.1, Rotor-Bearing Dynamics, The Vibration Institute, Illinois, 1982.
3. Morton P.G., Johnson J.H. and Walton M.H.: The Influence of Partial Grooving on the Dynamic Characteristics of Hydrodynamic Bearings, GEC Journal of Research, Vol.5, No.3, 1987.
4. Black H.F., Allaire P.E. and Barret L.E.: Inlet Flow Swirl in Short Turbulent Annular Seal Dynamics, 9th International Conference in Fluid Sealing, BHRA Fluids Engineering, Leeuwenhorst, The Netherlands, April 1981.
5. Ambrosch F. and Schwaebel R.: Method of and Device for Avoiding Rotor Instability to Enhance Dynamic Power Limit of Turbines and Compressors. U.S. Patent 4273 510 (1981).
6. Kirk R.G.: Labyrinth Seal Analysis for Compressor Design - Theory and Practice, International Conference of Rotor Dynamics I.F. To. MM. Tokyo, September 1988 pp 589-595.
7. Hart J.A. and Brown R.D.: The Use of Fluid Swirl to Control the Vibration Behaviour of Rotating Machines, International Conference on Vibrations in Rotating Machinery, I.Mech.E. Edinburgh, September 1988.
8. Smith D.M.: Journal Bearings in Turbomachinery, Chapman and Hall, London, 1969.
9. Vance J.M.: Rotordynamics of Turbomachinery, John Wiley, New York, 1988.

Stability Analysis For a Novel Design of Low Impedance Hydrodynamic Bearing

L K Lim, M J Goodwin, P J Ogrodnik and M P Roach

Department of Mechanical and Computer Aided Engineering
Staffordshire Polytechnic
Stafford, England

Abstract

An important problem in rotordynamics which researchers and bearing designers face is that of system stability. One of the causes of rotordynamic instability is oil whirl. This is an instability phenomena associated with the operating characteristics of the journal bearing and hence depends on the design of the bearings and the operating conditions in which it performs. This paper presents a theoretical stability analysis of a rigid rotor supported by a novel design of hydrodynamic oil film bearing. The novel bearing incorporates recesses in the bearing surface which are linked to a hydraulic accumulator and have been shown to offer superior unbalance response performance to conventional designs. However, their stability characteristics have not been investigated before. The paper describes a theoretical investigation using linearised oil film coefficients. Results are presented in the form of stability maps showing the variation of system stability with operating characteristics. The results obtained for the novel bearing are also compared with those obtained for conventional bearings.

Notation

c	Radial clearance.
C_{ij}	Bearing oil film damping force coefficients, force in i direction per unit velocity in j direction.
D	Bearing diameter.
e	Eccentricity.
K_{ij}	Bearing oil film stiffness force coefficients, force in i direction per unit displacement in j direction.
m	Rotor mass.
P	Stability operating parameter, $W/mc\omega^2$.
W	Static Bearing load.
ε	Eccentricity ratio, e/c.
ω	Angular speed (rad/sec).

Introduction

The phenomena of oil whip/ oil whirl has been under investigation for several decades. Despite this some of the factors which affect the stability of the rotor bearing system are not yet understood. Since the paper published by Rankine (1) in 1869, many researchers have contributed to the understanding of rotor bearing system dynamics and of oil whirl in particular. In 1924, Newkirk (2) utilised an empirical method to discover that whipping of the shafts ceased when the oil supply was halted. This discovery led him to reach a conclusion that the oil whip/ oil whirl phenomena is due to vibration within the oil film. Later, Newkirk et al (3) investigated three rotor-bearing models. He concluded, like other researchers, that eccentricity ratio is an important indicator of the stability of a rotor bearing system. He compared his results with Hummel and did not agree with Hummel's conclusion, as Hummel concluded in his work that a mild instability developed when the eccentricity ratio became less then 0.7. In fact Newkirk observed that instability of the system occurs with eccentricity ratios of up to 0.96. Newkirk also discovered that misalignment affects the stability of the rotor bearing system.

In 1960, Holmes (4) utilised Routh's Stability Criteria to examine the stability of a rigid rotor system. By using the first order Taylor's theorem, eight linearized oil film coefficients were computed. A stability map was drawn, indicating the system stability under various operating conditions, and it was shown that a system with an operating parameter $W/mc\omega^2$ less than 0.1 is unstable. Boroomand (5) in 1990 developed a novel type of hydrodynamic bearing. The design of this novel bearing is similar to that of a conventioanl partial arc bearing. However, unlike a conventional bearing, this bearing included a recess in the bearing surface, and a hydraulic accumulator was connected to the recess via a remote control valve. The function of the valve was to enable the accumulator to be isolated from the bearing oil film. Boroomand showed that the novel design had some superiority over conventional bearings: the unbalance response of a machine running in this new bearing design was found to be far superior to that for a machine running in conventional bearings, with the peak rotor amplitude very much lower than that for a system with conventional bearings. Although the superiority of Boroomand's novel bearing design in the area of unbalance response has been demonstrated, the stability of the bearing has not yet been assessed. The purpose of this paper is to examine the stability of a rigid rotor running in this novel bearing. The investigation also includes examining the stability of a similar bearing without the hydraulic accumulator in order to determine the effect of the the accumulator on the stability of the rotor bearing system.

Theory

The theory developed below is similar to that described in reference (6) and is used to assess the stability of a rigid rotor running in different types of bearing.

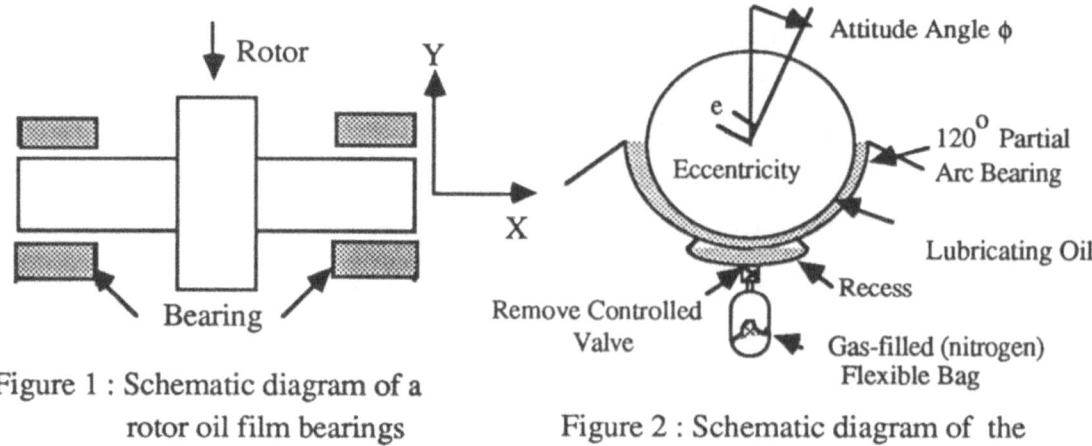

Figure 1 : Schematic diagram of a
rotor oil film bearings

Figure 2 : Schematic diagram of the
novel design of bearing

The analysis applies to a rotor mass mouted on a rigid shaft which runs in two identical hydrodynamic bearings as illustrated in Figure 1. Figure 2 shows the novel design of hydrodynamic journal bearing being investigated. The hydrodynamic bearing may be modelled in terms of eight linearised stiffness and damping coefficients. The relationship between the bearing forces and journal motion is given by the equations of motion for the journal movement in the horizontal and vertical directions which may be written as ;

$$K_{xx}x + K_{yy}y + C_{xx}\dot{x} + C_{xy}\dot{y} = -M\ddot{x} \qquad [1]$$

$$K_{yx}x + K_{yy}y + C_{yx}\dot{x} + C_{yy}\dot{y} = -M\ddot{y} \qquad [2]$$

where x and y are the horizontal and vertical displacements of the rotor, M is half the rotor mass, and K_{ij} is the oil film stiffness force in the i direction per unit displacement in the j direction, and C_{ij} is the oil film damping force in the i direction per unit velocity in the j direction. If the rotor is momentarily displaced from the equilibrium position by some random input, any subsequent free vibrations of the rotor in the horizontal and vertical direction will take the form

$$x = X_o e^{st} \qquad [3]$$

$$y = Y_o e^{(st + \Phi)} \qquad [4]$$

where X_o and Y_o are the vibration amplitudes in the horizontal and vertical directions respectively, and Φ is a phase lag angle. By differentiating equations [3] and [4], to give expressions for velocity and acceleration in the horizontal and vertical directions respectively, and substituting into equations [1] and [2], expressions for the ratio x/y may be obtained from each equation and combined to give

$$(M^2) s^4 + (MC_{xx} + MC_{yy}) s^3 + (MK_{xy} + MK_{yy} + C_{xx}C_{yy} - C_{xy}C_{yx}) s^2$$
$$+ (K_{yy}C_{xx} + C_{yy}K_{xx} - K_{xy}C_{yx} - C_{xy}K_{yx}) s + (K_{xx}K_{yy} - K_{xy}K_{yx}) = 0$$

which it has four roots of s. To test for system stability, one must examine the motion of the journal which follows a momentary displacement from its steady running position. If the journal were to return to a stable equilibrium position, then this would be characterised by values of x and y which decrease with time, that is by a negative value of α, the real part of s. For motion of the mass to be unstable, at least one value of α must be greater than zero. The value of the imaginary part of s, β, indicates the frequency of the resulting vibrations.

Results and Discussion

A computer program based on the theory presented above was written to investigate the stability of the rotor bearing system. This program was used to examine the stability of different bearing designs, including those with large and small recesses incorporated, and those with and without the hydraulic accumulator connected. Figure 3 shows the developed view of the bearing arc for the small recess bearing and figure 4 shows that for the large recess bearing. The stability of a conventional bearing was also examined in the investigation so that a comparison could be made. The bearing stiffness and damping coefficients used in the analysis were calculated from the dimensionless groups published by Boroomand (5) and by Bannister (7). Unless otherwise stated as a variable, the system parameters used were journal diameter 53.4 mm, bearing length/ diameter (L/D) ratio 0.75, radial clearance 50 μm, lubricant dynamic viscosity 0.05 N.s.m^{-2} and rotor mass 891 kg. Results are presented in the form of stability maps as indicated in Figures 5 to 8. The vertical axis of the stability map is an operating parameter P which may be used to describe the system characteristics as mentioned above, thus for a given rotor bearing system, operating parameter is proportional to $1/\omega^2$. The horizontal axis is eccentricity ratio ε_o. A system may be modelled by calculating ε_o and P with increasing speed and plotting their respective values on the figure. The locus in Figure 5 represents the corresponding P and ε values as speed ω is increases. Also the analysis described above was carried out for different P and ε values and the resulting values of α noted. Those points which resulted in values of $\alpha = 0$ are indicated by the line on the figure representing a stability boundary :- the limit of stable operation of the system. The figure indicates where the locus crosses the marked stability boundary from the stable region into the unstable regime and the corresponding values of P and ε can be noted. Figure 5 shows the stability map for the bearing with a small recess, both with and without the hydraulic accumulator connected. Similarly, figure 6 shows the stability map for the bearing with a large recess both with and without the hydraulic accumulator connected.

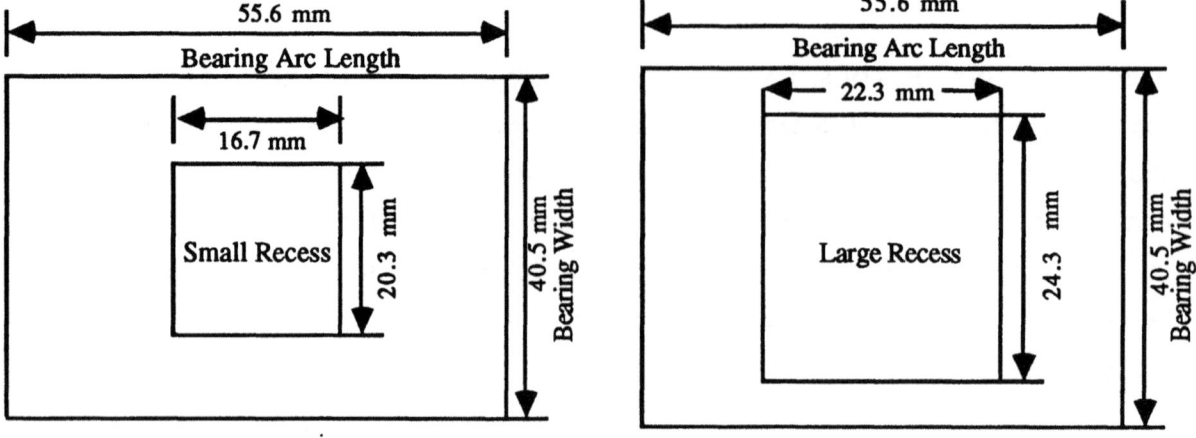

Figure 3   Dimension of the Small Recess Bearing

Figure 4   Dimension of the Large Recess Bearing

The stability limit for a conventional bearing is also shown in these two figures for comparison. It can be seen that the occurence of whirl onset for the conventional bearing corresponds to a much lower value of system operating parameter, which means that a much higher value of journal rotational speed is necessary for whirl onset. This means that the novel-design of bearing is not as stable as a conventional bearing.

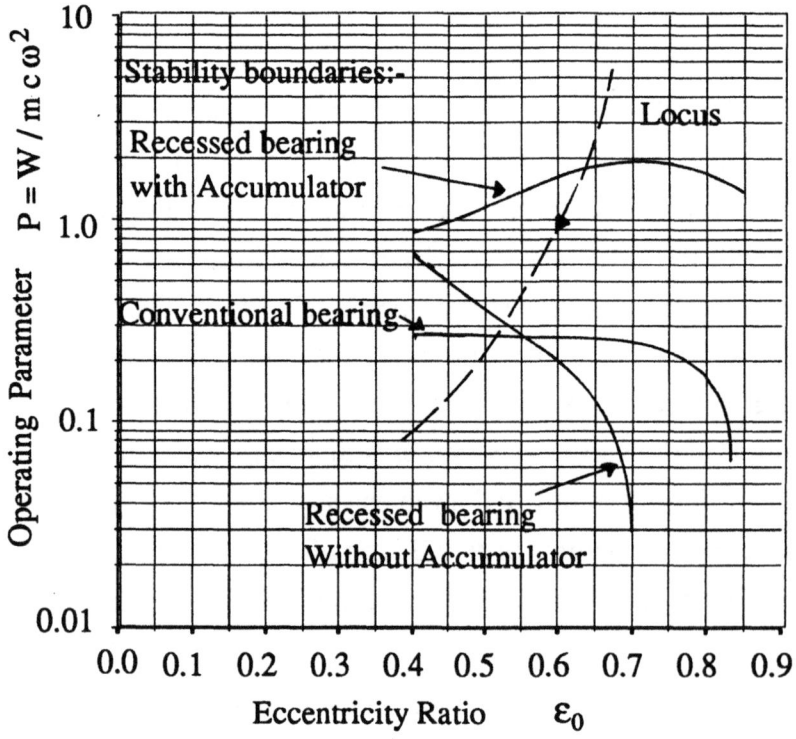

Figure 5: Stability Map for Small Recess Bearing

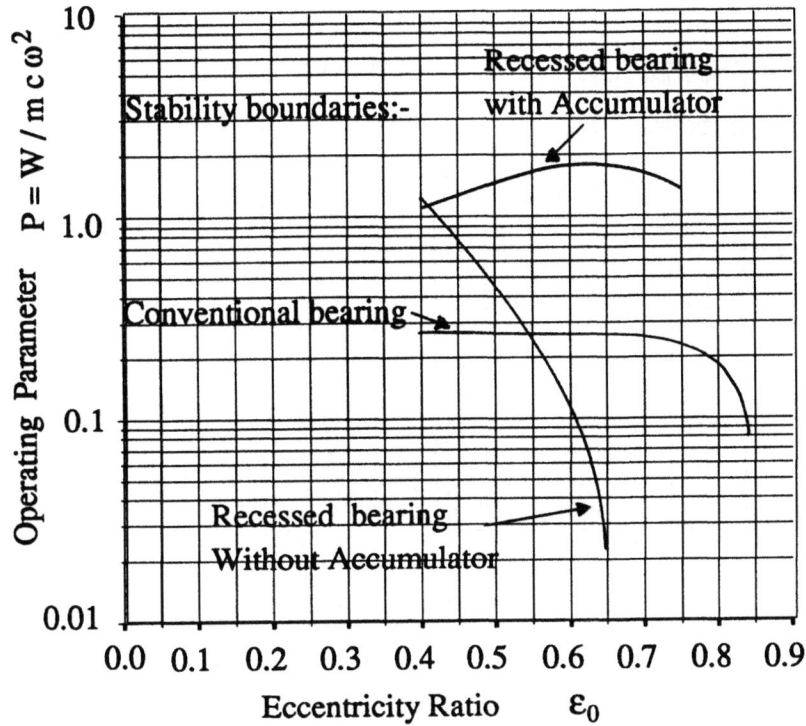

Figure 6: Stability Map for Large Recess Bearing

Figures 5 and 6 also show that the bearing with a recess but without the hydraulic accumulator is more stable than with the accumulator, especially when the eccentricity ratio is above 0.6. However it is still less stable than the conventional bearing.

Figure 7 shows the stability map for the bearing with large and small recesses with and without the hydraulic accumulator connected. It is clearly shown in this figure that for eccentricity ratios less than 0.5, the bearing with the small recess without hydraulic accumulator connected is more stable than the large recess bearing without hydraulic accumulator connected. However, when the eccentricity ratio of the bearing is greater than 0.5, the situation is reversed, and the large recess bearing is more stable than the small recess bearing. Figure 7 also shows the stability map of the novel design of bearing with both large and small recesses with the hydraulic accumulator connected. The two curves shown are similar and so it can be concluded that the stability characteristics of the two bearings are similar. This outcome has revealed that when the accumulator is connected to the bearing, the size of the recess does substantially not affect the stability characteristics of the bearing.

A possible explanation for the reduced stability characteristics of the new bearing with recesses connected to an accumulator is that because lubricant can flow freely to and from the accumulator, less of it is squeezed through the bearing clearance when journal

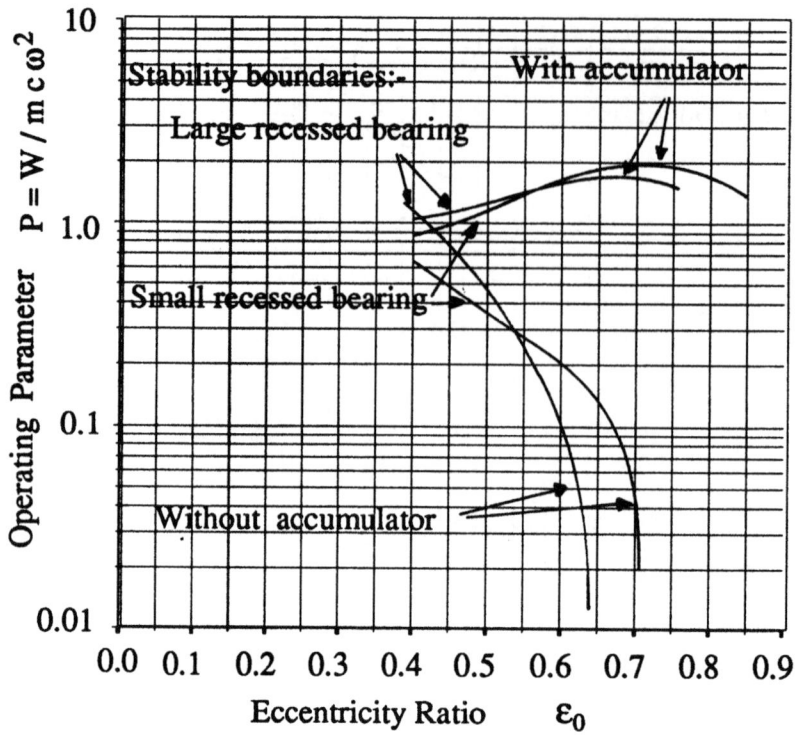

Figure 7: Stability Map for Large and Small Recess
Bearing with and without Accumulator

vibration occurs. This may give rise to a lower damping coefficient for the bearing, and therefore to reduced stability.

A similar explanation might also be used for the reduced stability levels, as compared with conventional bearings, in the absence of accumulators; in these instances it could be that the presence of a bearing recess reduces the squeeze area and squeeze film damping capacity of the bearing, and so again reduces the system stability. It is noteworthy that the results presented relate to a bearing with only a single recess. It is unclear from the results what the sensitivity of the system stability is to the number of recesses and their location. For example it might be that a bearing with two recesses, each with their centre located at 45° to the vertical, might have improved stability characteristics. More work is required to fully investigate this.

Conclusion

The work has shown that the new design of bearing that has been investigated has a lower stability threshold then has a conventional bearing. The lower stability may be associated with a lower oil film damping caused by both the bearing recess and by the presence of the accumulator. For the particular conditions investigated, the new bearing was always more unstable with the accumulator connected than when it was disconnected. Even with the accumulator disconnected, the stability is lower than it is for a conventional bearing. The size of the bearing recess does not appear to have a substantial influence on the system stability. For the two sizes investigated there were some conditions under which the smaller recess made the system more stable. This was the case both with and without the accumulator connected to the bearing.

References

[1] Rankine, W. J. M. " On the Centrifugal Force of Rotating Shaft ", *The Engineers, April 1869, pp 249*

[2] Newkirk, B. L. " Shaft Whipping ", *General Electric Review, Vol 27, 1924, pp 169-178*

[3] Newkirk, B. L. and Lewis, J. F. " Oil Film Whirl - An Investigation of Disturbances Due to Oil Films in Journal Bearing ", *Trans ASME, Jan 1956, pp21*

[4] Holmes, R. " The Vibrations of a Rigid Shaft on Short Sleeve Bearings ", *Jrnl. Mech. Eng. Sci., Vol. 2, No. 4, 1960, pp 331*

[5] Boroomand, T. " Variable Impedance Journal Bearings for Rotor Bearing Systems ", *PhD Thesis, Staffordshire Polytechnic, Feb 1990.*

[6] Goodwin, M. J. " Dynamics of Rotor-Bearing System ", *Published by Unwin Hyman Ltd, First Edition 1989*

[7] Bannister, R. H. " A Theoretical and Experimental Investigation Illustrating the Influence of Non-Linearity and Misalignment on the Eight Oil Film Force Coefficients ", *Conf. of. Vibration in Rotating Machinery, Churcil College Cambridge. IMechE 1976, pp 271-278.*

On the Performance and Stability
of Magnetic Bearings

D.K. Anand
Professor of Mechanical Engineering
and Systems Research Center

J.A. Kirk
Professor of Mechanical Engineering
University of Maryland
College Park, Maryland 20742

R. Zmood
RMIT
Melbourne, Australia

E. Rodriquez
Senior Engineer
Goddard Space Flight Center
Greenbelt, Maryland

ABSTRACT

The Advanced Design and Manufacturing Laboratory at the University of Maryland was awarded a contract by GSFC/NASA to design and prototype a 500 WH magnetically stabilized flywheel energy storage system based upon several new innovations. These innovations include composite materials, magnetic suspension, and permanent magnet ironless armature, brushless motor/generator technology.

The flywheel energy storage system was spun at over 5000 rpm (the first rigid body critical is around 4200 rpm) to yield stable performance. The successful completion by this project establishes a viable technology and engineering base for designing and construction of a prototype magnetic bearing flywheel energy storage system. This paper summarizes the research and development work done and includes the final design of the magnetic bearings control system, and motor/generator; construction of the prototype system consisting of the magnetic bearing stack, flywheel, container and display module; and conclusions and recommendations for future directions based upon the results of laboratory testing.

INTRODUCTION

The magnetic bearing developed as a part of this contract is unique in that it combines the usage of the permanent magnet and electro-magnetic coils (1-5). The bearings are used to center the flywheel and support the flywheel weight at high speed. The bearing stator has 4 permanent magnets (PMs), 8 electromagnetic coils (EMs), 2 magnet plates, 2 control flux plates and 8 control pins. The rotor contains a return ring and a flywheel, which spins at high speed and stores energy.

The system has an active positioning control in the radial direction and a passive support capability in the axial direction. The bearing is divided into four quadrants to decouple the flux produced by adjacent quadrants. The flux paths are separated into two independent axes, east-west and north-south, and each axis has its own independent control system. Figure 1 shows the cross section of a pancake magnetic bearing. If the flywheel is not in the center, the permanent magnets will cause a destablizing force to pull it farther from the center. A control system responds to the motion by sending a current through the electro-magnet which results in additional corrective flux. The corrective flux adds to the bias flux on the small gap side and subtracts on the large gap side to produce a net restoring force. This bearing has undergone considerable research and development, at this laboratory, and has been reported over the years in fifty papers and theses listed in ref. 5.

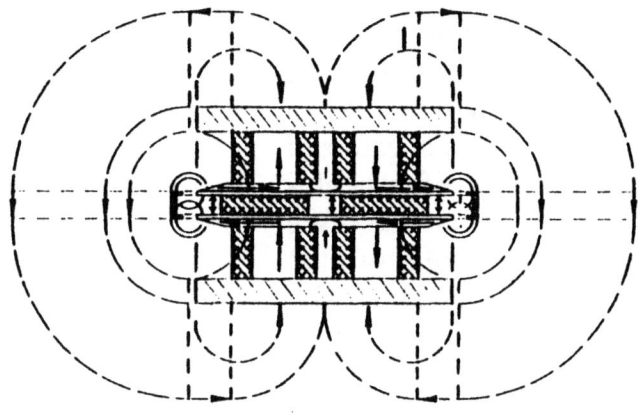

Fig. 1 Cross-Sectional View of Flux Paths Showing
Magnetic Flux Paths of Bearing

DESIGN CONSIDERATIONS

In the axial direction, the bearing passively supports the axial load of the flywheel by virtue of the fact that the flux lines bend sufficiently to support the weight. In the radial direction, the flywheel is affected by two forces; the destablizing force from the permanent magnets (PM) and the corrective force from the electromagnetic (EM) coils. The combined (restoring) radial force of bearing is

$$F_{rad} = K_i I - K_x X$$

where K_i is the force/current stiffness of EM coils, I is the control current, K_x is the radial stiffness of the permanent magnet, and X is the displacement of flywheel from the center. Figure 2 shows the radial forces of the bearing including the destablizing, corrective, and the combined forces. When the restoring force is zero, the corrective force equals the destablizing force. The range between these points is defined to be the maximum stable range (X_{stb}). The stable range of flywheel is related to the maximum control current of the power amplifier and the characteristics of PM and EM coils but independent of the control system. The maximum displacement range of the flywheel without current saturation is called the linear range, X_{lin}. In the linear range, K_a is the active stiffness of the magnetic bearing which can be represented as

$$K_a = CK_i - K_x .$$

For a magnetic bearing, a larger linear range, a higher active stiffness and a greater maximum restoring force are desired. Design algorithms were developed for the system to satisfy these requirements and were used iteratively to obtain the final design. This final design of the magnetic bearing is shown in Figure 3. A detailed methodology of the design procedure and the manufacture of the bearing is given in ref. 5.

CONTROL SYSTEM

The pancake bearing flywheel system shown in Figure 1 has two degrees of freedom in the radial direction. The motion of the system is represented by 2 translation dynamic equations in two orthogonal axes. The system has a negative spring constant due to bias flux of the permanent magnet and the flywheel is symmetric without mass imbalance. We assume that the translation motion in the X and Y direction are independent. The stack bearing flywheel system has 4 degrees of freedom, which represents the displacements sensed by the position transducers at either the inside radius or the periphery of the flywheel. The flywheel spins about the axial direction (Z axis) at speed of Ω rad/sec. The two bearings are assumed to be identical and have the same passive radial stiffness.

A proportional and derivative control system is used for the magnetic

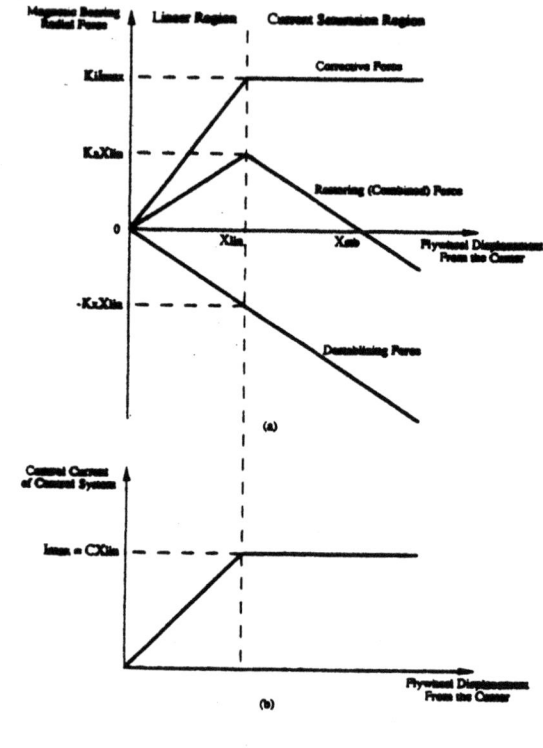

Fig. 2 Radial Forces

bearing flywheel system and the block diagram for one axis of the system is shown in Figure 4. Four control systems are used in the stack bearing flywheel system and are independently controlled. The linear model can be considered as a single input and single output (SISO) system. This has been extensively simulated and results of this design are discussed in detail in (1-4).

A nonlinear model was built to study the self-suspension and limit cycle oscillation phenomenon observed in experiment. When the power wasconnected, the flywheel not only failed to self-suspended but often broke into self-sustaining oscillations unless the mechanical touchdown gap was well adjusted. Also, the magnetic bearing flywheel system broke into limit cycle oscillation due to a large disturbance. Analysis and simulation of the nonlinear model has shown that reduction of the inductance of the electromagnetic coils will improve the robustness of the control system and relax the adjustment of the controller coefficients. It has been proven, by experimental and simulation results, that the combination of the power amplifier saturation and a large inductance of the EM coils causes the uncontrollable oscillations.

For a better understanding of the control system, the nonlinear model is simplified with only the current and displacement feedback loops. In addition, it is assumed the bearing actuator can be approximately modelled by a first order differential equation. For certain values of the gains (5), it can be shown that the phase plane trajectory either converges to the origin or executes a limit cycle on the phase plane. The conditions which need to be satisfied to ensure the magnetic bearing phase plane trajectories converge to the origin, and are thus stable, can be obtained from this simplified model.

These conditions are shown in Fig. 5, using the parameters for the experimentally tested bearings. It is noted that if the gain is chosen to stabilize the bearing control system for some nominal values of the parameters K_x/K_i but the coil inductance is large, then only very small variations in K_x and K_i can be tolerated if stability is to be maintained. The uncertainty in the values of these parameters however is usually quite large and will vary with changes in the rotor position, so if the coil inductance is too large then it may become impossible to stabilize the bearing control systems. When the coil inductance was reduced by a factor of 1:4 by re-winding the coils a marked improvement in bearing stability, and the ability of self-suspend, was observed.

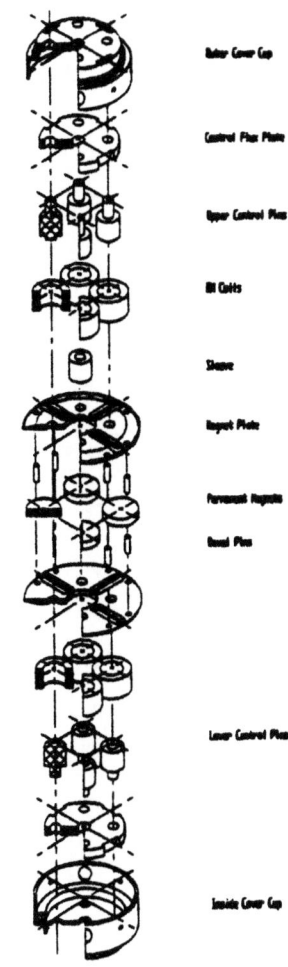

Fig. 3 Final Design of Magnetic Bearing

The final design goals of the control system are that the bandwidth of the control system must be greater than the first critical frequency (around 4000 r.p.m.); maximum excursion of the flywheel should be less than the mechanical touchdown bearing gap (6 mils); and good transient response with the balance of minimum energy and error.

A magnetic bearing stack is defined as being stable if it satisfies two conditions. One, it should be stable from the control point of view (see previous discussion on stability limits) and two, the maximum excursions of the flywheel should be less than 0.152 mm (6 mils) due to the presence of stops. If the first condition is satisfied then the second can be applied to determine the maximum forces and torques that can be applied on the flywheel. These maxima depend on the magnetic bearing parameters, flywheel rpm and value of control system gain. The choice of gain is especially important. A low value of gain allows the bearing to be spun to a higher rpm, but at the cost of lowering the static stiffness of the bearing.

Stability can be determined by observing the eigen values of the system matrix A. This essentially determines a range of values of K for which the system is stable. It is seen that the system is stable as long as the lower limit of gain is constant and the upper limit decreases with flywheel speed. The study established that the system will be stable up to a flywheel speed of 60,000 rpm.

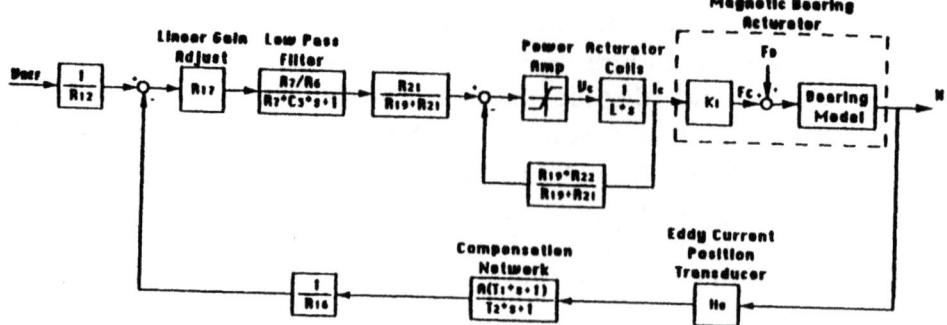

Fig. 4 Nonlinear Control System Model

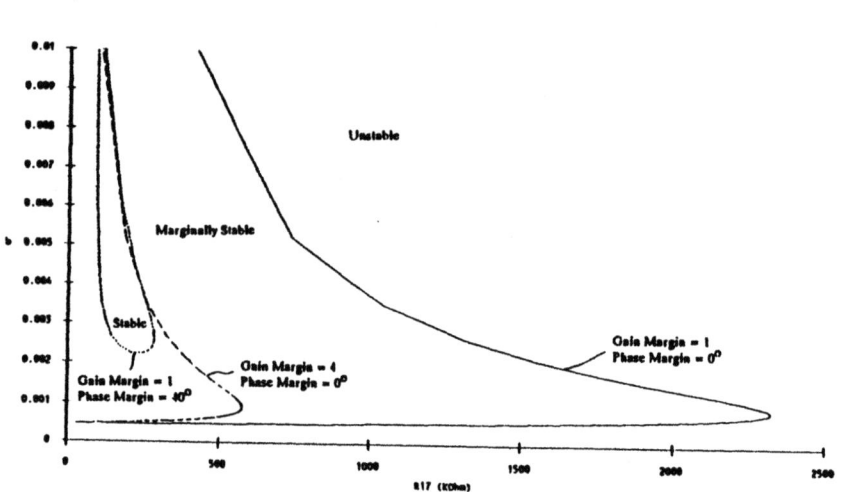

Fig. 5 Gain Margin and Phase Margin for Control System
with Adjustable Variables

The maximum force/torque specifications for the modified 102 mm (4") magnetic bearing stack is generated by simulating its response on EASY5, as shown in Fig. 6. The maximum force that can be applied at the center of mass of the flywheel, say along the x-axis, depends on the response at x_1 and x_2 which is independent of the flywheel speed. A torque, say along the y-axis, results in displacements at x_1 and x_2 and also at y_1 and y_2 due to the effects of gyroscopic coupling. Also the maximum values of the displacement recorded at x_1, y_1, x_2 and y_2 varies with the flywheel speed. It is important to keep this in mind when determining the maximum torque on the flywheel.

<u>TESTING</u>

During and after the construction phase of the prototype, the following tests were conducted:

- measurement of the various parts of the magnetic bearing system
- determination of K_x, K_i and K_A of individual bearings
- determination of the linear and stable regions of individual bearings
- suspension of individual bearings
- determination of K_T and K_I of motor
- spin testing of motor
- suspension and testing of the entire stack.

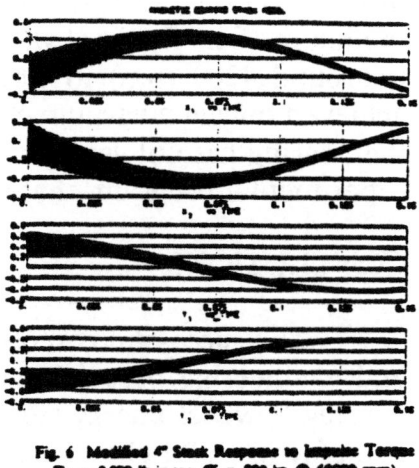

Fig. 6 Modified 4" Stack Response to Impulse Torque
T_o = 0.832 lb.in.sec. (K = 500 in @ 60000 rpm)
Units: x-axes (sec), y-axes (mils)

The measured and design parameters of the physical bearing and associated system is given in Table 1. It is noted that the manufactured hardware is well within the tolerances specified. Typical plots of results of testing individual bearings is shown in Fig. 7. This information was

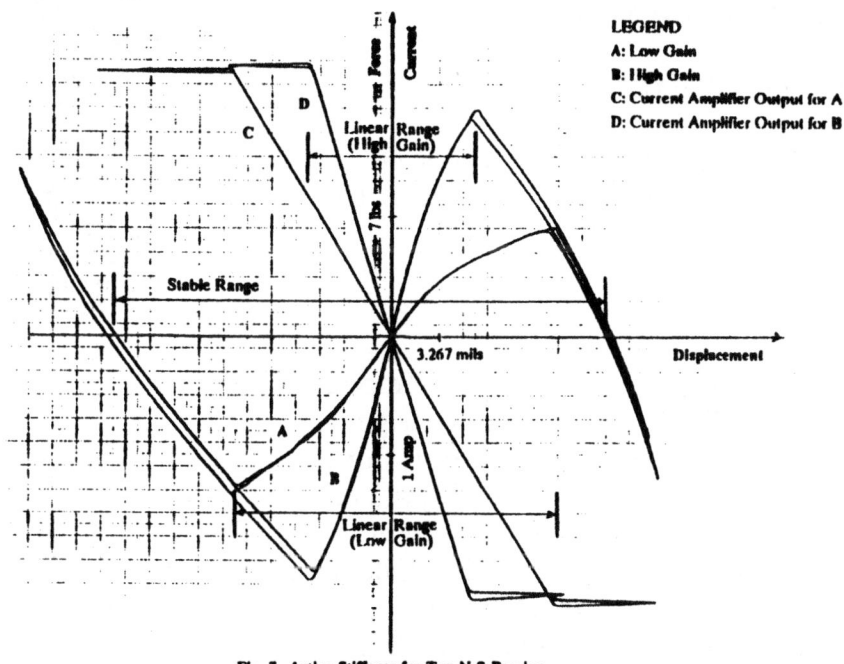

LEGEND
A: Low Gain
B: High Gain
C: Current Amplifier Output for A
D: Current Amplifier Output for B

Fig. 7 Active Stiffness for Top N-S Bearing

used to derive K_x, K_i, and K_A which is shown in Table 2. This table also shows the linear and stable ranges. The linear range increases when the gain is decreased. This means that the stiffness is decreased and the resulting bearing has a larger linear range but is soft. We note that a linear range of around 0.004" can be achieved and is necessary for self-capture. The bearings were individually suspended and spun to over 1000 rpm. The results were quite robust although the gain was on the low side, thereby having a relatively soft suspension. The testing of the motor/generator has been discussed at length in ref. 5. We simply note that K_T and K_I values were fairly close to those that were predicted.

138

The stack configuration, shown in Figure 8, once connected to the display/control panel, was suspended so that the bias current in the coils was a minimum. The flywheel was spun at over 5000 rpm (the first rigid body critical is around 4200 rpm) to yield stable performance. Higher spin rates can be achieved only in a vacuum chamber. Although such a chamber was constructed, the bearing has not yet been tested in a vacuum owing to time constraints. The experimental program, that will continue, will cover this testing phase.

CONCLUSIONS

The successful completion by this project establishes a viable technology and engineering base for designing and construction of a prototype magnetic bearing flywheel energy storage system. Specifically, we have proven the manufacture of tight tolerance bearings, stability and spin above first critical, use of sensors inside the bearing thereby eliminating runout problems, successful integration of surface wound brushless dc motor with EM/PM magnetic bearings, and finally established the analytical basis of magnetic bearing design including nonlinearities and saturation.

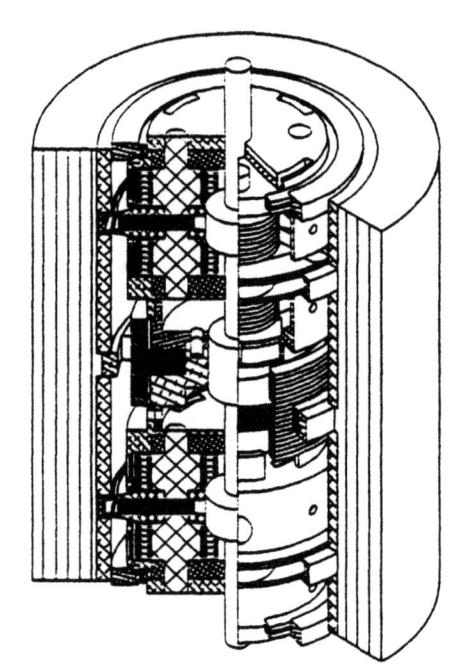

Fig. 8 Prototype Structure

ACKNOWLEDGEMENTS

This research was supported by NASA through a GSFC Contract No. NAS5-30091.

REFERENCES

1. Anand, D.K., Kirk, J.A., Zmood, R.B., Studer, P.A. and Rodriguez, G.E., "System Considerations for a Magnetically Suspended Flywheel," Proceedings of the 21st Intersociety Energy Conversion Engineering Conference, August 25-29, 1986, San Diego, California, pgs. 2.449-2.453.

2. Kirk, J.A., Anand, D.K., Evans, H.E. and Rodriguez, G.E., "Magnetically Suspended Flywheel System Study," NASA Conference Publication 2346, "An Assessment of Integrated Flywheel System Technology, Dec. 1984, pgs. 307-328.

3. Anand, D.K., Kirk, J.A. and Iwaskiw, P., "Magnetically Suspended Stacks for Inertial Energy Storage Flywheel," Proceedings of the 22nd Intersociety Energy Conversion Engineering Conference, August 10-14, 1987, Philadelphia, Pa., Vol. 2, pgs. 769-774.

4. Kirk, J.A. and Anand, D.K., "Satellite Power Using a Magnetically Suspended Flywheel Stack," Journal of Space Power, Vol. 22, Issue 3&4, March 1988.

5. "Magnetically Suspended Flywheels for Inertial Energy Storage", GSFC/NASA Contract NAS5-30091, January 1991.

	E-W		N-S	
	Top	Bottom	Top	Bottom
Spring Constant (lb/in)	-1538		-1545	
Electromagnetic Coil (lb/Amp)	14.1	12.3	13.1	11.3
System Gain (Amp/in)	185	375	196	399
Active Stiffness (lb/in)	1067	3349	1337	3962
Max Amplifier Current (Amp)	1.84	1.85	2.16	2.2
Max Force (lb)	6.4	13.4	8.9	15.3
Linear Range (mil)	9.1	4.6	10.7	5.3
(Stable Range (mil)	14	13.9	17.1	17

Table 1: Final Test Results

Identification of Friction Factors for Modelling the Exciting Forces caused by Flow in Labyrinth Seals

G. Thieleke and H. Stetter
Universität Stuttgart

Abstract

The interaction of the flow through the labyrinth seals with the shaft of the rotor can affect the stability of turbomachines. Exciting forces, so–called cross forces or non–conservative forces, arise, and act perpendicular to the rotor eccentricity. This effect is caused by an unsymmetrical pressure distribution within the labyrinth cavities.

In order to describe the compressible, turbulent flow in a labyrinth seal, *bulk–flow–theories* predicting the rotordynamic coefficients for labyrinth seals, are often applied. A friction factor model is used to describe the wall shear stresses acting on the rotor and the stator of the seal. The friction factors can be identified by experimental stationary–rotor flow tests.

Test results are presented for the friction factors of a staggered labyrinth. These labyrinths are widely used on the tip shrouding of bladings in steam or gas turbines. Based on the identified friction factors, the stiffness coefficients are determined using an One–volume–bulk–flow–theory similar to Childs theory. The theoretical results are compared with experimental results and show good agreement.

Nomenclature

c	Stiffness–term
c	Circumferential velocity of flow
C_R, Δr	Nominal clearance
c_{uo}	Inlet swirl
d	Damping–term
D	Diameter
$E_0^* = \frac{1}{2}\rho_0 c_{u0}^2 / \Delta p$	Flow characteristic
F	Force
f	Cross-sectional area of control volume
H	Local radial clearance
h_d	Hydraulic diameter
K_Q	Cross force spring-coefficient
m	Inertia–term
m	Number of cavities
$m_{R;S}, n_{R;S}$	Coefficients of friction factor
$\dot{m}_i$	Leakage flow per circumferential length
p	Pressure
Δp	Pressure difference at the seal
R	Gas constant

Re	Reynolds number
R_R	Shaft radius
T	Temperature
t	Time
L	Pitch of seal strips
u_w	Rotor peripheral velocity
U	Length on which shear stress acts
λ	friction factor
μ	Flow coefficient
ν	Kinematic viscosity
ρ	Density of fluid
π	Pressure ratio (p_a/p_0)
φ	Peripheral angle
σ	Shear stress
ω	Shaft angular velocity

subscripts

i	i-th chamber value
R	rotor
S	stator
u	circumferential direction
y	y-direction
z	z-direction
$-$	mean-value
$*$	dimensionless value

Not presented notations are declared in text.

1 Introduction

In order to determine the rotordynamic coefficients of labyrinth gas seals using an *One-volume-bulk-flow model*, it is essential to know the behaviour of the flow coefficient μ and the friction factors λ along the seal. However, there are nearly no data available for the friction factors of labyrinth gas seals. With the flow measurings of the circumferential velocity, it is possible to identify these coefficients from experimental investigations for the centered rotor position. The flow coefficient and the friction factors are different for each labyrinth geometry, and they must be known for predicting the rotordynamic coefficients of labyrinth gas seals.

For small motion about a centered position the force-motion equations for a labyrinth seal are assumed to be of the form:

$$\begin{pmatrix} F_y \\ F_z \end{pmatrix} = \begin{pmatrix} c_{yy} & c_{yz} \\ c_{zy} & c_{zz} \end{pmatrix} \begin{pmatrix} y \\ z \end{pmatrix} + \begin{pmatrix} d_{yy} & d_{yz} \\ d_{zy} & d_{zz} \end{pmatrix} \begin{pmatrix} \dot{y} \\ \dot{z} \end{pmatrix} + \begin{pmatrix} m_{yy} & m_{yz} \\ m_{zy} & m_{zz} \end{pmatrix} \begin{pmatrix} \ddot{y} \\ \ddot{z} \end{pmatrix} \qquad (1)$$

where y, z are components of the rotor displacement relative to the stator; and (F_y, F_z) are the components of the force acting on the rotor; and (c,d,m) are stiffness, damping and inertia coefficients respectively. They are called the rotordynamic coefficients of labyrinth seals.

2 Seal Analysis

The presented seal analysis uses an one-volume-bulk-flow model similar to Childs theory [1]. The basic equations for compressible flow in a labyrinth seal are derived for a single cavity control volume as shown in figure 1.

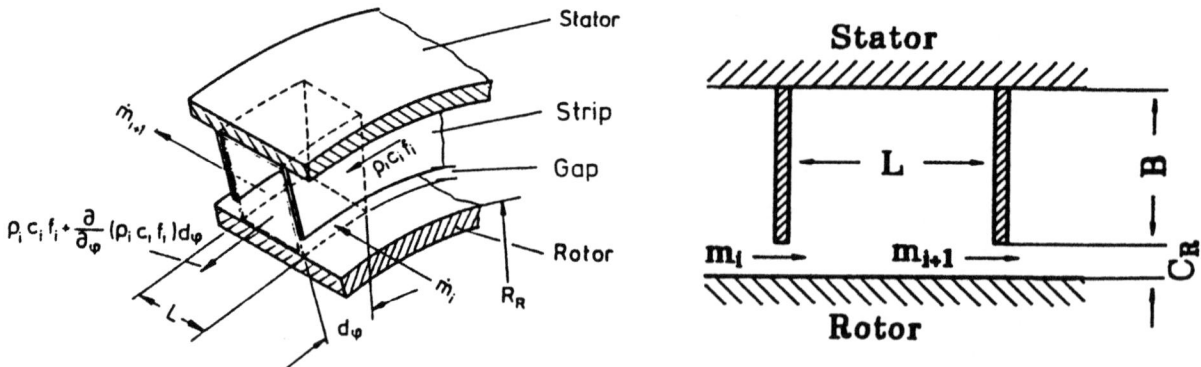

Figure 1: Cavity control volume

The governing equations are:

- Continuity equation:

$$\frac{\partial}{\partial t}\left(\rho_i f_i\right) + \frac{1}{R_R}\frac{\partial}{\partial \varphi}\left(\rho_i f_i c_i\right) + \dot{m}_{i+1} - \dot{m}_i = 0 \tag{2}$$

- Momentum equation:

$$\frac{\partial}{\partial t}\left(\rho_i f_i c_i\right) = -\frac{1}{R_R}\frac{\partial}{\partial \varphi}\left(\rho_i f_i c_i^2\right) - \frac{f_i}{R_R}\frac{\partial p_i}{\partial \varphi} + \dot{m}_i c_{i-1} - \dot{m}_{i+1} c_i$$
$$-\sigma_{t,Si} U_{Si} + \sigma_{t,Ri} U_{Ri} \tag{3}$$

- Leakage equation:

$$\dot{m}_i = \mu_i H_i \sqrt{\frac{p_i^2 - p_{i+1}^2}{RT}} \tag{4}$$

with leakage mass flow per circumferential length

- Friction equation:

$$\begin{aligned}
\sigma_{t,Si} &= \frac{1}{2}\rho_i c_i^2 \lambda_{Si} = \frac{1}{2}\rho_i c_i^2 n_{Si} Re_{Si}^{m_{Si}} \\
&= \frac{1}{2}\rho_i c_i^2 n_{Si}\left(\frac{|c_i| h_{di}}{\nu}\right)^{m_{Si}} \text{sign}(c_i) \\
\sigma_{t,Ri} &= \frac{1}{2}\rho_i (R_R\omega - c_i)^2 \lambda_{Ri} = \frac{1}{2}\rho_i (R_R\omega - c_i)^2 n_{Ri} Re_{Ri}^{m_{Ri}} \\
&= \frac{1}{2}\rho_i (R_R\omega - c_i)^2 n_{Ri}\left(\frac{|R_R\omega - c_i| h_{di}}{\nu}\right)^{m_{Ri}} \text{sign}(R_R\omega - c_i)
\end{aligned} \tag{5}$$

with h_d as hydraulic diameter.

- Ideal gas law:

$$\frac{p_i}{\rho_i} = RT \tag{6}$$

In the friction equation (5) λ_{Si} and λ_{Ri} represent the friction factors relative to the stator and the rotor respectively. Hirs turbulent bulk–flow model [2] assumes that these friction factors can be written as a function of the constants n, m and the Reynoldsnumber Re relative to the surface on which the shear stress acts. The friction parameters n, m are generally empirically determined and describe the influence of relative roughness and the geometry of the labyrinth cavity. The friction parameters n, m for smooth annular seals are determined from axial pressure drop experiments along the seal [3]. The same procedure is used to determine the friction factors for honeycomb–seals —smooth rotor and roughened stator— [4], [5]. Based on the assumption that the behaviour of the friction factors for seals and annular surfaces over the Reynoldsnumber is similar, Serkov [6] determines a correction factor Ψ in order to calculate the friction factors with the following equation:

$$\lambda_{laby} = \Psi \lambda_{annular\ seal} \quad \text{with} \quad \lambda_{annular\ seal} = nRe^m \ . \tag{7}$$

Thereby, the correction factor Ψ represents an integral value over the seal length.

3 Identification Procedure

The basic equation for identification of the friction factors λ_{Si} and λ_{Ri} is the momentum equation for the centered rotor position (zeroth–order momentum equation) which is formed as follows:

$$2\frac{RT}{p_i}\frac{(c_{i-1} - c_i)}{c_i^2}\frac{\dot{m}}{U_{Si}}sign(c_i) = \lambda_{Si} + \left(\frac{u_w - c_i}{c_i}\right)^2 \frac{U_{Ri}}{U_{Si}}\lambda_{Ri}sign(c_i - u_w)sign(c_i) \tag{8}$$

Based on the measurements pressure p_i, circumferential velocity c_i, flow rate $\dot{m}$ and peripheral velocity of rotor u_w for the centered rotor position, it is possible to identify the friction factors λ_{Ri} and λ_{Si}. Equation (8) is linear (equation of straight line) and can be written as follows:

$$h = \lambda_{Si} + \lambda_{Ri}x \tag{9}$$

with

$$h \quad = \quad 2\frac{RT}{p_i}\frac{(c_{i-1} - c_i)}{c_i^2}\frac{\dot{m}}{U_{Si}}sign(c_i) \tag{10}$$

$$x \quad = \quad \left(\frac{u_w - c_i}{c_i}\right)^2 \frac{U_{Ri}}{U_{Si}}sign(c_i - u_w)sign(c_i) \tag{11}$$

For a constant Reynoldsnumber for stator and rotor Re_{Si} and Re_{Ri}, the unknown λ_{Ri} and λ_{Si} are fitted for different values of h and x according to equation (9). In the test–facility, the values h, x and also the Reynoldsnumber Re_{Si}, Re_{Ri} are varied by the test parameters pressure ratio at the seal $\pi = p_a/p_0$, inlet swirl c_{u0} and peripheral velocity of rotor u_w.

In the second step of the identification procedure, the friction parameters n, m can be found by a least square fitting for different Reynoldsnumber Re_{Si}, Re_{Ri} and friction factors λ_{Si}, λ_{Ri} respectively as follows:

$$\lambda = nRe^m \tag{12}$$

For some special cases, it is possible to examine the identified friction factors λ_{Si}, λ_{Ri} according to equation (9). These special cases are:

- **Special case 1:** No shear stress acting on the rotor. This means that the circumferential velocity c_i is equal to peripheral velocity of rotor u_w. Thus, equation (8) reduces to:

$$2\frac{RT}{p_i}\frac{(c_{i-1} - c_i)}{c_i^2}\frac{\dot{m}}{U_S}sign(c_i) = \lambda_{Si} \tag{13}$$

In this case the friction factors λ_{Si} are determined directly from measurements without the searching for equal Reynoldsnumbers Re_{Si}, Re_{Ri}.

- **Special case 2**: Reynoldsnumber of rotor Re_{Ri} equal Reynoldsnumber of stator Re_{Si}. This special case appears for $u_w = 0$ and for $u_w = 2c_i$, i. e. for no rotor rotation and for u_w which is double the circumferential velocity of the flow. The value x only depends on the geometry of the labyrinth seal. If the friction factors λ_{Si} are known (special case 1), it is possible to determine the friction factors of rotor λ_{Ri} from measurements with no rotor rotation:

$$\lambda_{Ri} = (h - \lambda_{Si})\frac{U_{Ri}}{U_{Si}} \tag{14}$$

- **Special case 3**: h = 0 — that means $\dot{m} = 0$ or $c_{i-1} = c_i$ —. For h = 0, there is a correlation between the friction factors of rotor and stator:

$$\frac{\lambda_{Si}}{\lambda_{Ri}} = -\left(\frac{u_w - c_i}{c_i}\right)^2 \frac{U_{Ri}}{U_{Si}} sign(c_i - u_w)sign(c_i) \tag{15}$$

The ratio $\lambda_{Si}/\lambda_{Ri}$ only depends on the velocities c_i, u_w and the geometry data U_{Ri} and U_{Si}.

4 Experimental Investigations

In order to identify the friction factors, it is essential to measure the circumferential velocity of the flow c_i in each cavity. The pressure distribution along the seal, the flow rate $\dot{m}$, the temperature T and the peripheral velocity of rotor u_w are also measured. The measuring of the resulting circumferential component of the labyrinth flow in each cavity is made with three–hole-cylinder probes specially developed for this purpose (see figure 2). Since the probes are also used in the area of the labyrinth strips and near the rotor surface, the wall effect on the calibration factors, defined in figure 2, must be tested and taken into consideration.

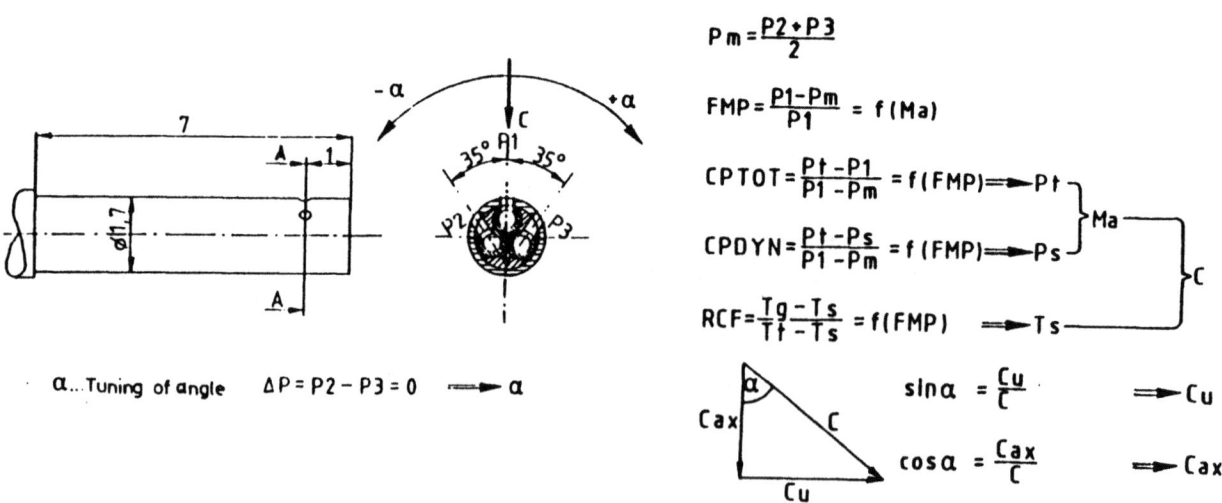

Figure 2: Three–hole-cylinder probe

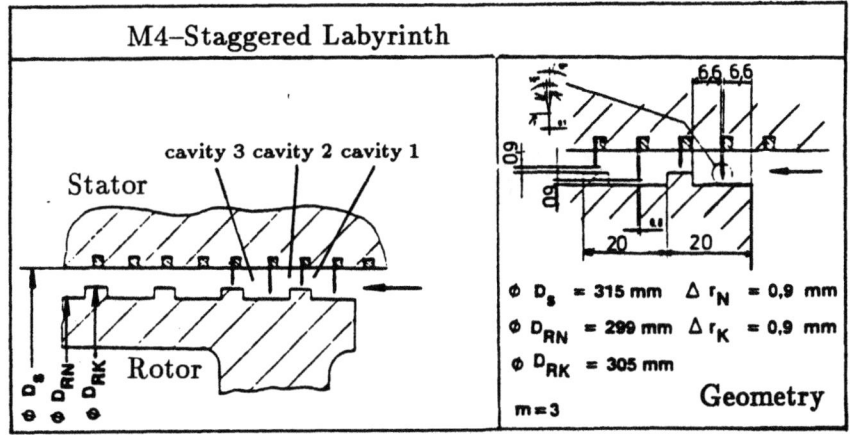

Figure 3: Geometry of labyrinth seal

The investigations are carried out on a labyrinth test–facility at the 'Institut für Thermische Strömungsmaschinen und Maschinenlaboratorium' of Stuttgart University. Using the test–facility, the flow forces in labyrinth seals (stiffness coefficients) can be determined experimentally [7].

The geometry of the investigated labyrinth seal is shown in fig. 3. It represents a staggered labyrinth with grooved rotor and strips on the stator. This labyrinth seal can be found on the tip shrouding of bladings in steam or gas turbines.

5 Results

Fig. 4 shows the characteristic of the friction factors of stator $\lambda_{S1,3}$ (cavity 1 and 3) versus the Reynoldsnumber Re_S. The cavities 1 and 3 have the same geometry with a transition from rotor groove to rotor serration. The results are independent of the used identification algorithm (eq. (9) or eq. (13)). The friction factors of stator λ_S are nearly constant over the range of the Reynoldsnumber from $10 \cdot 10^3$ to $120 \cdot 10^3$. The value of λ_S is 0.02 approximately. Fig. 4 shows that for the same geometry of labyrinth cavity the characteristic of the friction factors of stator is the same.

Fig. 5 shows the characteristic of the friction factors of stator λ_{S2} (cavity 2). The cavity 2 consists of a transition from rotor serration to rotor groove. The value of λ_{S2} is smaller compared to $\lambda_{S1,3}$. The experimental results show that the friction factors of stator depend on the geometry of the labyrinth seal. In contrast to the friction factors of stator, the friction factors of rotor do not depend on the labyrinth geometry. The approximated curves according to equation (12) are also depicted in fig. 4, 5. The identified friction parameters for the staggered labyrinth are:

$n_{Si} = 0.185$	$m_{Si} = -0.207$	for	$10^4 < Re < 12 \cdot 10^4$	$i = 1,3$
$n_{S2} = 0.164$	$m_{S2} = -0.306$			$i = 2$
$n_{Ri} = 0.239$	$m_{Ri} = -0.298$			$i = 1,2,3$

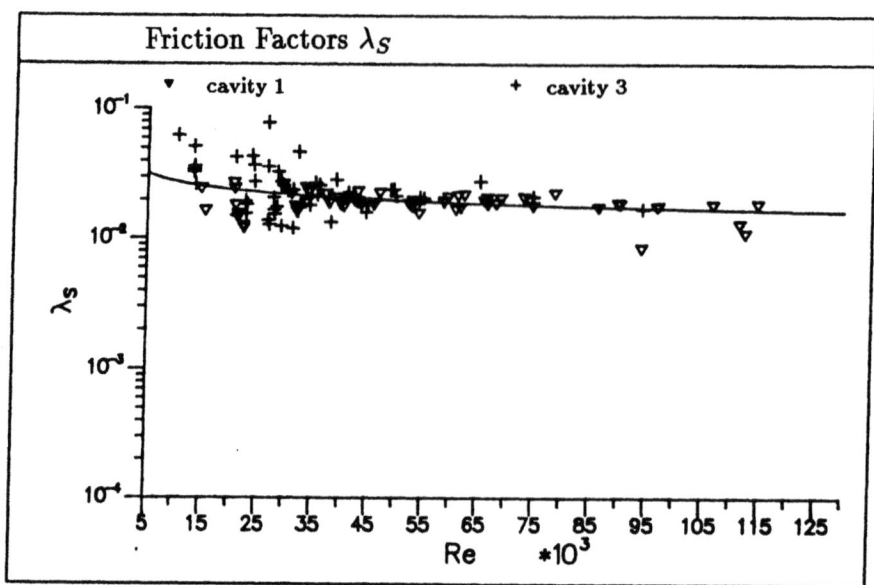

Figure 4: Friction factors of stator $\lambda_{S1,3}$, cavity 1 and 3

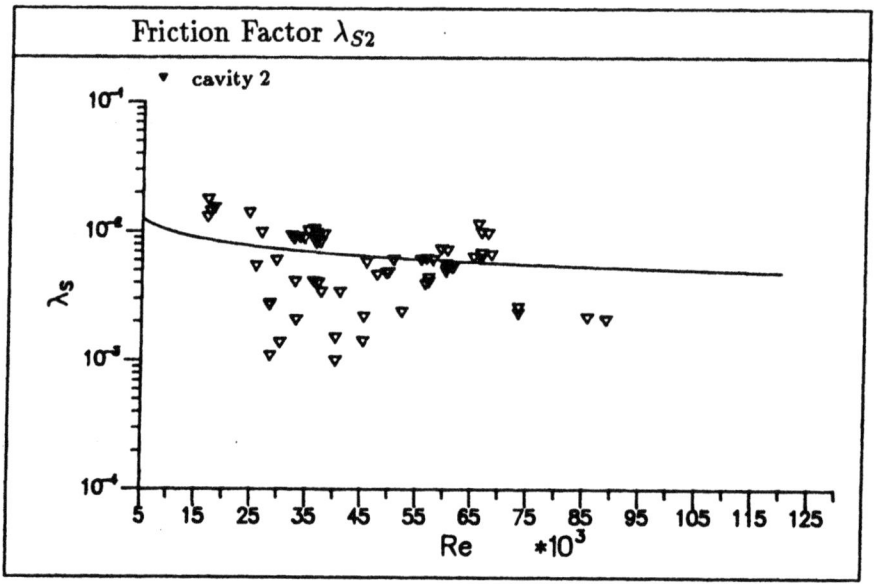

Figure 5: Friction factors of stator λ_{S2}, cavity 2

Fig. 6 shows summarized the results of the identified friction factors. In this figure there is also depicted the characteristic of friction factors of smooth [3], [8] and honeycomb–seals [5]. It is noteworthy that the friction factors of honeycomb–seals are smaller compared to the $\lambda_{S1,3}$-values of the staggered labyrinth. For the roughened honeycomb–seal a greater friction factor would be expected.

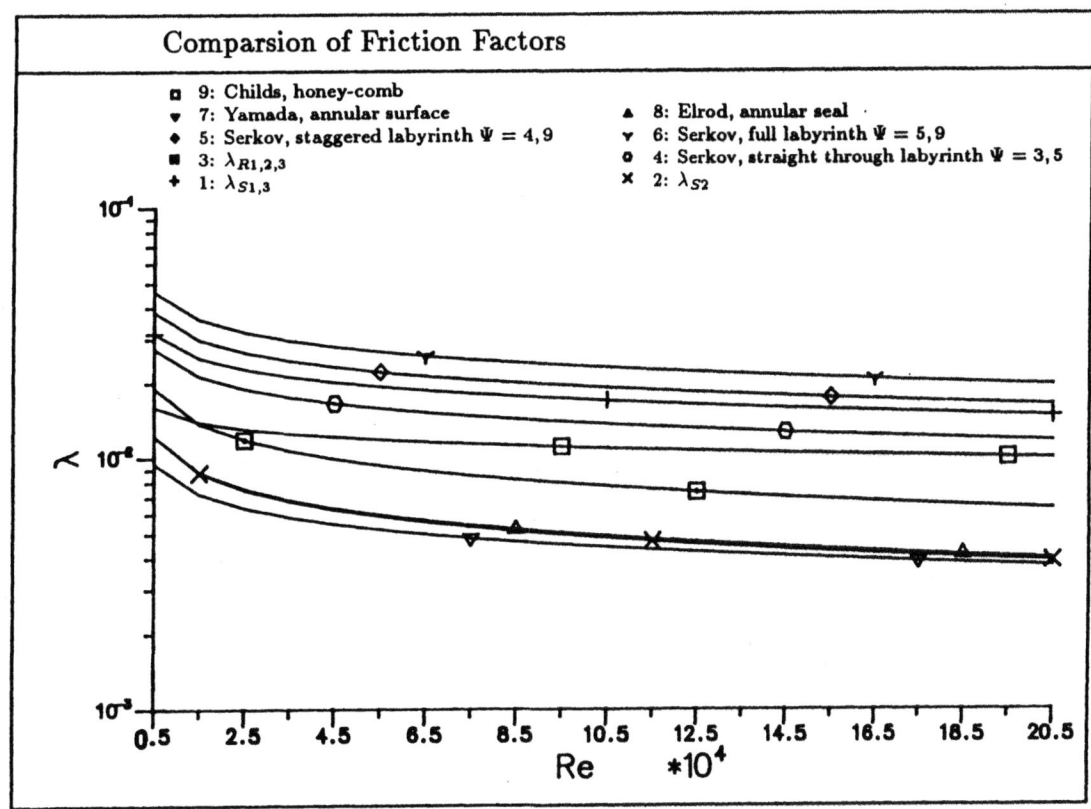

Figure 6: Comparison of friction factors

6 Comparison of Experimental and Theoretical Results

Using the governing equations of the bulk–flow–theory and the identified friction factors, the rotordynamic coefficients can be determined. The procedure of the solution described in [7] is similar to the perturbation analysis used by Childs [9]. Comparisons of theoretical and experimental results of the exciting cross–coupled stiffness coefficients c_{yz} respectively spring coefficients $\overline{K_Q^*}$ are made for the staggered labyrinth with 3 cavities. The correlation is as follows:

$$\overline{K_Q^*} = -\frac{c_{yz}C_R}{\Delta p_{st}R_R mL} \tag{16}$$

The results are presented for $\overline{K_Q^*}$ as a function of the flow characteristic E_0^*. The flow coefficient μ_i which depends on the geometry of the labyrinth seal, is also derived from experimental results [7].

The results are in good agreement with the spring coefficients $\overline{K_Q^*}$ (see figures 7, 8).

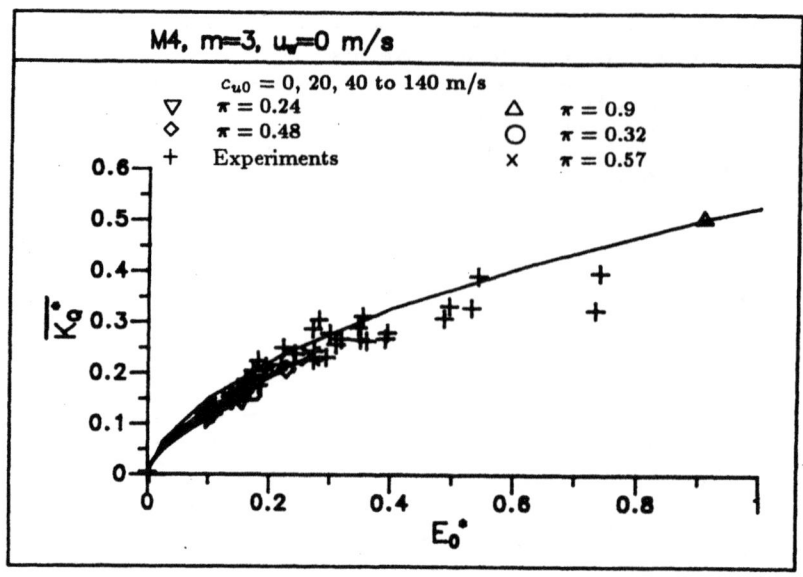

Figure 7: Theory versus experiment of spring coefficients $\overline{K_Q^*}$ for $u_w = 0$ m/s

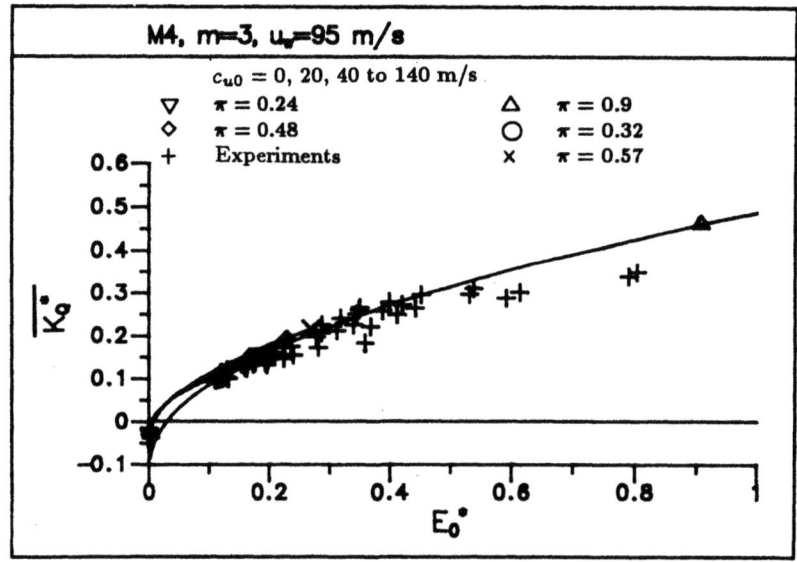

Figure 8: Theory versus experiment of spring coefficients $\overline{K_Q^*}$ for $u_w = 95$ m/s

7 Conclusions

Experimental results have been presented for the nearly unknown friction factors of a staggered labyrinth seal. Based on the one–volume–bulk–flow model, an identification algorithm is developed to determine the friction factors. Based on the measurings of the circumferential velocity in each labyrinth cavity, the identification of the friction factors is possible.

The results of this study support the following conclusions:

- The identification of friction factors of labyrinth seals is feasible.

- The friction factors depend on the geometry of the labyrinth seal.

- Using the identified friction factors, the results of the bulk–flow theory are in good agreement with the experimental results.

References

[1] Childs, D. W., Scharrer, J. K.: An Iwatsubo–Based Solution for Labyrinth Seals: Comparison to Experimental Results. Trans. ASME, Journal of Engineering for Gas Turbines and Power, 1986, Vol. 108, S. 325–331.

[2] Hirs, G. G.: A Bulk Flow Theory for Turbulence in Lubricant Films. Trans. ASME, Journal of Lubrication Technology, Apr. 1973, S. 137–146.

[3] Yamada, Y.: Resistance of Flow Through an Annulus with an Inner Rotating Cylinder. Bull. JSME, Vol. 5, No. 18, 1962, S. 302–310.

[4] Nelson, C. C.: Analysis for Leakage and Rotordynamic Coefficients of Surface-Roughened Tapered Annular Gas Seals. Trans. ASME, Journal of Engineering for Gas Turbines and Power, Oct. 1984, Vol. 106, S. 927–934.

[5] Childs, D., Ramsey, C.: Seal Rotordynamic Coefficient Test Results for a Model SSME ATD–HPFTP Turbine Interstage Seal with and without a Swirl Brake. NASA Conference Publication, Proceedings of a workshop 'Rotordynamic Instability Problems in High–Performance Turbomachinery', Texas A&M University, College Station Texas, Mai 1990.

[6] Serkov,S.A.: Die Bestimmung von aerodynamischen Kräften in Turbinendichtungen zur Erhöhung der Grenzstabilität. Wärme, Vol.92, 1986, No.4-5, S.79-83.

[7] Thieleke,G., Stetter,H.: Experimental Investigations of Exciting Forces caused by Flow in Labyrinth Seals. Proceedings of a workshop at Texas A&M University 21-23 May 1990, entitled Rotordynamic Instability Problems in High Performance Turbomachinery, pp. 103-128.

[8] Elrod, D., Childs, D. W., Nelson, C.: An Annular Gas Seal Analysis Using Empirical Entrance and Exit Region Friction Factors. Trans. ASME, Journal of Tribology, Apr. 1990, Vol. 112, S. 196–204.

[9] Childs, D. W., Kim, C. H.: Analysis and Testing for Rotordynamic Coefficients of Turbulent Annular Seals with Different, Directionally-Homogeneous Surface Roughness Treatments for Rotor and Stator. NASA Conference Publication 2338, Rotordynamic Instability Problems in High-Performance Turbomachinery 1984.

Physical Modelling and Data Analysis of the Dynamic Response of a Flexibly Mounted Rotor Mechanical Seal

An Sung Lee and Itzhak Green
Woodruff School of Mechanical Engineering, Georgia Institute of Technology
Atlanta, Georgia 30332

ABSTRACT

The dynamic behaviour of mechanical face seal has been an active area of research for the past three decades. Analytical and experimental work was exclusively devoted to the flexibly mounted stator seal (FMS). Recent theoretical work on the dynamics of the noncontacting flexibly mounted rotor (FMR) seal proved that it is superior in every aspect of dynamic behaviour compared to the FMS seal. The FMR seal is inherently stable regardless of operating speed, the maximum relative misalignment is smaller, and the critical stator misalignment is larger. These are measures of superior performance. Yet, no experimental investigation of the noncontacting FMR seals has been reported. This work describes the design and construction of a noncontacting FMR mechanical face seal test rig. Features of the rig design and data analysis will be introduced. Finally, experimental results will be compared with theory.

NOMENCLATURE

C	= seal clearance	r_o	= outer radius
d_r	= total displacement of rotor	R_i	= dimensionless inner radius, r_i/r_o
d_{rl}	= displacement of rotor to γ_{ri}	R_m	= dimensionless mean radius, $(1+R_i)/2$
d_{rn}	= displacement of rotor to γ_s	t	= time
D	= $D_f + D_s$	β^*	= coning
D_f	= fluid film damping	β	= dimensionless coning, $\beta^* r_o/C$
D_s	= support damping	γ_r	= total rotor response
I_p	= polar moment of inertia	γ_{ri}	= initial rotor misalignment
I_t	= transverse moment of inertia	γ_{rl}	= rotor response to γ_{ri}
K	= $K_f + K_s$	γ_s	= fixed stator misalignment
K_f	= fluid film stiffness	γ_{rn}	= rotor response to γ_s
K_s	= support stiffness	ξ,η,ζ	= inertial coordinate systems
p_w	= water pressure	γ_ξ,γ_η	= rotor misalignment in the $\xi\eta\zeta$-system
Q^*	= leakage, Equation (11)	μ	= viscosity
Q	= dimensionless leakage	ψ_s	= phase angle of γ_s with x-axis
r_i	= inner radius	ω	= shaft angular velocity

INTRODUCTION

Conventional contacting seals are load unbalanced. The unbalanced load causes seal faces to be in mechanical contact. As speeds, pressures, and temperatures in modern high performance rotating machinery increase, the wear due to contact between mating seal faces becomes one of the main causes of seal failure. A possible solution to the problem is the use of a noncontacting seal. In the noncontacting seal a thin fluid film lubricates the seal faces, thus reducing friction loses and wear. A good understanding of the dynamic behaviour of the seal is essential for reliable noncontacting operation with minimum leakage. Seal dynamics has been an active area of research (Etsion, 1982, 1985, and 1991, and Allaire, 1984). Etsion and Burton (1979) tested a face seal model consisting of a rigidly mounted rotor and a flexibly mounted stator under eccentric loading, i.e., initial stator misalignment. Self-excited oscillations in the form of combined precession and nutation of the stator were observed. The hydrostatic, hydrodynamic, and squeeze effects in mechanical face seals with diametral tilt and coning were analysed by Etsion and Sharoni (1980), Sharoni and Etsion (1981), and Etsion (1980). Metcalfe (1981, and 1982) analysed the dynamic tracking ability of a flexibly mounted face to an angular misalignment of a fixed face, and experimentally observed the dynamic whirl to be close to half of the shaft speed in a well-aligned mechanical face seal. Sehnal et al. (1983) experimentally investigated the effects of a face coning on the seal performance by comparing torque, face temperature, leakage, and wear of a conventional flat-face seal with three coned face seals. Etsion and Constantinescu (1984) experimentally observed the dynamic behaviour of a noncontacting flexibly mounted stator (FMS) mechanical face seal. They showed that the stator misalignment and its phase shift are time dependent. Green and Etsion (1985) analysed the dynamic behaviour of a noncontacting FMS mechanical face seals and the analysis was shown to be valid in many practical cases. Green and Etsion (1986) experimentally measured the axial stiffness and damping coefficients of elastomeric O-rings that commonly serve as secondary seals.

Theoretical work on the dynamics of a noncontacting flexibly mounted rotor (FMR) mechanical face seal was performed by Green (1989, and 1990). That work showed that the FMR seal is inherently stable regardless of operating speed, provided that the inertia ratio, I_t/I_p (transverse moment of inertia over polar moment of inertia), is less than one, which is a practical ratio in most mechanical face seals. Further, the FMR seal was found to be a better design than the FMS seal in terms of various seal performance criteria, i.e., the total relative misalignment, critical stator misalignment, and threshold speed of instability in the case where the inertia ratio is greater than one. Analytical steady-state solution was given in terms of transmissibilities (amplitude ratio of response over forcing input). However, no experimental investigation of the noncontacting FMR seal has been reported. An experimental investigation remains to be performed in order to evaluate the theoretical results and to verify the advantages of the FMR seal. In this work, features of a noncontacting FMR mechanical face seal test rig design and data analysis will be introduced, and experimental results will be compared with theory.

Equations of Motion and Steady-State Responses of FMR Seal

The equations of motion for a noncontacting FMR mechanical face seal were derived by Green (1990). The equations of motion in the angular mode are presented here in the inertial $\xi\eta\zeta$-system (see the nomenclature for definitions)

$$I_t\ddot{\gamma}_\xi + I_p\omega\dot{\gamma}_\eta + (D_s+D_f)\dot{\gamma}_\xi + (D_s+\frac{1}{2}D_f)\omega\gamma_\eta + (K_s+K_f)\gamma_\xi$$

$$= \gamma_s(K_f\cos\psi_s + \frac{1}{2}D_f\omega\sin\psi_s) + K_s\gamma_{ri}\cos\omega t$$

$$I_t\ddot{\gamma}_\eta - I_p\omega\dot{\gamma}_\xi + (D_s+D_f)\dot{\gamma}_\eta - (D_s+\frac{1}{2}D_f)\omega\gamma_\xi + (K_s+K_f)\gamma_\eta$$

$$= \gamma_s(K_f\sin\psi_s - \frac{1}{2}D_f\omega\cos\psi_s) + K_s\gamma_{ri}\sin\omega t$$

$$(1)$$

The fixed stator misalignment and initial rotor misalignment, each is defined with respect to the axis of shaft rotation, γ_s and γ_{ri}, respectively, are always present due to the manufacturing and assembly tolerances. These misalignments act as forcing inputs upon the rotor. A closed form solution for the steady-state response of Equation (1) was obtained in terms of transmissibilities by Green (1989) as follows:

$$\frac{\gamma_{rs}}{\gamma_s} = \frac{\sqrt{K_f^2 + \frac{1}{4}D_f^2\omega^2}}{\sqrt{K^2 + (D_s\omega + \frac{1}{2}D_f\omega)^2}}$$

$$\frac{\gamma_{rl}}{\gamma_{ri}} = \frac{K_s}{\sqrt{[(I_p-I_t)\omega^2 + (K_s+K_f)]^2 + (\frac{1}{2}D_f\omega)^2}}$$

$$(2)$$

where γ_{rs} and γ_{rl} are the rotor responses to γ_s and γ_{ri}, respectively. These transmissibilities will be compared with the experimental ones.

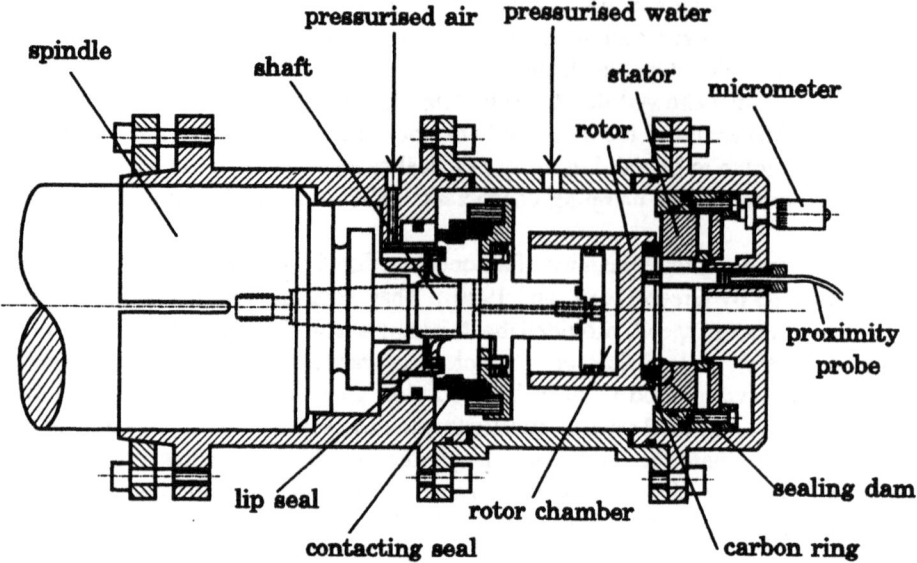

Fig. 1. Noncontacting flexibly mounted rotor seal test rig

TEST RIG

A noncontacting FMR mechanical face seal test rig is shown in Fig. 1. A sealing dam is formed between the faces of the stator and the carbon graphite ring that is attached to the rotor. Separation of the two faces is achieved by utilising the hydrostatic effect as it prevails in a converging gap between the flat-face carbon graphite ring and the coned-face stator. The rotor is flexibly mounted on the shaft, by means of an elastomeric Nitrile (Buna N) O-ring, to allow its tracking to the stator misalignment. Various seal clearances at the sealing dam can be obtained by changing air pressure in the rotor chamber. Pressurised air is supplied through holes in the shaft and housing. The pressurised air and pressurised water are separated by a contacting mechanical seal at one end, whereas the pressurised air is sealed by a lip seal at the other end. The shaft is screwed into a precision spindle which is driven by a motor mounted on a separate structure through a wafer spring coupling. For convenience of machining, maintenance, and adjusting of the rig, the housing is made of three parts. All possible leakage paths are sealed by O-rings. The stator misalignment is adjusted by three micrometers, and the dynamic behaviour of the rotor is detected by three eddy current proximity probes whose signals are sampled by a data acquisition system. Leakage is measured by a flow meter placed on the supply line of pressurised water.

Rotor (primary ring) and Support System
The 0.5198 kg rotor is made of AISI 4140 Steel. The polar moment of inertia, I_p, is 4.1619×10^{-4} kg $\cdot$ m^2, and the transverse moment of inertia, I_t, is 2.8032×10^{-4} kg $\cdot$ m^2 (I_t/I_p, 0.67353). The carbon graphite ring has an O.D. of 50.8 mm and an I.D. of 40.64 mm (radius ratio, r_i/r_o, of 0.8).

Series of test runs indicated that one O-ring support was not sufficient to ensure small misalignments of the rotor. Therefore, a shaft head which has the support system of two O-ring and one spring was designed (Fig. 2), and the rotor response was dramatically improved.

Stator (mating ring)
The stator is made of 440C stainless steel. The stator face was heat-treated and hardened to about 62 Rockwell C to reduce damage due to abrasive wear. Finally, polished to the flatness of two Helium light bands. The coning, which is important for the axial and angular stiffnesses of the fluid film, was mechanically induced into the stator face (Fig. 3). Various coning angles can be obtained by adjusting the tightening of the bolts distributed circumferentially throughout the rear holder of the stator.

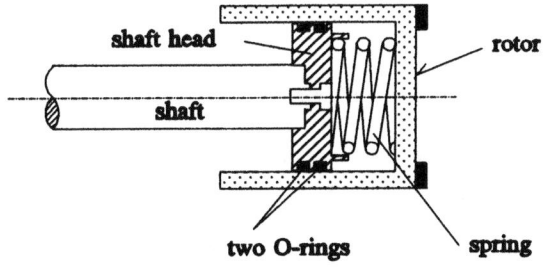

Fig. 2. Shaft head design

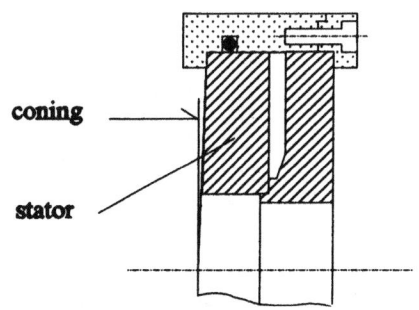

Fig. 3. Coning mechanism of the stator

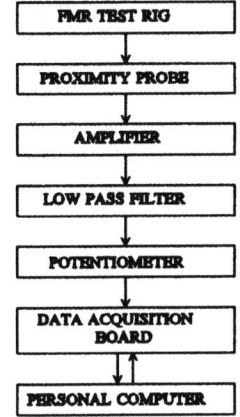

Fig. 4. Flow of signals

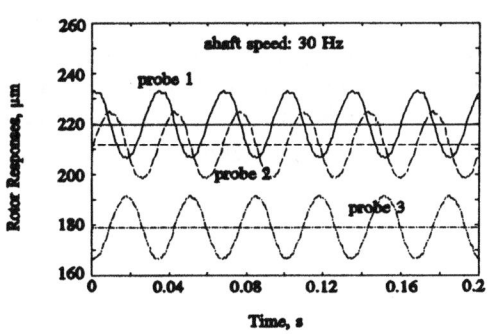

Fig. 5. Rotor Responses without the stator and water

Data Acquisition

The flow of signals in test set-up is shown in Fig. 4. A low pass filter with a cut-off frequency of 1000 Hz was used to eliminate high frequency cross-talk noises among the probes. A potentiometer was used to drop the amplified voltage output. The data acquisition system consists of a personal computer and a data acquisition board of maximum 100 KHz sampling rate. The board was driven by a real time data acquisition software.

DATA ANALYSIS

Reference Plane, Rotor and Stator Misalignments

A rotor reference plane perpendicular to the shaft axis needs to be defined to measure the rotor misalignment. Operation of the rig without the stator and water showed that the rotor responses were sinusoidal and synchronous with the shaft speed (Fig. 5). The mean values of the rotor responses defined the rotor reference plane. Then, the initial rotor misalignment, γ_{ri}, was measured by the probes with respect to the rotor reference plane under the same conditions, except that the shaft was stationary that time. The fixed stator misalignment, γ_s, was measured with respect to the rotor reference plane by the following procedure: First, the stator was placed in the test rig. Then, full contact between the rotor and stator was imposed by pressurising the rotor chamber. At that state, the rotor misalignment as measured by the probes assumed the same misalignment of the stator, which was γ_s.

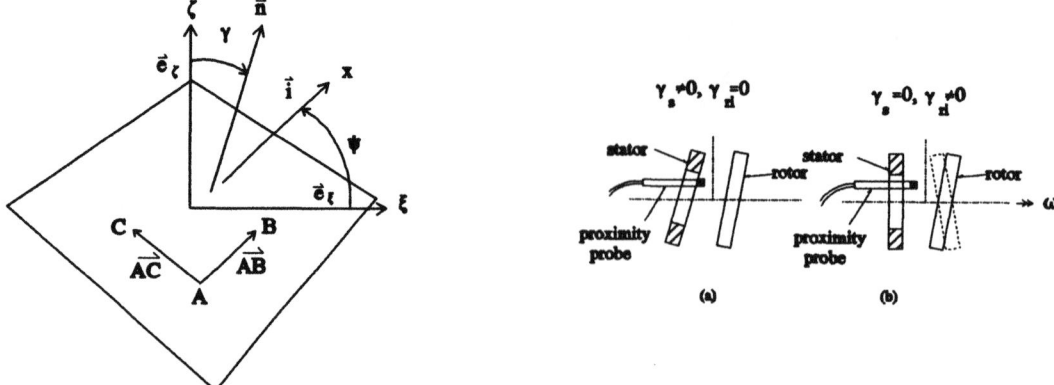

Fig. 6. Misalignment and precession of a plane

Fig. 7. Models of rotor responses to the misalignments

Calculation of Misalignment and Precession

Referring to Fig. 6 and based on vector algebra, misalignment and precession of the rotor can be found. ξ and ζ are the inertial reference axes where the latter coincides with the axis of shaft rotation. A, B, and C are points in the rotor plane as measured by the three probes. A unit out-normal vector of the plane, $\vec{n}$, is defined by

$$\vec{n} = \frac{\vec{AB} \times \vec{AC}}{|\vec{AB} \times \vec{AC}|} \tag{3}$$

The misalignment of the plane, γ, is obtained by

$$\vec{e}_\zeta \cdot \vec{n} = \cos\gamma \quad ; \quad \therefore \gamma = \cos^{-1}(\vec{e}_\zeta \cdot \vec{n}) \tag{4}$$

The nutation axis, x, is defined by

$$\vec{i} = \frac{\vec{e}_\zeta \times \vec{n}}{|\vec{e}_\zeta \times \vec{n}|} \tag{5}$$

The precession angle, ψ, is the angle between ξ and x, and is obtained by

$$\vec{e}_\xi \cdot \vec{i} = \cos\psi \quad ; \quad \therefore \psi = \cos^{-1}(\vec{e}_\xi \cdot \vec{i}) \tag{6}$$

Separation of Total Rotor Misalignment into Components

The total rotor misalignment vector, $\vec{\gamma}_r$, is the vector sum of two components

$$\vec{\gamma}_r = \vec{\gamma}_{rs} + \vec{\gamma}_{rl} \tag{7}$$

In order to compare the experimental rotor responses with the theoretical transmissibilities of Equation (2), separation of $\vec{\gamma}_r$ into its components, $\vec{\gamma}_{rs}$ and $\vec{\gamma}_{rl}$, is necessary. A model of the rotor response, γ_{rs}, to γ_s in the absence of γ_{ri} is illustrated in Fig. 7a. Since γ_s is the constant magnitude space-fixed angle, γ_{rs} is also of the same nature, i.e., constant magnitude space-fixed angle. A model of the rotor response, γ_{rl}, to γ_{ri} in the absence of γ_s is illustrated in Fig. 7b. Since γ_{ri} is rotating at a shaft speed and is of constant magnitude, γ_{rl} is also of the same nature, i.e., rotating at the shaft speed and of constant magnitude. It can be recognised that the total displacements, $d_{r,j}$ (j = 1, 2, 3), of the rotor at the three probes positions are the sums of constant displacements, $d_{rs,j}$, to γ_s and oscillating displacements, $d_{rl,j}$, to γ_{rl};

$$d_{r,j}=d_{rs,j}+d_{rl,j}=d_{rs,j}+|d_{rl}|\cos(\omega t-\phi_j) \ , \ j=1,2,3 \tag{8}$$

where ϕ_j is a phase angle. The mean values of $d_{r,j}$ are $d_{rs,j}$;

$$d_{rs,j}=\frac{1}{N}\sum_{k=1}^{N} d_{r,jk} \ , \ j=1,2,3 \tag{9}$$

where N is the number of data points in one shaft revolution. The subtracted values of $d_{r,j}$ by $d_{rs,j}$ are $d_{rl,j}$;

$$d_{rl,j}=d_{r,j}-d_{rs,j} \ , \ j=1,2,3 \tag{10}$$

Then, γ_{rs} and γ_{rl} can be calculated from $d_{rs,j}$ and $d_{rl,j}$, respectively, using the radial locations of the probes.

Seal Clearance

Since the fluid film stiffness and damping coefficients of the FMR seal are functions of the seal clearance, the dynamic behaviour of the FMR seal will be affected by it. Thus, controlling the seal clearance at a prescribed value during experiment is important. Etsion and Constantinescu (1984) used a simplified leakage equation to calculate the seal clearances of the noncontacting FMS seal. That expression is valid for the FMR seal as well. Hence,

$$Q^*=\frac{\pi p_w}{6\mu}C^3 Q \ ; \ \ Q=\frac{R_m}{1-R_i}[1+\frac{3}{2}\beta(1-R_i)] \tag{11}$$

where Q^* is the leakage, p_w is the water pressure, C is the seal clearance, and Q is the dimensionless leakage. Even if a relatively large error of Q is present, the error will affect much less the value of C when Equation (11) is used. Equation (11) will be used to indirectly calculate the seal clearances of the noncontacting FMR seal test rig.

RESULTS AND DISCUSSIONS

Operation of the test rig was performed at a water pressure of 0.2068 MPa, and a coning angle of 0.0112 rad, and shaft speeds of 600, 900, 1200, 1500, and 1800 rpm. The leakage was monitored by the flow meter and kept at 3 l/h (0.833 cc/s) by controlling air pressure in the rotor chamber. The seal clearance was found to be 3.75 μm by using Equation (11). The misalignment inputs, γ_s and γ_{rl}, were 7.1883×10^{-4} and 5.2156×10^{-4} rad, respectively.

A time response of the rotor, $d_{rl,j}$, at one probe position due to γ_{rl} is shown in Fig. 8. $d_{rl,j}$ was obtained after subtracting the mean value, $d_{rs,j}$, due to γ_s from $d_{r,j}$. Its power spectral density (Fig. 9) has a big spike at the shaft speed of 30 Hz (1800 rpm), and has negligible values at other frequencies. The corresponding trajectory of the rotor misalignment vector, $\vec{\gamma}_{rl}$, as a function of time is shown in Fig. 10. It was obtained after filtering out high frequency components. In Fig. 10, the distance from an origin to a point on the trajectory is the magnitude, and the angle between the line connecting the origin to the point and the ξ-axis is the precession. The order of γ_{rl} was 10^{-4} rad and this justifies the treatment of γ_{rl} as the vector. From Figs. 9 and 10, it can be noticed, respectively, that the experimentally obtained $\vec{\gamma}_{rl}$ has the same rotating speed as the shaft speed and constant magnitude. This is exactly what was predicted by the theoretical analysis. The experimental static (γ_{rs}/γ_s) and dynamic (γ_{rl}/γ_{rl}) transmissibilities at different shaft speeds are summarised and compared with the theoretical counterparts in Table 1.

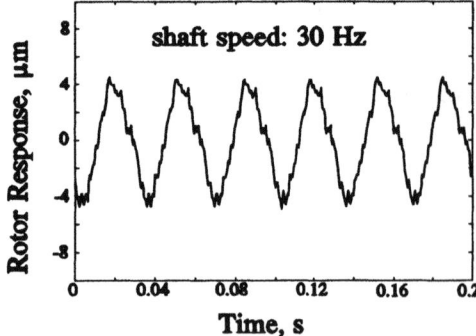

Fig. 8. Rotor response to the initial rotor misalignment

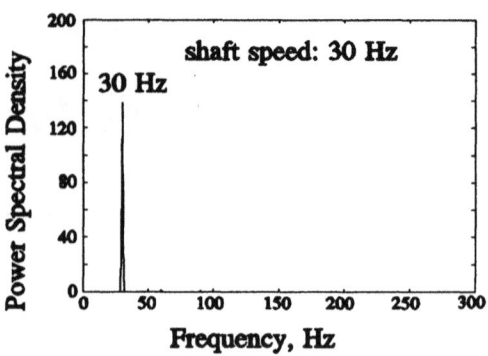

Fig. 9. Power spectral density of the rotor response

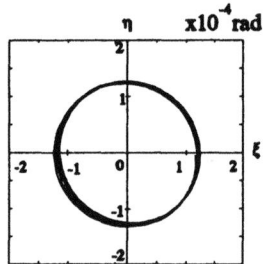

Fig. 10. Trajectory of the rotor misalignment vector

Table 1. Experimental and theoretical transmissibilities at p_w = 0.2068 MPa and C = 3.75 μm

Shaft Speed (rpm)	γ_{rs}/γ_s		γ_{rl}/γ_{ri}	
	Exp.	Theo.	Exp.	Theo.
600	0.6916	0.7602	0.2762	0.2392
900	0.6769	0.7595	0.2323	0.2398
1200	0.7013	0.7593	0.2459	0.2397
1500	0.7519	0.7593	0.2340	0.2394
1800	0.6676	0.7594	0.2442	0.2388

CONCLUSIONS

In order to experimentally investigate the dynamic behaviour of the noncontacting FMR mechanical face seal, a test rig was designed and built. Features of the rig design and methods required to analyse experimental data were introduced. Operation of the rig and data analysis indicated that the qualitative and quantitative behaviours of the FMR seal agreed well with theoretical predictions. Since the inertia ratio of the rotor was less than one, the dynamic instability was not a problem (Green, 1990).

The theoretical model of the FMR seal considered by Green (1989 and 1990) was based on the assumption that the center of mass of the rotor is fixed. However, the rotor of the test rig was mounted on the shaft and thus, its center of mass cannot be considered fixed anymore. In order to estimate the effect of the shaft dynamics on the rotor response, coupling of the seal and shaft dynamics needs to be performed. The coupling should include the fluid film and support rotor dynamic properties, and the misalignment forcing inputs.

ACKNOWLEDGMENT

This work was supported in part by the National Science Foundation under Grant Number MSM-8619190. This support is gratefully acknowledged. The authors would like to thank Mr. Ariel Trau and Dr. Scott Bair who helped in the design and construction of a prototype test rig. Mr. Trau spent his sabbatical leave from the Israel Armament Development Authority at Georgia Institute of Technology.

REFERENCES

Allaire PE (1984) Noncontacting Face Seals for Nuclear Applications-A Literature Review. Lubrication Engineering, Vol. 40, 6, pp 344-351.

Etsion I, Burton RA (1979) Observation of Self-Excited Wobble in Face Seals. ASME Trans. Journal of Lubrication Technology, Vol. 101, 4, pp 526-528.

Etsion I, Sharoni A (1980) Performance of End-Face Seals with Diametral Tilt and Coning--Hydrostatic Effects. ASLE Trans., Vol. 23, 3, pp 279-288.

Etsion I (1980) Squeeze Effects in Radial Face Seals. ASME Trans. Journal of Lubrication Technology, Vol. 102, 2, pp 145-152.

Etsion I (1982) A Review of Mechanical Face Seal Dynamics. The Shock and Vibration Digest, Vol. 14, 4, pp 9-14.

Etsion I (1985) Mechanical Face Seal Dynamics Update. The Shock and Vibration Digest, Vol. 17, 4, pp 11-15.

Etsion I (1991) Mechanical Face Seal Dynamics 1985-1989. The Shock and Vibration Digest, Vol. 23, 4, pp 3-7.

Etsion I, Constantinescu I (1984) Experimental Observation of the Dynamic Behaviour of Noncontacting Coned-Face Mechanical Seals. ASLE Trans., Vol 27, 3, pp 263-270.

Green I, Etsion I (1985) Stability Threshold and Steady-State Response of Noncontacting Coned-Face Seals. ASLE Trans., Vol. 28, 4, pp 449-460.

Green I, Etsion I (1986) Pressure and Squeeze Effects on the Dynamic Characteristics of Elastomer O-Rings Under Small Reciprocating Motion. ASME Trans. Journal of Tribology, Vol. 108, 3, pp 439-445.

Green I (1989) Gyroscopic and Support Effects on the Steady-State Response of a Noncontacting Flexibly Mounted Rotor Mechanical Face Seal. ASME Trans. Journal of Tribology, Vol. 111, pp 200-208.

Green I (1990) Gyroscopic and Damping Effects on the Stability of a Noncontacting Flexibly-Mounted Rotor Mechanical Face Seal. Dynamics of Rotating Machinery, Hemisphere Publishing Company, pp 153-173.

Metcalfe R (1981) Dynamic Tracking of Angular Misalignment in Liquid-Lubricated End-Face Seals. ASLE Trans., Vol. 24, 4, pp 509-506.

Metcalfe R (1982) Dynamic Whirl in Well-Aligned, Liquid-Lubricated End-Face Seals with Hydrostatic Tilt Instability. ASLE Trans., Vol. 25, 1, pp 1-6.

Sehnal J, Sedy J, Zobens A et al. (1983) Performance of the Coned-Face End Seal with regard to Energy Conservation. ASLE Trans., Vol. 26, 4, pp 415-429.

Sharoni A, Etsion I. (1981) Performance of End-Face Seals with Diametral Tilt and Coning--Hydrodynamic Effects. ASLE Trans., Vol. 24, 1, pp 61-70.

The relevance of the dynamic behaviour of the supporting structure in calculating critical speeds of multistage centrifugal pumps with interstage seals

N.Bachschmid *, K.L.Cavalca **, F.Cheli *
(*) Dipartimento di Meccanica, Politecnico di Milano
(**) PhD student at Dipartimento di Meccanica, Politecnico di Milano

ABSTRACT

The dynamic behaviour of a six-stage centrifugal pump rotor is calculated with different models of the supporting structure. The differences in the obtained results show that an accurate modelling of the supporting structure is necessary if the dynamic properties of the seals are considered.

1. INTRODUCTION

This paper deals with the well-known problem of the influence of the seals on the rotor dynamic behaviour (see i.e. [1], [2], [3]).

In rotor-supporting structure systems with oil-film bearings and without intermediate seals, the rotor supporting structure interaction, generally, leads to a more or less consistent lowering of the critical speed values with respect to the values calculated by neglecting the influence of the supporting structure and, sometimes, also in the appearing of the so-called "foundation critical speeds" (see i.e. [4]).

In standard critical speed calculations generally the presence of the supporting structure is considered by introducing in the rotor model or an appropriate spring stiffness or an appropriate spring-damper-mass (1 d.o.f.) system in correspondence of each bearing: the results of these calculations are, in most cases, satisfactory.

In the case of a rotor with intermediate seals, a dynamic connection between rotor and stationary parts (casing and foundation) exists not only at bearing locations (in position close to the two ends of the shaft) but also at each intermediate seal location. The interaction between rotor and stationary parts is therefore much stronger.

The aim of this paper is to show, by means of a numerical simulation, that the dynamic behaviour of the rotor can be realistically predicted only if the supporting structure is accurately modelled. Instability problems are neglected and only forced vibrations are calculated: a maximum in the frequency response curve represents a critical speed.

A series of numerical results are presented and analyzed. Firstly the behaviour of a centrifugal pump rotor supported by two multilobe bearings and a foundation modelled by two single d.o.f. systems is simulated in order to observe the rotor behaviour. Afterwards the behaviour of the same rotor, supported by the two bearings with 13 cylindrical plain seals between it, is simulated in two cases: modelling the foundation with 13 more one d.o.f. systems in correspondence of the seals and modelling the foundation by a finite element method and a modal reduction.

Significant changes in the system behaviour are observed in presence of the seals and an accurate modelling of the supporting structure is shown to be very important.

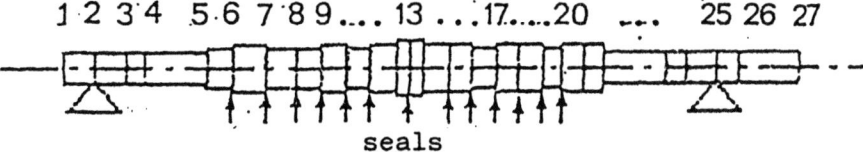

Fig.1 Scheme of the rotor, bearings and seals.

2. DESCRIPTION OF THE PUMP AND THE FOUNDATION

A six stage axially-split pump is supported by two multilobe bearings: 13 cylindrical plain seals between the bearings will be also considered. The rotor shaft is modelled by finite element method utilizing cylindrical beam elements. All other parts of the pump fixed on the shaft are considered like additional masses on the elements, neglecting their stiffness and considering only their inertial mass effect.

Fig.1 shows a scheme of the rotor supported by the two bearings and the seals.

Stiffness and damping characteristics of the oil-film are evaluated for different speeds of rotation, taking into account the load on each bearing. Some results can be seen in fig.2.

The seal stiffness are calculated considering the rotor perfectly centred with respect to the seal which is, therefore, unable to carry any load. Short seal theory is adopted, for its simplicity, and the method introduced by Allaire [5] is used for calculating the stiffness coefficients. Inlet whirl is not taken into account. The pressure drop, which is variable with the rotational speed of the rotor, is here, for the sake of simplicity, considered constant. The damping factors of the seals are neglected: fig.3 shows a scheme of the seals. The stiffness curves of seal n.5 are shown in fig.4.

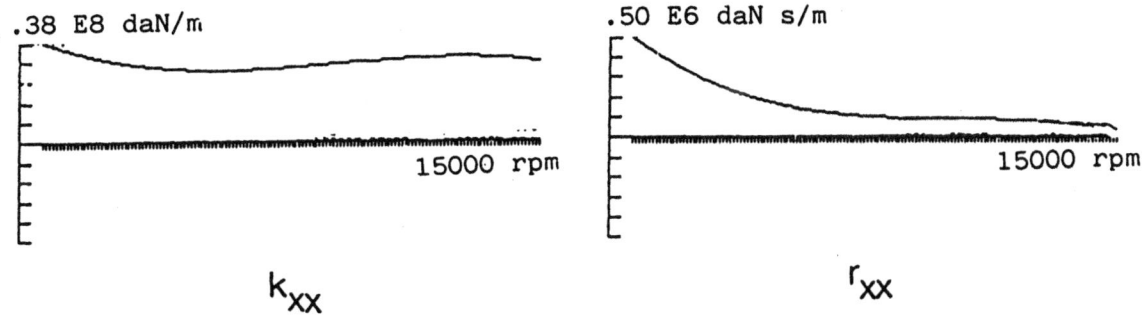

Fig.2 Oil-film direct stiffness and damping coefficients versus speed (bearing n.1, diameter D=90mm, width L=20mm).

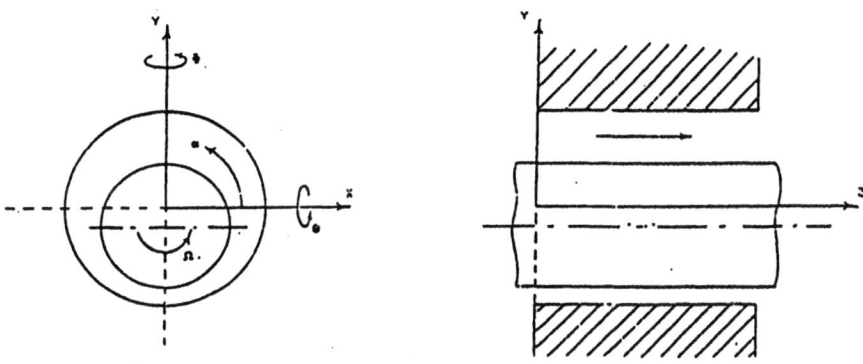

Fig.3 Scheme of the seals.

160

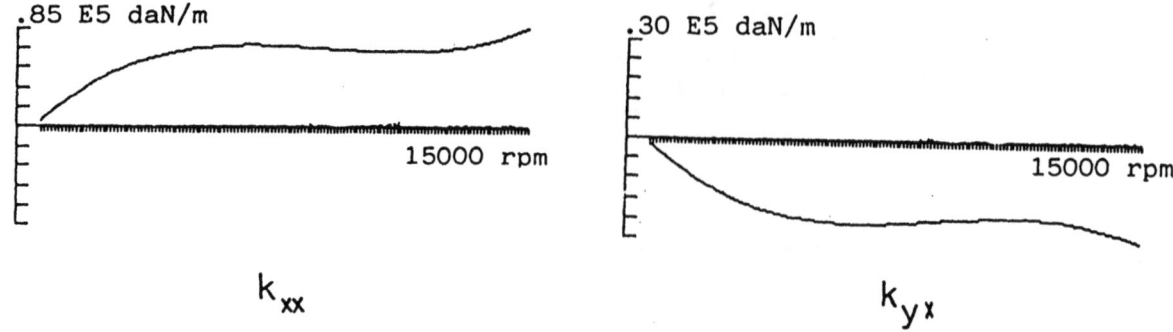

Fig.4 Direct and cross-coupled stiffness coefficients of the seal n.5 versus speed: diameter D = 65 mm, lenght L = 45 mm, radial clearance δ=270 μm, pressure drop p=900 kPa.

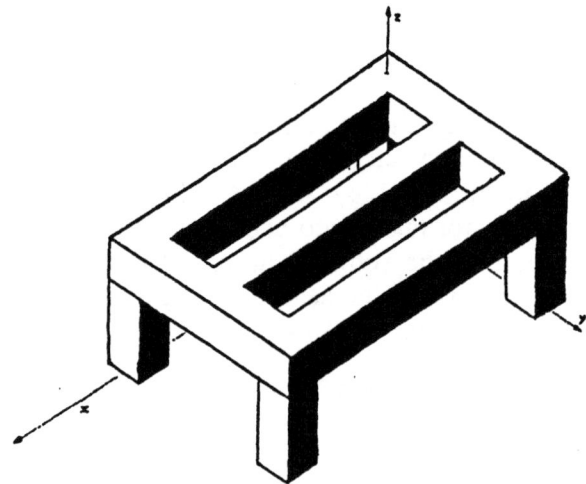

Fig.5 The foundation.

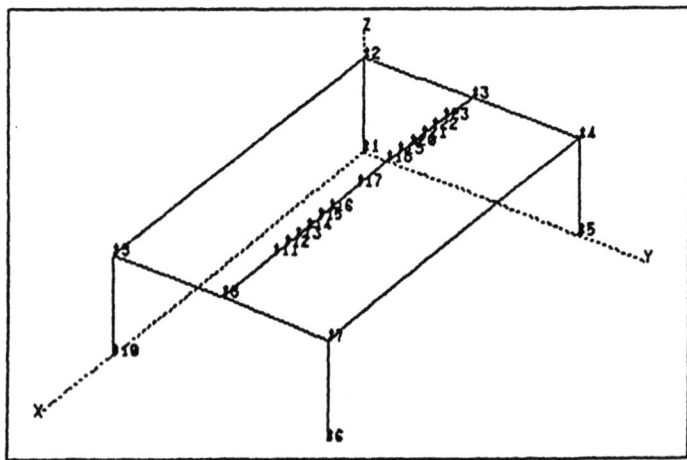

Fig.6 Foundation finite beam element model.

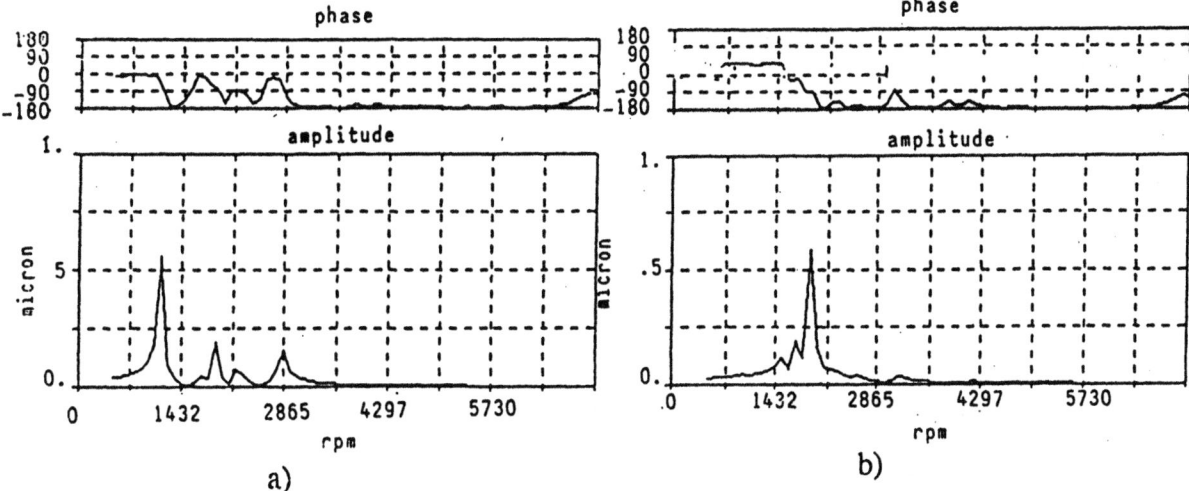

Fig.7 Frequency response on the node 17: (a) horizontal and (b) vertical direction.

Horizontal and vertical direct coefficients are equal, due to the symmetry and null-load on the seal. The cross-coupled coefficients are equal and opposite in sign.

An internal damping of the rotor is also considered and a damping coefficient matrix equal to 0.0005 of the rotor stiffness matrix is assumed.

The foundation is shown in fig.5. The f.e. model has 24 beam elements and 23 nodes as shown in fig.6. The model, taking into account also stiffness and masses of the pump casing, represents the complete supporting structure.

The frequency response curves, corresponding to an excitation by means of unity forces (10 N), applied, in horizontal and vertical directions, on node 17, evaluated at the node 17 of the foundation in horizontal (Y) and vertical (Z) directions, are plotted in fig.7.

The minimum dynamic stiffness, at resonance, is about 0.2×10^8 N/m and outside the resonance region, the dynamic stiffness is above 2.0×10^8 N/m, both in horizontal and vertical directions. As can be seen, although the foundation is extremely simple, the frequency response is similar to real foundation-casing systems in terms of natural frequencies and dynamic stiffness.

The dynamic effects of the foundation on the complete rotor-bearings-seals-foundation system is taken into account by a modal truncation method (described in [6]). We consider, as independent variables, a set of hybrid coordinates: the rotor node displacements $\underline{X}_r$ and the foundation modal co-ordinates $\underline{q}$: the equation of motion become:

$$\begin{bmatrix} [M_r] & 0 \\ 0 & [m_f] \end{bmatrix} \begin{Bmatrix} \ddot{X}_r \\ \ddot{q} \end{Bmatrix} + \left(\begin{bmatrix} [R_r] & 0 \\ 0 & [r_f] \end{bmatrix} + [R_{roq}] \right) \begin{Bmatrix} \dot{X}_r \\ \dot{q} \end{Bmatrix} +$$

$$+ \left(\begin{bmatrix} [K_r] & 0 \\ 0 & [k_f] \end{bmatrix} + [K_{roq}] \right) \begin{Bmatrix} X_r \\ q \end{Bmatrix} = \begin{Bmatrix} E_e \\ 0 \end{Bmatrix} \qquad (1)$$

where $[M_r]$, $[K_r]$ and $[R_r]$ are mass, stiffness and damping matrices of the rotor, $[m_f]$, $[k_f]$, $[r_f]$ are mass, stiffness and damping diagonal matrices related to the considered model of the foundation (containing modal parameters) and $[K_{roq}]$ and $[R_{roq}]$ are equivalent stiffness and damping matrices of the linearized oil-film and seal forces, defined by means of the corresponding mode shapes. Eq.(1) is solved in frequency domain, being the external rotor forces $\underline{E}_e = \underline{E}_{eo} e^{i\Omega t}$, and Ω the rotor speed velocity.

The modal reduction of the foundation results in 3 vertical and 4 horizontal modes as follows:

Vertical modes (rpm)	Horizontal modes (rpm)
1200	1200
2000	2000
2400	2300
	3000

Tab.I Vertical and horizontal natural frequencies of the foundation.

3. FREQUENCY RESPONSE OF THE ROTOR

The frequency response due to unbalance forces is calculated. Different situations are considered:

3.1 Rotor without seals

In this case, the rotor is supported by the multilobe bearings and the foundation is modelled like a 1 d.o.f. system which presents mass, stiffness and damping characteristics as follows:

mass	$M = 10\,Kg$		
vertical stiffness	$K_{xx} = 1.0 \times 10^9\ N/m$	vertical damping	$R_{xx} = 1.0 \times 10^5\ Ns/m$
lateral stiffness	$K_{yy} = 0.5 \times 10^9\ N/m$	lateral damping	$R_{yy} = 0.5 \times 10^5\ Ns/m$

Table II: 1 d.o.f. foundation scheme.

Above values can be considered as standard assumptions in the absence of detailed calculations or experimental results.

The first and second critical speeds are visualized in fig.8. The considered unbalances on the rotor are following: 1.25 Kgmm on node 6 and 2.50 Kgmm on node 20 with the same phase.

As can be noticed, the first critical speed is at approximately 2500 rpm and the second critical speed is positioned around 14000 rpm. The dynamic magnification factors are rather high (4÷5) and the critical speeds are clearly recognizable.

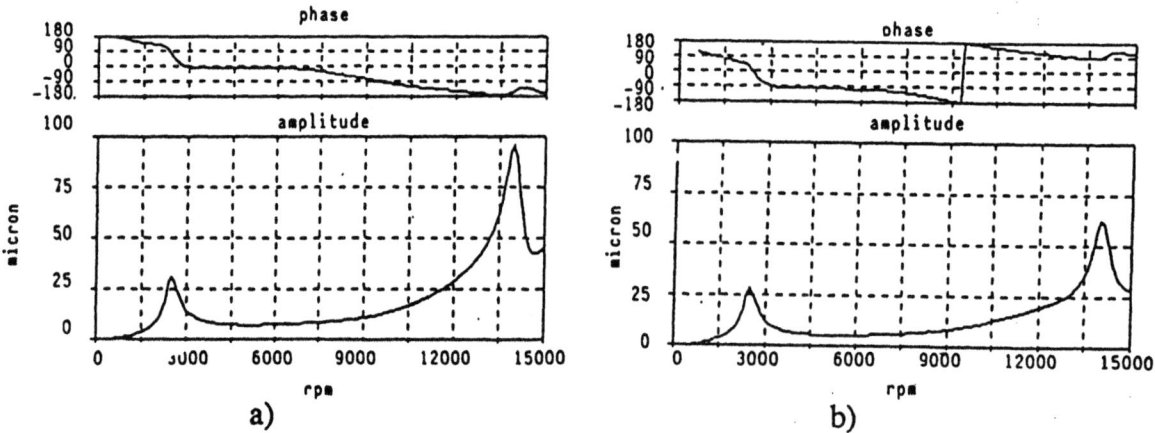

Fig.8 Rigid foundation, no seals: frequency response at the bearing 1: (a) horizontal and (b) vertical direction.

3.2 Rotor frequency response with seals

This case is similar to the previous one with respect to the foundation. The only difference is the presence of the seals. In correspondence of each seal a 1 d.o.f. system is introduced, with the same parameter numerical values as those of the bearings.

Fig.9a, relative to first bearing of the pump, shows a sharp peak in correspondence of 14000 rpm: the corresponding mode shape (fig.9b) has a rather high amplitude at the overhung coupling hub and one node between the bearings, and is, therefore, in some way, similar to the second mode without seals.

A further very smooth peak can be noticed just below 6000 rpm at the first bearing: the corresponding mode shape is similar to the first mode. But if we look at the displacements at the first seal, given in fig.10a and at the second bearing in fig.11, the only maximum below 14000 rpm is found at 7000 rpm in the seal; the corresponding deflection shape (fig.10b) is still similar to the first mode.

The first critical speed is therefore not clearly recognizable and seems to be or shifted towards high rotating speeds or completely cancelled.

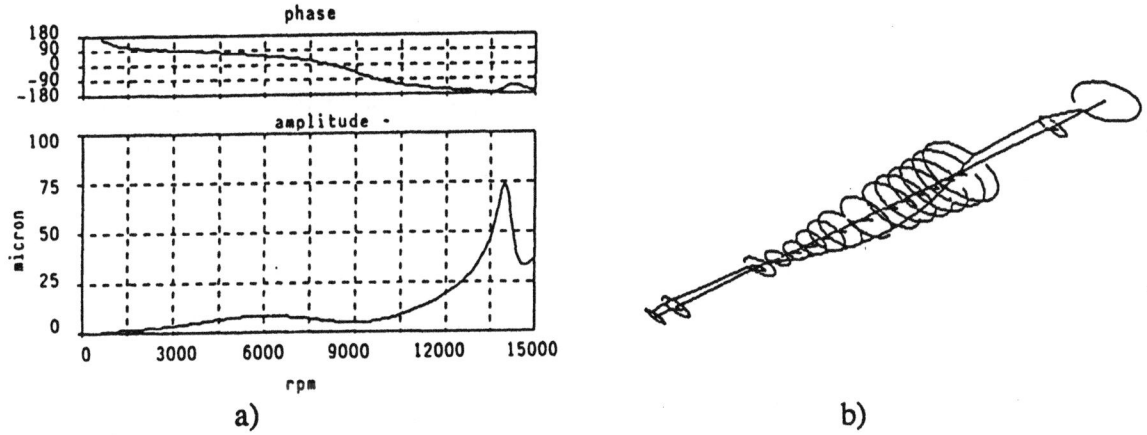

a)   b)

Fig.9 Frequency response of the rotor with seals at the first bearing: (a) horizontal amplitude, (b) deflection shape at 14000 rpm.

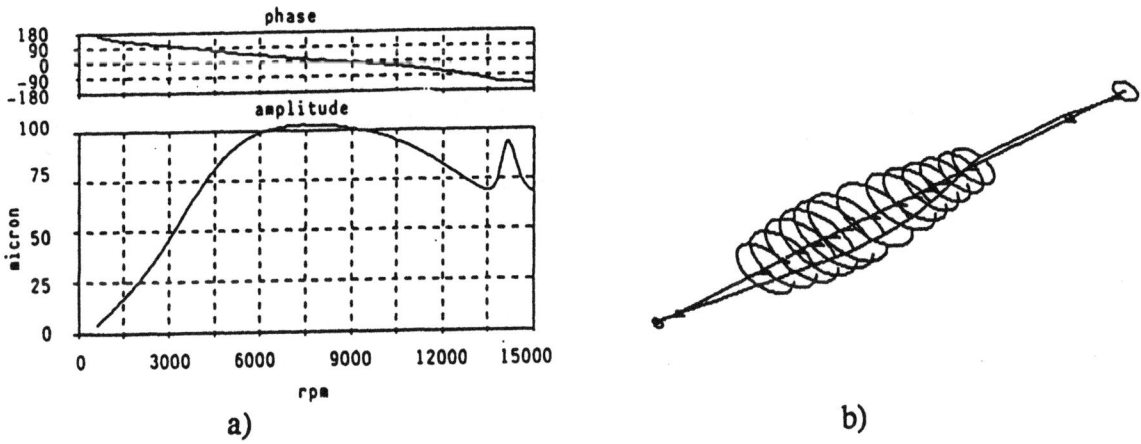

a)   b)

Fig.10 Frequency response of the rotor at the first seal: (a) horizontal amplitude, (b) deflection shape at 7000 rpm.

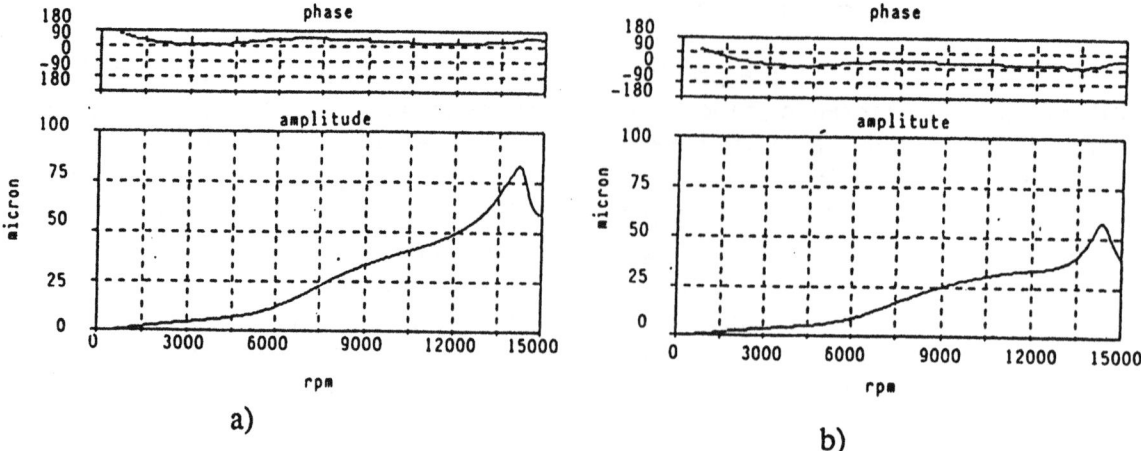

a)

b)

Fig.11 Frequency response of the rotor at the second bearing: (a) horizontal amplitude, (b) vertical amplitude.

If we consider now the rotor with seals and with the flexible continuous foundation, modelled as a 7 d.o.f. system by a modal analysis, we obtain more realistic frequency response diagrams.

In fig.12 the vibrations at bearing 1 are represented where almost three peaks are recognizable: the last one at about 14000 rpm is the II mode critical speed at same speed as before but with a very clear mode shape (fig.13a). The first peak at 1100 rpm in horizontal direction is a pure "foundation critical speed" as shown by the deflection mode represented in fig.13b; the first peak at 1900 rpm in vertical direction could be considered as a mixed mode (fig.13c). The deflection shape of the rotor is already similar to the first mode shape, but is influenced by the foundation mode at same speed. The classical first mode shown in fig.13d corresponds to the critical speed of 2700-2900 rpm which is evidenced in the frequency response curve by a peak.

Summarizing, it can be noticed that the first critical speed exists and is split into two modes (at 1900 and 2900 rpm) both not very far from original critical speed of 2500 rpm. Furthermore, if we look at the relative displacements in the seals, we get completely different values in the two cases: with 1 d.o.f. foundation models we have in the first seal about 100 μm at 7500 rpm and 93 μm at 14000 rpm, with the flexible foundation we have respectively 52 μm and 374 μm. So with the poor 1 d.o.f. model a critical situation may be suspected more likely at 7500 rpm than 14000 rpm, where instead a heavy rub could occur according to the flexible foundation results.

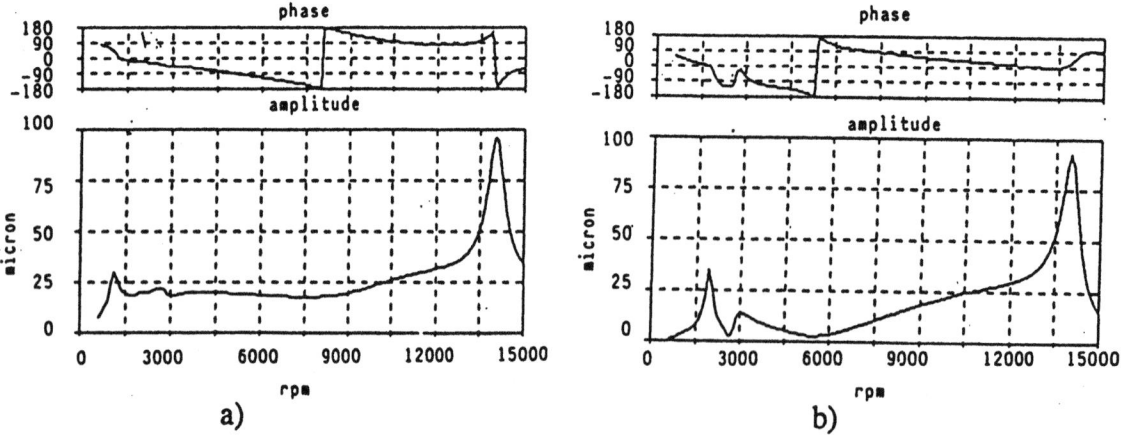

a)

b)

Fig.12 Frequency response of the rotor with seals (flexible foundation) at the first bearing: (a) horizontal and (b) vertical direction.

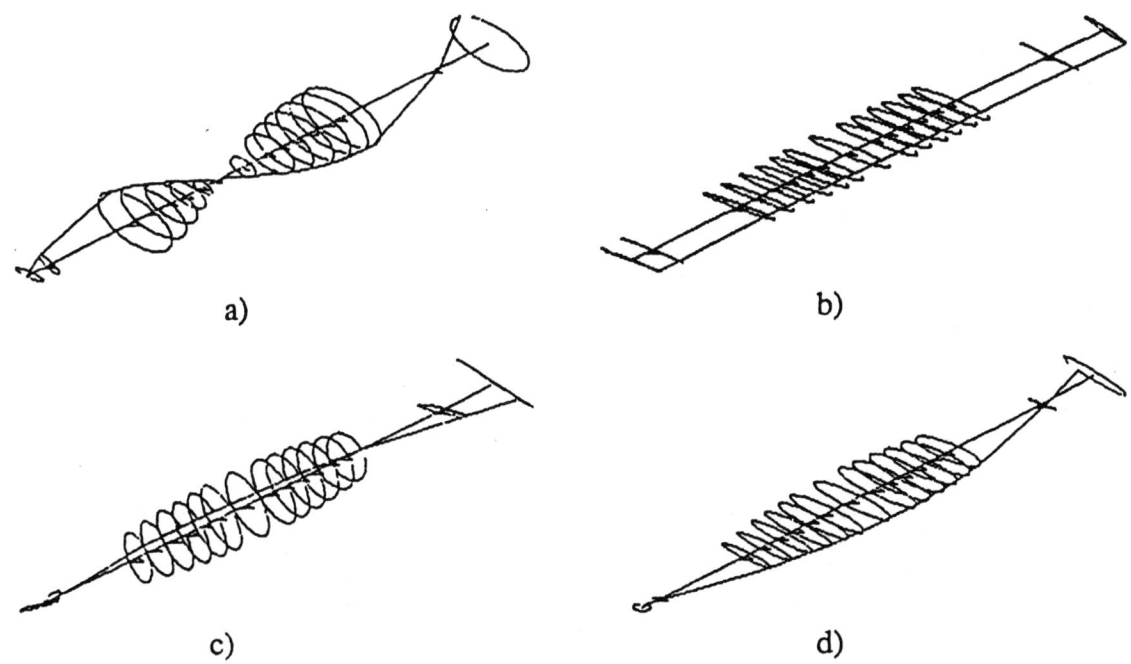

a)   b)

c)   d)

Fig.13 Frequency response with seals (flexible foundation): critical modes: a) at 14000 rpm, b) at 1100 rpm, c) at 1900 rpm, d) at 2900 rpm.

4. CONCLUSIONS

The numerical simulation shows that in case of rotors with many seals between the bearings, which is common in multistage centrifugal pumps, an accurate modelling of the supporting structure (casing+foundation) is extremely important in order to recognize critical speeds and calculate frequency responses due to unbalance. Although stability problems were not considered in this paper, it is reasonable to suppose that above conclusions are valid also in case of instability analysis.

ACKNOWLEDGMENTS

This study received a grant by MURST (Italy), by CNR (Italy) and CNPq (Brazil).

REFERENCES

[1] Nordmann R., "Seal properties", Rotordynamics 2 - Problems in turbomachinery, Springer Verlag, 1988, pp.153-173.
[2] Black H.F., "Effects of hydraulic forces in anular pressure seals on the vibrations of centrifugal pump rotor", JMES, Vol.II, N.2, 1969, pp.206-213.
[3] Childs D.W. "Finite length solutions for rotordynamic coefficients of turbulent anular seals", ASME 82-Lub-42.
[4] Bachschmid N., Pizzigoni B., DiPasquantonio F., "A method for investigating the dynamic behaviour of a turbomachinery shaft on a foundation", ASME paper 77-DET-16, DET Conference, Chicago (USA), Sept. 1977.
[5] Allaire P.F., Lee C.C., Gunter E.J., "Dynamics of short excentric plain seals with high axial Reynolds number ", Journal of Spacecraft and Rockets, Vol.15, N.6, Nov-Dec 1978, pp.342-347.
[6] Cheli F., Lucchesi Cavalca K., Dedini F., Vania A., "Dynamical behaviour analysis of rotor structure systems by a modal truncation method", 9th International Modal Analysis Conference IMAC, Florence, 1991.

The FEM Development of a General, Hollow Conical Shaft Element for Use in Transfer Matrix Analysis of Rotor-Bearing Systems Part I: Formulation of the Model

Mark S. Darlow and Thomas J. Bievenue

Technion - Israel Institute of Technology

ABSTRACT

Traditional beam and shell deformation models have been found to be inadequate for the analysis of shafts with tapered sections. Consequently, a procedure has been developed wherein a static finite element formulation is used to extract flexibility terms which can be used in transfer matrix based dynamic analysis. Such an approach is more computationally efficient than utilising a full, dynamic finite element analysis.

Part I of this paper presents the formulation of the Finite Element Method (FEM) model used for static deformation analysis of hollow, conical sections. The model proved to be non-dimensional in a geometrical sense as well as in a material sense, except for the small effect of varying Poisson's ratio. The parameters used are a non-dimensional thickness, length and taper angle. Element bending flexibility coefficients were calculated as functions of the geometrical parameters. The relationships between these flexibility values and transfer matrix analysis flexibility terms were established and the corresponding flexibility values were calculated. The model and procedure for extracting these flexibility values is described in sufficient detail for the reader to recreate the model and calculate the flexibility coefficients for any given set of geometrical parameters.

The FEM modelling procedure was used to generate flexibility coefficients for a wide range of geometrical parameters. These results are tabulated in Part II of this paper, which also presents experimental correlation results.

INTRODUCTION

Currently, there are many different methods for handling conical sections in rotordynamic analysis programs. The most commonly used method is to break the conical section into a series of cylindrical sections of increasing (or decreasing) diameter. Other methods include the use of analytical theories based on beam and/or shell analysis. Four methods were compared in this work; the stepped model, the shell membrane theory by Kinerk [1], the transfer matrix beam theory by Darlow [2] and the static finite element approach introduced by Vest and Darlow [3] and extended herein. The derivations of these methods and additional details of the results of this work can be found in Bievenue [4].

These four methods will each yield different flexibilities and consequently different predicted critical speeds will result. Previous work by Vest [5] and Vest and Darlow [3,6] concluded that the finite element approach produced flexibilities that greatly improved the accuracy of the predicted critical speeds as compared to the other methods. In this approach, a static finite element analysis is conducted to extract equivalent transfer matrix flexibility terms for a given conical geometry. All dynamic analyses can then conducted using a transfer matrix analysis code which is much more computationally efficient.

The procedure for modelling a conical section, for a FEM analysis, and extracting the flexibility terms is presented herein. The results already generated for a large number of geometries are presented in Part II of this paper. In this way, the reader can use previously generated flexibility values, or interpolate from these, if the geometry of interest is sufficiently close to one of the given models. Alternatively, the reader can construct and evaluate a FEM model based on the geometry of interest. The modeling used here and the results generated are non-dimensionalised so as to maximise their applicability in practical situations.

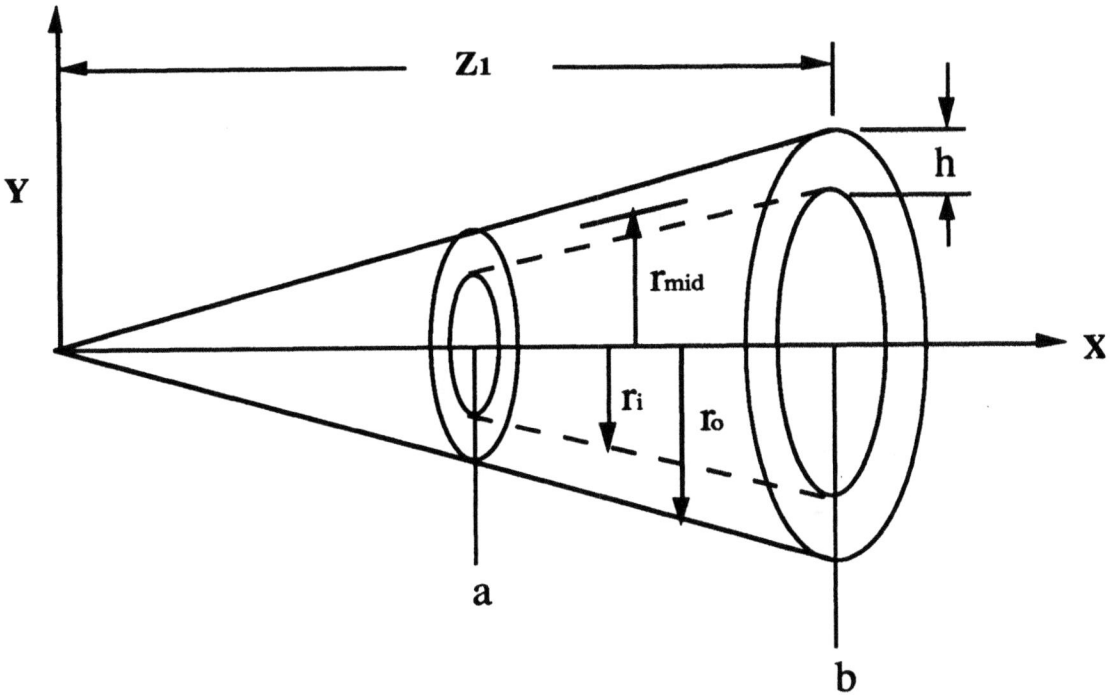

Fig 1. Conical section diagram

CONICAL SECTION GEOMETRICAL DEFINITIONS

The geometry of a hollow, conical section is specified as shown in Fig. 1. The non-dimensional parameters used were η, λ, and θ. r_{mid} is the radius of the conical section halfway between the ends and halfway between the outer and inner radii of the conical section. θ is the taper angle of the conical section. η and λ are defined as the normalised conical thickness and length, respectively. Note that r_{mid} is a dimensional quantity. r_{mid} is the only dimensional quantity needed to create a physical model. The flexibilities are dependent on the value of r_{mid}, but in a known manner (inverse proportionality). Given a set of non-dimensional parameters η, λ, and θ, different models can be generated by changing the value of r_{mid}. The flexibilities for all of these models will be related by the value of r_{mid}. Eqs. (1) and (2) show the relationships for the non-dimensional parameters.

$$\eta = \frac{h}{r_{mid}} \tag{1}$$

$$\lambda = \frac{L}{r_{mid}} \tag{2}$$

Given the four parameters, η, λ, θ and r_{mid}, a conical model can be generated. Using these parameters, Vest [5] examined models for η=0.0625, .125, .25, .5, 1.0, and θ=10, 20, 30, 40 degrees. λ was kept constant at a value of 1.0. r_{mid} was set equal to 25.4 mm (1.0 in.) to take advantage of its relationship with the flexibility coefficients. This set of values yielded 20 combinations of η, λ, and θ which were used to create 20 different models. Bievenue [4] incorporated different values of λ are and considered taper angles up to 80 deg. In all, 134 different models were generated. These models and their corresponding flexibilities are presented in Part II of this paper.

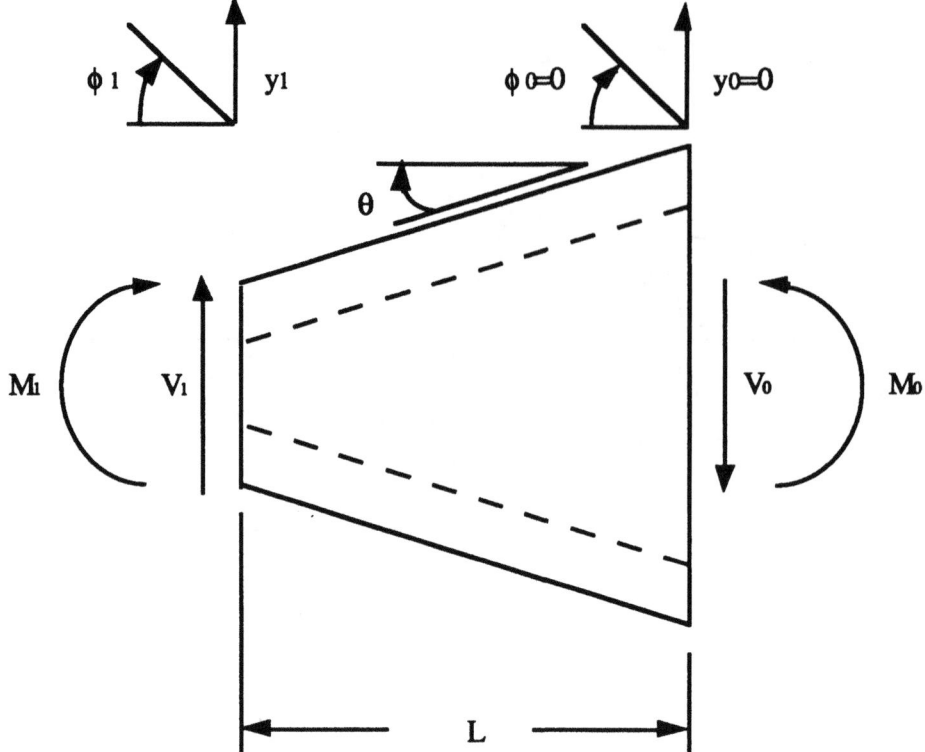

Fig. 2. Conical section free body diagram

FLEXIBILITY REPRESENTATION

Fig. 2 shows a cantilevered, conical-section, free-body diagram with applied moments and shear forces. In a transfer matrix analysis, the state variables at the left end are evaluated as a function of those at the right end. For a lumped-mass representation, this relationship is given by Eq. (3).

$$\begin{Bmatrix} y_1 \\ \phi_1 \\ M_1 \\ V_1 \end{Bmatrix} = \begin{bmatrix} 1 & L & K_4 & K_3 \\ 0 & 1 & K_1 & K_2 \\ 0 & 0 & 1 & L \\ 0 & 0 & 0 & 1 \end{bmatrix} \begin{Bmatrix} y_0 \\ \phi_0 \\ M_0 \\ V_0 \end{Bmatrix} \tag{3}$$

where these K terms represent the deflection and angle at the left (free) end as a function of the moment and shear at the right (constrained) end. Flexibility terms are most easily extracted from the FEM results by evaluating the deflections and angles at the free end for unit values of moment and shear, separately, also at the free end. These flexibilities are represented by:

S_{11} - deflection as a function of shear
S_{12} - deflection as a function of moment
S_{21} - angle as a function of shear
S_{22} - angle as a function of moment

The S terms can be related to the K terms by

$$\begin{Bmatrix} y_1 \\ \phi_1 \end{Bmatrix} = \begin{bmatrix} -(K_3 - LK_4) & K_4 \\ -(K_2 - LK_1) & K_1 \end{bmatrix} \begin{Bmatrix} V_1 \\ M_1 \end{Bmatrix} = \begin{bmatrix} S_{11} & S_{12} \\ S_{21} & S_{22} \end{bmatrix} \begin{Bmatrix} V_1 \\ M_1 \end{Bmatrix} \tag{4}$$

where L is the axial length of the conical section.

CONICAL ORIENTATION

The FEM models were all generated (except for a few test cases) for conical sections with a positive slope. That is, the conical section gets larger in diameter from left to right. For the negative slope case, flexibility terms can be generated from the results of an equivalent positive slope case using the following relationships [4].

$$K_1^* = K_1 \tag{5}$$

$$K_2^* = LK_1 - K_2 \tag{6}$$

$$K_3^* = K_3 \tag{7}$$

$$K_4^* = K_2 \tag{8}$$

$$S_{11}^* = S_{11} - LS_{12} - LS_{21} + L^2 S_{22} \tag{9}$$

$$S_{12}^* = LS_{22} - S_{12} \tag{10}$$

$$S_{21}^* = LS_{22} - S_{21} \tag{11}$$

$$S_{22}^* = S_{22} \tag{12}$$

where the asterisked parameters are those for the negative slope case.

FINITE ELEMENT MODEL

Fig. 3 illustrates the Finite Element model that represents the conical section used herein. Taking advantage of symmetry, only half of the conical section is used. This model is, in most regards, the same as that used by Vest [5] except for a few minor differences. These differences will be pointed out after discussing the attributes of the model.

The model consists of four main sections; two cylindrical sections, a conical section and an end face. Vest concluded that the cylindrical sections were necessary to provide an appropriate elastic boundary condition for the ends of the conical sections. This boundary condition allows the conical ends to warp, as would be expected in realistic situations. This warping allows S_{12} and S_{21} to be different values whereas in the shell and beam theories these two values are equal, due to the assumption that the interface deforms in a planar fashion. Through Vest's finite element work, he found that an appropriate length for the cylindrical sections is the average radius of the interface; the average radius being the average of the inner and outer radii of the cylindrical-conical interface. The end face, shown in Fig. 3, attached to the smaller end of the model is used to provide a stiff loading face. It is necessary to create a stiff face so that when the model is loaded, the end of the cylindrical section does not warp. This does not, however, prevent the ends of the conical sections from warping.

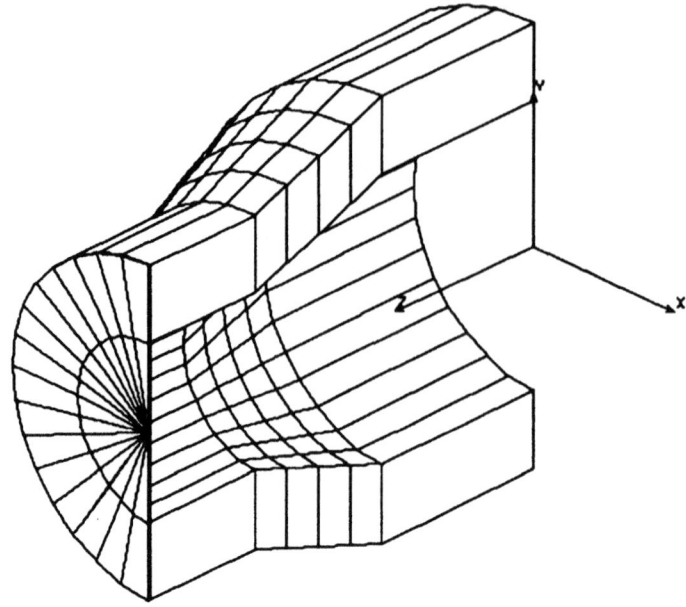

Fig. 3. Finite Element model

Vest used two different types of elements in his model. For the conical and cylindrical sections of the model, 20 noded parabolic solid elements were used; and for the end face, eight noded shell elements were used. For the work described herein, a 20 noded isoparametric solid element was used for the entire model.

The only difference between Vest's model and the model used here is in the construction of the end face. In order to provide a stiff end face using shell elements, Vest made the thickness of these elements equal to 100 times that of the thickness of the rest of the model. To create a similar stiff end face using solid elements it was necessary to choose a proper thickness and an appropriate Modulus of Elasticity. After a certain amount of convergence testing, it was concluded that by making the end face 0.25 mm (0.01 in.) thick and using a Modulus of Elasticity equal to 10^{13} MPa (10^{15} psi) the model yielded similar results to that of Vest's model.

Model Loading

As mentioned above, unity shear and moment loadings are applied to the model. However, since the model is cut in half, the actual shear and moment loads are also cut in half. The deflections and rotations extracted from the deformed model are then the actual S's as given above. The loads are applied to the left end of the model on the end face shown in Fig. 3. It should be noted that the desired load conditions are based on the loads being applied to the end of the conical section, but in this model the loads are actually applied to the end face. This creates a slight problem for the shear load case. Applying a shear load at the end face creates a moment at the cylindrical-conical interface which is not desired. For this reason, a counter-acting moment must be applied to the end face.

Flexibility Calculations

The flexibilities for any given model will simply be the deflections and rotations of the model due to the two load cases. These deflections and rotations are for the left end of the conical section with respect to the right end. The right end of the conical section will also have deflections and rotations since it is not constrained, but attached to the cylindrical section. It is therefore necessary to subtract the right end deflections and rotations from the left end deflections and rotations. The warping at the ends of the conical sections made the calculations sensitive to which nodes were used to calculate the deflections and rotations. The answers were different if the inner nodes were used as opposed to the outer nodes. Vest experimented with different methods of determining the appropriate deflections and rotations, and decided that the best approach was to use the deflections and rotations of the mid-nodes at the ends of the conical sections.

The rotations were determined by averaging the rotations of each mid-node around the circumference of the conical section. The rotation of each mid-node is found by dividing its axial deflection by its corresponding radius, assuming that for small angles $\sin(\theta) \approx \theta$. The total rotation of the left end with respect to the right end is found by subtracting the average rotation of the right end from the average rotation of the left end. Eq. (13) shows the proper method of extracting the rotations.

$$\phi_1 = \left[\sum_1^N \frac{Z_{disp}}{(N)\text{Radius}} \right]_{\text{Left End}} - \left[\sum_1^N \frac{Z_{disp}}{(N)\text{Radius}} \right]_{\text{Right End}} \tag{13}$$

The deflections are found in a similar manner. The vertical deflections of each mid-node around the circumference of the conical section are averaged, and the right end averaged deflection is subtracted from the left end averaged deflection. It is also necessary to subtract out the rigid body deflection of the conical section. This rigid body deflection is simply the average rotation of the right end times the length of the conical section. Eq. (14) shows the proper method of extracting the deflections.

$$y_1 = \left[\sum_1^N \frac{Y_{disp}}{N} \right]_{\text{Left End}} - \left[\sum_1^N \frac{Y_{disp}}{N} \right]_{\text{Right End}} - L \left[\sum_1^N \frac{Z_{disp}}{(N)\text{Radius}} \right]_{\text{Right End}} \tag{14}$$

Using Eqs. (13) and (14) and the flexibility relationships, the flexibilities can be determined for each load case. The value of N in Eqs. (13) and (14) is equal to the number of nodes around the circumference of the conical section.

Convergence of Model

Vest found that only four elements along the conical length are needed for the results to converge. Fig. 3 shows that there are 16 elements around the circumference of the model, four elements along the conical length, one element in the radial direction and only one element along the cylindrical lengths. Vest found that the flexibilities converged for the models he analysed. Since this work is expanding on the range of λ and θ it was necessary to check for the appropriate number of elements needed for convergence. The range of values for η was not going to be expanded so the number of elements in the radial direction did not need to be examined.

Convergence was checked for each change in θ. The most flexible model at each angle was chosen to determine the proper number of elements. The smallest value of η and the largest value of λ would yield the most flexible case and thus thus the corresponding combination of parameters for each angle was evaluated for convergence. Four different cases were examined. Models were created with 4, 8, 12, and 16 (and, in one case, 20) elements along the conical length. The flexibilities of each case were checked to see if they converged.

Table 1 shows the number of elements necessary for a 5% convergence for each taper angle. If too few elements are used the resulting flexibility terms are generally stiffer than they should be.

Table 1 Number of elements for convergence

Taper Angle	# Elements
10	4
20	4
30	12
40	12
50	12
60	16
70	16
80	20

NON-DIMENSIONALISATION OF MODEL RESULTS

While the basic geometric parameters are non-dimensional, the FEM model is still a function of E and r_{mid}. It is desirable to have a fully non-dimensionalised model that can be scaled in size and extended for the use of alternate materials. Thus, it was necessary to establish the relationships between the flexibilities, and E and r_{mid}. For both the beam and shell theories, the flexibilities are inversely proportional to E. Models were built with values of E that were double and half that of the original value of E. The results were as expected. If E was doubled the flexibilities should be cut in half, and the results from the finite element model showed this to be the case. By halving E the corresponding flexibilities should double, and again the finite element model showed this relationship. This demonstrates that the Modulus of Elasticity could be factored out of the flexibility results.

Another model was created to check the effect of doubling the value of r_{mid}. Eqs. (15)-(17) show the relationships that the flexibility terms are expected to have with r_{mid}.

$$S_{11} \propto \frac{1}{r_{mid}} \tag{15}$$

$$S_{12} \text{ \& } S_{21} \propto \frac{1}{r_{mid}^2} \tag{16}$$

$$S_{22} \propto \frac{1}{r_{mid}^3} \tag{17}$$

If r_{mid} is doubled then it is expected that S_{11} would be halved, S_{12} and S_{21} would be quartered, and S_{22} would be eighthed. A model was created with r_{mid}=50.8 mm (2.0 in.) to compare to the model created using r_{mid} of 25.4 mm (1.0 in.) The non-dimensional parameters were kept the same for each model. In this way the relationships given in Eqs. (15)-(17) could be checked. The flexibilities for the two models were calculated and it was found that these relationships were valid. This demonstrates that it is reasonable to factor out r_{mid} as well as E from the flexibility terms. It is then also reasonable to run all of the FE models using one value of r_{mid} and E and then relate the resulting flexibility terms to any value of r_{mid} and E needed. Thus the model is fully non-dimensional.

While Poisson's Ratio is dimensionless, it is also material dependent. Therefore, a study was conducted [4] to evaluate the sensitivity of the flexibility results to changes in Poisson's Ratio. The effects were found to be small; less than 2% for a range of Poisson's Ratio values from 0.29 to 0.33 which encompasses most common shaft materials (i.e., aluminium and steel). Thus , the results can be generally taken to be material independent. However, if a FEM model is to be created for a specific shaft design, it would be advisable to include the correct material properties for that design.

SUMMARY AND CONCLUSIONS

An approach for using static FEM modeling of general, hollow conical sections for use in transfer matrix analysis of rotor-bearing systems has been presented in sufficient detail for the reader to apply this method using commercially available Finite Element and transfer matrix analysis codes. The resulting procedure is much more computationally efficient than conducting dynamic FEM analyses. It is also more accurate than using conventional beam or shell models available in current commercial rotordynamic analysis codes; whether they be based on FEM or transfer matrix approaches. This is because conical sections do not, in general, behave precisely according to either beam or shell theory.

For the reader who cannot, or chooses not, to develop a FEM model for his application but is dealing with a conical geometry sufficiently close to one of those studied by Bievenue [4], a full compilation of numerical results is presented in Part II of this paper, along with the results of a series of experiments conducted to verify the validity of this analytical technique.

REFERENCES

1. Kinerk, H., "Deflection of Cones under Lateral Load and Moment", GE TIS No. R58AAGT708, Sept., 1958.
2. Darlow, M. S., Murphy, B. T., Elder, J. A., and Sandor, G. N., "Extension of the Transfer Matrix Method for Rotordynamic Analysis to Include a Direct Representation of Conical Sections and Trunnions", ASME Transactions, Journal of Mechanical Design, Vol. 102, No.1, Jan. 1980, pp. 122-129.
3. Vest, Todd A. and Darlow, Mark S., "A Modified Conical Beam Element Based on Finite Element Analysis: Experimental Correlations," ASME Transactions, Journal of Vibration and Acoustics, Vol. 112, pp. 350-354, July 1990.
4. Bievenue, T., "Finite Element Modeling of Hollow Conical Sections for Critical speed Analysis", Master Thesis, Department of Mechanical Engineering, Rensselaer Polytechnic Institute, May 1991.
5. Vest, T. A., "Dynamic Modeling of Linear Beam-Like Structures Using Finite Element Analysis and Measured Vibration Characteristics", Doctoral Thesis, Department of Mechanical Engineering, Rensselaer Polytechnic Institute, December 1989.
6. Vest, T. A., Darlow, M. S., "An Evaluation of Beam and Shell Models for Hollow Conical Shafts", Proceedings of the 6th International Modal Analysis Conference, pp. 201-209, Kissimee, FL, February 1988.

The FEM Development of a General, Hollow Conical Shaft Element for Use in Transfer Matrix Analysis of Rotor-Bearing Systems Part II: Analysis Results and Experimental Correlation

Mark S. Darlow and Thomas J. Bievenue

Technion - Israel Institute of Technology

ABSTRACT

Traditional beam and shell deformation models have been found to be inadequate for the analysis of shafts with tapered sections. Consequently, a procedure has been developed wherein a static finite element formulation is used to extract flexibility terms which can be used in transfer matrix based dynamic analysis. Such an approach is more computationally efficient than utilising a full, dynamic finite element analysis. The formulation of the Finite Element Method (FEM) model and a description of the modelling procedure was presented in Part I of this paper.

Reported in this part is a continuation of this effort in which 134 conical models were created and analysed using a parametric study of the geometry of hollow conical sections. The parameters used are a non-dimensional thickness, length and taper angle. These parameters were varied to include a wide range of geometrical models and the resulting flexibility coefficients are tabulated herein..

These flexibility terms were subsequently used in a transfer matrix, critical speed program and compared to similar results from analytically derived beam and shell models. The critical speeds predicted using these three methods were also compared to critical speeds obtained using the stepped cylinder approximation, which is typically used in engineering practice to model conical sections. Of the four methods, the finite element method proved to be the most accurate in predicting the actual resonant frequencies of an experimental test shaft.

INTRODUCTION

Currently, there are many different methods for handling conical sections in critical speed programs. The most commonly used method is to break the conical section into a series of cylindrical sections of increasing (or decreasing) diameter. Other methods include the use of analytical theories based on beam and/or shell analysis. Four methods were compared in this work; the stepped model, the shell membrane theory by Kinerk [1], the transfer matrix beam theory by Darlow [2] and the static finite element approach introduced by Vest and Darlow [3] and extended herein. The derivation of these methods and additional details of the results of this work can be found in Bievenue [4].

The procedure for modeling a conical section, for a FEM analysis, and extracting the flexibility terms was presented in Part I of this paper. The results generated for a large number of geometries are presented herein. With these results, the reader can use previously generated flexibility values, or interpolate from these, if the geometry of interest is sufficiently close to one of the given models. Alternatively, the reader can construct and evaluate a FEM model based on the geometry of interest, using the approach described in Part I of this paper. The modelling and the results generated are non-dimensionalised so as to maximise their applicability in practical situations.

The ultimate goal of this research is to provide a tool for accurate, efficient dynamic analysis of rotor-bearing systems which incorporate conical shaft sections. In order to establish whether this goal has been achieved, the appropriate flexibility coefficients were used in a transfer matrix, undamped critical speed program for correlation with the results of a series of resonance tests conducted by Vest [5]. Corresponding flexibility coefficients based on analytically-derived beam and shell models, as well as a stepped cylinder approximation, were also used in the critical speed program, for comparison.

The geometry of a hollow, conical section is specified in Part I of this paper and is not repeated here, for the sake of brevity. Recall that the non-dimensional parameters used are η, λ, and θ; where θ is the taper angle of the conical section and η and λ are defined as the normalised conical thickness and length, respectively. Recall also that the analysis results were found to bear consistent relationships to Young's Modulus (E) and the average cone radius (r_{mid}).

FINITE ELEMENT RESULTS

Table 1 shows the parameter values used in this study, which covers a very wide range of conical geometries. While there are 240 possible combinations of the η, λ, and θ parameters, many of these combinations do not form proper conical sections. There are actually 134 physically realisable combinations of these parameters. Models for these 134 combinations were created and analysed in this work. Table A.1 in Appendix A shows the parameters used for each model and the corresponding flexibilities. For all of these models the Modulus of Elasticity was 210 GPa (30×10^6 psi), r_{mid} was 25.4 mm (1.0 in.), and Poisson's Ratio was 0.3. The flexibilities given in Table A.1 are the flexibilities with E and r_{mid} factored out (S_{11}'', S_{12}'', S_{21}'', and S_{22}''). Bievenue [4] also presented the corresponding flexibilities calculated from the shell and beam theories, but these are not presented here for the sake of brevity. The flexibilities provided in Table A.1 are for cones with a positive taper angle. The corresponding values for an equivalent geometry with a negative taper angle can easily be determined from these values, as described in Part I of this paper.

Table 1. Values of non-dimensional parameters

θ	10	20	30	40	50	60	70	80
λ	0.25	0.5	1.0	2.0	3.0	4.0		
η	0.0625	0.125	0.25	0.5	1.0			

The flexibilities resulting from the FE method are for specific geometrical models. A question arises as to what to do if the model being studied has non-dimensional parameters that fall in between those listed in Table A.1. It is possible that an interpolation scheme could be used to extract reasonable flexibility coefficients for these circumstances. It may also be possible to fit some equation to the data in Table A.1. The best approach, but not always possible, is to create a FE model for the given parameters, including the correct material properties, analyse it, and extract the exact flexibilities.

CRITICAL SPEED PREDICTION

The ultimate purpose of this effort is to improve the dynamic analysis, including the prediction of critical speeds, for shafts that include hollow conical sections. The flexibility terms for the FE, shell and beam methods have been used in a transfer matrix analysis, undamped critical speed program. The results from an experimental shaft built and tested by Vest [5] was used for comparison.

The shaft, shown in Fig. 1, was made out of aluminium and included two hollow conical sections of different orientation. The conical sections of the shaft were machined to different wall thicknesses to create more than one test case. Three cases were examined by Vest [5], with wall thicknesses of 1.59, 3.18, and 6.35 mm (0.0625, 0.125, and 0.25 in.). Vest [5] experimentally determined the first five non-rotating resonances for these three cases. The third case, where the thickness is 6.35 mm (0.25 in.), will be considered here first.

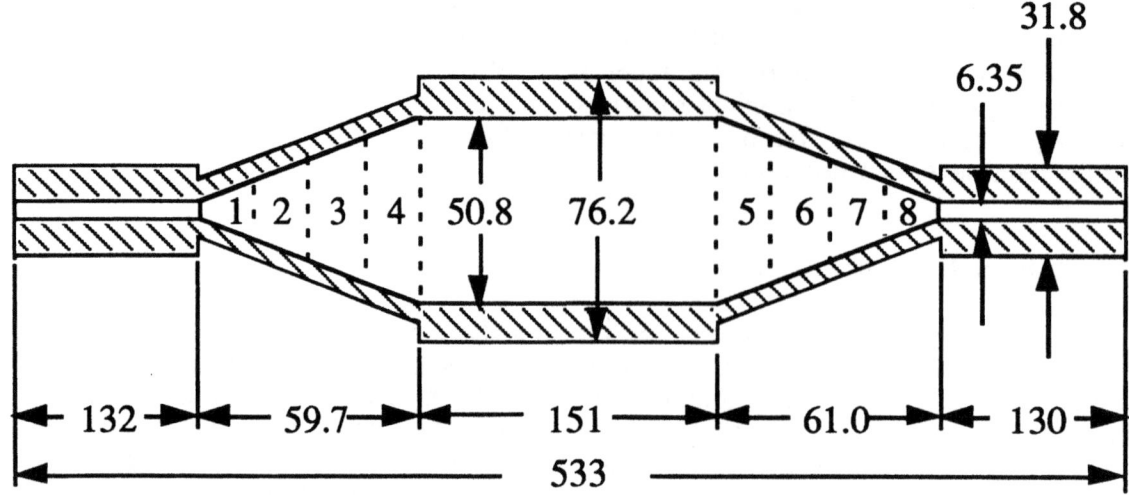

Figure 1. Test shaft dimensions (mm)

The overall dimensions for the shaft are given in Fig. 1. The conical sections were broken down into four sections each and the dimensions of these sections were then measured. Once the dimensions were known, the non-dimensional parameters were determined and the flexibilities from these values were found. For the shell and beam methods the flexibilities could be calculated using the shell and beam equations [5]. Since the non-dimensional parameters fall in between the models given in Table A.1, two methods were used to determine the flexibilities using the FE method. The first method was to use straight line interpolation using the data given in Table A.1. The second method was to create finite element models for each of the sections, analyse them and extract the exact FE flexibilities. The non-dimensional parameters for each section are given in Table 2. They are listed, as shown in Fig. 1, with sections numbering from one through eight. Some of the first four sections are not identical to the last four sections due to machining differences, and thus have different parameters.

Table 2. Non-dimensional parameters for experimental shaft

Section #	η	λ	θ
1	0.7692	1.4589	18.75
2	0.4641	1.0127	18.13
3	0.3385	0.7385	18.13
4	0.2696	0.5882	16.51
5	0.2304	0.5882	16.51
6	0.3385	0.7385	18.13
7	0.4641	1.0127	18.13
8	0.7294	1.5915	17.56

Table 3. S_{22}'' for experimental shaft sections

Section #	S_{22}''(FE)	S_{22}''(FE-int)	S_{22}''(Shell)	S_{22}''(Beam)
1	0.7820	0.9570	1.0237	0.5830
2	0.9530	1.0605	1.0710	0.6940
3	0.9936	1.1778	1.0430	0.6940
4	0.9581	1.0392	0.9681	0.6920
5	1.1600	1.1380	1.1328	0.8140
6	0.9581	1.1778	1.0430	0.6940
7	0.9530	1.0605	1.0710	0.6940
8	0.8670	1.0998	1.1298	0.6830

Using these parameters the flexibilities for each method were determined. These flexibilities are shown in Table 3. Only S_{22}'' is shown in Table 3 for the sake of brevity. It can be seen that there is quite a difference between the flexibility terms for each method. The interpolated values for the FE method are up to 27% different than the exact terms. It should be mentioned that the conical section on the right side of the shaft is opposite in orientation from the left side. For this case the orientation equations given in Part I of this paper were used to determine the appropriate flexibilities. This does not effect S_{22}'', however.

The resonance frequencies were determined using input files which represented the FE method, the FE method using interpolated values, the shell method, the beam method and the stepped approach. There were also comparisons made of critical speeds with all of the flexibility terms used, as opposed to the critical speeds obtained when using only S_{22}'' while setting the other flexibility terms to zero. While the complete set of results is presented by Bievenue [4], only the FE, shell and beam method results are presented here, for brevity.

S_{22}'' Domination

To determine if S_{22}'' was really the dominant term in calculating the critical speeds, as indicated by Vest [5], six different cases were compared. The FE method was used with and without the other flexibility terms and compared to the shell and beam methods under the same conditions. Table 4 shows the results. The cases that used all of the flexibilities are marked with an *. Table 5 shows the % difference between each method and the experimental data.

Table 4. Critical speeds with and without S_{11}'', S_{12}'', and S_{21}''

Mode	Exp.	FE *	FE	Shell *	Shell	Beam *	Beam
1	750	708.2	716.9	660.3	670.0	799.4	788.6
2	1305	1271.6	1273.6	1186.0	1190.1	1394.2	1386.3
3	3900	3968.4	3965.0	3933.0	3950.5	4211.5	4052.3
4	5225	5674.0	5153.0	5686.1	5110.5	5990.6	5344.3
5	7456	7685.7	7620.3	7567.9	7522.4	7828.5	7742.9

Table 5. % error from Table 4

Mode	Exp.	FE *	FE	Shell *	Shell	Beam *	Beam
1	750	-5.6	-4.4	-12.0	10.7	6.5	5.1
2	1305	-2.6	-2.4	-9.1	8.8	6.8	6.2
3	3900	1.8	1.7	0.85	1.3	8.0	3.9
4	5225	8.6	-1.4	8.8	2.2	14.6	2.3
5	7456	3.1	2.2	1.5	0.9	5.0	3.9

In all three methods, the cases where only S_{22}'' was used yielded better results. The difference between using the other three flexibility terms and not using them changed the critical speeds by no more than 1.2% for the FE method except for mode 4. In all three methods mode 4 was improved dramatically if only S_{22}'' was used. The predicted mode shapes are quite different for the cases where all of the terms are used and where only S_{22}'' was used. When only S_{22}'' was used the mode shapes were symmetric, as would be expected. When all of the flexibility terms were used the mode shapes showed many irregular inflection points. Although the inclusion of all of the flexibilities (for the FE method) has little effect on the critical speeds, it does seem to affect the mode shapes. These odd inflection points are inconsistent with what would be expected. It is expected that the shaft should behave in a symmetrical manner since the geometry of the shaft is symmetric. Due to the odd mode shapes, and including the fact that the predicted critical speeds are better using only S_{22}'', it is recommended that the other three flexibilities not be used. It should be noted that the orientation change will then have no effect since S_{22}'' is not affected by a change in orientation.

Critical Speeds of the Other Two Cases

As mentioned above, there were three cases of different wall thicknesses for which resonance frequencies were measured. Table 4 shows the critical speeds for the third case where the thickness was 6.35 mm (0.25 in.). Table 6 shows the critical speeds calculated for the first case where the thickness was 1.59 mm (0.0625 in.). The critical speeds for the second case, where the wall thickness is 3.18 mm (0.125 in.), are shown in Table 7. These calculations were made using only the S_{22}'' term. From Tables 6 and 7 it is clear that the FE method predicted the critical speeds better than the shell or beam methods.

Table 6. Critical speeds for a wall thickness of 1.59 mm (0.0625 in.)

mode	Exp.	FE	% diff	Shell	% diff	Beam	% diff
1	229	230.9	0.82	254.1	11.0	298.1	30.2
2	530	531.6	0.3	590.3	11.4	689.1	30.0
3	2693	2747.7	2.0	3177.2	18.0	3472.2	28.9
4	3675	3745.9	2.0	4254.8	15.8	4614.3	25.6
5	7825	8082.4	3.3	8378.3	7.1	8575.1	9.6

Table 7. Critical speeds for a wall thickness of 3.18 mm (0.125 in.)

mode	Exp.	FE	% diff	Shell	% diff	Beam	% diff
1	366	404.0	10.4	406.5	11.1	481.1	31.5
2	745	771.2	3.5	772.1	3.6	911.7	22.4
3	3343	3576.1	7.0	3652.3	9.3	3822.2	14.3
4	4418	4586.5	3.8	4705.0	6.5	4983.9	11.7
5	7675	8167.8	6.4	8205.6	6.9	8372.9	9.1

SUMMARY AND CONCLUSIONS

The purpose of this effort was to determine if the FE approach was a valid method in determining the static flexibilities of hollow conical sections for use in critical speed programs. The FE method was used to create a non-dimensional model from which flexibility coefficients were extracted and used in a transfer matrix critical speed program. Kinerk's shell membrane theory and Darlow's exact beam theory were also used to develop flexibility coefficients that were used in the critical speed program to compare the critical speeds using these methods to the critical speeds resulting from the FEM approach.

Experimental verification showed that the FE approach is a viable candidate for critical speed analysis. The FE method, in general, predicted the first five critical speeds more accurately than the other methods when compared to the results of an experimental test shaft. It was shown that S_{22}'' was a very dominant term in the prediction of these critical speeds. The other three flexibility terms actually made the predictions worse and affected the mode shapes in a peculiar manner. For these reasons S_{11}'', S_{12}'', and S_{21}'' were omitted from the critical speed prediction calculations.

REFERENCES

1. Kinerk, H., "Deflection of Cones under Lateral Load and Moment", GE TIS No. R58AAGT708, Sept., 1958.
2. Darlow, M. S., Murphy, B. T., Elder, J. A., and Sandor, G. N., "Extension of the Transfer Matrix Method for Rotordynamic Analysis to Include a Direct Representation of Conical Sections and Trunnions", ASME Transactions, *Journal of Mechanical Design*, Vol. 102, No.1, Jan. 1980, pp. 122-129.
3. Vest, Todd A. and Darlow, Mark S., "A Modified Conical Beam Element Based on Finite Element Analysis: Experimental Correlations," ASME Transactions, *Journal of Vibration and Acoustics*, Vol. 112, pp. 350-354, July 1990, ASME Paper No. DE-Vol. 18-4, pp. 349-353, Montreal, September 1989.
4. Bievenue, T., "Finite Element Modeling of Hollow Conical Sections for Critical speed Analysis", Master Thesis, Department of Mechanical Engineering, Rensselaer Polytechnic Institute, May 1991.
5. Vest, T. A., "Dynamic Modeling of Linear Beam-Like Structures Using Finite Element Analysis and Measured Vibration Characteristics", Doctoral Thesis, Department of Mechanical Engineering, Rensselaer Polytechnic Institute, December 1989.

APPENDIX

Table A.1. Flexibility coefficients from FEM modelling of conical sections

η	λ	θ	S_{11}''	S_{12}''	S_{21}''	S_{22}''
0.0625	0.25	10	3.08179	-0.90693	-0.32141	1.63684
0.125	0.25	10	1.68873	-0.20057	-0.21113	0.73225
0.25	0.25	10	0.86017	-0.04497	-0.11212	0.34190
0.5	0.25	10	0.39805	-0.00499	-0.04579	0.15689
1.0	0.25	10	0.16048	0.00173	-0.01021	0.06610
0.0625	0.5	10	5.46646	-1.76020	-0.07388	3.38214
0.125	0.5	10	3.07593	-0.45377	-0.15409	1.54269
0.25	0.5	10	1.62900	-0.08653	-0.11102	0.71433
0.5	0.5	10	0.78067	0.00410	-0.04791	0.32508
1.0	0.5	10	0.32169	0.01404	-0.00559	0.13493
0.0625	1.0	10	11.77727	-0.59614	0.98133	6.28778
0.125	1.0	10	6.12668	-0.08065	0.40189	3.05688
0.25	1.0	10	3.19094	0.09196	0.14702	1.46182
0.5	1.0	10	1.57403	0.11757	0.06758	0.67114
1.0	1.0	10	0.67561	0.07901	0.05086	0.27586
0.0625	2.0	10	30.31854	5.34485	6.23303	12.47660
0.125	2.0	10	15.27137	2.75928	3.04349	6.17948
0.25	2.0	10	7.67690	1.44888	1.46240	3.01552
0.5	2.0	10	3.74897	0.76758	0.70125	1.40461
1.0	2.0	10	1.66542	0.37564	0.33434	0.57286
0.0625	3.0	10	61.17678	15.32229	15.63966	19.97905
0.125	3.0	10	30.58740	7.68714	7.76347	9.93082
0.25	3.0	10	15.22236	3.87104	3.80620	4.86029
0.5	3.0	10	7.37506	1.93458	1.83176	2.26107
1.0	3.0	10	3.29289	0.88429	0.82729	0.90901
0.0625	4.0	10	109.38000	29.67362	29.52650	29.83928
0.125	4.0	10	54.61458	14.84941	14.70262	14.82945
0.25	4.0	10	27.08374	7.41656	7.25140	7.24739
0.5	4.0	10	13.06537	3.636780	3.48916	3.34497
1.0	4.0	10	5.82034	1.60763	1.53418	1.31380

Table A.1. Flexibility coefficients from FEM modelling of conical sections (cont'd.)

η	λ	θ	S_{11}''	S_{12}''	S_{21}''	S_{22}''
0.0625	0.25	20	3.07643	-2.10786	-0.76018	2.89070
0.125	0.25	20	1.70811	-0.50149	-0.50933	1.05500
0.25	0.25	20	0.86769	-0.13167	-0.27105	0.43620
0.5	0.25	20	0.39952	-0.02910	-0.11364	0.18167
1.0	0.25	20	0.16085	-0.00468	-0.02930	0.07106
0.0625	0.5	20	5.28363	-4.17407	-0.55368	6.25904
0.125	0.5	20	3.02747	-1.24312	-0.57824	2.48447
0.25	0.5	20	1.61701	-0.32890	-0.37797	1.00765
0.5	0.5	20	0.77504	-0.06461	-0.17330	0.40744
1.0	0.5	20	0.32026	-0.00341	-0.04400	0.15292
0.0625	1.0	20	11.27862	-3.30379	-0.09131	10.57267
0.125	1.0	20	5.85356	-1.26098	-0.26770	4.92617
0.25	1.0	20	3.05063	-0.37466	-0.26734	2.20341
0.5	1.0	20	1.50920	-0.03722	-0.14408	0.91708
1.0	1.0	20	0.65695	0.03802	-0.02020	0.33636
0.0625	2.0	20	28.33854	2.43468	4.00103	22.43174
0.125	2.0	20	14.21685	1.37598	1.82237	10.98621
0.25	2.0	20	7.08476	0.85559	0.80429	5.16312
0.5	2.0	20	3.42335	0.55470	0.39622	2.22673
1.0	2.0	20	1.54869	0.31045	0.22629	0.79678
0.0625	3.0	20	57.72981	14.14439	14.03783	48.79503
0.125	3.0	20	28.76981	7.28443	6.92882	23.80591
0.25	3.0	20	14.16732	3.83678	3.43557	11.11627
0.5	3.0	20	6.74679	2.00346	1.72780	4.62598
0.0625	4.0	20	112.20633	41.93481	38.95284	138.27411
0.125	4.0	20	55.75383	21.65595	19.57982	65.52837
0.25	4.0	20	27.22211	11.11383	10.03915	28.76674
0.0625	0.25	30	3.157183	-4.046296	-1.094713	6.190949
0.125	0.25	30	1.800489	-.9135869	-.8410377	1.804175
0.25	0.25	30	.9082337	-.2346722	-.4557718	.6342412
0.5	0.25	30	.4122668	-.0562173	-.1915715	.2319215
1.0	0.25	30	.1643354	-.0117383	-.0508968	.0809231
0.0625	0.5	30	5.464565	-7.459964	-.7585356	13.33458
0.125	0.5	30	3.15328	-2.3195	-.9819293	4.794189
0.25	0.5	30	1.695071	-.6274378	-.6841326	1.675184
0.5	0.5	30	.8052817	-.1420967	-.3181201	.5834762
1.0	0.5	30	.3298644	-.0219718	-.0869363	.1899141
0.0625	1.0	30	12.0015	-6.935786	-.8384542	22.34959
0.125	1.0	30	6.166869	-2.665841	-.9201164	9.776401
0.25	1.0	30	3.166819	-.8970184	-.7150807	4.06615
0.5	1.0	30	1.549119	-.1966365	-.3730443	1.50273
1.0	1.0	30	.675382	-.0003485	-.0934056	.4710452
0.0625	2.0	30	31.82771	-1.273809	2.849597	67.08393
0.125	2.0	30	15.7297	.2004688	.816011	30.7316
0.25	2.0	30	7.64236	.6366522	.3058457	13.20013
0.5	2.0	30	3.579366	.589743	.2800163	4.889745
0.0625	3.0	30	81.63316	36.22445	28.30956	566.159
0.125	3.0	30	39.63663	20.82369	16.46648	234.9312
0.0625	0.25	40	3.491135	-7.197884	-1.342119	14.30648
0.125	0.25	40	2.023333	-1.548604	-1.267148	3.503515
0.25	0.25	40	1.005038	-.3746633	-.7009859	1.033515
0.5	0.25	40	.4437814	-.0907070	-.2911689	.3284562
1.0	0.25	40	.1729809	-.0202879	-.0775346	.0994250
0.0625	0.5	40	6.426979	-11.87152	-1.016165	29.43473
0.125	0.5	40	3.588508	-3.906941	-1.481308	10.33081
0.25	0.5	40	1.911686	-1.061013	-1.107044	3.204923
0.5	0.5	40	.8909118	-.2456549	-.5123119	.9546413
1.0	0.5	40	.3568998	-.0444939	-.1405421	.2631547

Table A.1. Flexibility coefficients from FEM modelling of conical sections (cont'd.)

η	λ	θ	S_{11}''	S_{12}''	S_{21}''	S_{22}''
0.0625	1.0	40	14.70507	-11.68185	-2.541177	54.39634
0.125	1.0	40	7.339257	-4.547387	-2.032707	23.12136
0.25	1.0	40	3.637264	-1.572442	-1.359744	9.118
0.5	1.0	40	1.729381	-.3866886	-.6601499	2.993986
0.0625	2.0	40	46.8587	.0821764	-3.520591	537.8501
0.125	2.0	40	22.53825	3.552784	.2357719	217.2241
0.0625	0.25	50	4.301141	-13.491	-1.387454	37.66038
0.125	0.25	50	2.495852	-2.811274	-1.898625	8.168906
0.25	0.25	50	1.215929	-.6103008	-1.087327	1.944804
0.5	0.25	50	.5139369	-.1431274	-.4381925	.5294405
1.0	0.25	50	.1921767	-.0320869	-.1143637	.1363745
0.0625	0.5	50	8.774644	-19.15576	-2.040299	72.16558
0.125	0.5	50	4.590368	-6.755789	-2.312877	26.08593
0.25	0.5	50	2.381743	-1.855383	-1.814926	7.454237
0.5	0.5	50	1.083145	-.4194677	-.8247193	1.862474
1.0	0.5	50	.4187516	-.0774524	-.2184864	.4239034
0.0625	1.0	50	20.9277	-20.61343	-8.17082	184.3872
0.125	1.0	50	10.16485	-7.758286	-4.907073	75.98711
0.25	1.0	50	4.796905	-2.49754	-2.4756	27.98586
0.5	1.0	50	2.181477	-.5945798	-1.000564	7.994161
0.0625	0.25	60	6.505805	-27.79935	-1.840291	119.8955
0.125	0.25	60	3.584203	-6.099998	-3.129908	25.91065
0.25	0.25	60	1.710317	-1.139321	-1.870654	4.709243
0.5	0.25	60	.6839754	-.2475729	-.7088799	1.044417
1.0	0.25	60	.2383001	-.0522501	-.1749143	.2236118
0.0625	0.5	60	14.39799	-35.05427	-7.523112	233.3883
0.125	0.5	60	7.012428	-12.81323	-4.780609	87.40136
0.25	0.5	60	3.449373	-3.669731	-3.326863	24.5223
0.5	0.5	60	1.532762	-.7982261	-1.450112	4.964546
1.0	0.5	60	.5684088	-.1401582	-.3594627	.8790427
0.0625	1.0	60	40.1175	-39.83031	-41.67092	2182.013
0.125	1.0	60	18.99492	-9.0838	-12.20627	748.5374
0.25	1.0	60	8.552913	-.2283742	-1.251468	202.9755
0.0625	0.25	70	13.67559	-65.4084	-11.79299	531.8389
0.125	0.25	70	6.73101	-17.47135	-7.787985	138.4404
0.25	0.25	70	3.121125	-2.90099	-4.217132	19.74319
0.5	0.25	70	1.210894	-.5659825	-1.428463	3.137126
1.0	0.25	70	.3815382	-.1043128	-.3115508	.5200892
0.0625	0.5	70	30.81442	-80.55015	-42.60354	1670.436
0.125	0.5	70	14.55359	-27.27561	-17.17515	614.6411
0.25	0.5	70	6.668323	-7.608674	-6.883269	172.6719
0.5	0.5	70	2.893044	-1.679053	-2.845006	27.36207
0.0625	0.25	80	49.99622	-229.2213	-137.462	10386.16
0.125	0.25	80	23.4041	-66.78418	-44.46315	3184.845
0.25	0.25	80	10.71572	-12.52832	-16.51675	441.0124
0.5	0.25	80	4.374857	-2.331468	-4.778488	37.56271

Response of a Discontinuously Nonlinear Rotor System

D. Gonsalves, R. D. Neilson, A. D. S. Barr
Department of Engineering, University of Aberdeen. U.K.

ABSTRACT

This paper looks at the study of a rotor system with bearing clearance effects, with particular emphasis on the possibility of chaotic motion. The equations of motion are presented and used in numerical simulations for different combinations of experimental parameters. The results are evaluated using various chaos and spectral analysis techniques and compared to those obtained from an experimental rig.

INTRODUCTION

Severe nonlinearities, in the form of a discontinuous stiffness effect, can be introduced into rotor systems as a result of clearances.

Systems exhibiting this effect have been studied by several investigators. Ehrich investigated the subharmonic vibration of rotors in a bearing clearance [1] and Bently explained the occurrence of subharmonics in his experimental results by using a simple horizontal model with a bearing clearance [2].

Two of the authors, Neilson and Barr, investigated, both numerically and experimentally, bearing nonlinearities in a four degree of freedom rotor system [3], while Ehrich showed the occurrence of higher subharmonics in a high speed rotor system with a bearing clearance, using numerical integration [4].

The above research concentrated on the periodic and aperiodic motions. However, Kim and Noah [5] investigated the response and stability of a modified Jeffcott rotor system with a bearing clearance, (using a Harmonic Balance/Alternating Frequency/time technique) where they observed the presence of chaotic motion.

THEORETICAL MODEL

The system studied here comprises a rotor, which is elastically supported by a symmetrical stiffness k_1, and mounted in a radial clearance g, provided by a snubber ring. The ring is also elastically supported, but by stiffer springs k_2 ($>>k_1$). The secondary stiffness k_2 comes into play when the rotor traverses the radial clearance and

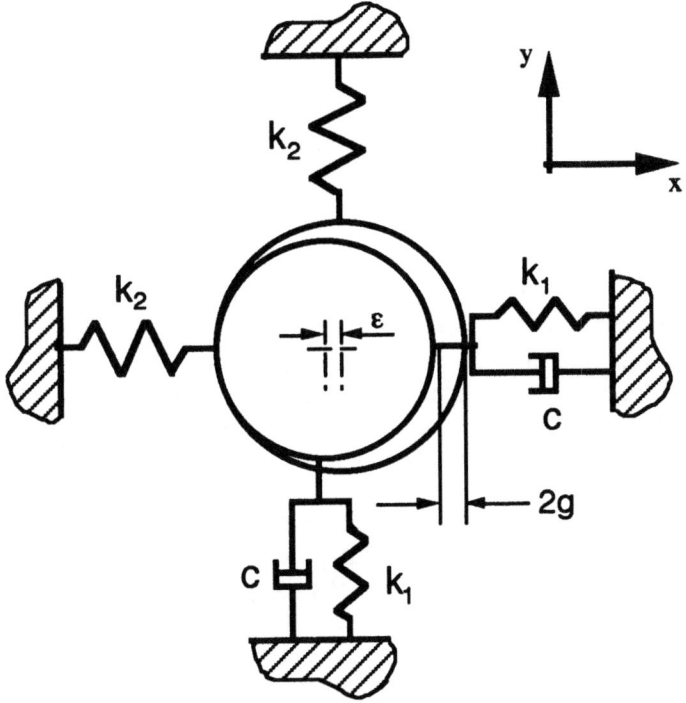

Fig. 1 Diagrammatic model of the system.

comes into contact with the snubber ring. The rotor may be positioned eccentrically within the ring, by an amount ε, and the situation where the rotor is just touching one side of the bearing statically, i.e. $\varepsilon = g$, is of particular interest here (Fig. 1). An out of balance provides the excitation force.

Equations of motion for $(x^2 + y^2)^{1/2} > g$.

In the x direction :

$$\ddot{x} + 2v\omega_1\dot{x} + \omega_1^2 x + \omega_2^2 \left((x - \varepsilon)(1 - \frac{g}{R})\right) = \Omega^2\rho \cos \Omega t$$

In the y direction :

$$\ddot{y} + 2v\omega_1\dot{y} + \omega_1^2 y + \omega_2^2 \left(y(1 - \frac{g}{R})\right) = \Omega^2\rho \sin \Omega t$$

where

$$\omega_1 = \sqrt{\frac{k_1}{M}}, \qquad \omega_2 = \sqrt{\frac{k_2}{M}}, \qquad \rho = \frac{m\rho_1}{M}, \qquad R = \sqrt{(x - \varepsilon)^2 + y^2}$$

M is the mass of the rotor
$m\rho_1$ is the out of balance
Ω is the shaft speed
v is the damping ratio

Discontinuity functions which have zeros when the rotor contacts the ring, are used to detect when a discontinuity condition occurs. The displacement discontinuity function δ is given by :

$$\delta = \sqrt{(x - \varepsilon)^2 + y^2} - g$$

The equations of motion were transformed into first order differential equations and solved numerically, using a piecewise solution. The method, developed by Borthwick [6], is based on the classical fourth order Runge-Kutta algorithm. This and the detection and interpolation procedures used were the same as those described and later used by Neilson and Barr [3].

A spectral analysis was carried out on the resulting computed responses, applying the Hanning window prior to using standard Fast Fourier Transform techniques to produce waterfall plots.

Simulations were also run in order to obtain bifurcation diagrams. These are diagrams used in the study of chaotic motion and are plots of some measure of the motion of a system as a function of a system parameter. Sampling occurs at the same phase, once every forcing cycle. In this case the measure of motion is the amplitude of the response in the x direction and the rotational speed of the rotor is the system parameter. The number of different points at each value of the system parameter gives the period of the motion. When the bifurcation diagram loses continuity it can mean either quasi-periodic, aperiodic or chaotic motion.

A separate integration, having zero initial conditions, was run for each value of forcing frequency, with the response of the system being allowed to settle for 150.25 cycles before data was recorded.

Spatial orbits and Poincaré maps were produced for periodic, quasi-periodic and chaotic motions. Poincaré maps are sampled in the same way as for a bifurcation diagram, except the points are plotted in the phase space.

EXPERIMENTAL RIG

To allow comparison with the theoretical results, an experimental rig was designed and constructed, with a radial gap of 0.5 mm, a damping ratio of 0.085 and a spring stiffness ratio, k_2/k_1, of 36.6. An eccentrically mounted mass on either end of the rotor provided the excitation force, with a ρ value of 0.06 mm.

The response was monitored by two non-contacting eddy current probes. A pair of light emitting diode opto-switches together with a disc with holes drilled in it were used to monitor the speed of the rotor and the position of the imbalance mass.

The rotational speed, Ω, was increased from 10 - 50 Hz (600 - 3000 rpm) and the response investigated using some of the same techniques used on the computational results.

RESULTS AND DISCUSSION

Initially the system parameters of the model were set up to emulate the rig. Figs. 2 and 3 show the waterfall plot and bifurcation diagram produced from the model over the same speed range, 10 - 50 Hz. For increasing shaft speed the response builds up to resonance, with a peak at approximately 23 Hz. The response then reduces in amplitude and enters into chaotic motion, characterised by a broad band of amplitudes. Two more chaotic bands exist separated by periodic motion. Comparison with the spectral content of the response from the rig, Fig. 4, indicates good correlation with the only major discrepancy being the existence of sidebands about the half subharmonics at higher speeds.

Figs. 5 - 12 show the spatial plots (orbits) and Poincaré maps from the theoretical model and the experimental rig, respectively, for period three motion, at 35 Hz, and chaotic motion, at 46 Hz. The periodic case, Figs. 5 - 8, shows good correlation with the form of the orbits and maps being fairly similar. The very narrow band in which this period three could be elicited is consistent between the model and the rig. At 46 Hz the wide-band, continuous frequency content (Figs. 2 and 4), the apparent randomness of the orbit, Figs. 9 and 11, together with the obvious structure of the Poincaré map is a strong indicator of chaotic motion. Unfortunately the maps do not correlate quite so well.

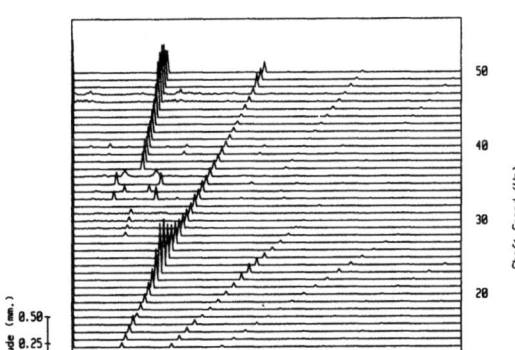

Fig. 2 Spectral response of the model

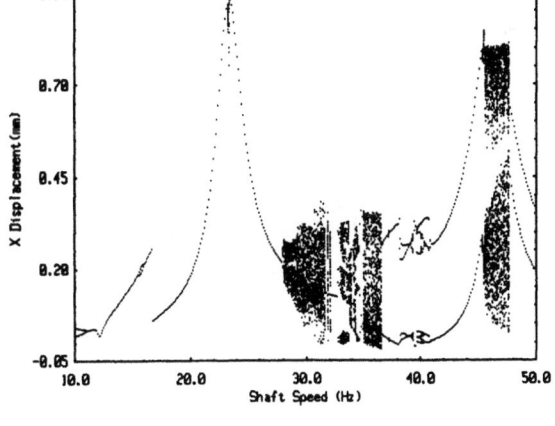

Fig. 3 Bifurcation diagram of response of the model

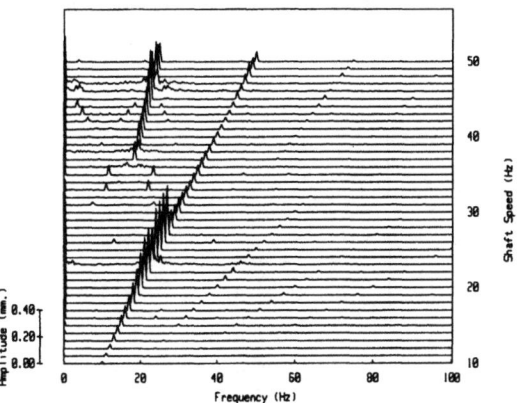

Fig. 4 Spectral response of the experimental rig

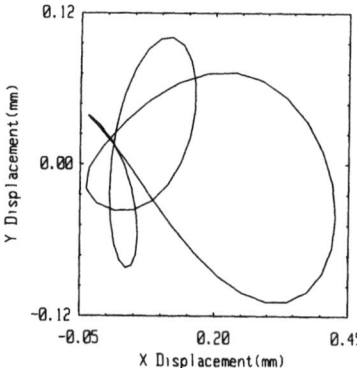

Fig. 5 Spatial plot of periodic motion response of the model

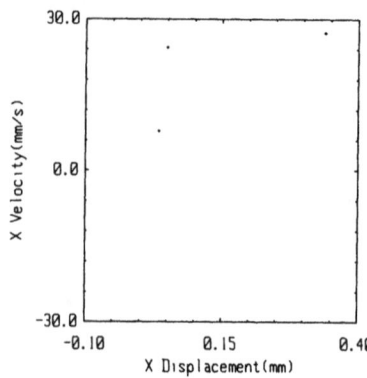

Fig. 6 Poincaré map of periodic motion response of the model

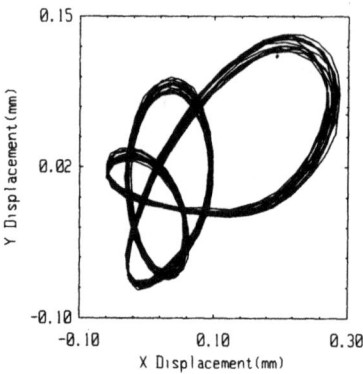

Fig. 7 Spatial plot of periodic motion response of the rig

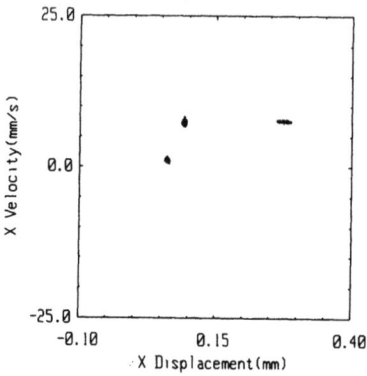

Fig. 8 Poincaré map of periodic motion response of the rig

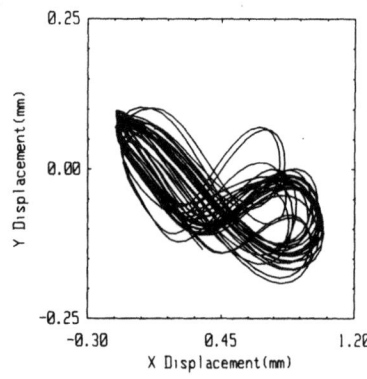

Fig. 9 Spatial plot of chaotic motion response of the model

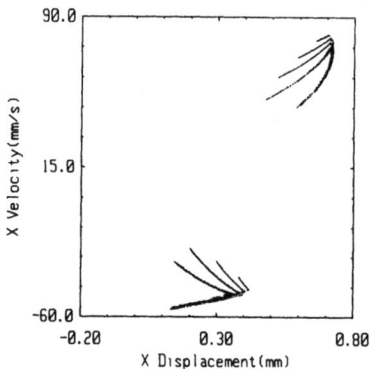

Fig. 10 Poincaré map of chaotic motion response of the model

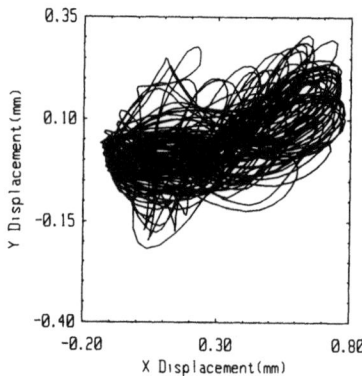

Fig. 11 Spatial plot of chaotic motion response of the rig

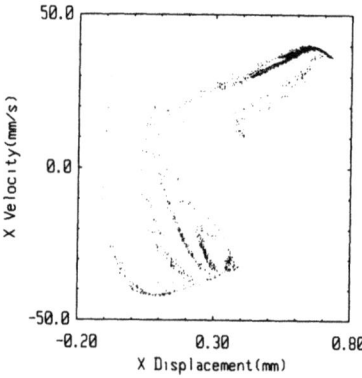

Fig.12 Poincaré map of chaotic motion response of the rig

A spatial plot and Poincaré map for a quasi-periodic response of the model are also produced, Figs. 13 and 14. The continuous closed orbit in Fig. 14 is characteristic of quasi-periodic motion.

The region between the shaft speeds 32.5 Hz and 48 Hz was examined in more detail for the case above. The periodic motion can be seen more clearly in the bifurcation diagram, Fig. 16, with periods one, two, three, four and eight being very apparent.

Two other cases for a different value of damping ratio and then a different clearance were simulated.

The effect of changing the level of damping to 0.125, Figs. 17 and 18 (the damping ratio Shaw [7] used in his study of a single degree of freedom motion), showed that an increase in damping changes the response at the lower shaft speeds to a quasi-periodic motion and causes the chaotic motion at the higher speeds to disappear. This agrees with previous results [5].

An increase in gap and eccentricity to 10 m, with the damping level remaining at 0.085, (Figs. 19 and 20) simulates a discontinuity confined to motion in the x direction, as the rotor comes into contact with a more or less flat surface. Again, the response becomes more periodic with the bifurcations being more evident. The chaotic motion that existed between the speeds 45 Hz and 47.5 Hz, when the radial clearance was 0.5 mm, has also disappeared.

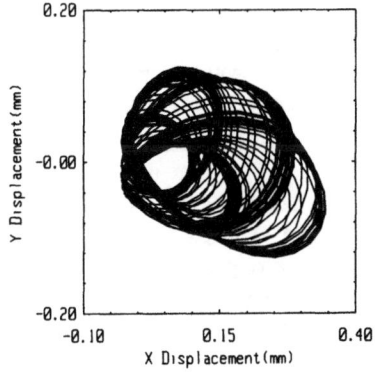

Fig. 13 Spatial plot of quasi-periodic response of the model

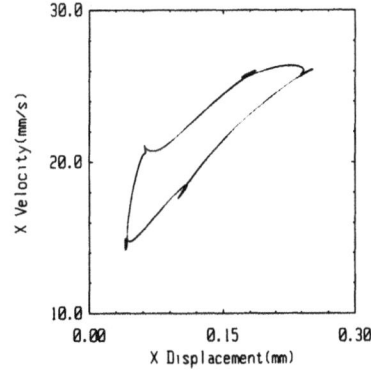

Fig.14 Poincaré map of quasi-periodic response of the model

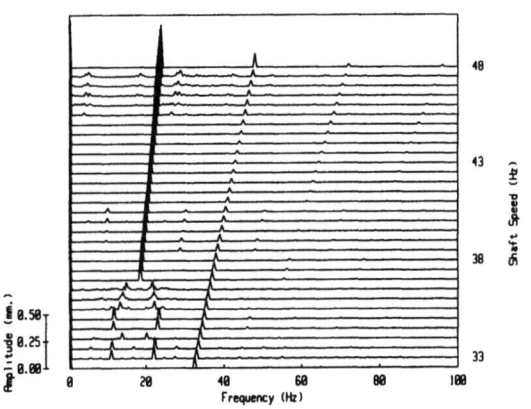

Fig. 15 Spectral response of the model

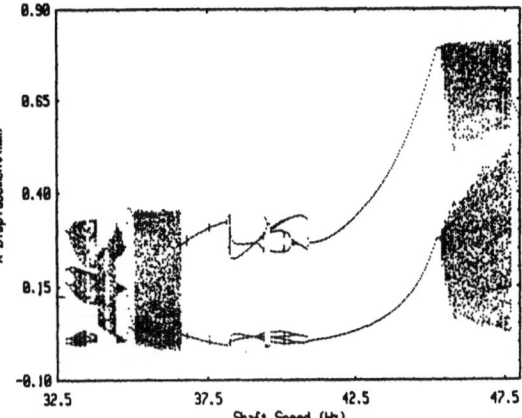

Fig.16 Bifurcation diagram of response of the model

188

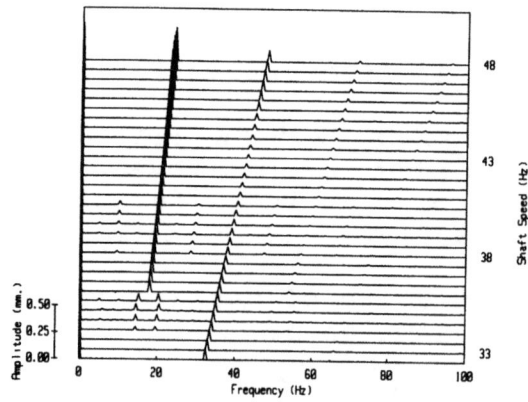

Fig. 17 Spectral response
of the model for ν = 0.125

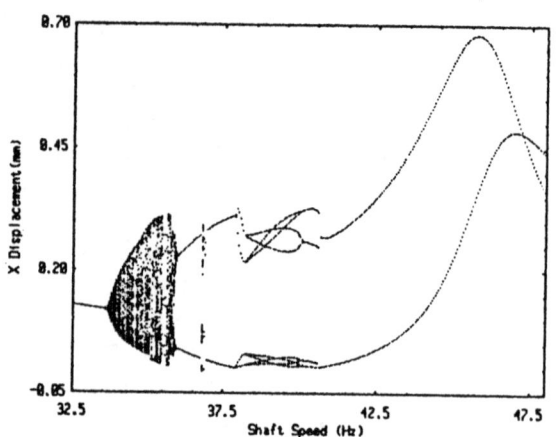

Fig. 18 Bifurcation diagram
response of the model for
ν = 0.125

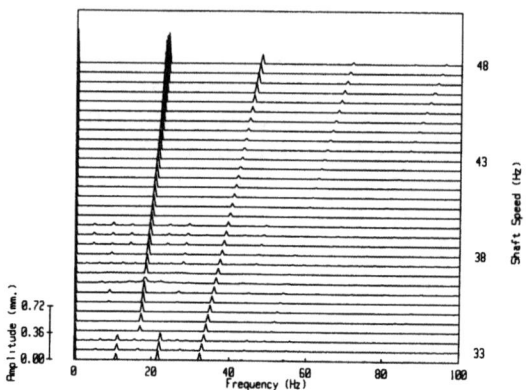

Fig. 19 Spectral response of
the model for g = 10 m

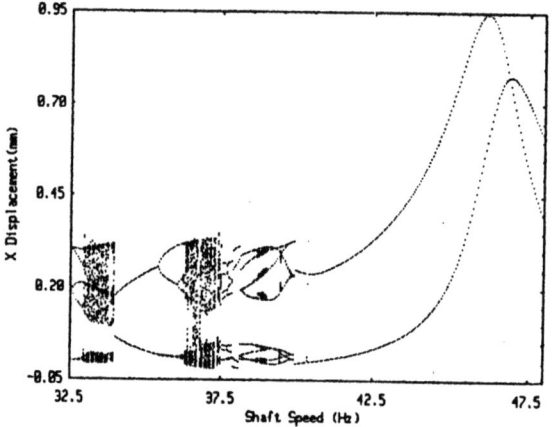

Fig. 20 Bifurcation diagram of
response of model for g = 10 m

The combination of a damping ratio of 0.125, a stiffness ratio of 15 and a clearance and eccentricity of 10 m produces a bifurcation diagram where the period doubling route to chaos can be observed quite clearly. The bifurcation points correlate quite well with Shaw's results [7].

CONCLUSIONS

The results obtained from both the experimental rig and the numerical simulation display good correlation and show that a physical system of this type can exhibit chaotic motion.

The simulations indicate that, in certain regions, small changes in damping or clearance result in major changes in the form of the response of the system, from periodic to chaotic or quasi-periodic motion and vice versa. They also confirm Kim and Noah's result that an increase in damping eventually eliminates any subharmonic motion.

ACKNOWLEDGEMENTS

The authors would like to acknowledge the financial support of the Science and Engineering Research Council for this work.

REFERENCES

1. Ehrich FF. Subharmonic vibration of rotors in bearing clearance. ASME Conference and Show on Design, 1966, Paper 66-MD-1:1-4
2. Bently DE. Forced Subrotative Speed Dynamic Action of Rotating Machinery. ASME. Petroleum Engineering Conference, Dallas, 1974, Paper 74-PET-16:1-8
3. Neilson RD, Barr ADS. Dynamics of a rigid rotor mounted on discontinuously nonlinear elastic supports. Proceedings of the I. Mech. E. 1988;202 (C5):369-376.
4. Ehrich FF. High Order Subharmonic Response of High Speed Rotors in Bearing Clearance. J Vibration, Acoustics, Stress, and Reliability in Design 1988;110:9-16
5. Kim YB, Noah ST. Bifurcation Analysis for a Modified Jeffcott Rotor with Bearing Clearances. Nonlinear Dynamics 1990;1:221-241
6. Borthwick WKD. The numerical solution of discontinuous structural systems. Second International Conference on Recent Advances in Structural Dynamics, University of Southampton 1984:307-316
7. Shaw SW, Holmes PJ. A Periodically Forced Piecewise Linear Oscillator. J Sound and Vibration 1983;90(1):129-155

A PC-Based Evaluation and Monitoring System For Rotating Machinery

Marcello Typrin, Eric C. Pawtowski, R. G. Kirk
VIRGINIA POLYTECHNIC INSTITUTE AND STATE UNIVERSITY

ABSTRACT

Using a data acquisition system controlled by a personal computer to monitor rotating machinery has the benefits of improved accuracy, consistency in analysis procedure and fast performance. The Virginia Tech Rotor Dynamics Laboratory has developed such a system in the Windows 3.0 operating environment; it is called the Intelligent Monitoring System (IMS). This data acquisition system has a modular software design in which specific tasks are controlled by independent and interchangeable software modules. The system has been successfully used to read vibration data from a small test rig and evaluate the rig's performance.

INTRODUCTION

Compared to the first pc introduced ten years ago, today's pc has more memory, its information storage capacity is greatly increased, and it is much faster. Given these improvements in performance, in addition to the abundance of software and special purpose boards, the pc is now being used in more sophisticated and demanding industrial applications.

One such application is the Intelligent Monitoring System (IMS). IMS is a monitoring and evaluation system for rotating machinery that can be operated independently or as the data acquisition module for an on-line rotating machinery expert system. The entire project (monitoring system and expert system) is the result of a two year long project named The Wizard Of Tech (WOT) that was developed at the Virginia Tech Rotor Dynamics Laboratory [1,2,3]. The focus of this paper will be on the monitoring and evaluation capabilities of IMS.

Industries that use rotating machinery such as turbines, compressors, fans and pumps often rely on plant engineers or technicians to manually collect and compile large quantities of data and then return a diagnosis of the problem. The quality and speed of the analysis is often compromised due to the tedious and repetitive nature of the task and the voluminous amounts of data required for a thorough analysis. IMS automates the data collection and analysis process, thus making it faster and more accurate than its manual counterpart.

What makes IMS truly exceptional, however, is its flexibility. IMS is a modular system which means that a specific task, such as the acquisition of vibration data or the evaluation of this data, is controlled by program sections called modules. These modules can be written by the user in any language and customized for his particular application or he can use the ready-to-run modules provided by IMS. This flexibility means that any given rotating machine can be monitored by a variety of hardware by simply including the appropriate modules.

SYSTEM OPERATION

IMS runs in the Windows 3.0 multitasking environment. Under this multitasking environment, several programs called modules run simultaneously. Each module is a DOS application dedicated to performing a specific task: acquiring temperature data or vibration data, performing an FFT, opening and closing relays, interpreting and evaluating data, and displaying data and results of the evaluations.

Modules are organized in groups according to the type of data which they acquire and process. In the prototype, for example, the FFT group is comprised of the module that acquires the FFT data and the module that evaluates the FFT data. A DVF-3 group also exists and it is comprised of the module that controls the operation of the DVF-3 (Bently Nevada Corporation) and another module that evaluates the data obtained from the DVF-3. As other types of data, such as temperatures and flow rates, are required, groups containing the necessary modules could be created by the user.

Note that communication between a group's modules is necessary to transfer data between the acquisition module and the evaluation module and to synchronize the operation of the modules within a group; i.e., the DVF-3 evaluation module must only performs its task after the DVF-3 acquisition module has finished acquiring the data. This communication is accomplished by reading and writing files. For example, when the DVF-3 acquisition module has finished collecting data, it creates a file containing that data. The DVF-3 evaluation module attempts to read that file. If the file is not found, then the acquisition module has not yet finished its task and the evaluation module will attempt to read the data file a short time later. Attempts to read the data file are continued until the evaluation module successfully reads the data file. Once it succeeds, the evaluation module deletes the data file so that it will not read and process the same data again. Refer to Fig. 1 for a schematic of how modules within a group interact.

Each group runs independently of all other groups. This is a desirable feature since if a single group were to intentionally or unintentionally stop running, the entire system would not be forced to stop as well. Instead a warning message would appear on the monitor stating that a particular group has failed.

Although a given group runs independently of other groups, coordination between at least some, if not all, groups is necessary to insure that relays stay closed long enough to insure that all data acquisition modules have enough time to collect the data. This is required only when a specific signal is processed by more than one acquisition module. In the prototype, for example, the FFT group's acquisisiton module and the DVF-3's acquisition module both process the same signal. In the FFT acquisition module, data is processed in 10 seconds whereas 30 seconds are required by the DVF-3 acquisition module. Thus if the relay is switched to the next monitoring location after 10 seconds, the DVF-3 acquisition module would not have finished its task. A switch control module, called the control node, prevents these conflicts. The control node will open the current relay and close the next one only after the slowest acquisition module (DVF-3) has finished acquiring data. The control node communicates with the acquisition modules with files using the same principle as depicted in Fig. 1.

Another module, called the display/failure node, is required to check if a group has failed. This node will continuously check the output file containing the evaluation of every group's evaluation module. If a particular evaluation module does create an output file within a user prescribed time, then the system will report that group as failed. A schematic showing how the nodes interact with the groups is given in Fig. 2.

The display/failure node, aside from checking on the status of every group, will also display the evaluations and data associated with each group. In the prototype, several screens exists; Figs. 3,4,5. The master screen, Fig. 3 shows each monitored location in a color coded box. The color of each box indicates the status of that location; i.e., poor, satisfactory or good. Specific information such as the frequency spectrum, Fig 4, or phase, gapvolts, speed, and overall vibration content, Fig 5 is available in the FFT and General Data screen respectively.

CREATING A MODULE

IMS comes with modules that control the DVF-3 and the Rapid System's FFT Spectrum Analyzer and their respective evaluation modules. Also included are the two nodes. Since the acquisition modules control specific instruments, the user may be interested in creating modules for other instruments. The user may also be interested in personalizing the evaluation modules to meet specific needs of an application which the IMS evaluation modes cannot meet. Customizing an application is quite simple and can

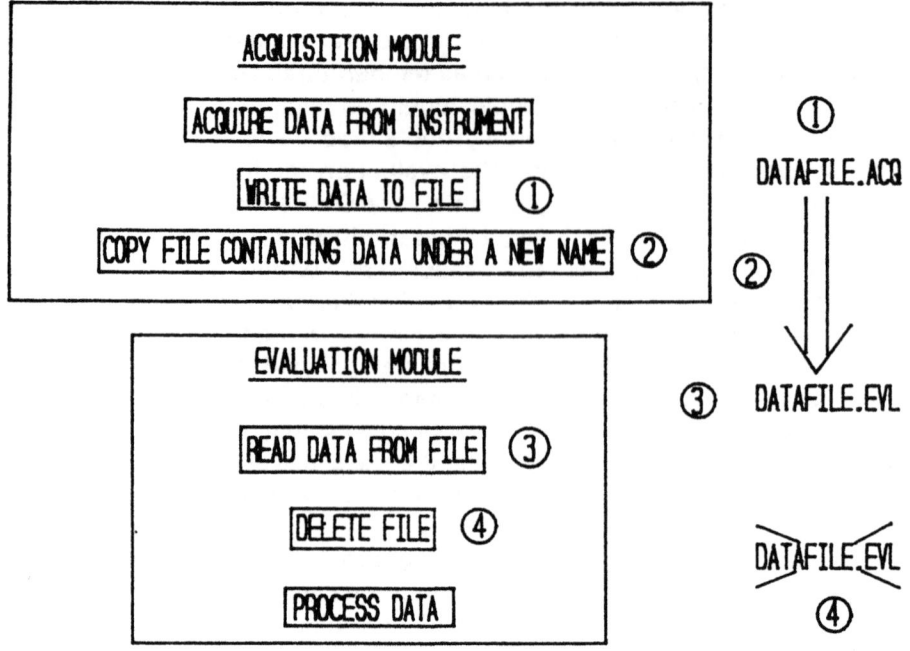

Figure 1. Interaction of modules within a group

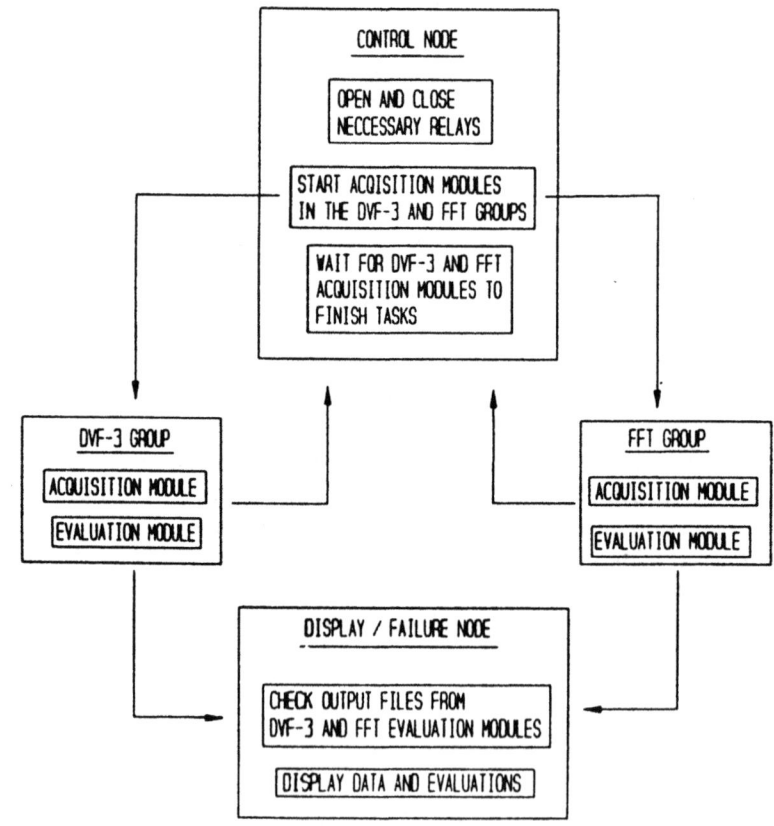

Figure 2. Interaction of nodes with groups

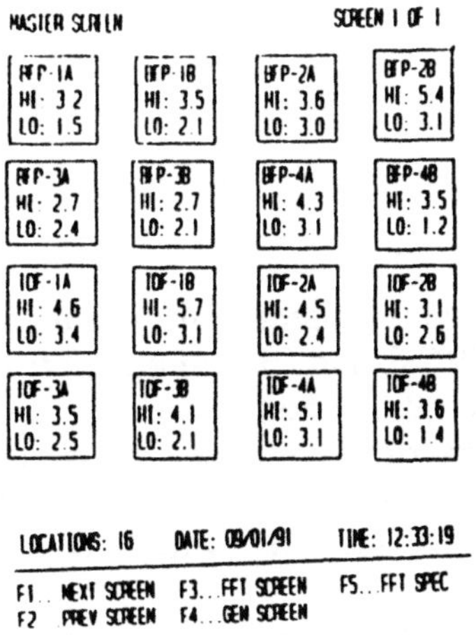

```
MASTER SCREEN                    SCREEN 1 OF 1

┌────────┐ ┌────────┐ ┌────────┐ ┌────────┐
│ BFP-1A │ │ BFP-1B │ │ BFP-2A │ │ BFP-2B │
│ HI: 3.2│ │ HI: 3.5│ │ HI: 3.6│ │ HI: 5.4│
│ LO: 1.5│ │ LO: 2.1│ │ LO: 3.0│ │ LO: 3.1│
└────────┘ └────────┘ └────────┘ └────────┘

┌────────┐ ┌────────┐ ┌────────┐ ┌────────┐
│ BFP-3A │ │ BFP-3B │ │ BFP-4A │ │ BFP-4B │
│ HI: 2.7│ │ HI: 2.7│ │ HI: 4.3│ │ HI: 3.5│
│ LO: 2.4│ │ LO: 2.1│ │ LO: 3.1│ │ LO: 1.2│
└────────┘ └────────┘ └────────┘ └────────┘

┌────────┐ ┌────────┐ ┌────────┐ ┌────────┐
│ IDF-1A │ │ IDF-1B │ │ IDF-2A │ │ IDF-2B │
│ HI: 4.6│ │ HI: 5.7│ │ HI: 4.5│ │ HI: 3.1│
│ LO: 3.4│ │ LO: 3.1│ │ LO: 2.4│ │ LO: 2.6│
└────────┘ └────────┘ └────────┘ └────────┘

┌────────┐ ┌────────┐ ┌────────┐ ┌────────┐
│ IDF-3A │ │ IDF-3B │ │ IDF-4A │ │ IDF-4B │
│ HI: 3.5│ │ HI: 4.1│ │ HI: 5.1│ │ HI: 3.6│
│ LO: 2.5│ │ LO: 2.1│ │ LO: 3.1│ │ LO: 1.4│
└────────┘ └────────┘ └────────┘ └────────┘

LOCATIONS: 16    DATE: 09/01/91    TIME: 12:33:19

F1   NEXT SCREEN    F3...FFT SCREEN    F5...FFT SPEC
F2   PREV SCREEN    F4...GEN SCREEN
```

Figure 3. Master screen

FFT SCREEN

BFP-1A #1 IB BRG

RPM RANGE	MAX COMP [MILS]	STATUS
0-1000	1.7	GOOD
1001-2000	2.9	POOR
2001-3000	2.1	FAIR
3001-4000	3.5	FAIR
4001-5000	1.1	GOOD

SHAFT SPEED: 3600 RPM

DATE: 09/03/91   TIME: 09:12:45

F1...OTHER LOCATION   F3...MASTER
F2...SPECTRUM   F4...GENERAL SCREEN

Figure 4. FFT screen

GENERAL SCREEN

BFP-1A #1 IB BRG

	OVERALL [MILS]	1X	PAHSE [DEG]	1X AND PHASE	GAPVOLTS
CURRENT VALUE	3.5	2.5	325	*	7.8
STATUS	FAIR	FAIR	*	NOMINAL	ACCEPTABLE
CURRENT-REF	0.1	-0.2	-74	*	1.1
STATUS	GOOD	GOOD	WARNING	WARNING	POOR

SHAFT SPEED: 3600 RPM

DATE: 09/03/09   TIME:09:12:51

F1...OTHER LOCATION   F3...MASTER SCREEN
F2...SPECTRUM   F4...FFT SCREEN

Figure 5. General Data screen

be done by following the format given in the shell program. The shell program shown in Fig. 6 insures that a customized module can communicate with the existing modules.

STANDARDS

Although it is possible to create a customized evaluation module, it is probably unnecessary. IMS allows the DVF-3 and FFT evaluation modules to be changed by using the standards editor. A standards editor changes the parameters that a module uses to evaluate the performance of a machine. These parameters are found in the user-defined standards. For example, if vibration as function of shaft speed requires evaluation, then the vibration and speed data would be referenced to the AMPLITUDE vs. SPEED standard.

There are three types of standards; scaler, vector, and graphic. The scaler standards are the simplest. They involve only a piece of data such as speed or gap voltage. The standard is set by defining an acceptable median value for the data, and a series of one or more deviations by which the data's value may vary around the median. In the prototype system, a scalar standard for gap volts was developed. The median value was defined to be 8.0 volts. Any value within 0.5 volts of the median (7.5 volts to 8.5 volts) was considered "nominal". Any value greater than 0.5 volts but less than 1.0 volt from the median (7.0 volts to 9.0 volts) was considered to be in a "warning" condition. Deviations greater than 1.0 volt were in the "alarm" condition.

Vector standards are slightly more complex: they involve two variables that can be used to define a two dimensional vector, such as vibration amplitude and phase angle. These standard are set by defining the nominal magnitude and phase angle of the vector formed by the two variables, and a series of one or more allowable deviations by which the magnitude of the actual vector is allowed to vary from the nominal one. An illustration of this type of standard for synchronous vibration phase for the rotor kit is given in Fig 7. The nominal vector is 2 mils [50.8 μm] at a phase of 40 degrees. Actual vectors whose tips fall to within a 1 mil [25.4 μm] circle of this vector are considered "nominal". Those greater than 1 mil [25.4 μm] but less than 2 mils [50.8 μm] are marginal, greater than 2 mils [50.8 μm] and less than 3 mils [76.2 μm] are taken as a "warning", and vectors greater than 3 mils [76.2 μm] are in "alarm" conditions.

The final type of standard used employs a graphic format of one variable vs. another. These graphs are divided into operational regions, similar to traditional amplitude vs. speed plots. A machine's status is found by determining which of the plot's regions contain its operating parameters.

For more information on these standards, see reference [5].

HARDWARE

A schematic of IMS's hardware is shown in Fig. 8. It includes a 16 MHz -80386 pc, a Bernoulli Box, the Rapid System's FFT Spectrum Analyzer, Bently Nevada's DVF-3, and Metrabyte's 32 channel relay board [MEM-32], 20 channel thermocouple board [THERM-20], and 16 channel A/D converter board [MAI-16].

The pc must have a 80386 processor or better to insure operation of the Windows 3.0 operating system. Furthermore, at least 5 megabytes of random access memory is required to support the memory requirements of Windows 3.0, the modules, a ramdrive and cache. Note that a cache is not required, but its exclusion would noticeably increase operating time.

The DVF-3 can process two channels simultaneously. It provides machine operating speed, overall and synchronous vibration levels, phase, and gap volts which is a measure of static shaft location. Harmonic and subharmonic vibration content is provided by the FFT Spectrum Analyzer. It has the ability to process four channels in

```
'***************************************************************************
'
'        Shell program written in Microsoft Quick BASIC, Version 4.50
'
'        This shell is for an evaluation module.  A shell for the acquisition
'        module has a similar structure.
'
'***************************************************************************

ON ERROR GOTO ERRORHANDLER:

continue=1           'WHILE loop allows the module to run indefinitely'
WHILE continue=1

  OPEN  "path:filename.xxx" FOR INPUT AS #1
  CALL READDATA( parameter list ) '<<< user defined subroutine'
  CLOSE #1
  KILL "path:filename.xxx"
  CALL EVALUATE_DATA( parameter list ) '<<< user defined subroutine'
  OPEN "path:filename.yyy" FOR OUTPUT AS #1
  CALL SEND_EVALUATIONS_TO_FILE( parameter list ) '<<< user defined subroutine'
  CLOSE #1
  NAME "path:filename.yyy" AS "path:filename.zzz"

WEND

ERRORHANDLER:
 IF (ERR=53) OR (ERR=58) THEN RESUME

'ERR = 53 means that the file does not exist'
'ERR = 58 means that the fIle already exists'

'for ERR=53 or ERR=58, the same line that generated the
'error is executed again

 IF (ERR<>53) AND (ERR<>58) THEN
  PRINT "UNEXPECTED ERROR. ERROR CODE IS: ";ERR
 END IF

END
```

Figure 6. Shell program

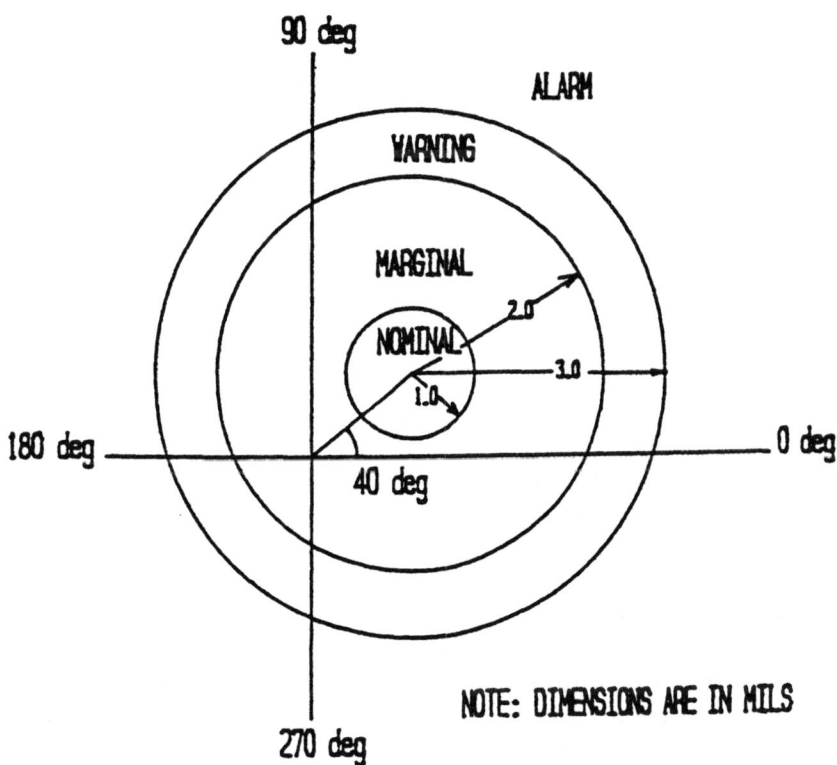

Figure 7. Vector standard

196

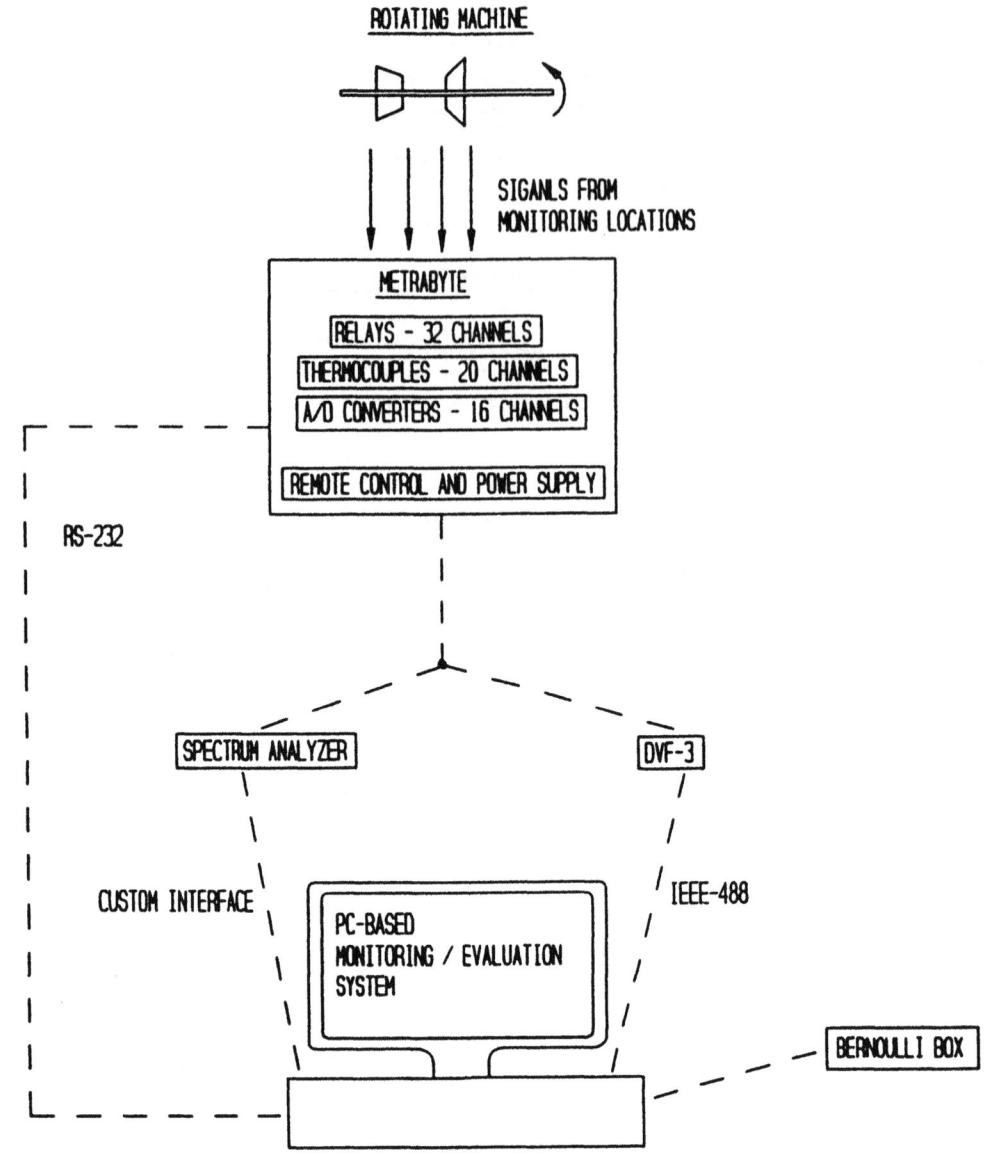

Figure 8. Hardware

succession, but is configured to process only two so as to maintain compatibility with the DVF-3. A standard IEEE-488 card is the programmable interface between the pc and the DVF-3. The Rapid Systems FFT Spectrum Analyzer uses a custom programmable interface. Temperatures are acquired via the Metraybyte THERM-20 board and can handle up to 20 J, K, T, R S, B, or E type thermocouple. Switching is performed by the Metrabyte MEM-32 relay board. The MAI-16 board is intended to handle the pressure and flow rate signals. The MAI-16, THERM-20 and MEM-32 are interfaced with the pc via the REM-64 controller which in turn is connected to the pc through the RS-232 serial port in the pc.

The Bernoulli box is a large storage medium used to hold logged data on removeable 20 megabyte disks.

CONCLUSION

IMS has successfully monitored and evaluated the performance of a rotating machine in the Windows 3.0 multitasking environment. This environment provides for a flexible system that supports a variety of monitoring instruments and can be applied to all types of rotating machinery. The flexibility is further given by allowing the user to customize the evaluation (standards) for these specific machines.

ACKNOWLEDGEMENTS

This work has been sponsored by a joint industry/Virginia Center for Innovative Technology Grant No. CAE-88-011-01, with additional support from Amoco Research, Bently Nevada, Dresser-Rand, DuPont, Elliott Company, Ingersoll-Rand, and Virginia Power. The authors are especially grateful for the support given by Dr. Ira Jacobsen, Director of the Institute of Computer Aided Engineering, Charlottesville, Virginia and for the technical assistance given by Mr. Roy Mondy, Rotating Equipment Specialist, Virginia Power, Richmond, VA.

REFERENCES:

1. Kirk, R. G., Hoglund, J., and Mondy, R. E., "Development of a PC-based Off-line Expert System Evaluation Of Turbomachinery Response," Proceedings of the 1st International Monitoring & Diagnostics Conference, Las Vegas, Nevada, Sept 11-14, 1989, pp. 439-444

2. Kirk, R. G., Hoglund, J., and Keesee J., "Application of Artificial Intelligence For Rapid Evaluation of Turbomachinery Dynamic Response And Stability," Proceedings of the JSME International Symposium on Advanced Computers For Dynamics and Design, Tsuchiura, Japan, Sept. 6-8, 1989, pp. 149-154.

3. Kirk, R. G., Hoglund, J., and Mondy, R. E., "Development of a Unique PC-based Turbomachinery Expert System," Proceedings of IASTD Conference, Zurich, Switzerland, June 24-26, 1989.

4. Kirk, R. G., Pawtowski, C. E., Typrin, M. "An Intelligent Standards Filter For a PC-based Expert System," Proceedings of 45th Meeting of the Mechanical Failures Prevention Group, Annapolis, Maryland, April 9-11, 1991.

SOME EFFECTS OF THE EMERGING SCIENCE OF DESIGN ON THE DESIGN OF TURBOMACHINERY USING A COMPUTER-BASED DESIGN MODEL IN CONJUNCTION WITH A KNOWLEDGE-BASED EXPERT SYSTEM

Eur Ing Professor Dr. Douglas Muster
Department of Mechanical Engineering
University of Houston
Houston, Texas 77204-4792

ABSTRACT

Engineers are at a turning point with respect to how they view their design problems. They are moving from a world view based in the Newtonian concepts of reductionism and mechanism and closed systems in equilibrium isolated from their environments to another based in the holism of systems thinking and systems open to their environments. The effects of this Kuhnian paradigm shift are discussed here in terms of the state of the art and state of the science of design, especially as they apply to the design of turbomachinery.

INTRODUCTION

Modern computers provide designers with a unique, integrating medium capable of both representing and storing knowledge and using it to conduct the processes of design. In this paper and its companion (Muster, 1992), two related aspects of the use of knowledge-based expert systems in the computer-based design of turbomachinery are discussed. The specific problems associated with organizing and using knowledge-based expert systems are embedded inextricably in the larger and more general design problem of using metalevel, discipline-independent approaches and techniques in conjunction with discipline-dependent knowledge bases and expert systems (Muster et al., 1989). In turn, the state of the latter topic is directly related to the state of the emerging science of design.

There is a direct relationship between the scope and complexity of a design problem and the form and depth of the knowledge bases and expert systems used by designers to establish a family of satisficing solutions to it. Satisficing --- even, optimal --- solutions for simple, single-discipline design problems can be established using well-known computer-aided-design (CAD) methods. The sequential, iterative nature of CAD methods requires that the factual information input to the computer be proper and correct, nothing more. Springs, bearings, simple structural elements and their connectors and single-purpose components (such as electric motors and generators) can be designed in this way. Usually, the data base used in conjunction with a CAD method is a computer-based archive of facts alone, not facts and rules. The method itself includes all of the rules needed to establish a solution. Decisions internal to the method are made using clearly stated decision trees (not expert systems).

When the system being designed is relatively complex and its emergent properties and behavior are characterized by single-cause-multiple-effect laws, often CAD methods are simply not appropriate, except for the design of relatively simple subsystems whose characteristics and environment can be defined in isolation from the system considered as a whole. Under these circumstances, satisficing --- surely not optimal --- families of solutions can be established using computer-based systems approaches in which the decisions of a designer are supported by the interactive use of knowledge bases and expert systems. A gas turbine and its auxiliary equipment is an example of such a system; its bearings could be selected (that is, designed) using a CAD method. The design of a gas turbine would be made more difficult if the designer attempted to take into account such down-the-road considerations as manufacturing and maintaining the gas turbine system. It is this last case with which we are concerned here.

In the sections that follow, we discuss the state of the art and the state of the science of design, particularly as they relate to the design of systems. A discussion of the knowledge bases and expert systems that can be used in modern, decision-based design approaches is given in the companion paper referred to earlier (Muster, 1992).

THE STATE OF THE ART AND SCIENCE OF DESIGN

From many sources, there is strong evidence to suggest that the future in engineering and, particularly, in engineering design is being shaped by the synergism occurring between three interrelated influences, namely, systems thinking, computer technology and a systems-based paradigm that uses the concept of defining problems in terms of the transdisciplines in which they are embedded. This last permits an engineer to use the newly developed systems-based approaches to design discussed elsewhere (Beitz, 1987) (Hubka, 1982) (Muster and Mistree, 1988, 1990) (Muster et al., 1989) (Pahl and Beitz, 1984).

Systems thinking and computer technology exist in an overlapping world of synergistic action. In this combined world, interactive changes occur almost too swiftly to be recognized as occurring separately. The clearly delineated bounds of the traditional, Newtonian-science-based disciplines can no longer serve as the universally accepted and sole means for defining engineering problems. Instead, the broadening influence of systems thinking encourages engineers to approach a design problem that spans portions of several traditional disciplines in terms of a new paradigm. Using this paradigm, engineers define their design problems and, simultaneously and consciously, they define the laws and relationships that characterize the body of knowledge, the transdiscipline, in which the problem is embedded.

A REVIEW OF THE ART AND SCIENCE OF DESIGN

In the rationale above, we have stated the assumptions on which we have based the review of the state of the art and state of the science of design given below. From the perspective of this view of design, we can develop a statement that describes the contemporary roles of systems thinking and computer technology in negotiating satisficing solutions to design problems arising in transdisciplinary bodies of knowledge.

The Passage of Design from an Art to a Science: When electronic computers were introduced, about five decades ago, they were used simply as the preferred means for making the iterative, number-crunching processes more efficient. Design itself and the cleverness with which the computer expertise of the designer could be deployed remained arts. Design theory --- an early euphemism for the science of design --- was a backwater of research ignored by most engineers and scientists.

Finally, change came. In engineering institutions, there was a swing from emphasis on pragmatism and the art of engineering to emphasis on the engineering sciences and especially to research in mechanics, the thermal sciences and metallurgy, which were identified closely with the practice of engineering (Simon, 1981). Little research was directed towards establishing the science of design and there was little incentive to seek new design paradigms. Instead, much effort was spent on improving the efficiency of the analytically based, consecutively applied, incremental methods which had been used since the start of the Industrial Revolution.

The Science of Design Today: Recent work by members of a relatively broad-based community of researchers in design theory suggests that the development of a transdisciplinary body of knowledge which supports the science of design can be and is being realized. However, it is probably a realistic assessment of the situation to state that, at this time, the science of design is in a pretheory stage (Dixon, 1987) (Simon, 1981, p. 155 et seq.). The body of work developed thus far has been largely uncoordinated and, in the main, reflects the research interests of relatively narrowly oriented individuals. However, a cohesive, world-wide, goal-directed community of researchers in design theory --- whose origins and principal interests are in engineering --- is developing with concentrations of activity in Australia, Denmark, Germany, Great Britain, Japan, the Netherlands and the United States.

The community-at-large of design theorists includes many nonengineering researchers from such diverse disciplines as architecture, economics, systems theory, management science and even the social and behavioral sciences. Over a decade ago, Simon (1981, p. 129) stated, "Everyone designs who devises courses of action aimed at changing existing conditions into preferred ones." The general nature of design and the catholic interests that motivate designers are discussed by Coyne et al. (1990, p. 5). In their words, "Design is a purposeful activity ... [that] involves a conscious effort [on the part of the designer] to arrive at a [desired] state of affairs in which certain characteristics [that satisfy a perceived pattern of needs] are evident."

Systems Thinking, Systems Practice: The science of design being developed in the studies referred to above is based in systems thinking and its applications in systems practice. When it is applied to the design of a system, it emphasizes both the emergent properties of the system as an entity and the separate and collective properties of the system and its subsystems in their intrinsic environments. It is a "paradigm shift" of the kind defined by Kuhn (1970). It is the antithesis of the traditional approach to design that requires the harmonious coupling of a designer's experience-based intuition and his skills in performing analyses which emphasize both reductionism and mechanism and the design of the components of a system in isolation from the influences of their environments. An implied goal of the studies is to move engineering design from its historical position as the premier mechanical art towards a new, unique and singular position at the interface between art and the natural sciences (Muster and Mistree, 1989).

Prior to about 1970, the common theme of most design-related publications was the development of computer-based techniques for improving the efficiency of existing traditional iterative design methods. A few publications were concerned with how best to introduce creativity into the design process. Apparently, relatively few researchers were interested in developing the science of design. In the seventies, as large-capacity computers became accessible to engineers, the interest of researchers shifted to broad design methods and computer-oriented design tools, such as CAD, CAD/CAM and CAE.

Then, in the eighties, there was a rebirth of interest in the science of design as we approached the turning point described by Capra (1982). Now, as we enter the last decade of the twentieth century, the principal interest in design research appears to be the development of the transdisciplinary bodies of knowledge that would support a definitive theory of design and the development and ap-

plication of computer-based tools to support the decisions of designers.

We believe it is important for engineers to recognize the
open nature of design methods, computer-based or otherwise. Recall that the design methods described in the
literature are guidelines, not protocols. Guidelines do
not require strict adherence to procedure, protocols do.
In this sense, the term "design protocol" is a near-oxy-
moron. Whether they are performing simple titration
tests or complex experiments in quantum mechanics, scientists follow prescribed protocols with robot-like precision and rigor. A measure of their success and skill may
well be that they can describe and perform a protocol so
precisely that, later, other scientists can perform the
same test and expect to obtain exactly the same results.
Recall the recent contretemps involving cold fusion.

Not so in design. There are no universal design protocols followed by all engineers. Just the opposite is
true. There is a key element in the science of design
that requires human intervention and interaction in the
design process. It is central to understanding the
structure and application of the science of design (Muster and Mistree, 1988 and 1989; Muster and Revans, 1989).
As a consequence, we strive to include sufficient flexibility in our guidelines for designers that they can make
good use of their experience-based intuition.

<u>Some Recent Critical Reviews of the Design Literature</u>:
There are several critical reviews of the relatively recent design literature, including that of Shupe (1988)
whose review is summarized in Table 1. These include
those by Pahl and Beitz (1984) (who give a comprehensive
review of the evolution of design in Germany), Hubka and
Schregenberger (1987) (who provide a short review of
trends in the science of design in Europe and in the
United States), Andreasen (1987) (who gives an overview
of the state of art in Europe), Brei et al. (1989) (who
made a survey of research methods to study design), Cross
(1989) (who has written an easy to read review of design
methods), De Boer (1989) (who has reviewed and evaluated
several general-problem-solving and design methods) Finger and Dixon (1989a, 1989b) (who wrote a two-part review
of the state of the art of mechanical engineering design)
and Muster and Mistree (1990) (who summarized recent contributions to the state of the art).

CONCURRENT ENGINEERING DESIGN FOR THE LIFE CYCLE

Some of the same researchers who are investigating the
foundations of the science of design are interested in
another significant engineering problem, the <u>concurrent
engineering design of a product (or system) for its life
cycle (CED/LC)</u>. In the recent past, many facets of the
subject have been studied exhaustively, in particular by
industrial and governmental agencies with a vested inter-

est in being able to purchase reliable, long-service products that are relatively easy to manufacture and maintain (Calkins et al., 1989). The primary goal of concurrent engineering design for the life cycle is to minimize costs to the user over the complete life cycle of a product while maximizing its quality and performance (Winner et al., 1988). This last statement is a paraphrase of the substance of a definition of the processes associated with CED/LC given by Winner.

It is outside the scope of this review to cover in detail the concurrent engineering design of a product for its life cycle; however, in order to make this section sensibly self-contained, we summarize the scope of the problems represented by concurrent engineering design for the life cycle of, say, a gas turbine and how it may be affected by systems thinking and computers.

The Process of Concurrent Engineering Design for the Life Cycle: Designing a system (or product) for its life span involves more than simply creating an artifact capable of accomplishing specific functions. In addition to this, the system must be available to perform its mission reliably for its anticipated service life with minimal downtime for scheduled servicing and maintenance. It must be capable of being produced in quantity using materials and production-line techniques appropriate to the market for which it is intended. It must be made of materials and in a manner which makes service and maintenance relatively easy. Under normal conditions, it should require minimal amounts of consumables, such as lubricants, finishes (say, paints and polishes) and small, expendable parts (say, spark plugs, filters and fuses). Finally, for all of its life span, the product must be safe when it is used by an ordinary user in a manner intended by the designer. In addition and only infrequently, other more specialized considerations may have to be taken into account. These could include such diverse considerations as requiring that the product be resistant to corrosion by a constituent of its environment, that it be capable of being used in conjunction with specific items of equipment, or that it operate more quietly than might normally be expected. Past efforts to introduce these generic and product-specific, life-span considerations into the design process have been limited in nature and have been attempted only for specific products whose service life conditions are known relatively completely or can be foreseen with reasonable certainty.

Three Subcultures in Engineering: The situation has been complicated further by the way in which engineers and managers usually view the spectrum of activities encompassed by a product's life span. In the view of many engineers, in all but extraordinary circumstances, a product's life span is considered in terms of three separate, essentially distinct and nonoverlapping phases, the trichotomous, sequential grouping of design, manufacturing and maintenance.

This process of segmenting the events and activities that comprise the life span of an industrial product, has had the corollary effect of creating and nurturing three sub-cultures in engineering with sometimes conflicting goals and value systems. In most industrial organizations, the three cultural systems exist contemporaneously and in close physical proximity to each other. In the main, communication between the three groups is controlled explicitly by the hierarchical structure of the organization in which they exist as administrative equals. Perhaps, more importantly, it is often the case that effective informal, direct communication between individuals in the groups is not encouraged. In fact, it may be discouraged implicitly by the different intellectual and skill profiles of the individuals recruited for each group. In many organizations, it is not unusual to find that there is little interest among the members of a group in the problems and methods of those in another group.

The bodies of knowledge which support the activities of each group have been developed separately, much in the manner of different specializations in the sciences and mathematics, without significant interaction or interchange of ideas and methods among the practitioners in each field. Often, in academic institutions, the trichotomous relationship among the three cultures is accentuated by placing the faculty and students interested in them in separate academic units, sometimes even in different colleges.

Design and design methods are studied with equal fervor in engineering institutions and in industry but with different emphases. Only in the recent past has the study of manufacturing machinery, techniques and methods emerged as a respectable academic subdiscipline of engineering. Before this shift in emphasis, manufacturing was studied primarily by engineers in industry and not in faculties of engineering and design. Finally, engineers in academic institutions have virtually ignored maintenance of machinery as a subject suitable for academic inquiry. At the same time, engineers in industry usually treat it as a necessary --- but often neglected --- post-manufacturing constraint on the operation of the products they design and manufacture. These attitudes and considerations have stifled attempts to develop a generalized, domain-independent theory within which events which may occur over the life span of an engineering artifact can be taken into account. Without such a theory, no systematic method can be developed for introducing the generic and product-specific, life-span considerations cited above into the early stages of design.

CLOSURE

We have discussed the confluence of circumstances that support our conclusion that design is being shaped by a synergism occurring between three interrelated influences, namely, systems thinking, computer technology and a systems-based paradigm that uses the concept of defining problems in terms of the transdisciplines in which they are embedded. A corollary effect of the interaction among these factors is the creation of a body of knowledge that supports the emerging science of design. The bounds and unique combination of interactive subdisciplines that comprise the new science of design are becoming more clearly defined, permitting engineering designers to organize and use knowledge-based expert systems in the computer-based design of relatively complex mechanical systems, including concurrent engineering design for their life cycles.

REFERENCES

Andreasen, M. M. (1987), "Design Strategy", _Proc., 1987 Intl. Conf. Engrg. Des._, v1 pp 171-178.

Beitz, W. (1987), "General Approach of Systematic Design --- Application of VDI Guideline 2221", _Proc., 1987 Intl. Conf. Engrg. Des._, v1 pp 15-20.

Brei, M. L., D. A. Dierolf, and K. J. Richter (1989), _A Survey of Research Methods to Study Design_, IDA Paper P-2155, Institute of Defense Analyses, Alexandria, VA 23211.

Calkins, D. E., R. S. Gaevert, F. J. Michel and K. J. Richter (1989), _Aerospace System Unified Life Cycle Engineering_, IDA Paper P-2151, Institute for Defense Analyses, Alexandria, VA 22311.

Capra, F. (1982), _The Turning Point_, Simon & Schuster, New York, pp 516.

Coyne, R. D., M. A. Rosenman, A. D. Radford, M. Balachandran, and J. S. Gero (1990), _Knowledge-based Design Systems_, Addison-Wesley Publishing Co., New York, pp 567.

Cross, N. (1989), _Engineering Design Methods_, John Wiley & Sons, New York.

De Boer, S. J. (1989), _Decision Methods and Techniques in Methodical Engineering Design_, Academisch Boeken Centrum, De Lier, The Netherlands.

Dixon, J. R. (1987), "On Research Methodology towards a Scientific Theory of Engineering Design", _Artificial Intelligence for Engrg. Des._, v1n3 pp 145-157.

Finger, S., and J. R. Dixon (1989a), "A Review of Research in Mechanical Engineering Design. Part I: Prescriptive, Descriptive and Computer-based Models of Design Processes", _Research in Engineering_

206

Design, v1 pp 51-67.

Finger, S., and J. R. Dixon (1989b), "A Review of
Research in Mechanical Engineering Design. Part II:
Representations, Analysis and Design for the Life
Cycle", Research in Engineering Design, v1 pp 121-
137.

Hubka, V. (1982), Principles of Engineering, Butterworth,
London, pp 352.

Hubka, V., and J. Schregenberger (1987), "Paths towards
Design Science", Proc., 1987 Intl. Conf. Engrg.
Des., v1 pp 3-14.

Kuhn, T. S. (1970), The Structure of Scientific
Revolutions, University of Chicago Press, Chicago,
IL, pp

Muster, D. (1992), "The Organization and Use of
Knowledge-based Expert Systems in the Computer-based
Design of Turbomachinery", Proc., Turbodynamics '92,
pp.

Muster, D., and F. Mistree (1988), "The Decision Support
Problem Technique in Engineering Design", Intl.
Jour. Appl. Engrg. Educ., v4n1 pp 23-33.

Muster, D., and F. Mistree (1989), "Engineering Design
as it Moves from an Art towards a Science: Its
Impact on the Education Process", Intl. Jour. Appl.
Engrg. Educ., v5n2 pp 239-246.

Muster, D., and F. Mistree (1990), "Issues in Engineering
Design Research as Engineering Design Moves from an
Art towards a Science", ASEE Jour. Engrg. Educ.,
v80n8 pp 1014-1016.

Muster, D., F. Mistree and J. K. Allen (1989),
Partitioning and Planning for Unified Life Cycle
Engineering, DM&A, Inc., Houston, Texas, USAF
Contract FY 1457-88-05215, Sep 1989, pp 104/5 disks.

Muster, D., and R. W. Revans (1989), "Action Learning:
An essential element in the emerging science of
design", Proc., 33rd Annual Mtg. Intl. Soc. Sys.
Sci., vII pp 56-64.

Pahl, G., and W. Beitz (1984), Engineering Design, Ed.,
K. M. Wallace, The Design Council, London, pp 450.

Shupe, J. A. (1988), Decision-Based Design: Taxonomy and
Implementation, Ph. D. Dissertation, University of
Houston, Department of Mechanical Engineering,
Houston, Texas 77204-4792, pp 188.

Simon, H. A. (1981), The Sciences of the Artificial,
2nd ed., The MIT Press, Cambridge, MA, pp 247.

Winner, R. I., J. P. Pennell, H. E. Bertrand and M. M. G.
    Slusarczuk (1988), "The Role of Concurrent
    Engineering in Weapons Systems Acquisition", IDA
    Report R-338, Institute for Defense Analyses,
    Alexandria, VA 22311.

Yoshikawa, H., and T. Tomiyama (1984), "Knowledge
    Engineering and CAD", Proc. Intl. Symposium on
    Design and Synthesis, Tokyo, Japan.

THE ORGANIZATION AND USE OF KNOWLEDGE-BASED EXPERT SYSTEMS IN THE COMPUTER-BASED DESIGN OF TURBOMACHINERY

Eur Ing Professor Dr. Douglas Muster
Department of Mechanical Engineering
University of Houston
Houston, Texas 77204-4792

ABSTRACT

System thinking and computer technology have grown explosively in the past three decades and in the process have given engineering designers a new world view with which they can frame their problems and new computer-based tools with which they can negotiate families of satisficing solutions to them. A key element in these new design processes and the body of knowledge which supports them is the organization and deployment of knowledge by means of modern computers that are capable of providing designers with a unique, integrating medium capable of representing and storing knowledge and using it to conduct the processes of design. Thus, the organization and use of both general archival and system-specific knowledge bases are of essential importance to designers.

INTRODUCTION

The explosive synergism that has occurred between systems thinking and computer technology in the past three decades and its effects on design, particularly engineering design, is discussed in a companion paper (Muster, 1992) and several other recent publications Capra (1982), Checkland (1981), Coyne et al. (1990), Dixon (1987), Muster et al. (1989) and Simon (1981).

Capra (1982) attributes the "turning point" occurring in science and technology in the eighties to the ".. dramatic change of concepts and ideas that has occurred in physics during the first three decades of the century, and is still being elaborated in our current theories of matter." In engineering, these changes have had the particular effect of moving designers from a world view based in the Newtonian concepts of reductionism and mechanism and closed systems in equilibrium isolated from their environments towards another based in the holism of systems thinking and systems open to their environments. The effects of this Kuhnian paradigm shift are discussed in the companion paper (Muster, 1992) in terms of their effects on the state of the art and state of the science of design.

We note in our companion paper that modern computers provide designers with a unique, integrating medium capable of representing and storing knowledge and using it to conduct the processes of design. In turn, the specific problems associated with organizing and using knowledge-based expert systems (Coyne et al., 1990) are embedded inextricably in the larger and more general design problem of using metalevel, discipline-independent approaches and techniques in conjunction with discipline-dependent knowledge bases and expert systems (Kamal et al., 1987). The latter topic is directly related to the state of the emerging science of design, the subject of the companion paper.

KNOWLEDGE BASES: THEIR ORGANIZATION AND USE BY DESIGNERS

The general knowledge base available to a designer prior to the time the goals of a specific design project have been established is archival in nature. Usually, it is relatively extensive and unstructured. It contains data, information, rules and standards related to design methods (from simple, experienced-based, analytical procedures to relatively sophisticated and formal CAD methods), materials (for structural purposes and for use as lubricants and coolants and in special-purpose components), complete systems (from central-station power plants to gas-turbine power packages) and the small systems and components (such as air compressors, pumps, blowers, generators, fuses, batteries and gaskets) used in virtually every engineering system. Significant portions of it are related to the engineering sciences, mathematics, the computer sciences and other general technological subjects. It represents the general archival knowledge source from which a designer can assemble the knowledge base required by a specific design project.

Three concepts on which to base a knowledge-based model of design are given in Coyne et al. (1990, Ch. 1). Their approach is based on the three concepts: "(1) _representation_, how we represent information in a computer system; (2) _reasoning_, what we understand by design reasoning; (3) _syntax_, the role of syntactic knowledge in design and computer systems." In addition, they present an argument "... for a model of design in which two separable tasks are operative: _interpretation_, the mapping between design descriptions and their performance requirements, and _generation_, the composition of designs." [Emphases shown above are in the original.]

Once the decision has been made to proceed with the design of an engineering system, data, information, rules and standards related to the design, manufacture and operation of such a system or of systems like it are assembled and collated by searching the general knowledge base. This specialized knowledge base represents the state of the art in the specific field that supports the design of the engineering system under consideration at the time the project is initiated. In part, it is related to the design problem at hand and its position relative to the state of the art. The archival knowledge base may be only marginally useful in the original design of a product which will be at the cutting edge in a new field of technology. For example, we cannot design the means to use the new superconducting materials under development today. As a conceptual system, we can conceive of the general means required to cool the superconducting materials and use them in bar or wire form in a generator, but we cannot manufacture a real-world, concrete system until the state of the art has progressed to the point where the bars and wires can be fabricated. For a variant or adaptive design project, the archival general knowledge base may contain truly up-to-date information from which the designer can form a useful system-specific knowledge base. The development of a typical knowledge base of this type is shown in Figure 1. It is shown relative to the spectrum of activities that comprises the life cycle of a product. As information is received from monitoring the operation of prototypical and production versions of the product system, the system-specific, state-of-the-art knowledge base can be refined and augmented continuously. In Figure 2, we show a representation of this process for a gas-turbine propulsion system.

A conceptual representation of the ideal processes by means of which a state-of-the-art knowledge base is organized and refined is shown in Figure 1 as it progresses from its initial state (derived from archival sources) to a fully developed state (derived from information obtained by monitoring the performance of the specific system created and manufactured by the designer (Figure 2) and reports from the current literature in the fields allied to it (the latter are not shown in Figure 2)).

The knowledge base to which reference is made in Figure 2 could be organized in several ways. We show that the general class of the system to be "Rotating Machinery", its class "Turbines" and its subclass "Gas Turbines". Finally, after the system is built and in operation, a collection of files would be created in which the performance of "Gas Turbine 1", "Gas Turbine 2", etc., each a particular gas turbine power plant installed and operating in a specific vehicle, would be recorded.

In another part of the knowledge base there would be information related to the small systems and components used in the gas turbines (such as pumps, blowers and compressors, rolling-element bearings, journal bearings, gears, fasteners, vibration isolation mountings, shafts, lubricants and accessories). Eventually, when the system is in operation and its performance is being monitored, information on performance-related factors (such as wear of individual components and vibration amplitudes measured at specific locations) will be collected in files assigned to "Gas Turbine x". This data may be com-

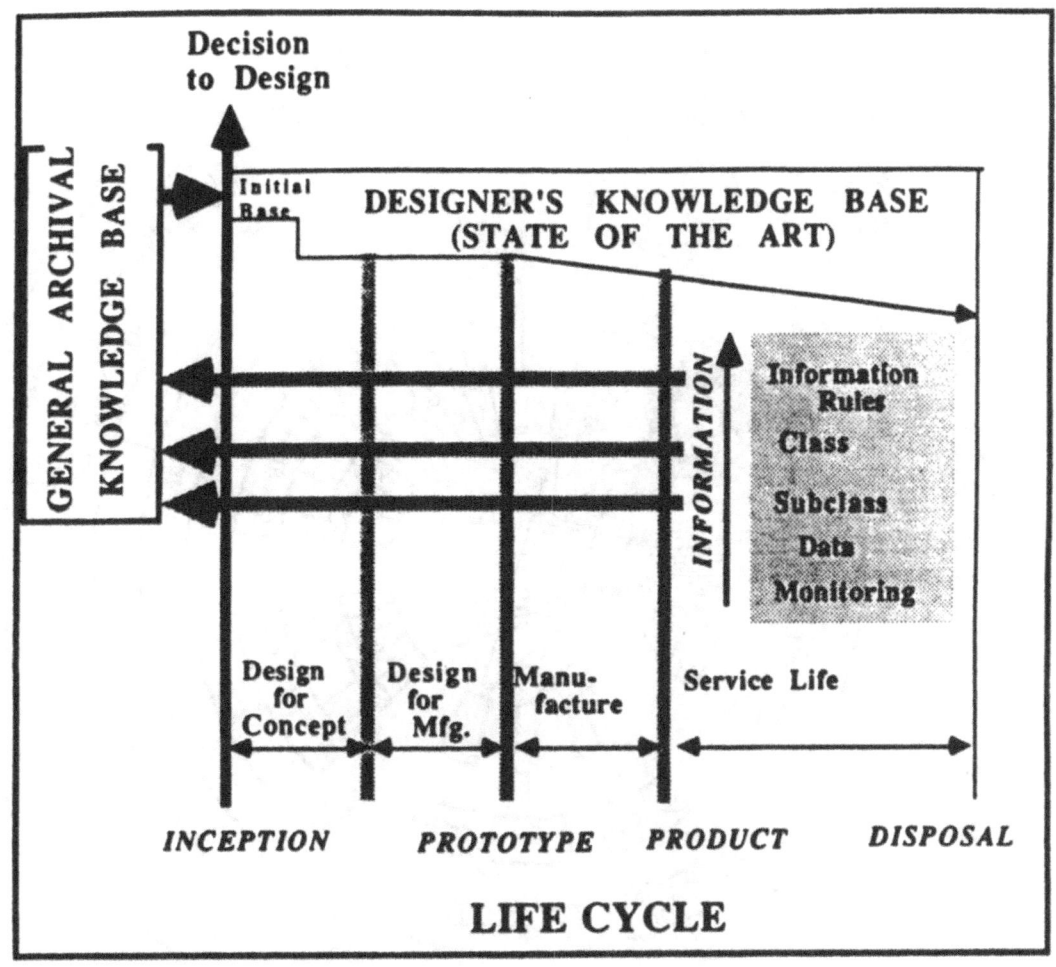

Figure 1
Constructing and using a state-of-the-art knowledge base.

pressed, reduced or treated in other ways; it may be related to specific performance parameters. Later, the
data or a summary of its principal characteristics may be
transferred to comparable files which deal with gas turbines with similar characteristics (speed, size, power,
etc.). Still later, the data in this subclass file can
be summarized and sent to the class files on "Turbines"
and then on to the files that deal with even more general
classes of machines.

In general, information in the knowledge bases we have
just described is moved outward in a hierarchy of files
arranged so as to accept detailed data and information
related to individual machines only (Figure 2). As the
data and information are processed and moved outward, its
identity is changed from the particular to the general
and, ultimately, it may be transformed into proprietary
(company), industry (SAE), national (ANSI) or international (ISO or IEC) standards representative of a broad
class of machines. Except for already-known, state-of-
the-art data and information, the files are constructed

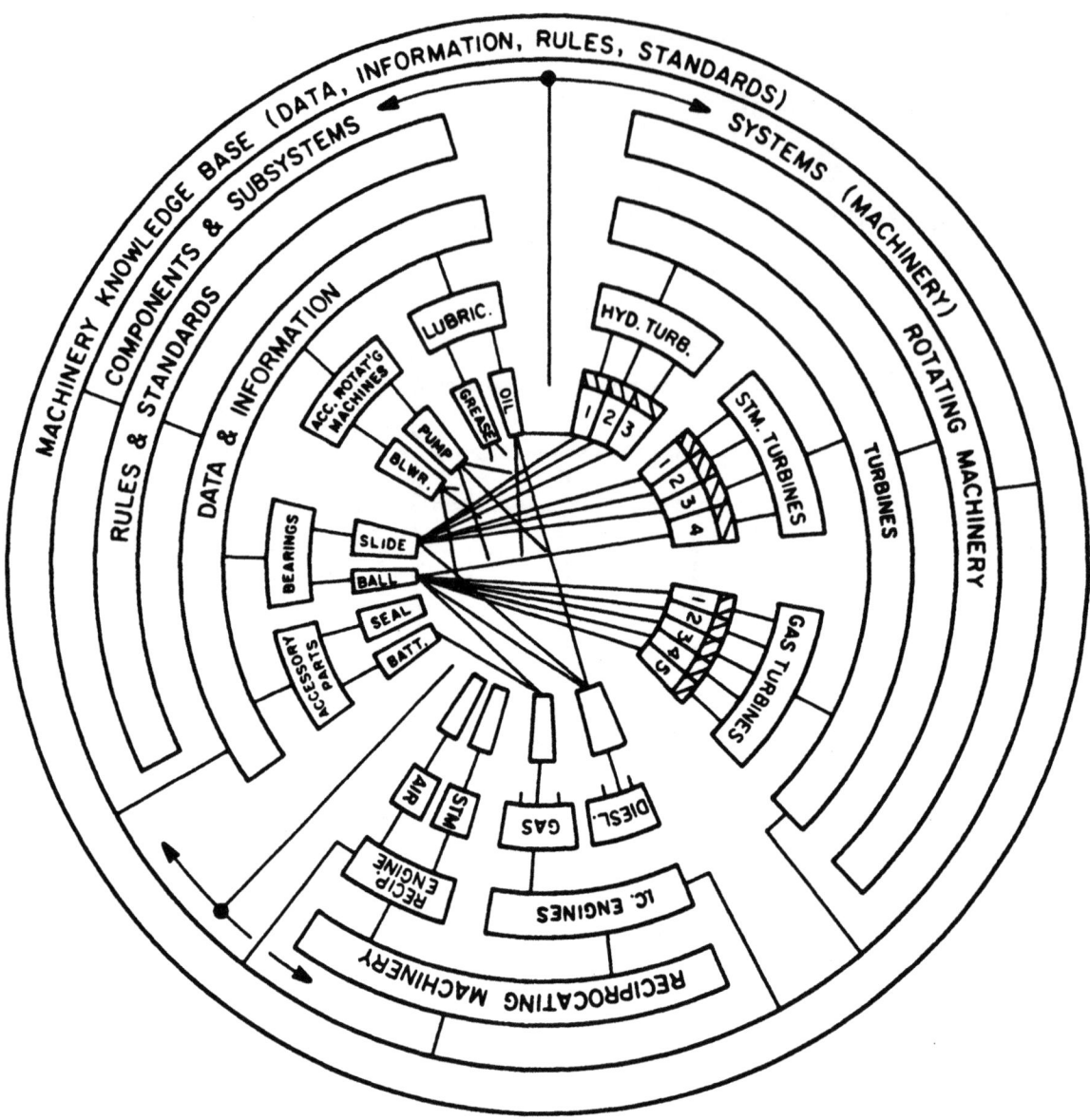

Figure 2
Evolution of a state-of-the-art knowledge base
developed for gas turbines.

from the inside of the circular representation outward
but are used by designers from the outside towards the
center.

System Time: In a physical system, such as a gas turbine
and its auxiliaries, system time can be monitored and
measured in two ways, in physical units of time (say,
seconds), or in event-based generalized-time units related to factors by which the performance of the system can
be measured, such as the on-line use of the system or the
wear patterns of its parts (Figure 3).

Physical time can be used in an engineering system only
if it can be related to the system's processes, or to
events in the system's environment.   This can occur only

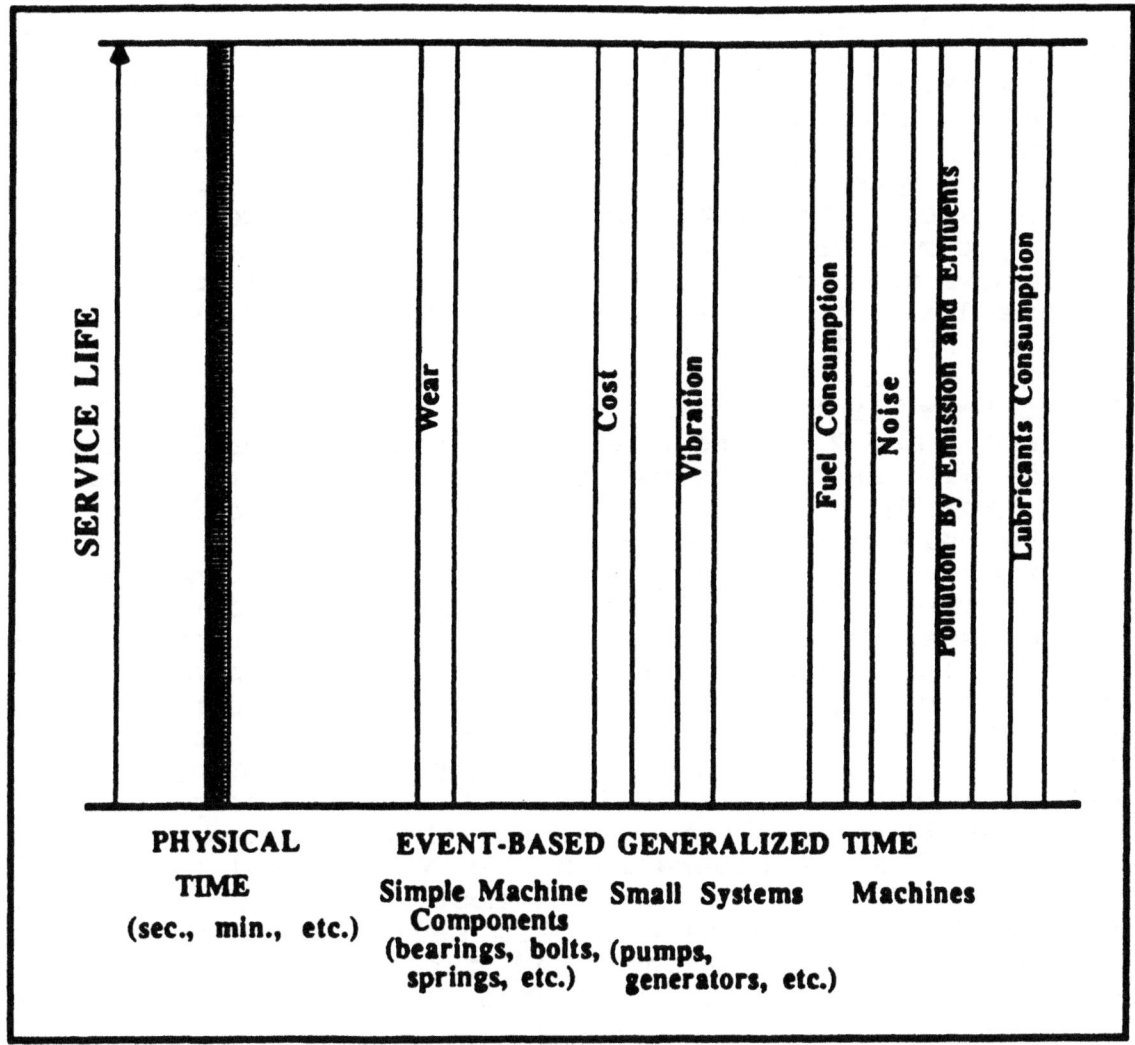

Figure 3
The relation of physical time and event-based generalized
time in the service-life of the life cycle of a product.

as the result of a deliberate, a priori choice made in
the early stages of design or as a modification of the
original design which stems from information obtained by
monitoring the system when it is operating. In either
case, knowledge of the system's processes and their relationship in physical time or event-based generalized time
can be used by the designer to cue the response of the
subsystem that controls the system's processes.

Event-based generalized time in a physical system (for
example, a gas turbine or a similar item of high-speed
rotating machinery) can be related to such performance-
related factors as on-line use, cumulative wear or the
rate of wear (as measured directly or in the detection
and analysis of wear particles in the machine's lubricants), changes in temperature or in the rate of flow of
process fluids in an absolute sense or of one location
relative to another, chemical changes, and the frequency
and amplitude of vibration (as measured at critical loca-

214

tions in the system)(Muster and Muster, 1990). In effect, the subsystem that controls the system's processes is programmed to respond with a specified action (for example, an audio or visual warning signal to the system's operator) when the system, or one or more of its elements reaches a predetermined threshold value of a selected performance-related factor. In this case, knowledge of the system's processes and their relation to, say, the amplitude of vibration measured at a bearing can be used at one level to cue a warning to the operator and at a greater level to shut down the system.

Event-based information taken at any of the levels in Figures 1 and 2 can be used to construct generalized-time scales such as those shown in Figure 3. It is logical to ask, "Are the event-based, generalized-time scales of the figure related to physical time?" In theory, at least, the answer is yes. Given enough data and information, we could devise transfer functions for an individual machine (or classes of machines or components common to all machines in a class) by means of which the two scales could be related. However, in order for such transfer functions to be useful, they would have to account properly for the periods when a machine simply sits idle, for the effect of the separate reliabilities of its components on the overall reliability of the machine taken as a whole and other like considerations. As a practical matter, it is not feasible to do this for most machines. It can be and is done for certain small, self-contained systems (such as electric motors and generators, batteries, pumps, etc.) and individual machine components (such as machine screws which are subject to repetitive loads and rolling-contact bearings).

Standards --- company proprietary, industry, national and international --- have been developed for the use of designers who select such components for use in their designs. However, engineers have been unable to devise practical, suitably precise transfer functions that relate physical time and significant events in the service lives of relatively large, complex systems. In these cases --- and there are many --- we are denied the benefits associated with being able to use the precision with which we can measure physical time. For these reasons, we have concluded that, in general, it is better to use event-based generalized time than physical time as a basis for devising appropriate schedules for monitoring the performance of most machines or classes of machines. Further, time-in-service (in hours or months) can be used as a single-figure- of-merit criterion for rating small components, but it is inappropriate to use such a criterion to estimate the service life of relatively large, expensive complex systems.

In summary, the information obtained from manufacturing and operational experience (as monitored by the event-based factors cited above) can be used by designers in concurrent engineering design for the life cycle of a product. The updating, expansion and refinement of the knowledge and data bases that contain this information is a continuous and, where feasible, automated process.

CLOSURE

In the past five decades, the design of lightweight, high-speed turbomachinery has paralleled closely developments in computer technology and systems thinking. The metalevel, systems-based, discipline-independent, design approaches that we are learning to use today in conjunction with discipline-dependent knowledge bases and expert systems have contributed significantly towards improving the quality of our designs. In addition, they have provided engineers with a rationale for making choices in the design of complex systems, say, a gas turbine and its auxiliary equipment. We have learned to monitor the performance and operating parameters associated with manufacturing and maintaining turbomachinery. We have improved significantly our ability to reduce the metaphorical distances between designing for concept, designing for manufacture and designing for maintenance.

REFERENCES

Capra, F. (1982), The Turning Point, Simon & Schuster, New York, pp 516.

Coyne, R. D., M. A. Rosenman, A. D. Radford, M. Balachandran, and J. S. Gero (1990), Knowledge-based Design Systems, Addison-Wesley Publishing Co., New York, pp 567.

Dixon, J. R. (1987), "On Research Methodology towards a Scientific Theory of Engineering Design", Artificial Intelligence for Engrg. Des., v1n3 pp 145-157.

Kamal, S.Z., H. M. Karandikar, F. Mistree and D. Muster, "Knowledge Representation for Discipline-independent Decision Making", Expert Systems in Computer-Aided Design, Ed., J. Gero, Elsevier Science Publishers, North-Holland, The Netherlands, pp 289-321.

Muster, D. (1992), "Some Effects of the Emerging Science of Design on the Design of Turbomachinery Using a Computer-Based Design Model in Conjunction with a Knowledge-based Expert System", Proc., Turbodynamics '92, pp

Muster, D., F. Mistree and J. K. Allen (1989), Partitioning and Planning for Unified Life Cycle Engineering, DM&A, Inc., Houston, Texas, USAF

Contract FY 1457-88-05215, Sep 1989, pp 104/5 disks.

Muster, G. L., and D. Muster (1990), "Some requirements

216

for automated data-reduction systems suitable for processing vibration measurements taken on high-speed rotating machinery for the support of operational and maintenance decisions", <u>Vibration and Wear in High Speed Rotating Machinery</u>, NATO ASI Series, Series E: Applied Sciences - Vol. 174, Ed., J. M. Montalvão e Silva and F. A. Pina da Silva, Kluwer Academic Publishers, Dordrecht, The Netherlands, pp 741-758.

Simon, H. A. (1981), <u>The Sciences of the Artificial</u>, 2nd ed., The MIT Press, Cambridge, MA, pp 247.

Optimising Parameters in Gear Hob Manufacturing to Reduce Gear Transmission Vibration and Wear

ZHAO ZHENGXU[*], R. W. BAINES[*], WANG ZHIHONG[**]

[*] Mechanical & Computer Aided Department, Staffordshire Polytechnic, Beaconside, Stafford, ST18 0AD, England.
[**] Department of Mechanical & Production, Shandong Polytechnic University, Jinan, Shandong, P. R. China.

ABSTRACT

This paper presents a new method to determine manufacturing parameters that will optimise tooth form and geometrical accuracy when manufacturing hobs. The method provides a higher profile accuracy and proper tooth form through transversely relieved grinding on conventional grinding machines. A model for the enveloping process of tooth profiles when hobs are ground is established to eliminate the needs for computation of complex transcendental equations. The grinding surface of the grinding wheel are well controlled by a new dressing method to generate concave cutting edges on the hobs. This results in the production of drum gear tooth form that provide a better meshing contact to achieve better working condition and transmission quality of subsequently generated gears or wormgears manufactured by the hob.

1. INTRODUCTION

The manufacturing parameters in gear hob manufacturing, especially the grinding wheel setting angles, the efficient diameter, the dressing path and associated angles, and the relative movements of the hob grinding machine directly affect the resulting tooth form, accuracy and hob life. These in turn affect the working condition and transmission quality of subsequently generated gears or wormgears manufactured by the hob. Currently, the determination of these parameters relies on two approaches. One is simply to select from available options and then modify the selected parameters[1]. Another approach is based on specially designed algorithms to find the suitable parameters. Those approaches however often require accessing a mainframe or large capacity computer to solve complicated transcendental equations. The computation involves tedious iteration processes and is proved to be prohibitive considering the time and the cost [2].

This paper reports a novel method to determine manufacturing parameters for gear hobs that will optimise tooth form and geometrical accuracy of hobs for both gears and wormgears. The method provides a hob grinding process model to eliminate the needs for the computation of transcendental equations. The grinding surface is controlled through the dressing process of the grinding wheel to generate concave cutting edges on the hobs. This results in the production of a drum gear tooth form that provides better meshing contact leading to gears with minimum transmission vibration and wear rates.

2. TRANSFORMATION EQUATIONS AND MATRICES

In the hob relieved grinding process, the spatial relationship between hob and grinding wheel is defined in terms of transformation matrices with respect to the body coordinate system of either the hob or the grinding wheel from a reference coordinate system. Because both the body coordinate system and the reference coordinate system are arbitrary, the overall coordinate system for a typical hob grinding process can be established as shown in Fig.1. Where $\{X_g, Y_g, Z_g\}$ and $\{X_h, Y_h, Z_h\}$ are body coordinate frames of the grinding wheel and the hob, respectively, and $\{X_r, Y_r, Z_r\}$ is the reference coordinate frame. The spatial relationships between the body frames $\{X_g, Y_g, Z_g\}$, $\{X_h, Y_h, Z_h\}$ and $\{X_r, Y_r, Z_r\}$ are uniquely defined by transformation equations (1) and (2).

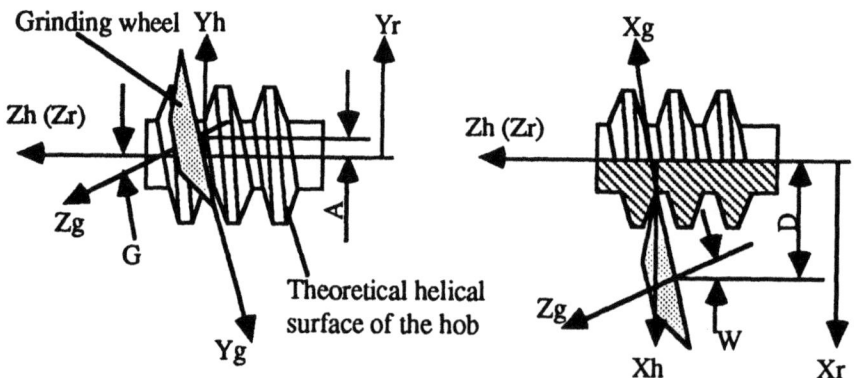

G = Setting angle (rad) of the grinding wheel measured in Yr Zr - Plane
A = Centre distance (mm) along Yr direction
D = Centre distance (mm) along Xr direction
W = Setting angle (rad) of grinding wheel measured in Xr Zr - Plane

Figure 1. Spatial relationships of grinding wheel and hob

$$\begin{bmatrix} Xh \\ Yh \\ Zh \\ Th \end{bmatrix} = Mhg \begin{bmatrix} Xg \\ Yg \\ Zg \\ Tg \end{bmatrix} \tag{1}$$

$$\begin{bmatrix} Xr \\ Yr \\ Zr \\ Tr \end{bmatrix} = Mrh \begin{bmatrix} Xh \\ Yh \\ Zh \\ Th \end{bmatrix} \tag{2}$$

where Mhg and Mrh are transformation matrices defined as follows

$$Mhg = \begin{bmatrix} -\cos W & 0 & -\sin W & D-QT \\ \sin G \sin W & -\cos G & -\sin G \cos W & A \\ -\cos G \sin W & -\sin G & \cos G \cos W & 0 \\ 0 & 0 & 0 & 1 \end{bmatrix} \tag{3}$$

$$Mrh = \begin{bmatrix} \cos T & \sin T & 0 & 0 \\ -\sin T & \cos T & 0 & 0 \\ 0 & 0 & 0 & -PT \\ 0 & 0 & 0 & 0 \end{bmatrix} \tag{4}$$

The spatial relationships of the grinding wheel and the hob are specified by two sets of variables {P, Q, T} and {A, D, G, W}. In the first set, T is the revolution angle of the hob. P and Q can be determined by equations (5) and (6).

$$P = \frac{zm}{2} \tag{5}$$

$$Q = \frac{f}{2\pi} \tag{6}$$

where z is the number of the starts of the hob.

m is the module of the hob (mm).

f is the transverse feed amount of the grinding wheel for each revolution of the hob (mm/rev).

The variables in the second set are setting parameters of the grinding wheel relative to the hob. They are of vital importance in the hob grinding process and directly affect the accuracy and the tooth form of the hob.

3. MODEL FOR HOB GRINDING PROCESS

The hob grinding process can be defined by a mathematical model in which the hob surface is enveloped by the grinding wheel. The grinding surface of the grinding wheel, as shown in Fig. 2, is defined using the parametric equations (7), (8) and (9).

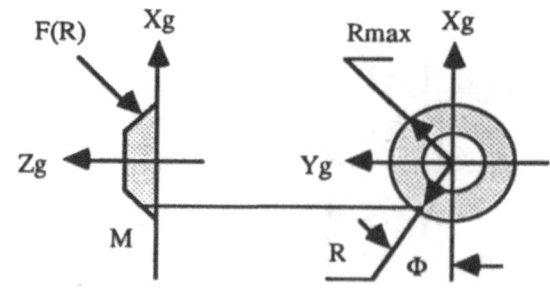

Figure 2. The grinding surface of grinding wheel

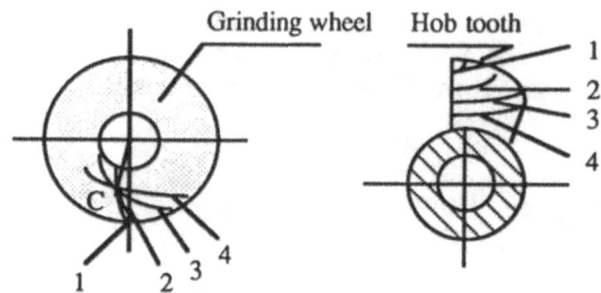

Figure 3. Instantaneous grinding curves

$$X_g = R \cos \Phi \tag{7}$$
$$Y_g = R \sin \Phi \tag{8}$$
$$Z_g = F(R) \tag{9}$$

where R and Φ are the parametric coordinates of the point M on the grinding wheel.

$F(R)$ represents the profile of the grinding wheel.

As shown in Fig.3, the final hob surface is actually enveloped by the grinding surface with different instantaneous grinding curves which are the instantaneous contact lines between the enveloped hob surface and the grinding surface [3].

Substituting X_g, Y_g, Z_g in Eq.(1) with Eqs.(7), (8) and (9), the instantaneous grinding curves can be obtained through the equation (10).

$$\begin{bmatrix} Q-Y_h, & X_h, & P \end{bmatrix} \begin{bmatrix} N_{xh} \\ N_{yh} \\ N_{zh} \end{bmatrix} = 0 \tag{10}$$

Where $\{N_{xh}, N_{yh}, N_{zh}\}$ represent the normal vector on the point M. For simplicity, Eq.(10) can be written as

$$\Psi(R, \Phi, T) = 0 \tag{11}$$

Replacing X_g, Y_g, Z_g in Eq. (1) with (7), (8), and (9) and then substitute (1) for (2), the hob grinding process model can be represented by combining Eq. (11) with relationships given in equations (12), (13) and (14).

$$X_r = \zeta \cos T + \chi \sin T \tag{12}$$

$$Y_r = - \zeta \sin T + \chi \cos T \tag{13}$$

$$Z_r = \lambda - PT \tag{14}$$

where $\zeta = D - QT - R \cos W \cos \Phi - F(R) \sin W$

$\chi = A - R \cos G \sin \Phi + R \sin G \sin W \, con \, \Phi$

$\lambda = - R \sin G \sin \Phi + F(R) \cos G \cos W - R \, con \, G \sin W \cos \Phi$

4. THE GRINDING SURFACE AND THE DRESSING PATH

The parametric equations (7), (8) and (9) have not defined exactly the particular required grinding surface. There are two geometrical factors which are of vital importance. The first one is the maximum diameter D_{gmax} of the grinding wheel and the second is the profile of the grinding surface in $X_g Z_g$ plane, i.e., $F(R)$ which is defined by Eq. (9). A larger grinding wheel will increase the productivity and the quality of the ground surface. However, D_{gmax} must be controlled to avoid excessive interference between the grinding wheel and the hob. The maximum diameter of the grinding wheel can be calculated using formulae (15a) to (15 j). In these formulae, Z_k represents the number of chip slots of the hob. δ_d, r_s, h_s, H, r_f and r_a are design parameters of the hob and are illustrated in Fig 4.

$$D_{gmax} = \frac{r_f \sin (\varepsilon - \sigma)}{\sin (\alpha_b + \delta - \varepsilon + \sigma) \cos (\varepsilon - \sigma - \delta)} \tag{15a}$$

where $\varepsilon = \dfrac{360}{Z_k}$ \qquad (15b)

$$\beta_s = \frac{\varepsilon - (H + r_s (ctg \, \frac{\delta_d}{2})) \, \frac{\delta_d}{r_a}}{2} \tag{15c}$$

$$r_d = r_a - \frac{k\beta_s}{\varepsilon} \tag{15d}$$

$$\alpha_d = tg^{-1} \frac{kZ_k}{2\pi r_d} \tag{15e}$$

$$\sigma = \beta_s - \frac{180 \, h_s \sin 2\alpha_d}{2\pi (r_d - h_s)} \tag{15f}$$

$$r_b = r_f - \frac{k\beta_s}{\varepsilon} \tag{15g}$$

$$\alpha_b = tg^{-1} \frac{kZ_k}{2\pi r_b} \tag{15h}$$

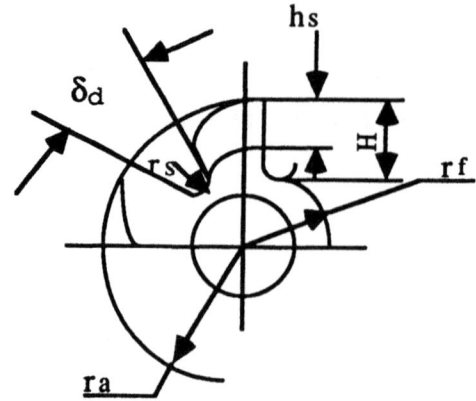

Figure 4. Design parameters of hob tooth

$$\delta = tg^{-1}(\frac{r_f}{r_b \sin (\epsilon - \sigma)} - ctg (\epsilon - \sigma)) \tag{15i}$$

$$k = \frac{f}{Z_k} \tag{15j}$$

The profile of the grinding surface $F(R)$ is achieved based on the criterion given in relation (16).

$$\frac{d^2(F(R))}{d^2R} = \frac{CE}{\sqrt[3]{(R^2 - E^2)^2}} > 0 \tag{16}$$

where C and E are constants which are associated with the dressing angle α_p and the position of the start point of the dressing path, see Fig.5. C and E can be determined by the relations (17), (18) and (19).

$$C = \frac{P_x \, tg\alpha_p}{\sqrt{R_{gmax}^2 - E^2}} \tag{17}$$

$$E = P_x \sin \alpha_p - P_y \cos \alpha_p \tag{18}$$

$$P_x^2 + P_y^2 \geq R_{gmax}^2 \tag{19}$$

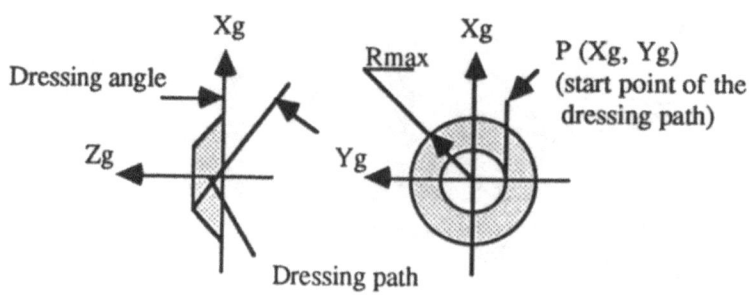

Figure 5. Dressing angle and dressing path of the grinding wheel

The dressing path of the grinding wheel can therefore be defined by parametric expressions given in equations (20), (21) and (22).

$$X_g = E \, (ctg \, \Phi \cos \alpha_p - \sin \alpha_p) \tag{20}$$

$$Y_g = E \, (ctg \, \Phi \sin \alpha_p + \cos \alpha_p) \tag{21}$$

$$Z_g = C \, (\sqrt{R_{gmax}^2 - E^2} - E \, ctg \, \Phi) \tag{22}$$

5. SETTING AND DRESSING PARAMETERS

The determination of the dressing path and dressing angle is dependent on the correct setting parameters A, D, G and W. These setting parameters in turn must be determined in order to achieve the correct tooth form and high accuracy. Theoretically, the relieved hob surfaces should be helical. It can be proved that the relieved ground hob surfaces represented by Eqs. (11), (12), (13) and (14) are non-helical surfaces [4] [5]. As shown in Fig. 6, the relieved ground hob surface and the theoretical helical surface are in contact with each other. The contact curve can be described as

$$r = \sqrt{r_b^2 + (r_1 - QT)^2} \tag{23}$$

where $r_b = R_c \, (\sin W \sin G \cos \Phi_c - \cos G \sin \Phi_c) + A - F(R_c) \cos W \sin G$.

r_1 is the radius of the pitch cylinder of the hob.

F and R are parametric coordinates of the intersect point C of the instantaneous grinding curves (see Fig. 3).

The angle η between the contact curve and the middle line of the theoretical hob surface is calculated by differentiating r with respect to T

$$tg \, \eta = \frac{dr}{r \, dT} \tag{24}$$

The correct tooth form, high accuracy and long hob life can then be achieved when the angle η is controlled to its minimum value. Based on relation (24), the optimal setting and dressing parameters of the grinding wheel can be obtained from equations (25) to (29).

$$tg \, \Phi_c = \frac{\sin G}{\sin W \cos G + \dfrac{P}{r_b} \cos W} \tag{25}$$

$$\Delta = (\frac{dF(R)}{dR}) \, R = R_c = \frac{\cos W \, tg \, G}{\sin \Phi_c - \sin W \, tg \, G \cos \Phi_c} \tag{26}$$

$$A = - (Q + \cos W \, tg \, G \, (F(R_c) - \frac{R_c}{\Delta}) + \frac{r_b}{\cos G}) \tag{28}$$

$$D = r_1 + R_c \cos \Phi_c + F(R_c) \sin W \tag{29}$$

224

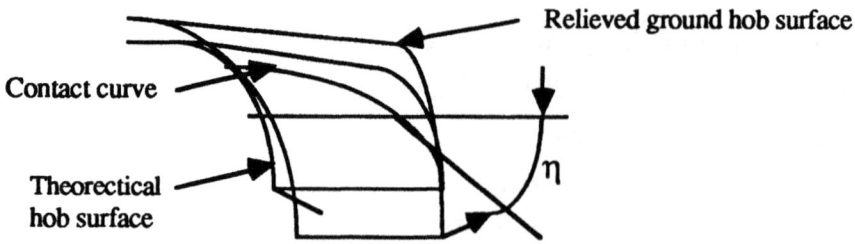

Figure 6. The spatial relation between ground and theoretical hob surfaces

6. CONCLUSIONS

Transversely relieved hob grinding involves a complex process of determining the setting and dressing parameters for the grinding wheel. The ground hob surfaces are non-helical surfaces and therefore complicated computation is required to solve transcendental equations. This paper has reported a new method to solve the problem by determining the manufacturing parameters for hobs used in both gear and wormgear cutting processes. The maximum deviation of the tooth profile ground with the reported method can be controlled within the range from 0.6 to 4 μm for the hobs with 1 to 4 starts. The method, especially the the dressing technique, utilises the features of the transversely relieved grinding process to control the tooth form of the hob with the optimal parameters. This will improve the working condition and transmission quality, increase the load capacity, and reduce the transmission vibration and wear rates by providing smoothness and quiet continuity of the meshing contact of the gears and wormgears.

REFERENCES

1. Paquet, R. M. Plastics gearing software program. Computer in Mechanical Engineering 1988, May/June: 24-29.
2. Orthwein, W. C. Helical and worm gear design. Computer in Mechanical Engineering 1988, January/February: 38-43.
3. Nakano, Y. Method of cutting tool profiles for toothed workpieces having complicated tooth forms. The International Symposium on Gear and Power Transmission 1981, Vol 1:199-204.
4. Zhao, Z-. X and Wang, Z-. H. Study on transversely relieved grinding hob for ZI worm wheel. Proceedings of Shandong Engineering Institute 1985, 3: 1-21.
5. Liang, X-, C. Tooth grinding methods. Proceedings of Mechanical Engineering 1983, Vol 19, No 2. P. R. China.

FINITE ELEMENT CONTACT ANALYSIS OF THE CONE ROLLER ENVELOPING WORM GEARING AND STRESS COMPUTATION

Long Hui   Zhang Guanghui

National Laboratory of Mechanical Transmission
Chongqing University
Chongqing, 630044, P.R. of China

ABSTRACT

The cone roller enveloping worm gearing with multi-tooth contact instantaneously is investigated by the finite element contact analysis approach combined with the automatic generation techniques of computational modeling and computer graphic algorithm. The loading distribution among contact gears, contact force distribution along contact line and stresses distribution of the whole finite element model are simulated by the computer procedure. The numerical simulation is important for the further research of the recently developed worm gearing transmission.

I. INTRODUCTION

The cone roller enveloping worm gearing is a new type of worm gearing transmission[1], which has the properties of multi-tooth contact instantaneously and complex engaging state when working. In the process of operation, the simulation of the worm and worm wheel is very important. In this paper, a three-dimensional elastic finite element contact technique is employed in the investigation of this type worm gearing with three-tooth contact instantaneously. The loading distributions of the worm and worm wheel tooth are simulated from the beginning to the end of engaging. Based on the computational results of loading functions among the engaging tooth and pressure force distribution on the contact surface of each tooth, the stresses on the worm and the worm wheel are calculated by mixed scheme of finite element analysis for three-dimensional elastic contact problems with friction[2], which has the high computational efficiency of iteration as compared with other methods.

II. ALGORITHM

Based on the mixed approach of finite element method for three-dimensional elastic contact problems with friction, the finite element contact analysis scheme is established. In this scheme, the flexibility matrix of contact region is calculated by means of a modified Cholesky elimination algorithm. In the process of solving contact problems, such as the engaged cone roller enveloping worm gearing, the elimination of stiffness matrix and the calculation of flexibility matrix of contact region can be conducted simultaneously, the computational efficiency of the iterations by this approach is much higher than that of general flexiblity method.

In this approach, the displacement compatibility condition, force equilibrium condition and Coulomb's law of friction on the normal and tangential directions are satisfied along the contact surface. In order to obtain the flexibility matrix equation of contact region, the boundary conditions on contact surface can be formulated into coefficient matrices, and a modified Closhy decomposition-elimination is adopted. The matrix equation of contact problem can be written as:

$$[K] \ \{U\} = \{P\} + \{R\} \tag{2-1}$$

where $[K]$、$\{U\}$、$\{P\}$、$\{R\}$ denote stiffness matrices, displacment vectors, loading vectors and contact force vectors respectively. The flexibility matrix equation of contact region can be written as:

$$[Fa] \ \{R\} = - \{\triangle P\} - \{\varepsilon_0\} \tag{2-2}$$

where $[Fa]$ denotes the flexibility matrix of contact region, $\{\triangle P\}$ is the relative displacement vector between the contact node pairs and $\{\varepsilon_0\}$ is the initial gop vector between the contact node pairs. Once the flexibility matrix equation (2-2) is obtained, the iterations are condensed to the contact region. After the force and displacment vectors of contact surface have been solved by iterations on contact region, the numerical simulation of contact problems can be obtained by back-substituting the matrix equation (2-1) with the results of contact surface.

III. FINITE ELEMENT MODEL

In this paper the rotating automatic mesh-generation algorithm[3] is applied in forming finite element mesh for the cone roller enveloping worm gearing. For the generation of finite element mesh of the worm surface, coordinate values of contact nodes on the worm surface are obtained by the gearing surface equation of worm (3-1).

$$X1 = \cos\psi_1\cos\psi_2\,(a\cos\psi_2 + z_0\cos\gamma - y_0\sin\gamma - a_2) + \cos\psi_1\cos\psi_2$$
$$(b_2 + y_0\cos\gamma + z_0\sin\gamma + a\sin\psi_2) - \sin\psi_1\,(c_2 + x_0)$$

$$Y1 = \sin\psi_1\cos\psi_2\,(a_2 + y_0\sin\gamma - z_0\cos\gamma - a\cos\psi_2) - \sin\psi_1\sin\psi_2 \quad (3-1)$$
$$(b_2 + y_0\cos\gamma + z_0\sin\gamma + a\sin\psi_2) - \cos\psi_1\,(c_2 + x_0)$$

$$Z1 = \sin\psi_2\,(a\cos\psi_2 + z_0\cos\gamma - y_0\sin\gamma - a_2) - \cos\psi_2\,(b_2 + y_0\cos\gamma$$
$$+ z_0\sin\gamma + a\sin\psi_2)$$

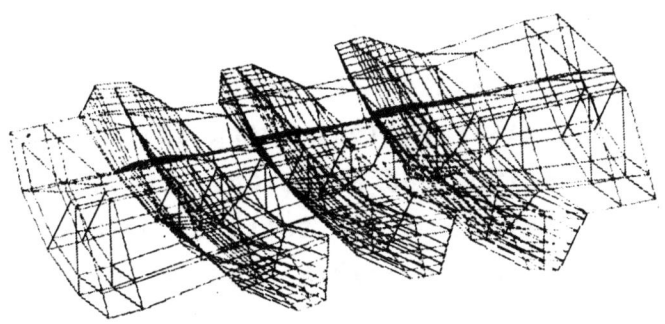

Fig. 1 Finite Element Mesh of Worm

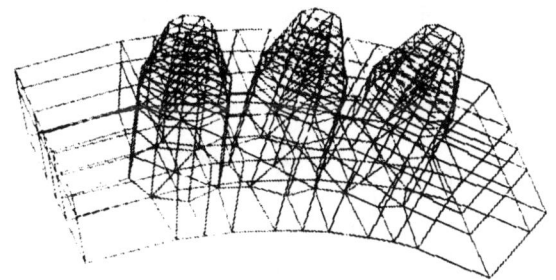

Fig. 2 Finite Element Mesh of Worm Wheel

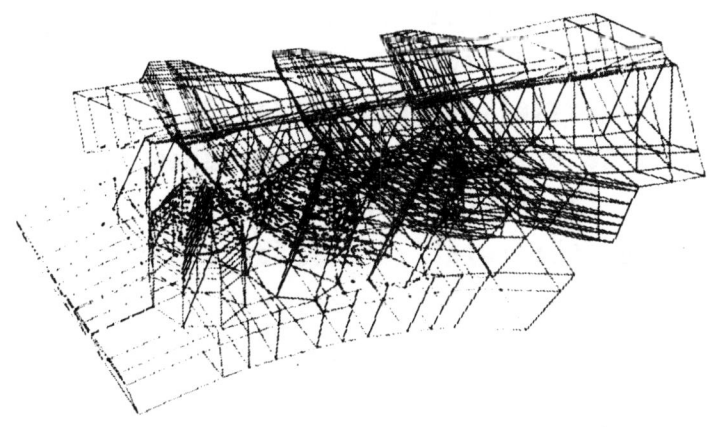

Fig. 3 Finite Element Model with Three-tooth Engaging Instantaneously

The finite element mesh of cone roller enveloping worm is shown in Fig 1 and Fig 2, respectively, and the finite element model of this kind of worm gearing with three-tooth contact instantaneously is shown in Fig 3. In an attempt to gain the accurecy of simulation, the torque acted on the worm can be simplified as the even distributed tangential forces acted on boundaries of the worm.

Ⅳ. COMPUTATIONAL RESULTS

4.1 Loading distribution among gears

Using the finite element contact analysis procedure, the numerical analytical results of the cone roller enveloping worm gearing in the instantaneous contact state are obtained after several times of iterations. The loading distributions among the three instantaneously contact teeth are 38.981%, 30.312% and 30.707%, respectively (Fig 4), which shows that the right tooth of worm wheel (from which the power is inputed) have undertaken the maximum loading and the medium tooth of worm wheel undertaken the minimum loading. The further investigation shows that the loading distribuations among instantaneously contact teeth are influenced by the gearing direction angle, the axial stiffness of worm and the helical angle.

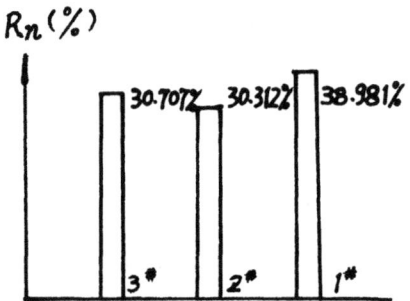

Fig. 4   Loading Distribution among Three Teeth

4.2 Contact force distribution

The normal and tangential contact forces (Rn, Rt, Rs) distributions along the contact line of three teeth are shown in Fig 5, Fig 6, Fig 7, respectively. The contact force distribution along contact line is influenced by the relative stiffness between worm and worm gear. If the difference of stiffness between the worm and worm gear is greater, and the contact force distributions are more centralized.

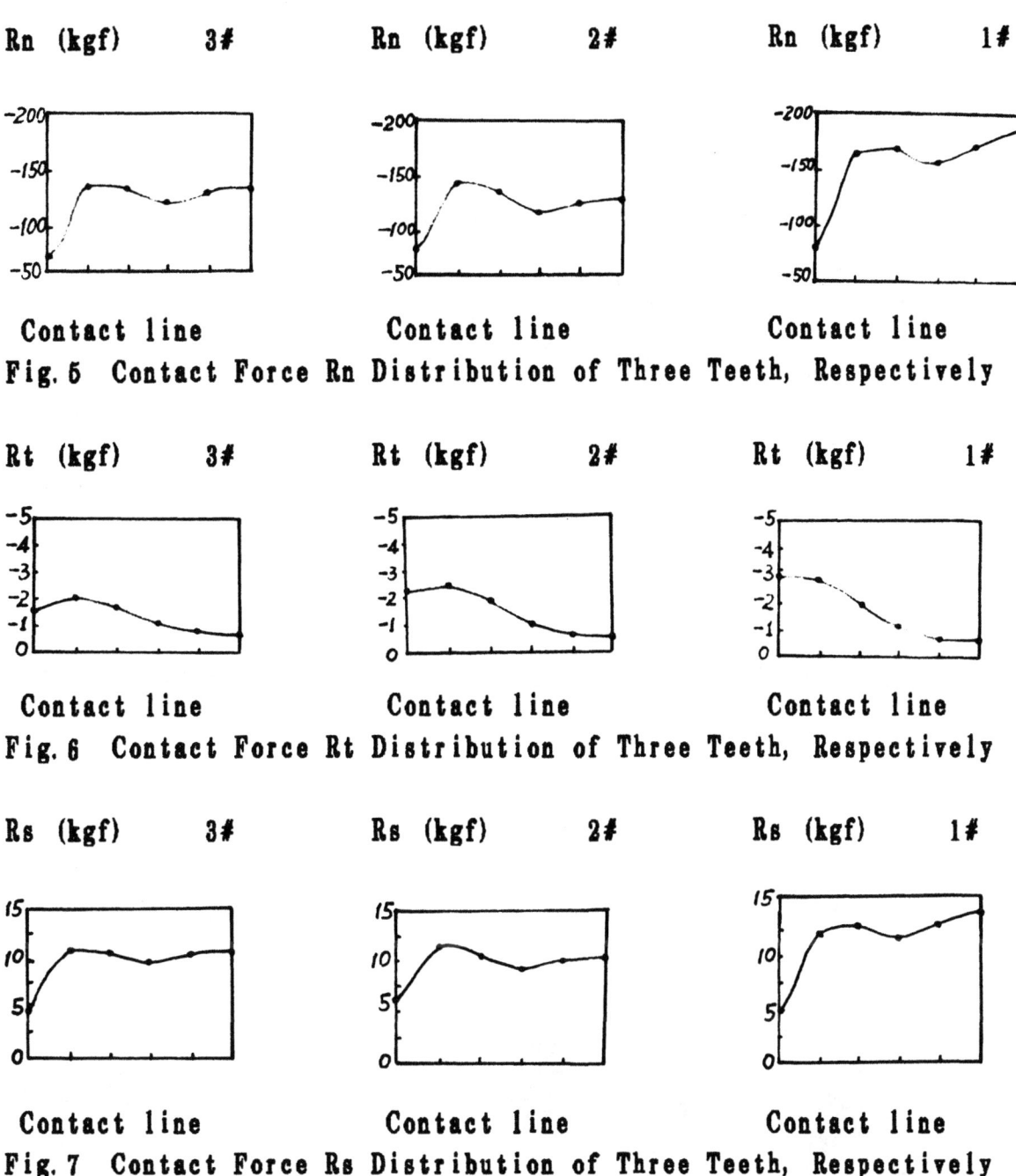

Fig. 5 Contact Force Rn Distribution of Three Teeth, Respectively

Fig. 6 Contact Force Rt Distribution of Three Teeth, Respectively

Fig. 7 Contact Force Rs Distribution of Three Teeth, Respectively

4.3 Stresses distribution

The contours of principal stresses σ_1, σ_3 on the engaging section of the cone roller enveloping worm gearing are shown in Fig8 and Fig9, respectively. As shown in the two figures, the maximum tension stresses appear in the root section of the loaded side, and the maximum compress stress appear in the contact surface of the worm wheel. However, the stresses of the worm wheel are higher than that of worm.

230

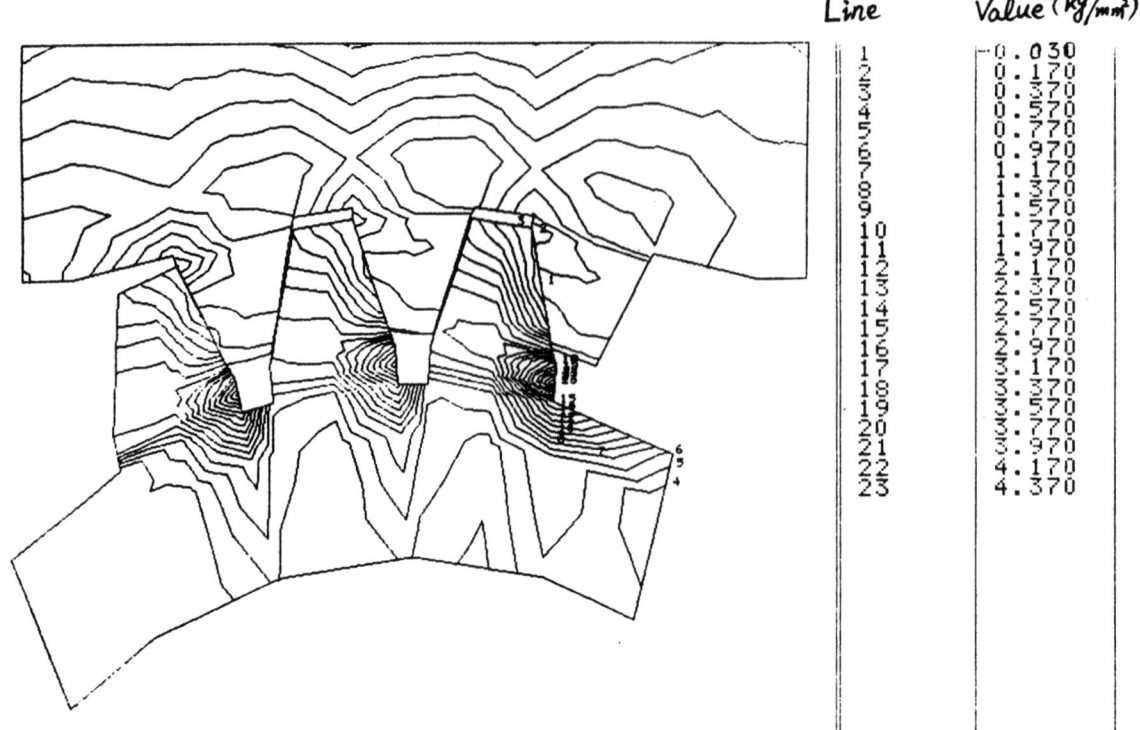

Line	Value (Kg/mm²)
1	-0.030
2	0.170
3	0.370
4	0.570
5	0.770
6	0.970
7	1.170
8	1.370
9	1.570
10	1.770
11	1.970
12	2.170
13	2.370
14	2.570
15	2.770
16	2.970
17	3.170
18	3.370
19	3.570
20	3.770
21	3.970
22	4.170
23	4.370

Fig. 8 Contours of Principal Stresses σ1 on the Engaging Section

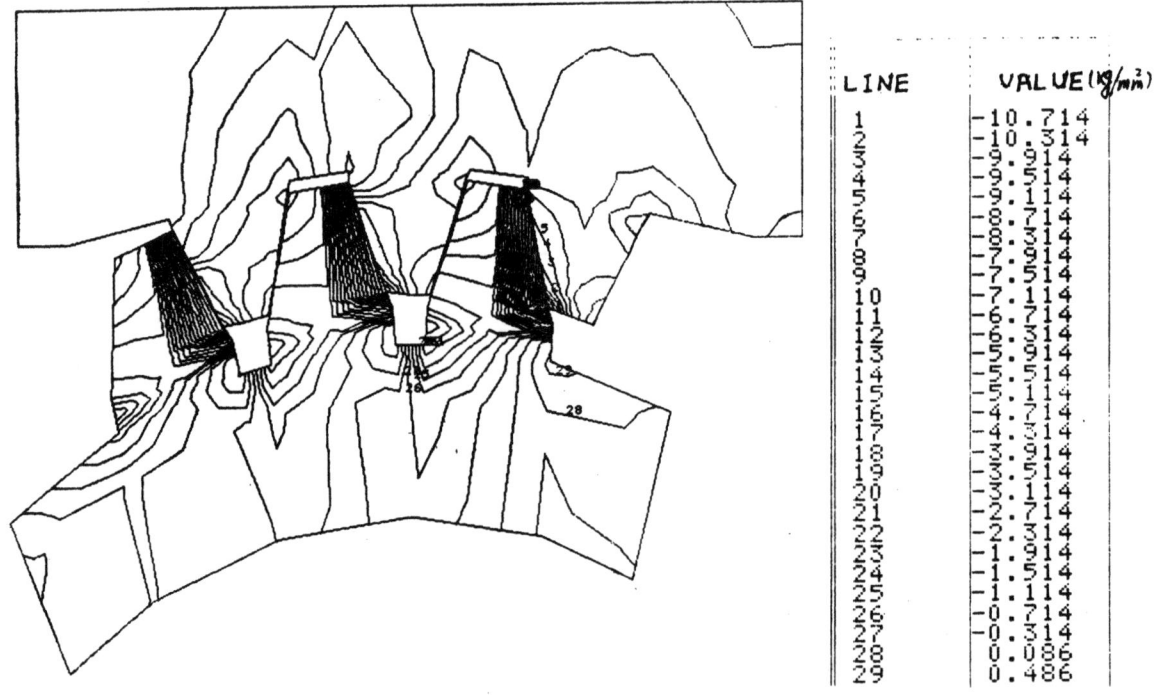

LINE	VALUE (Kg/mm²)
1	-10.714
2	-10.314
3	-9.914
4	-9.514
5	-9.114
6	-8.714
7	-8.314
8	-7.914
9	-7.514
10	-7.114
11	-6.714
12	-6.314
13	-5.914
14	-5.514
15	-5.114
16	-4.714
17	-4.314
18	-3.914
19	-3.514
20	-3.114
21	-2.714
22	-2.314
23	-1.914
24	-1.514
25	-1.114
26	-0.714
27	-0.314
28	0.086
29	0.486

Fig. 9 Contours of Principal Stresses σ3 on the Engaging Section

V. CONCLUSION

The finite element contact analysis approach is employed for the numerical simulation of the cone roller enveloping worm gearing, from which the worthwhile numerical results are obtained. It is of great importance in the further researches for a new type of worm gearing transimission. Since the type of cone roller enveloping worm gearing have many applications in industry, at present, we are developing a procedure system for simulation of this type of worm gearing for further researches.

VI. REFERENCES

[1] Long Hui, Zhang Guanghui, Strength Analysis of the Cone Roller Enveloping Worm Gearing, Research Report No. 2 of the Cone Roller Enveloping Worm Gearing in Chongqing University, January 1991

[2] Ou Hengan, Li Runfang, Gong Jianxia, An Improved Mixed Approach of Finite Element Method for Three-Dimensional Elastic Contact Problems With Friction, the 22nd Midwestern Mechanics Conference, October 1991

[3] Long Hui, Automatic Mesh-Generating Algorithm in Three-Dimensional Finite Element Analysis, Journal of Chongqing University, Vol. 11 (1988), 110-121

[4] T. D. Sachdeva, C. V. Ramakrishran, A Finite Element Solution for The Two-Dimensional Elastic Coctact Problems With Friction, Int. J. Num. Meth. in Engng., vol. 17 (1981), 1251-1271

Vibrational Behaviour of Rotors with Gear Couplings in Case of Insufficient Coupling Lubrication.

N. Bachschmid[*] - A. Curami[*] - F. Petrone[**]
(*) Professor - Dipartimento di Meccanica - Politecnico di Milano.
(**) Professor - Istituto di Macchine - Università degli Studi di Catania

ABSTRACT

An anomalous vibrational behaviour of a steam turbine driving a centrifugal compressor is described. The behaviour of the gear coupling between turbine and compressor is analyzed and a numerical simulation shows that most likely a coupling locking due to unsufficient lubrication occurred.

1. INTRODUCTION

Motors and operating machines are frequently connected with gear couplings especially when rotating speeds and powers are considerable.

In fact the gear coupling allows axial direction fairly wide relative movements and also permits small angular or radial misalignments.

The correct operation is strongly affected by an efficient lubrication of the teeth to allow a more uniform distribution of the contact forces and mainly to avoid phenomena of locking, due to the increase of the friction forces acting on the teeth in contact.

The need of providing an efficient lubrication is well known to coupling manufacturers, especially in case of high speed couplings, and different solutions are proposed.

A locked coupling represents a rigid linkage between the shafts, that imposes a whirling motion to them, in case of misalignment.

The aim of this paper is to describe the experimental results obtained measuring the anomalous vibrations of an air turbocompressor steam turbine, and to analyze the behaviour of the coupling.

The characteristics of these vibrations indicated clearly that the phenomena could not be related to instability problems due to oil film of the bearings, steam leakage, or to dynamic unbalance problems.

The hypothesis of a sudden lock-up of the coupling was then studied and the results of this analysis show that the malfunction could be due to this fault.

2. DESCRIPTION OF THE MACHINE

The high vibrations occurred on a steam turbine driving a double-flow, four-stage, centrifugal compressor (fig. 1).

The shaft of the compressor was supported on two bearings while the turbine had three bearings: two positioned at both extremities of the bladed body, while the third was allocated at the end of the extension shaft, near by the coupling; the extension shaft included between the second and the third bearing had a smaller diameter compared to the remaining part.

The two rotors were coupled by means of a gear coupling consisting of two hubs with external teeth that mesh with the internal teeth sleeves bolted with a spacer (fig. 2).

The normal operating speed range of the group was between 4300 rpm and 4600 rpm. The net turbine power was nearly 6 MW.

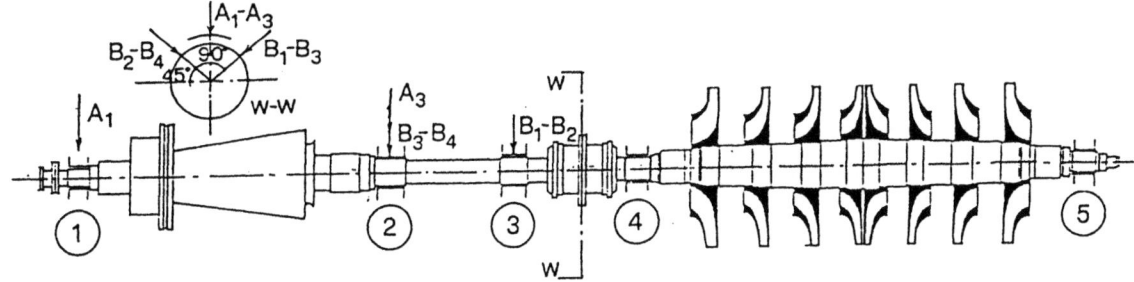

Fig. 1. The turbine-compressor rotor and proximitor locations.

3. EXPERIMENTAL BEHAVIOUR

The vibrations were measured by means of 4 proximitors mounted two by two at 90° on each of the two turbine bearings, placed near the coupling, evidencing the relative movement between the turbine shaft and the supporting case.

All the bearings of the turbine and of the compressor, included the axial bearings placed at the ends of the shafts, were also equipped with velocity probes.

A key-phasor for the 1 x rev reference was placed at the end of the turbine shaft.

Piezo-accelerometers were also used to measure the vibrations of the turbine case.

The accelerometers and proximitors signal outputs were sent to a RACALL multichannel tape-recorder as well as to a 8-channel ONO-SOKKI CF 880 spectrum analyzer.

Also some dial gauges were used to verify if during the performed tests relative displacements between turbine and compressor cases would appear.

From the amplitude peaks of the first diagram (fig. 3), carried-out during a run-up transient from 3750 rpm to 4500 rpm, it has been possible to identify a maximum of 1 x vibrations at 4200 rpm . As can be seen the dynamic magnification factor was very small with only little influence on first vibration mode and with a high value of damping ratio, due to oil film.

After the thermal transient the vibration amplitudes remained constant with a value of roughly 40 μm p-p on both bearings 2 and 3.

One day a sudden increase of the overall vibrations of bearing 3 occurred while the vibration level of bearing 2 remained almost the same (fig. 4). This new vibrational behaviour lasted for several weeks. During this period the group was operated at different speeds, and the 1x vibrations were recorded periodically showing no significant variations in the 4100-4500 rpm speed range, excluding therefore the effect of additional unbalance.

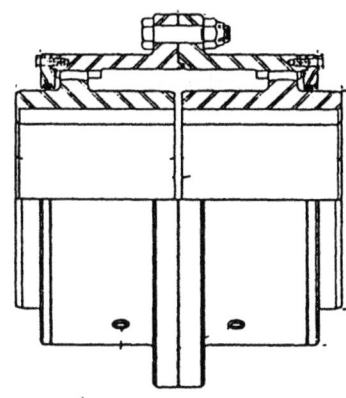

Fig. 2. The gear coupling

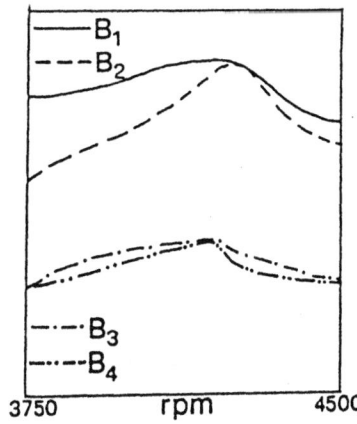

Fig. 3. First harmonic vibration components.

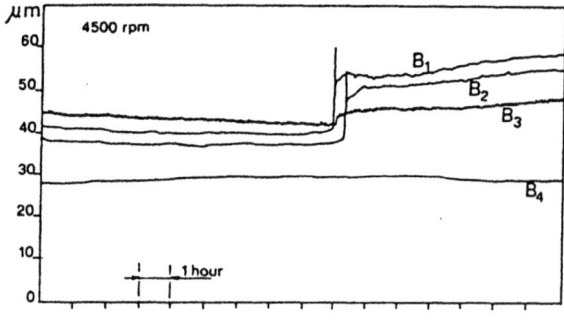

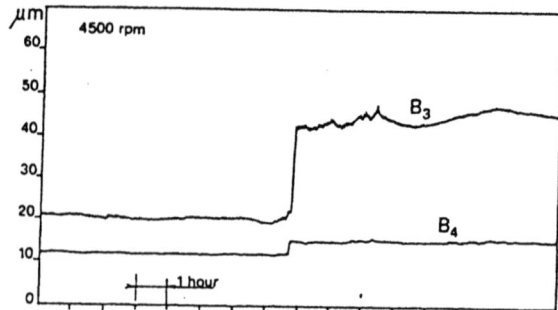

Fig. 4. Overall vibrations (1st increase)   Fig. 5. Overall vibration (2nd increase).

A similar but more dramatic vibration increase occurred sometimes later, approximately 40 hours after a run-up, which was noticeable also on bearing 2 as shown in fig. 5 (the vibrations on bearing 3 were not recorded due to overload condition in the board recording instruments).

The spectra of the proximitors on bearing 3 show that the level was of the same order of the radial clearance (figs. 6-7).

A verification of alignment between the turbine and compressor cases showed that a displacement occurred in the same time, probably due to different thermal expansions or sliding of the turbine on his foundation plate. During run-down tests the vibration levels decreased slightly and continuously with the speed , not showing any maximum at critical speed (figs. 8-9). The small increase in the 2x component was probably due to non-linear effects of oil film .

The accelerometers placed on the cases showed high vibrations with relevant higher harmonic components probably due to shock phenomena (fig. 10).

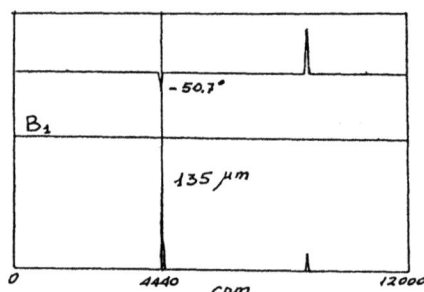

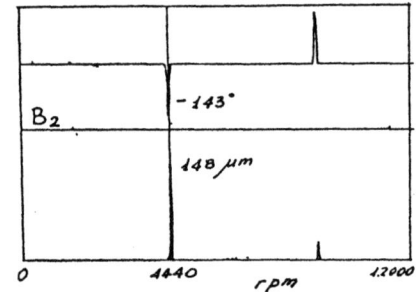

Fig. 6. Spectrum of proximitor B1   Fig. 7. Spectrum of proximitor B2.

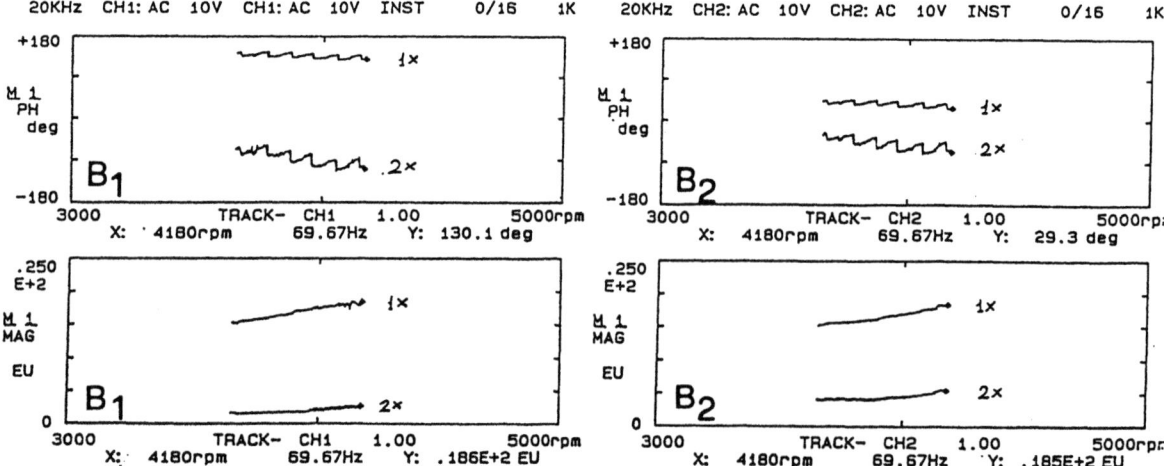

Fig. 8. Amplitudes and phases in bearing 3 versus speed during run down.

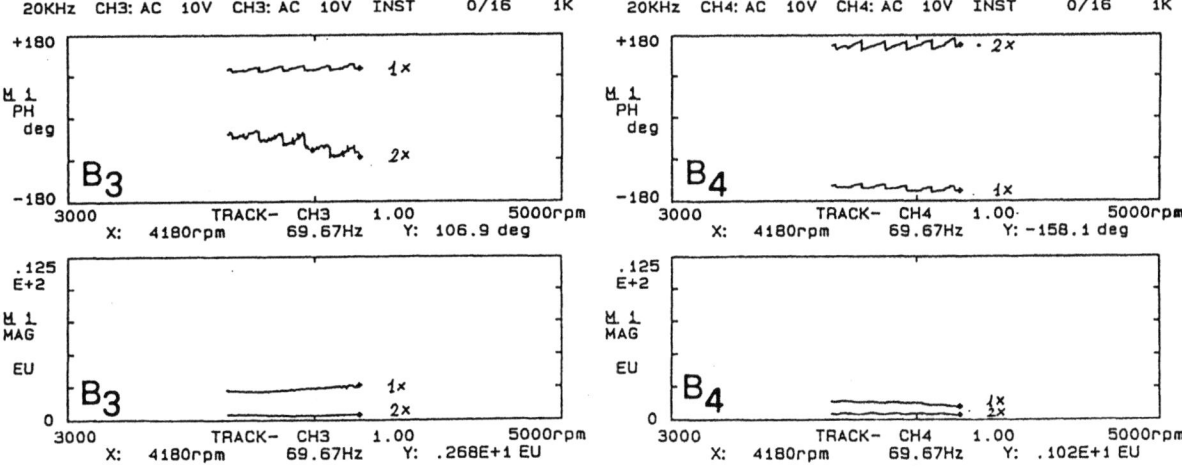

Fig. 9. Amplitudes and phases in bearing 2 versus speed during run down.

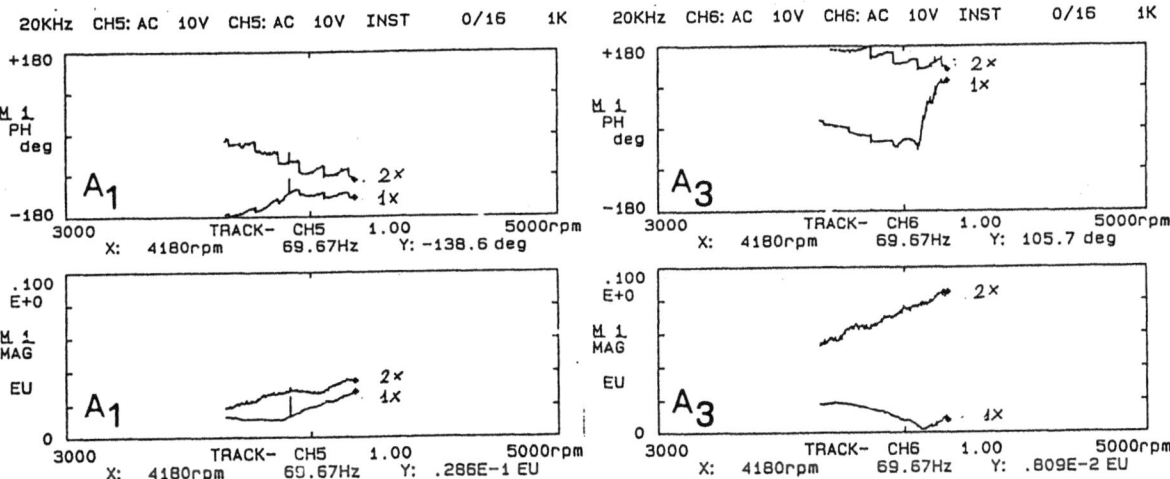

Fig. 10. Vibration accelerations of casing near bearings 1 and 3.

Afterwards a very slow run-down was made until the rotating speed got 2600 rpm. At this speed a sudden decrease of the vibrations of all the turbine bearings occurred; they came back to the values recorded during the normal operating conditions, and remained almost unchanged during the run-up to 4500 rpm.

The turbocompressor is now still running smoothly.

4. MALFUNCTIONING CAUSES

An analysis on the forces and couples acting in a gear coupling, shows that almost two different effects cause a deflection of the driving shaft:

a) the friction forces which arise in the presence of angular misalignment, due to the sliding action of the hub teeth with respect to sleeve ones, as explained in fig. 11, create a moment which deflects the end of the shaft; this effect is well-known [1];

if we suppose a transmitted torque of M = 12000 Nm, z teeth, a pitch-radius of r = 0.12 m, and considering a vanishing angular misalignment, we have on each tooth, in case of uniform distribution, a normal contact force of:

$$N_i = \frac{M}{r \, z \, \cos\theta} \tag{1}$$

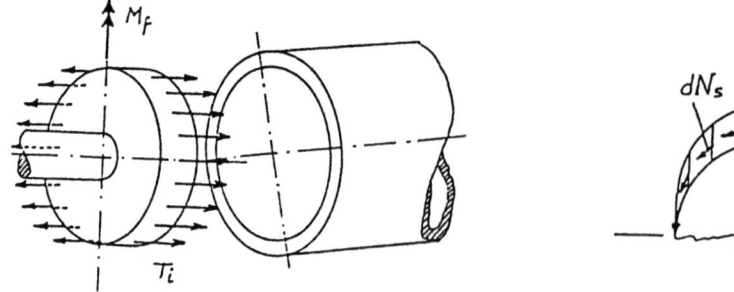

Fig. 11. Friction forces on gear coupling hub

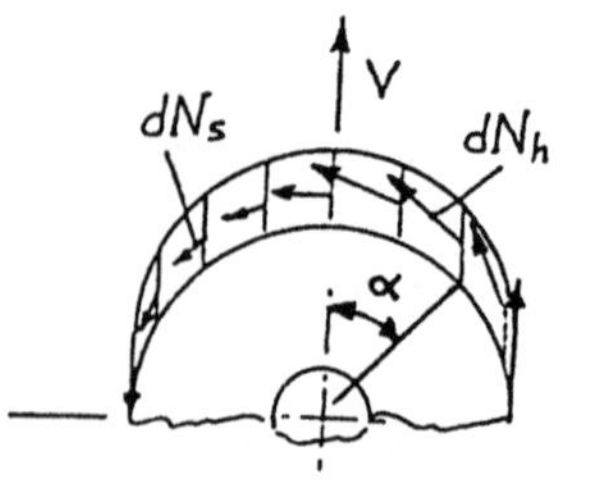

Fig. 12. Normal contact forces.

(where θ is the pressure angle), and a longitudinal friction force of $T_i = f\,N_i$ (where f is the friction coefficient).

The moment of above friction forces M_f can be calculated as follows:

$$M_f = \Sigma\, T_i\, r\, \text{sen}\, \alpha_i \tag{2}$$

where α_i is the angle formed by the radius passing through the contact point with respect to the plane containing the axes of the shaft and of the sleeve;

if we substitute a continuous distribution of contact forces on the pitch circle we get for the normal (dN_i) and the tangential (dT_i) elementary contact forces, the following:

$$dN_i = \frac{M\,r\,d\alpha}{r\,\cos\theta\,\,2\pi r} \tag{3}$$

$$dT_i = f\,dN_i \tag{4}$$

$$M_f = 2\int_0^{\pi} dT_i\, r\,\sin\alpha = \frac{2\,M\,f}{\pi\,\cos\theta} \tag{5}$$

in case of $f = 0.1$ and $\theta = 20°$ the value of M_f is 800 Nm; this moment seems not to be sufficient to generate shaft deflections as high as those measured;

b) uniform contact force distribution (i.e. equal force on each tooth) leads to a null resultant force; in the presence of tooth spacing errors, non uniform wear and so on, a non uniform contact force between teeth can arise; if these disuniformities have a random distribution, no significant resultant force arises; if more likely, they are stronger on half gear (and weaker on the other half gear) as shown in fig. 12, a resultant force arises which can be calculated in similar manner as above; in case of misalignment the disuniformity still increases;

if we suppose that the elementary contact forces on one half gear dN_h are higher by a percentage p of the mean value dN_i, and that they are smaller on the other half by the same amount, we have:

$$dN_h = dN_i\,(1+p) \tag{6}$$

$$dN_s = dN_i\,(1-p) \tag{7}$$

the resultant force in the vertical direction V is given by:

$$V = \int_0^{\pi} dN_i\,(1+p)\,\sin\alpha - \int_0^{\pi} dN_i\,(1-p)\,\sin\alpha = \frac{2\,p\,M}{\pi\,r\,\cos\theta} \tag{8}$$

in case of p = 0.15 and θ = 20° it results V = 10162 N; this radial rotating force is able to deflect the shaft by a considerable amount, as will be shown later, generating an angular "rotating" misalignment of the coupling.

Above considerations show moments and forces acting on the shaft extremity which are proportional to the driving moment; they cannot explain the jumps in the vibration amplitude.

A relative displacement of the turbine casing with respect to the compressor casing, due to different thermal expansions and/or faulty anchorage of the casings, cause a misalignment in the coupling, but no shaft deflection, if we neglect the friction force effect (a). But if, due to lack of lubrication, the friction forces increase strongly, a lock-up can occur and suddenly the shaft will be deflected and driven in a whirling motion with an amplitude equal to the total misalignment, if we suppose the other coupling end unaffected by this new force distribution.

This last hypothesis is supported by the fact that the first compressor bearing is rather stiff, and close to coupling end.

Once the coupling is locked-up a re-alignment of the casings do not change the situation of the whirling motion.

During the run-down of the group the transmitted torque M and also the contact forces decrease roughly with the square of the speed. The smaller friction forces cannot lock the coupling anymore, which suddenly starts again his slipping motion.

This mechanism could then completely explain the observed behaviour.

5. ANALYTICAL SIMULATION

On the basis of the dimensional data of the rotors, by means of a f.e.m. computing program [2], the behaviour of the two coupled rotors has been simulated.

First of all the flexural critical speeds of the two coupled rotors are obtained.

The turbine-compressor group has been modelled with 30 finite beam elements with 4 d.o.f. per node (fig.13). Supports and foundation are modelled with 1 d.o.f. mass-spring-damper systems. Oil films are modelled as usual by stiffness and damping coefficients. The numerical values are given in Table 1.

The 3rd bearing of the turbine has a rather small stiffness due to the fact that the load is very little.

The frequency response of this system has been calculated taking into account separately an unbalance of 1 kgm applied on:
a) the middle of the compressor (node 9);
b) the turbine, near the last stages (node 24);
c) the coupling (node 16).

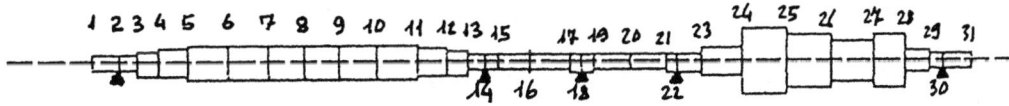

Fig. 13. Rotor model.

Table 1

Bearing n.		1		2		3	
Direction		vertical	horizontal	vertical	horizontal	vertical	horizontal
Support	stiffness	$1*10^{10}$	$5*10^9$	$1*10^{10}$	$5*10^9$	$1*10^{10}$	$5*10^9$
	damping	$1*10^5$	$5*10^4$	$1*10^5$	$5*10^4$	$1*10^5$	$5*10^4$
Oil film	stiffness	$1*10^9$	$5*10^8$	$1*10^9$	$5*10^8$	$8*10^7$	$4*10^7$
	damping	$1*10^6$	$5*10^5$	$1*10^6$	$5*10^5$	$8*10^4$	$4*10^4$

stiffness in N/m, dampings in Ns/m.

238

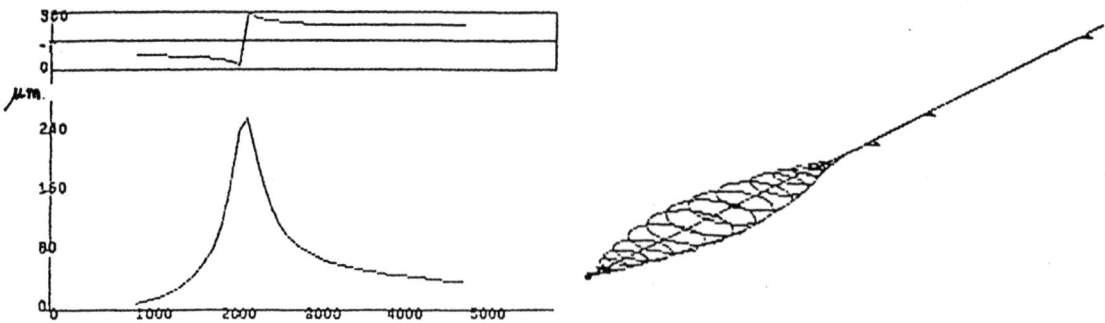

Fig. 14. Frequency response curve in compressor bearing (n.4) and deflection shape at critical speed.

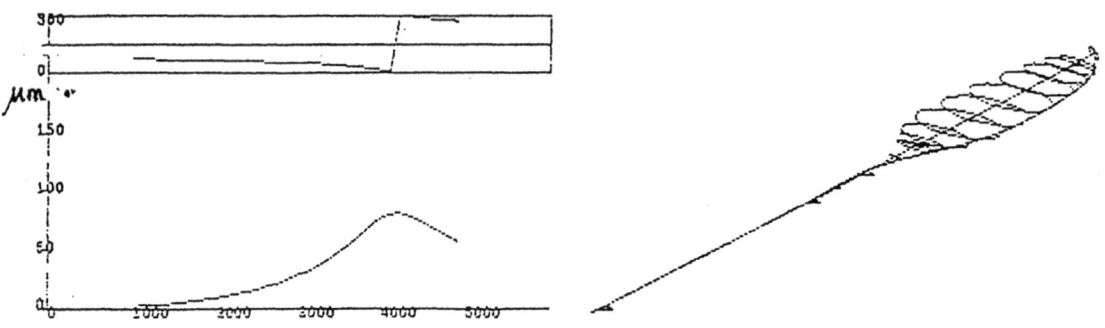

Fig. 15. Frequency response curve in turbine bearing (n.2) and deflection shape at critical speed

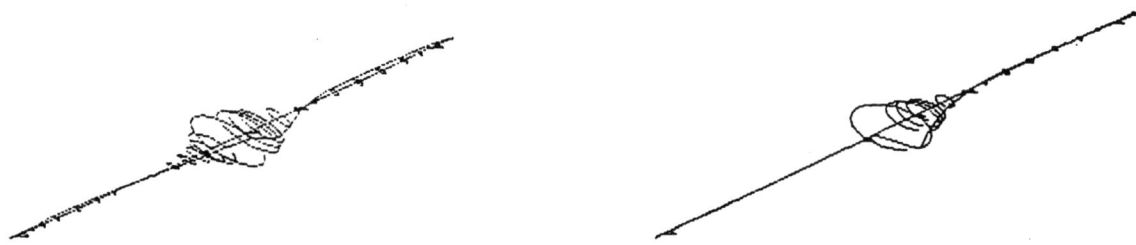

Fig. 16. Deflection shape of rotor at 4200 rpm with unbalance on coupling

Fig. 17. Deflection shape of turbine rotor in case of coupling misalignment.

In this manner the peaks in the response curves diagram could define the critical speeds.

The diagrams relatives to the (a) case (fig. 14) show that the first critical speed of the compressor is 2300 rpm; while the diagrams relatives to the (b) case (fig. 15) show that the first critical speed of the turbine is included between 4200 and 4900 rpm. The minor value might confirm the experimental one.

The deflection shapes relative to case (c) at 4200 rpm (fig. 16) show that the vibrations of the central body of the turbine, excited by unbalanced forces on the coupling, are very little, also at the turbine critical speed.

In order to simulate the behaviour of the coupling, a very thin beam element was introduced in the model between the two hubs, and a rotating force of 10000 N was applied to the turbine hub. In this way it was possible to calculate the effect described in point 4.b).

At normal operating speed of 4600 rpm the p-p amplitudes in μm, shown in table 2, were found.

The deflection shape (fig. 17) shows that the extension shaft only is affected by high vibrations.

Table 2

direction	bearing n.3	gear hub
horizontal	200	350
vertical	160	310

The amplitude values in correspondence of bearing 3 are close to the experimental ones.

Above calculation helps also to analyze the case of locked coupling: at 4600 rpm an amplitude of 160-200 μm in correspondence of bearing 3 can be generated by a misalignment of the coupling hub of 0.3-0.4 mm. This value is not unrealistic if we bear in mind that the total initial misalignment before the locking occurrence could be obtained by the sum of casing misalignment and of shaft deflection (4.a and 4.b).

6. CONCLUSIONS

An anomalous vibrational behaviour of a steam turbine driving an air compressor has been described.

The main characteristics of the observed phenomena indicated that a coupling malfunction could have been the origin.

A deeper analysis of the behaviour of the coupling, and the simulation made by means of a f.e. model of the rotors show that the locking of the coupling, due to insufficient lubrication, and the unlocking later-on at the decreasing of the transmitted torque, are able to generate in the turbine bearings a behaviour very similar to the observed one.

ACKNOWLEDGEMENTS

This study received a grant by MURST and by CNR.

REFERENCES

[1] J.R.Mancuso . Couplings and Joints. Mechanical engineering/45. M. Dekker (ed.), New York and Basel, 1986. pp.234-318.

[2] Diana G.,Curami A.,Pizzigoni B. "Computer analysis of rotor bearing systems. P.A.L.L.A. a package to analyze the dynamic behavior of a rotor -supporting structure system" Rotordynamics 2. Problems in Turbomachinery. 1988. Springer Verlag. pp. 191-260.

SESSION 10   ANALYTICAL METHODS II

ON DISCRETE-CONTINUOUS MODELLING OF THE ROTATING SYSTEMS FOR A NON-LINEAR TORSIONAL VIBRATION ANALYSIS

R. Bogacz[*)], H. Irretier[**)], T.Szolc[*)]

[*)] Institute of Fundamental Technological Research, Warsaw, Poland
[**)] Institute of Mechanics of the Kassel University, Kassel, Germany

ABSTRACT

In the paper there are investigated non-linear torsional vibrations in the drive systems of rotor machines. Considerations are performed using a discrete-continuous mechanical model, motion of which is described by classical wave equations. An application of the d'Alembert solutions of the motion equations leads to an appropriate system of linear and non-linear ordinary differential equations with a "shifted" argument. The shifted argument enables to solve this system of equations numerically in an appropriate sequence, which in comparison with coupled ordinary differential equations for an analogous discrete model, essentially increases numerical efficiency and accuracy of the proposed method. In the numerical examples there are investigated non-linear effects due to the elastic coupling, dry friction in the clutch and due to a non-linear character of damping.

1. INTRODUCTION

Torsional vibrations of the drive systems of machines and motor-vehicles driven by internal combustion reciprocating engines or electric motors are usually a source of severe dynamic loads of shaft segments, couplings, gears and other elements. Because of this reason, possibly accurate analysis of torsional vibrations of drive systems is extremely important for design optimization and to guarantee system quality and reliability. The drive systems of machines and motor vehicles are characterized by non-linear effects due to variations of mass moments of inertia of the system reciprocating parts, elastic couplings with non- linear characteristics, dry friction in clutches and vibration absorbers, backlashes in the gear stages and joints, variation of the gear stiffness as well as due to a non-linear character of damping [1,2]. For the non-linear torsional vibration analysis the considered drive systems are usually represented by discrete mechanical models, motion of which is described by appropriate systems of ordinary differential equations. Determination of drive system transient or steady-state non-linear torsional response usually reduces to computer simulations in the form of a direct integration of the motion equations [1,2]. Although very many advanced numerical methods have been developed so far, such computer simulations for relatively long time of calculations with a sufficiently small integration step still require a big computation effort. Moreover, there are some problems with a numerical stability and accuracy.

In order to overcome the mentioned above difficulties, in the paper an alternative method is proposed. This method is based on a discrete-continuous (hybrid) model of the drive system, for which the torsional wave propagation theory is applied for vibration analysis.

2. ASSUMPTIONS AND FORMULATION OF THE PROBLEM

In the paper a subject of considerations is the drive system of a rotor machine driven by the internal combustion reciprocating engine or by the electric motor. The discrete-continuous (hybrid) model of such system consists of $n+m+l+2$ rigid bodies of constant and variable mass moments of inertia I_j, $j=1,2,\ldots,n+m+l+2$, connected each other by means of cylindrical elastic elements with continuously distributed parameters as well as by means of massless torsional springs, Fig.1. The rigid bodies represent respectively mass moments of inertia of: a rotor of the electric motor or elements of the reciprocating engine auxiliary drives, n crank assemblies, flywheel, disks of the friction clutch with elastic coupling, gear wheels, Cardan universal joint elements, l rigid coupling disks and of the driven machine rotor. The elastic elements with distributed parameters of lengths l_i and constant torsional stiffnesses k_i, $i=1,2,\ldots,n+m+l+1$, represent shaft segments. However, the massless torsional springs of variable stiffnesses h_j, $j=n+2,n+3,\ldots,n+m+1$, represent flexibility of the elastic coupling, gear stages and joints, [1,2]. Internal and external damping in the system is represented by a linear model of the viscous type except the elastic coupling and the gear stages, for which non-linear damping terms are introduced [1]. The considered system is excited to vibrations by the active external torques $M_i(t)$, $i=2,3,\ldots,n+1$, produced by the reciprocating engine or, in a case of the electric motor, by the active external torque $M_1(t)$ where in the model one can assume n=0. On the system there are also imposed constant or variable passive external torques $M_k(t)$,

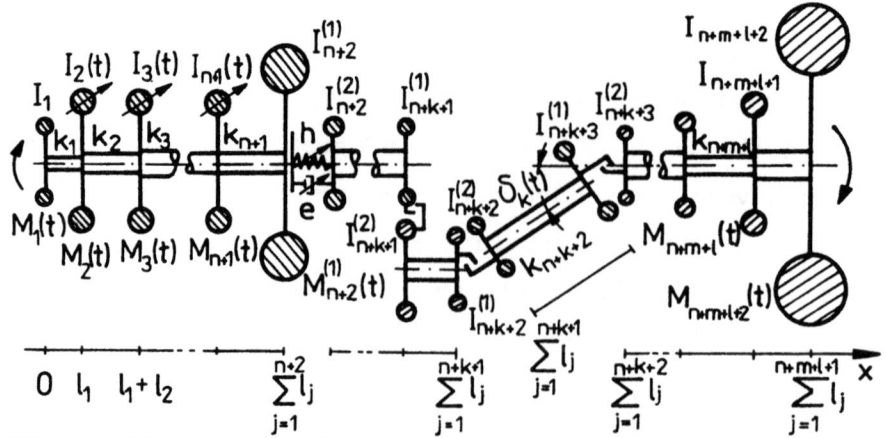

Fig.1. Discrete-continuous model of the drive system

$k=n+2, n+3, \ldots, n+m+l+2$. The constant components of mass moments of inertia of the hybrid model rigid bodies and the torsional stiffnesses of the elastic elements are determined using proper parameter identification procedures [3].

Equations of motion for angular displacements of the elastic element cross-sections are classical wave equations

$$a^2\theta_{i,xx}(x,t)-\theta_{i,tt}(x,t)=0, \qquad i=1,2,\ldots,n+m+l+1, \qquad (1)$$

where: $a^2=G/\rho$, t denotes time and x is a spatial coordinate parallel to the system rotation axis, Fig.1. These equations are solved with following boundary conditions

$$I_1\theta_{1,tt}+d_1\theta_{1,t}-c_1l_1\theta_{1,xt}-k_1l_1\theta_{1,x}=M_1(t) \qquad \text{for } x=0,$$

$$I_i(\theta_i)\theta_{i,tt}+[d_i+0.5\theta_{i,t}L_i(\theta_i)]\theta_{i,t}+c_{i-1}l_{i-1}\theta_{i-1,xt}-c_il_i\theta_{i,xt}+$$
$$+k_{i-1}l_{i-1}\theta_{i-1,x}-k_il_i\theta_{i,x}=M_i(t), \quad \theta_{i-1}=\theta_i, \quad i=2,3,\ldots,n+1, \text{ for } x=\sum_{j=1}^{i-1}l_j,$$

$$I_j^{(1)}\theta_{j-1,tt}+d_j^{(1)}\theta_{j-1,t}+c_{j-1}l_{j-1}\theta_{j-1,xt}+\kappa_j e_j(\Delta\theta_j(t))[\kappa_j\theta_{j-1,t}-\theta_{j,t}]+$$
$$+k_{j-1}l_{j-1}\theta_{j-1,x}+\kappa_j h_j(\Delta\theta_j(t))[\kappa_j\theta_{j-1}-\theta_j]=M_j^{(1)}(t), \quad \Delta\theta_j(t)=\kappa_j\theta_{j-1}-\theta_j,$$

$$\hspace{11cm}(2)$$

$$I_j^{(2)}\theta_{j,tt}+d_j^{(2)}\theta_{j,t}-c_jl_j\theta_{j,xt}-e_j(\Delta\theta_j(t))[\kappa_j\theta_{j-1,t}-\theta_{j,t}]-k_jl_j\theta_{j,x}-$$
$$-h_j(\Delta\theta_j(t))[\kappa_j\theta_{j-1}-\theta_j]=M_j^{(2)}(t), \quad j=n+2,n+3,\ldots,n+m+1, \text{ for } x=\sum_{i=1}^{j-1}l_i,$$

$$I_k\theta_{k,tt}+d_k\theta_{k,t}+c_{k-1}l_{k-1}\theta_{k-1,xt}-c_kl_k\theta_{k,xt}+k_{k-1}l_{k-1}\theta_{k-1,x}-k_kl_k\theta_{k,x}=M_k(t),$$
$$\theta_{k-1}=\theta_k, \quad k=n+m+2,n+m+3,\ldots,n+m+l+1 \text{ for } x=\sum_{i=1}^{k-1}l_i,$$

$$I_{nml2}\theta_{nml1,tt}+d_{nml2}\theta_{nml1,t}+c_{nml1}l_{nml1}\theta_{nml1,xt}+k_{nml1}l_{nml1}\theta_{nml1,x}=$$
$$=M_{nml2}(t) \hspace{6cm} \text{for } x=\sum_{i=1}^{nml1}l_i,$$

where $L_i(\theta_i)=dI_i(\theta_i)/d\theta_i$, $i=2,3,\ldots,n+1$, $nml2=n+m+l+2$, $nml1=n+m+l+1$ and

d_k, c_l are respectively constant external and internal damping coefficients, $k=1,2,\ldots,n+m+l+2$, $l=1,2,\ldots,n+m+l+1$. However, the functions $e_j(\Delta\theta_j(t))$ and $h_j(\Delta\theta_j(t))$, $j=n+2,n+3,\ldots,n+m+1$, denote non-linear damping and stiffness coefficients, respectively, for the elastic coupling, gear stages and joints [1]. Coefficients κ_j denote gear ratios, but for the elastic coupling, friction clutch and Cardan universal joints κ_j are assumed equal to unity. Superscripts (1) and (2) are assigned to the quantities corresponding respectively to the driving and driven elements in the system, Fig.1. The subscripts after commas denote partial differentiations.

Solutions of equations (1) are sought in the form of d'Alembert solutions

$$\theta_i(x,t)=f_i\left(at-x+\sum_{j=1}^{i-1}1_j\right)+g_i\left(at+x-\sum_{j=1}^{i-1}1_j\right), \quad i=1,2,\ldots,n+m+l+1. \quad (3)$$

The functions f_i and g_i in (3) represent torsional waves propagating in the elastic elements as a result of the external torque application. They are determined by the boundary and initial conditions [4]. Thus, substituting (3) into the boundary conditions (2) leads to following system of linear and non-linear ordinary differential equations with a "shifted" argument z for the functions f_i and g_i, $i=1,2,\ldots,n+m+l+1$,

$$r_{2,nml2}g''_{nml1}(z)+r_{1,nml2}g'_{nml1}(z)=M_{nml2}(z-1_{nml1})+s_{2,nml2}f''_{nml1}(z-21_{nml1})+$$

$$+s_{1,nml2}f'_{nml1}(z-21_{nml1}),$$

$$g'_i(z)=-f'_i(z-21_i)+f'_{i+1}(z-1_i)+g'_{i+1}(z-1_i), \quad i=1,2,\ldots,n,n+m+1,n+m+2,\ldots,nml1,$$

$$r_{21}f''_1(z)+r_{11}f'_1(z)=M_1(z)+s_{21}g''_1(z)+s_{11}g'_1(z),$$

$$r_{21}(z)f''_1(z)+r_{11}(z)f'_1(z)=M_1(z)+s_{21}(z)g''_1(z)+s_{11}(z)g'_1(z)+t_{21}f''_{i-1}(z-1_{i-1})+$$

$$+t_{11}f'_{i-1}(z-1_{i-1}), \quad i=2,3,\ldots,n+1, \quad (4)$$

$$\begin{bmatrix} p_{2,j-1} & 0 \\ 0 & r_{2,j} \end{bmatrix}\begin{bmatrix} g''_{j-1}(z+1_{j-1}) \\ f''_j(z) \end{bmatrix}+\begin{bmatrix} p_{1,j-1}(z) & -\kappa_j e_j(\Delta_j(z)) \\ -\kappa_j e_j(\Delta_j(z)) & r_{1,j}(z) \end{bmatrix}\begin{bmatrix} g'_{j-1}(z+1_{j-1}) \\ f'_j(z) \end{bmatrix}+$$

$$+\begin{bmatrix} \kappa_j^2 h_j(\Delta_j(z)) & -\kappa_j h_j(\Delta_j(z)) \\ -\kappa_j h_j(\Delta_j(z)) & h_j(\Delta_j(z)) \end{bmatrix}\begin{bmatrix} g_{j-1}(z+1_{j-1}) \\ f_j(z) \end{bmatrix}=\begin{bmatrix} M_j^{(1)}(z)+u_{2,j-1}f''_{j-1}(z-1_{j-1})+ \\ M_j^{(2)}(z)+s_{2,j}g''_j(z)+ \end{bmatrix}$$

$$+u_{1,j-1}(z)f'_{j-1}(z-1_{j-1})+\kappa_j e_j(\Delta_j(z))g'_j(z)+\kappa_j h_j(\Delta_j(z))[g_j(z)-\kappa_j f_{j-1}(z-1_{j-1})]$$
$$+s_{1,j}g'_j(z)+\kappa_j e_j(\Delta_j(z))f'_{j-1}(z-1_{j-1})-h_j(\Delta_j(z))[g_j(z)-\kappa_j f_{j-1}(z-1_{j-1})] \Bigg],$$

$$\Delta_j(z)=\kappa_j[f_{j-1}(z-1_{j-1})+g_{j-1}(z+1_{j-1})]-f_j(z)-g_j(z), \quad j=n+2,n+3,\ldots,n+m+1,$$

$$r_{2k}f''_k(z)+r_{1k}f'_k(z)=M_k(z)+s_{2k}g''_k(z)+s_{1k}g'_k(z)+t_{2k}f''_{k-1}(z-1_{k-1})+t_{1k}f'_{k-1}(z-1_{k-1}),$$

$$k=n+m+2,n+m+3,\ldots,n+m+l+1,$$

where: $r_{21}=c_1l_1+aI_1$, $r_{11}=l_s(k_1l_1+ad_1)/a$, $s_{21}=c_1l_1-aI_1$, $s_{11}=l_s(k_1l_1-ad_1)/a$,

$r_{2,nml2}=c_{nml1}l_{nml1}+aI_{nml2}$, $\quad r_{1,nml2}=l_s(k_{nml1}l_{nml1}+ad_{nml2})/a$,

$s_{2,nml2}=c_{nml1}l_{nml1}-aI_{nml2}$, $\quad s_{1,nml2}=l_s(k_{nml1}l_{nml1}-ad_{nml2})/a$,

$r_{2i}(z)=c_il_i+c_{i-1}l_{i-1}+aI_i(z)$, $r_{1i}(z)=l_s[k_il_i+k_{i-1}l_{i-1}+a(d_i+\Omega_i(z)L_i(z))]/a$,

$s_{2i}(z)=c_il_i-c_{i-1}l_{i-1}-aI_i(z)$, $s_{1i}(z)=l_s[k_il_i-k_{i-1}l_{i-1}-a(d_i+\Omega_i(z)L_i(z))]/a$,

$t_{2i}=2c_{i-1}l_{i-1}$, $t_{1i}=2k_{i-1}l_{i-1}l_s/a$, $\quad \Omega_i(z)=0.5[f_i'(z)+g_i'(z)]$,

$$i=2,3,\ldots,n+1,n+m+2,n+m+3,\ldots,n+m+l+1,$$

$p_{2,j-1}=c_{j-1}l_{j-1}+aI_j^{(1)}$, $\quad p_{1,j-1}(z)=l_s[k_{j-1}l_{j-1}+a(d_j^{(1)}+\kappa^2 e_j(\Delta_j(z)))]/a$,

$u_{2,j-1}=c_{j-1}l_{j-1}-aI_j^{(1)}$, $\quad u_{1,j-1}(z)=l_s[k_{j-1}l_{j-1}-a(d_j^{(1)}-\kappa^2 e_j(\Delta_j(z)))]/a$,

$r_{2,j}=c_jl_j+aI_j^{(2)}$, $\quad r_{1,j}(z)=l_s[k_jl_j+a(d_j^{(2)}+e_j(\Delta_j(z)))]/a$, $\quad j=n+2,n+3,\ldots$

$s_{2,j}=c_jl_j-aI_j^{(2)}$, $\quad s_{1,j}(z)=l_s[k_jl_j-a(d_j^{(2)}-e_j(\Delta_j(z)))]/a$, $\quad \ldots,n+m+1$,

and l_s is an arbitrary value. Using the Newmark method to solve (4) together with (3) one obtains system transient or steady-state dynamic response in the form of tangential stresses, torques, angular velocities and displacements of arbitrary cross-sections of the hybrid model elastic elements. The "shifted" argument in equations (4), which is a consequence of the wave interpretation of torsional vibrations, makes their right hand sides always known in each computation step. Thus, in contrary to coupled differential equations for an analogous discrete mechanical model, it is possible to solve equations (4) sequentially, one after another, in the presented order. This feature very essentially simplifies the numerical procedure making it much more efficient, stable and accurate, which will be confirmed by numerical examples.

3. NUMERICAL RESULTS

The proposed wave method of torsional vibration analysis was compared with traditional methods based on the discrete mechanical model. The analogous discrete model was characterized by the same number of the rigid bodies, identical values of the torsional stiffness coefficients, identical functions describing variation of the mass moments of inertia and the external torques, analogous damping terms, the same value of the total mass moment of inertia as well as by very similar values of first natural frequencies and mode shape functions as the corresponding hybrid model. Numerical calculations were performed for systems consisting only of the crank mechanisms of four and six cylinder in-line engines (i.e. for n=4 or n=6 and m=l=0) in steady-state operating conditions. Fig.2a presents plots of dynamic angular

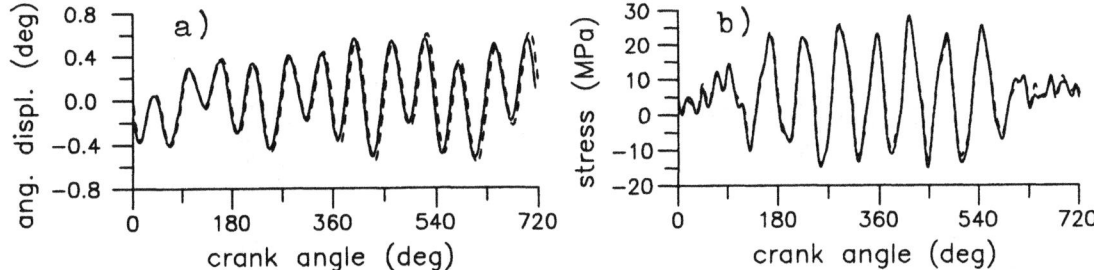

Fig.2. Steady-state response of the crank mechanisms of: (a) 4-cylinder,
(b) 6-cylinder engine using the wave (———) and the Hamming (---) method

displacements of the crankshaft free end of the four cylinder automobile
Diesel engine in resonant operating conditions. However, Fig.2b presents
plots of tangential stresses in the crankshaft 7th journal of the six
cylinder medium-speed Diesel engine at nominal operating conditions. The
obtained results of steady-state vibrations for the crank mechanisms of
the mentioned above engines, where for the discrete model the Hamming
method was applied, are characterized by very similar plot shapes. For
the investigated cases the greatest discrepancies of extreme values did
not exceed 6%-8%. But the wave method requires about 2.5-3.0 times
shorter computation time with excellent numerical stability in contrary
to the Hamming method. For the same cases the Runge-Kutta and the Taylor
method are characterized by better stability than the Hamming method but
require much longer computation time.

From the above comparisons it follows, that the wave method based on
the hybrid mechanical model is characterized by the greatest numerical
efficiency, very good stability and accuracy. Because of these
advantages, the proposed approach is the most convenient for relatively
long in time non-linear transient vibration analyses. As an example, a
run-up simulation was performed for the drive system of a rotor machine
driven by the six cylinder in-line Diesel engine by means of the elastic
coupling and two rigid couplings (i.e. for n=6, m=1, l=2). The
characteristic of the elastic coupling was assumed linear or progressive
and degressive hyperbolic sinusoidal. However, the damping coefficient
$e_8(\Delta\Theta_8(t))$ was described by the constant or parabolic function [1]. This
system was accelerated from the average rotational speed 500 [rpm] to
the nominal speed equal to 2800 [rpm]. Fig.3a presents plots of the
system response for constant h_8 and e_8 in the form of dynamic torque
transmitted by the drive shaft and in the form of vibratory angular
velocity of the machine rotor. However, Figs.3b and c present plots of
the mentioned quantities for variable h_8 and e_8, where Fig.3b
corresponds to the progressive coupling characteristic and Fig.3c
corresponds to the degressive one. Plots in Fig.3a are characterized by

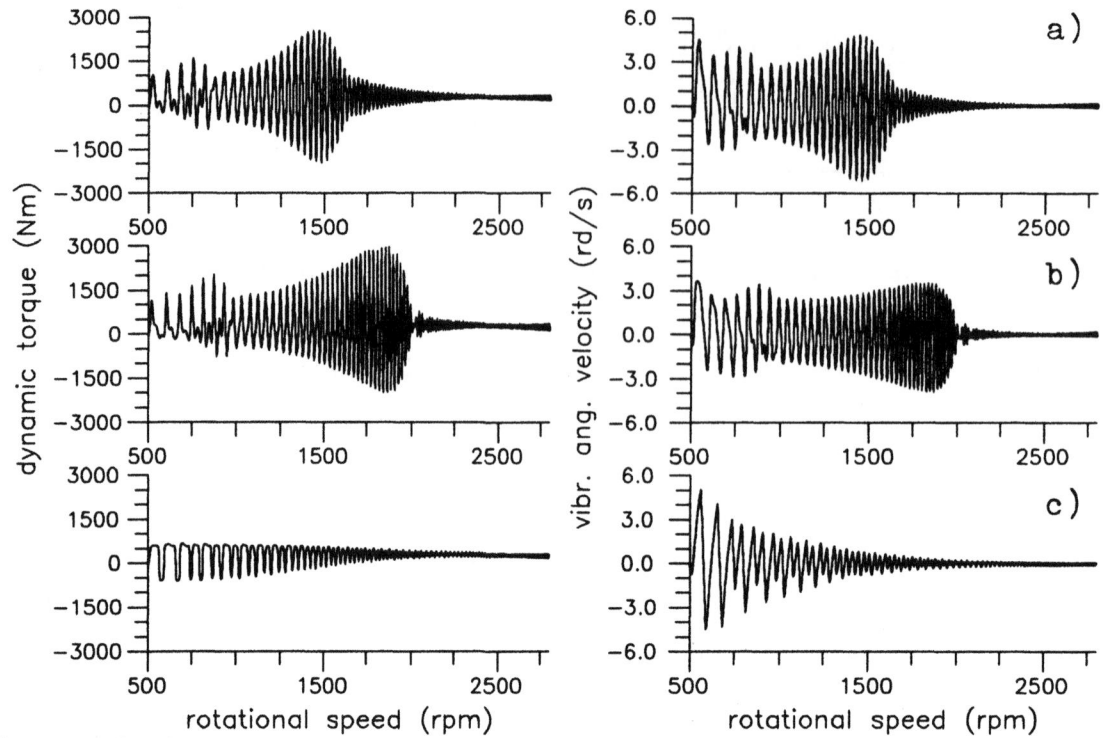

Fig.3. System transient response for various coupling characteristics

relatively fluent increase and decrease of local extremes occurring during a passage through the resonance zone. For the progressive characteristic of the coupling and the variable damping coefficient e_8 one obtains greater extreme values of the dynamic torque and smaller extreme values of the angular velocity than for constant h_8 and e_8, Fig.3b. Moreover, in this case the system transient response reaches maximum at higher value of the shaft average rotational speed and then decays rapidly. However, for the degressive coupling characteristic and variable e_8 the system response rises rapidly at the beginning of the run-up and then decays with an increase of the rotational speed, Fig.3c. The extreme values of the torque are much smaller but peaks of the angular velocity are slightly greater only at the beginning of the run-up. From this comparison it follows, that the elastic coupling with degressive characteristic seems to be the most optimum for the considered drive system.

In the case of coupling with progressive characteristic, in order to minimize the extreme values of the drive shaft transient response, a dry friction clutch of the overload type was applied, Fig.1. For the clutch there was assumed the Coulomb model of friction [1]. This clutch was expected to transmit maximal torque which value did not exceed 1.5 of the maximal average torque produced by the engine. For this case Fig.4 presents plots of the dynamic torque in the drive shaft and of the vibratory angular velocity of the machine rotor. The obtained curves

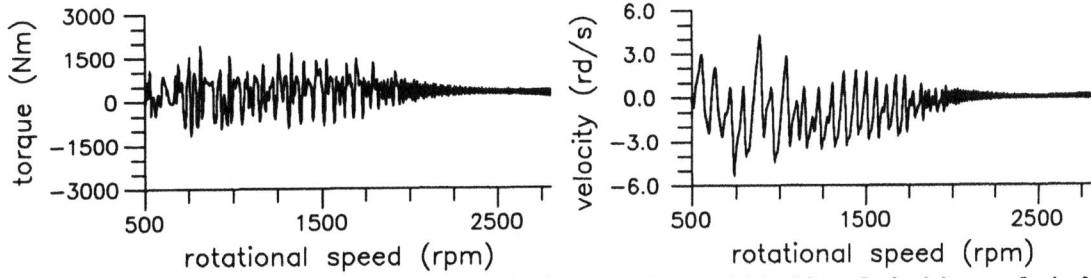

Fig.4. Transient response of the system with the friction clutch of the overload type

have a completely different character ·than the curves in Fig.3b. The overload friction clutch in the drive system caused an essential decrease of extreme values of the dynamic torque transmitted by the drive shaft.

4. FINAL REMARKS

In the presented paper non-linear torsional vibrations of the drive system were investigated. For this purpose a discrete-continuous (hybrid) model and the torsional wave propagation theory were applied. Because of a continuous distribution of mass in the elastic elements representing shaft segments, one can state that the discrete-continuous mechanical model better corresponds to reality than the discrete one. Moreover, the wave interpretation of torsional vibrations leads to ordinary differential equations with a "shifted" argument, which one can solve sequentially in an appropriate order. This feature of the proposed approach, in comparison with the classical methods based on the discrete mechanical model described by the coupled ordinary differential equations, causes an essential increase of a numerical efficiency of the presented procedure as well as yields very good numerical stability and accuracy. Thus, the wave method seems to be particularly advantageous from a practical standpoint.

REFERENCES

[1] Laschet A., "Simulation von Antriebssystemen", Springer-Verlag, 1988.

[2] Evans B.F., Smalley A.J., Simons H.R., "Startup of Synchronous Motor Drive Trains: The Application of Transient Torsional Analysis to Cumulative Fatigue Assessment", Trans.of the ASME, Paper No.85-DET-122.

[3] Bogacz R., Irretier H., Szolc T., "Analysis of Transient Torsional Vibrations of the Rotor Machine Using a Hybrid Model", Proc. of the 3rd Int. Conf. on Rotordynamics, IFToMM, Editions du CNRS, Lyon 1990, pp. 207-211.

[4] Bogacz R., Irretier H., Szolc T., "An Application of Torsional Wave Analysis to Turbogenerator Rotor Shaft Response", Trans. of the ASME, Journal of Vibration and Acoustics, 1991 (in print, ref. JVA-89-561).

THE ROTARY MOTION OF A THREE DEGREE OF FREEDOM MANIPULATOR LINKAGE

K L FAN, M J GOODWIN, M D Butler, D D L Risk
STAFFORDSHIRE POLYTECHNIC, U.K.
J E T PENNY
ASTON UNIVERSITY, U.K.

ABSTRACT

This paper describes a methodology for developing and solving the equations of motion of a three degree of freedom link mechanism used in many robot applications. A mathematical model, based on Lagrange's equation, describing the complete dynamics of the robot is obtained which allows for both the dynamic equations of motion of the manipulator linkage and the characteristics of its actuators. An example shows how the model can be used to determine the required actuator torques necessary to produce a given required motion. Simulation results are presented to illustrate the use of the model as a design tool.

INTRODUCTION

The equations of motion of a robot manipulator are a set of mathematical equations describing the dynamic behaviour of the individual links of the manipulator. More specifically, the dynamic equations relate the kinematic variables of the links(position, velocity and acceleration) to the forces or torques at the joints [1,2,3] which are required to produce and control the desired linkage motion. The Lagrangian formulation of the system's dynamic behaviour is described below in terms of work and energy using generalized coordinates. All the forces which are not related to work, and all constraint forces are automatically eliminated in this method. It has been shown that the resultant equations are generally compact and provide a closed form expression relating joint torques and joint displacements[2,4].

Earlier publications concerning robot arm modelling have tended to focus only on the arm dynamics, and have not shown how the equations of motion may be related to the arm kinematic analysis and subsequently used as a tool in robot arm design. Instead, the dynamic analysis, the kinematic analysis, and the robot design have tended to be treated in publications as relatively distinct considerations. This paper summarises the relationship between the kinematic and dynamic analysis and shows how both can be used in robot arm design.

MANIPULATOR DYNAMICS

The manipulator's equations of motion describe the relationship between the input joint torques and the output motion. In the <u>forward</u> dynamics analysis the joint torques, as functions of time are assumed to be known, and the resulting joint motion is calculated.

For the <u>inverse</u> dynamics analysis, the motions are the desired trajectories, described as time functions $\theta_i(t)$ through $\theta_n(t)$. The quantities that are calculated are the joint torques to be applied at each instant by the actuators in order to ensure that the required link trajectories are obtained. The relationship between the forward and inverse dynamics modelling is summarised in Figure 1 below.

In the <u>inverse dynamics</u> the joint torques are obtained by evaluating the right hand side of the closed form equation of motion, equation(1) below, using the specific trajectory data. The derivation of equation(1), using Lagrange's method, is given in reference[2,4].

$$\tau_i = \sum_{j=1}^{n} M_{ij}\ddot{\theta}_i + \sum_{j=1}^{n}\sum_{j=1}^{k} h_{ijk}\dot{\theta}_i\dot{\theta}_k + G_i \qquad \ldots\ldots\ldots\ldots(1)$$

For successive instants in time the joint velocities $\dot{\theta}_i$ and accelerations $\ddot{\theta}_i$ are computed from the given time functions, and then substituted into the right-hand side of equation(1). It must be noted that the coefficients, M_{ij}, h_{ijk} and G_i are all configuration dependent, that is, they depend upon the instantaneous robot arm position. Equation(1) is thus non-linear. It is also noteworthy that the torque required at one link is also dependent on the motion of other links. Thus the individual link equations of motion are coupled.

Equation(1) may be developed for all robot arm links, so that all link equations of motion may be written in matrix form as:

$$\vec{\tau} = M(\vec{\theta})\ddot{\vec{\theta}} + H(\vec{\theta},\dot{\vec{\theta}}) + G(\vec{\theta}) \qquad \ldots\ldots\ldots\ldots\ldots(2)$$

where; τ = applied torque vector
$\quad\quad \vec{\theta} = (\theta_1, \theta_2, \theta_i)^{\intercal}$
$\quad\quad M$ = Inertia matrix
$\quad\quad \underline{H}$ = nonlinear vector of centrifugal and coriolis forces
$\quad\quad \underline{G}$ = gravity loading vector

250

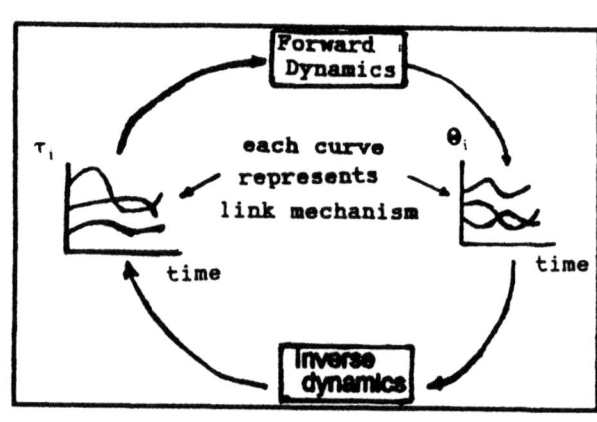

Fig.1 Showing a Complete
Dynamic Model

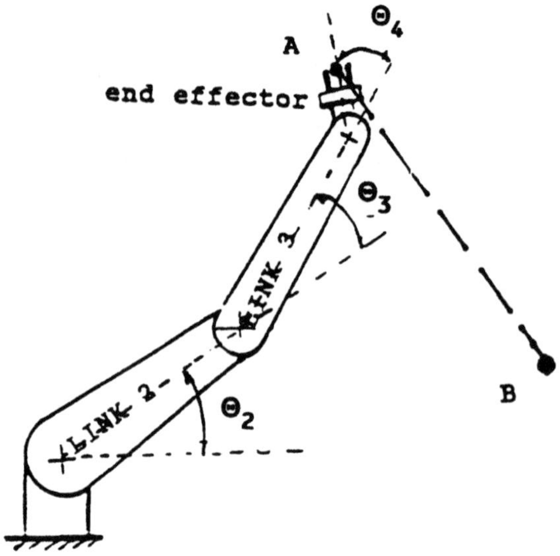

Fig.2 Manipulator
Configuration for a mot-
ion moving form A to B

Expanding equation(2), as shown in reference[4] for a 3
link robot and with the manipulator configuration as
Figure 2 gives

$$
M = \begin{bmatrix}
\begin{matrix} \frac{1}{3}m_2 d_2^2 \\ +2m_3(d_2\frac{2}{2}+\frac{d_3^2}{24}) \\ +2m_4(\frac{d_2^2}{2}+\frac{d_4^2}{24}) \end{matrix} & \begin{matrix} (m_3\frac{d_3^2}{12}+m_4\frac{d_4^2}{12}) \\ +(\frac{m_3}{2}+m_4)d_2 d_3 \cos\theta_3 \end{matrix} & \begin{matrix} (m_4 d_4 [\frac{d_2}{2}\cos(\theta_3+\theta_4) \\ +\frac{d_4}{12}]) \end{matrix} \\
\begin{matrix} m_3 d_2\frac{d_3}{2}\cos\theta_3+m_3\frac{d_3^2}{12} \\ m_4 d_2 d_3 \\ +m_4\frac{d_4^2}{12} \end{matrix} & \begin{matrix} \frac{1}{3}m_3 d_3^2 \\ +2m_4(\frac{d_3^2}{2}+\frac{d_4^2}{24}) \end{matrix} & \begin{matrix} m_4 d_3\frac{d_4}{2}\cos\theta_4 \\ +m_4\frac{d_4^2}{12} \end{matrix} \\
\begin{matrix} m_4 d_2 d_4 \cos(\theta_3+\theta_4) \\ +m_4\frac{d_4^2}{12} \end{matrix} & \begin{matrix} m_4 d_3\frac{d_4}{2}\cos\theta_4 \\ +m_4\frac{d_4^2}{12} \end{matrix} & \frac{1}{3}m_4 d_4^2
\end{bmatrix} \quad \cdots\cdots (3)
$$

where; m_2, m_3 and m_4 represent the masses of links 2,3 & 4
d_2, d_3 and d_4 represent the lengths of links 2,3 & 4
θ_2, θ_3 and θ_4 represent the angular movements among
joints 2, 3 & 4
$\dot\theta_2, \dot\theta_3$ and $\dot\theta_4$ represent the angular velocities of
links 2, 3 and 4
$\ddot\theta_2, \ddot\theta_3$ and $\ddot\theta_4$ represent the angular accelerations of
links 2, 3 and 4.

All the coriolis and centrifugal forces of links 2, 3 and
4 will be expressed in terms of H_2, H_3 and H_4 respectively

as;

$$H_2 = -(\frac{m_3}{2}+m_4) \, d_2 d_3 \dot\theta_3^2 \sin\theta_3 - \frac{m_4}{2} d_2 d_4 \dot\theta_4 \sin(\theta_3+\theta_4)(\dot\theta_3+\dot\theta_4) \ldots (4a)$$

$$H_3 = (d_3-d_4) \, m_4 d_2 \dot\theta_2 \dot\theta_3 \sin\theta_3 - \dot\theta_4^2 \sin\theta_4 \, (\frac{m_4}{2} d_3 d_4)$$
$$+ m_4 \dot\theta_2 \dot\theta_4 d_2 \frac{d_4}{2} \sin(\theta_3+\theta_4) \qquad \ldots \ldots (4b)$$

$$H_4 = -\frac{m_4}{2} d_2 d_4 \dot\theta_2 \dot\theta_3 \sin(\theta_3+\theta_4) \qquad \ldots \ldots (4c)$$

Also the gravitational effects on links 2, 3 and 4 are represented by the G_2, G_3 and G_4 terms which are also shown[4] to be:

$$G_2 = g[\,(\frac{m_2}{2}+m_3+m_4)\, d_2\cos\theta_2+(\frac{m_3}{2}+m_4)\, d_3\cos(\theta_3+\theta_4)$$
$$+(\frac{m_4}{2} d_4\cos(\theta_2+\theta_3+\theta_4)\,] \qquad \ldots \ldots (5a)$$

$$G_3 = g[\,(\frac{m_3}{2}+m_4)\, d_3\cos(\theta_2+\theta_3)+\frac{m_4}{2} d_4\cos(\theta_2+\theta_3+\theta_4)\,] \ldots \ldots (5b)$$

$$G_4 = \frac{m_4}{2} d_4 g\cos(\theta_2+\theta_3+\theta_4) \qquad \ldots \ldots (5c)$$

The **forward dynamics problem** is the calculation of the resulting robot motion when prescribed joint torques or forces are applied. This involves the use of a transposed form of equation(2) to obtain the manipulator's current position and velocity, θ and $\dot\theta$ for successive instants in time, given the vector of applied torques, τ. For each time instant considered the acceleration must then be integrated twice, using the Runge-Kutta(R-K) method[4,5] for example, to determine the resulting motion, that is velocity and position at the beginning of the next time step. If, for example, Euler's method is applied then the difference equation used to obtain θ is

$$\theta_{n+1} = \theta_n + h\Phi_n \qquad \ldots \ldots \ldots (6)$$

where, h is the step length (time interval), and

$$\Phi_n = \dot\theta_n \qquad \ldots \ldots \ldots (7)$$

Noting that, we can express equation(2) in the following form:

$$\ddot\theta = \dot\phi = M(\theta)^{-1}\,[\tau - H(\theta,\dot\theta) - G(\theta)] \qquad \ldots \ldots (8)$$

252

Therefore, applying Euler's method, we obtain

$$
\begin{pmatrix} \phi_2 \\ \phi_3 \\ \phi_4 \end{pmatrix}_{n+1} = \begin{pmatrix} \phi_2 \\ \phi_3 \\ \phi_4 \end{pmatrix}_n + h \begin{pmatrix} - & - & - \\ - & M & - \\ - & - & - \end{pmatrix}_n^{-1} \left\{ \vec{\tau} - H - G \right\}_n \quad \dots \dots (9)
$$

$$
\begin{pmatrix} \theta_2 \\ \theta_3 \\ \theta_4 \end{pmatrix}_{n+1} = \begin{pmatrix} \theta_2 \\ \theta_3 \\ \theta_4 \end{pmatrix}_n + h \begin{pmatrix} \phi_2 \\ \phi_3 \\ \phi_4 \end{pmatrix}_n \quad \dots \dots \dots (10)
$$

SIMULATION RESULTS

Consider the 3 degree of freedom robot moving from it's end effector position A to position B, as presented in figure 2.

In order to complete the whole movement within a given time, say 10 seconds, the robot is assumed to have a maximum acceleration and maximum deceleration to achieve this requirement as shown in the graphs of velocity and displacement versus time shown in figures 3 and 4.

The displacement-time relationship of the s-t curve shown in figure 4 has been plotted by using the constant area measurement[5] method from the v-t curve. The displacement was 0.3505m.

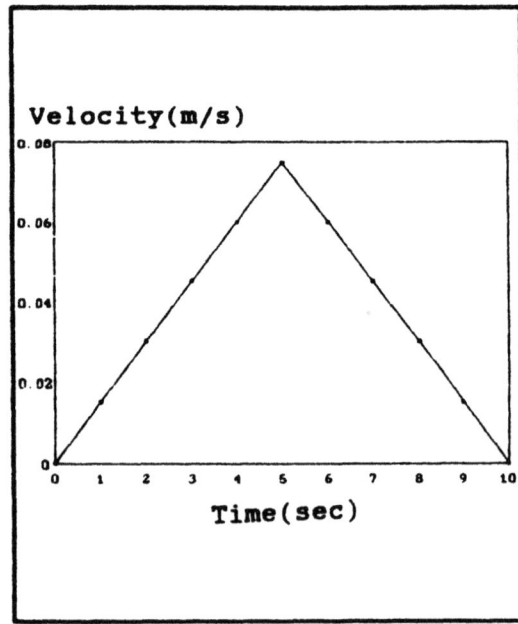

Figure 3 Variation of end effector velocity with time

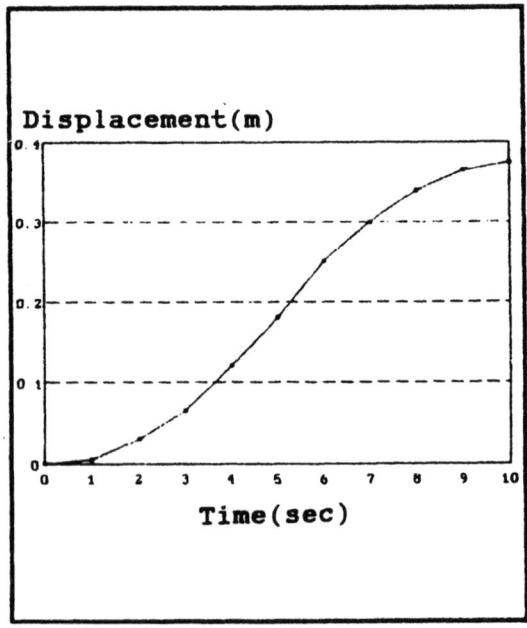

Figure 4 Variation of end effector displacement with time

Then, using the values of
displacement from the s-t
curve and given the initial
positions of the robot as
an input to an inverse
kinematic analysis[1,3]
program, a series of Θ_2, Θ_3
and Θ_4 values can be
generated as shown in
figure 5.

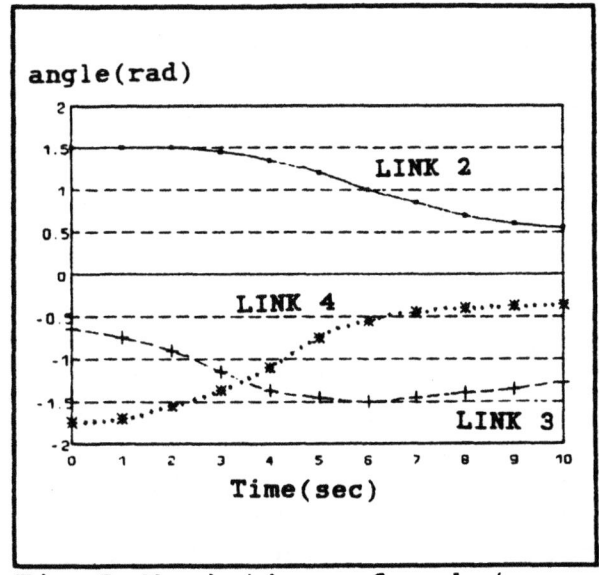

Fig.5 Variation of robot arm
joint angles with time

The finite difference
method was also
incorporated into the
inverse kinematic modelling
program[1,3], to work out
the corresponding angular
velocities and angular
accelerations at each
particular moment of time.
A sample of these results
is presented in figure 6 and 7.

Input of the Θ_i, $\dot{\Theta}_i$ and $\ddot{\Theta}_i$ values from the inverse
kinematics to the inverse dynamics model of equation(2),
enables the torques, τ_i, to be evaluated for every joint
at each different time interval. Figure 8 below shows
the variation of τ_i with respect to time.

Modelling of the robot motion under the action of known
applied actuator torques can be carried out using
equation(8).

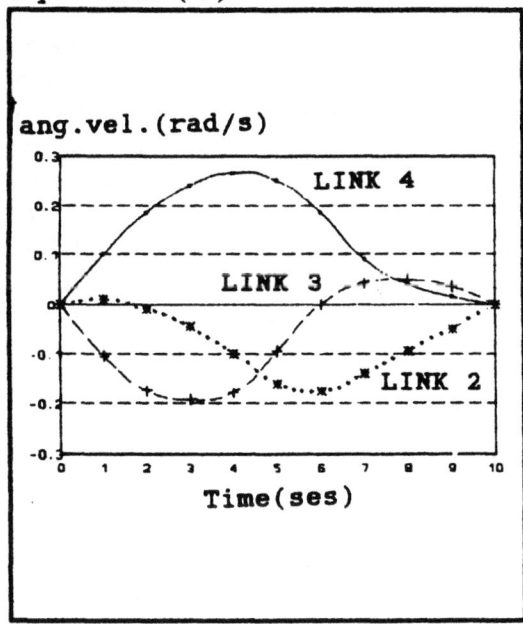

Fig. 6 Variation of robot
joint angular velocities
with time

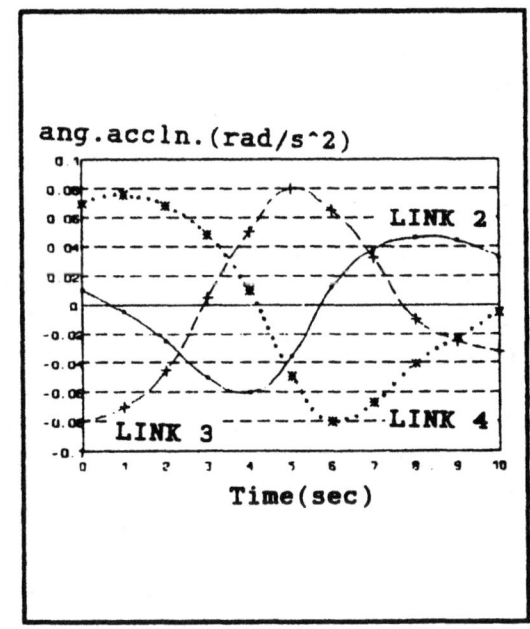

Fig. 7 Variation of robot
joint angular acceler-
ations with time

That is, knowing what torques are applied by the joint actuators enables the resulting link motion to be determined. This means of modelling the robot dynamic response was used by the authors as a means of checking the analysis and software used for the inverse dynamics modelling described above. By inputting the above torques shown in figure 8 to the forward dynamics modelling program based on equation(8), the resulting robot motion including

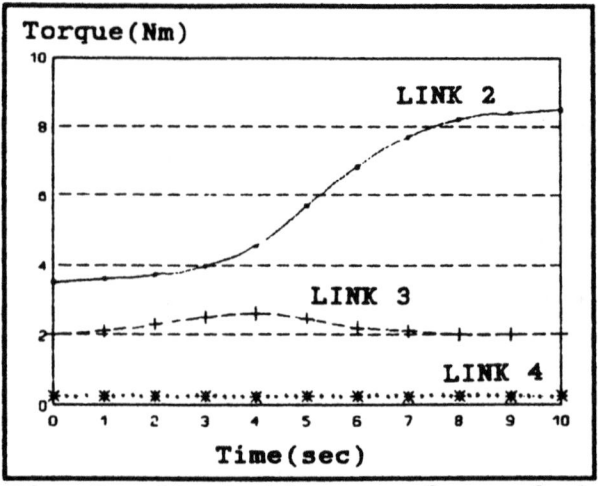

Fig.8 Variation of robot joint torques with time

positions of every link, θ_i, were shown to be the same values originally input to the inverse dynamic modelling program. In other words, both the forward and inverse dynamic modelling program were validated because both of the values of θ_i and τ_i generated or used in both the inverse and forward dynamics modelling programs matched each other.

RESULTS OF SIMULATION TESTS AND DISCUSSION

Several simulation tests have been carried out, with different displacements between position A and B, but with the same time span. A series θ-t, $\dot{\theta}$-t, $\ddot{\theta}$-t and τ-t curves have been plotted. It was found that they were similar in form to those shown above. τ_i always had the greatest value (around 10Nm maximum for the particular robot arm mass distributions used). The forces acting on link-2 were influenced not only by its movement but also by those of the other two links. It can also be seen in figure 8 that for the particular robot motion examined τ_i and τ_i need to have maximum values of about 3Nm and 0.5Nm respectively. This is the type of information required by the engineer when designing or selecting a robot for carrying out a particular task within a known time interval.

The derivation of the dynamic model of a 3 degree of freedom robot manipulator based on the Lagrange method is simple and systematic compared with other methods. It is noteworthy however that the method is more cumbersome to employ for systems with higher numbers of degrees of freedom. The resulting dynamic equations of motion, excluding the dynamics of the electronic control device, gear friction, and backlash, are a set of second order coupled, nonlinear, differential equations. Each equation contains a large number of torque or force terms

classified into four groups. They are:-
(a) inertial torque/force due to the links;
(b) reaction torques/forces generated by acceleration at other joints;
(c) velocity generated(centrifugal and coriolis) reaction torques/forces between joints; and
(d) torque/force due to gravity loading on the link.

In general, each inertial and gravity torque or force term depends on the instantaneous position or configuration of links while the velocity generated reaction torque or force terms depend both on the instantaneous velocity and on the configuration of the links. For most of the cases, (d) is a dominant factor on the equation of motion of each link.

CONCLUSION

The work presented here illustrates how the dynamic equations of motion in explicit symbolic form, in conjunction with kinematic analysis, may be derived and used for actuator force computation in robotic manipulators.

It has been demonstrated how the equations of motion as derived from the Lagrangian formulation are applied to a robot manipulator with 3 degrees of freedom. The method has been shown to be comparatively simple for systems with only three degrees of freedom, but is clearly more cumbersome with larger systems.

REFERENCES:

[1] K L FAN, "Second Progress Report on Kinematics Analysis of Robot Arm", August 1990. Internal report, Dept. of Mech & C/A Engg. Staffordshire Polytechnic.

[2] C S G Lee, "Tutorial on Robotics, IEEE Computer Society Press, 1983.

[3] N I Smith, "An Iterative Method for the Inverse Kinematics Problem of Robotics", Journal of Electrical & Electronic Engg., Australia, March 1986, vol.6, No.1, p50-53.

[4] K L FAN, "Third Progress Report on Dynamic Modelling of a Robot Arm", March 1991. Internal report, Dept. of Mech & C/A Engg. Staffordshire Polytechnic.

[5] W Vetterling, W Press & et al, "Numerical Recipes", 2nd Edition, p547-561.

PHYSICAL INTERPRETATION AND SOME PROPERTIES OF COMPLEX EIGENVECTORS IN ROTORDYNAMICS BY A TIME DOMAIN APPROACH

H.R.T. de Azevedo
CEPEL - Electrical Energy Research Center, Rio de Janeiro, Brasil
H.I. Weber
UNICAMP - Campinas State University, São Paulo, Brasil

ABSTRACT

This paper presents, at first, the equations of eigenvalue problems concerning non self-adjoint systems, having real matrices, for the case of rotating machines. Fixed frame approach is used, in which the bearing element matrices are not necessarily symmetric. Viscous damping can be also taken into account in the bearings.

It's presented by the authors the mathematical development concerning the relationships envolved in the physical interpretation of complex eigenvectors associated to modal properties in Rotordynamics. It's used a time domain approach in order to relate the nodal orbits described by these eigenvectors with the general equation of planar ellipses.

Some properties associated with invariants of the orbits are also shown in the paper. Particular cases of trajectories (linear or circular) are studied in the methodology, in order to obtain a systematic way to deal with graphical routines in computacional programs.

Finally, it's presented some graphical examples obtained from a Finite Element Modelling Package (Program ROTMEF) in which graphical routines based on the present methodology were implemented. Some comments concerning the influence of physical properties of the model (non-symmetries and damping) on the graphical output of the eigenvectors are also made.

INTRODUCTION

The study of mechanical behaviour concerning lateral oscilations in rotating machines, by a modal approach, envolves the treatment of pairs of complex eigensolutions (eigenvalues/eigenvectors). Some of its mathematical properties such as orthogonality, diagonalization of the system's matrices, etc, are well known and quite used. Nevertheless, the little available litterature covering physical interpretation or graphical visualization of these eigenvectors was not considered applicable for the purposes of the authors' study. The main objective of this study was, firstly, the development of a software package for rotor modelling by the Finite Element Method (Program ROTMEF).

The methodology presented in this paper was developed in a systematic basis in order to correlate the mathematical complex eigenvectors issued from modelling with the natural vibration modes (real quantities) measured on an actual machine.

Some results of these investigations are being introduced in conjunction with modal updating techniques, in order to detect parameter changes in the measured machine. We intend to use these results as a step to the implementation of Predictive Maintenance Procedures in Large Hidrogenerating Units of the Brazilian Electric Network.

SYSTEM'S EQUATIONS

Finite Element Method was adopted in order to discretize the continuous system. Rotating axis-symmetric shaft and disc elements associated with bearing elements (linearized) not necessarily symmetric can be used to create de model. The shaft and disc elements are considered conservative, nevertheless viscous damping properties of the bearings are possible to be taken into account. Its equations are obtained, for a fixed frame approach, as better described by Ref. (1). The system's general equation, of order N, is presented in matricial form as:

$$[M]\,\ddot{\Delta}(t) + [C + G]\,\dot{\Delta}(t) + [K]\,\Delta(t) = \mathbf{F}(t) \tag{1}$$

The solution of the homogeneous system is obtained by means of the state space formulation:

$$\begin{bmatrix} C + G & : & M \\ \cdots & : & \cdots \\ M & : & O \end{bmatrix} \begin{Bmatrix} \dot{\Delta}(t) \\ \ddot{\Delta}(t) \end{Bmatrix} - \begin{bmatrix} -K & : & O \\ \cdots & : & \cdots \\ O & : & M \end{bmatrix} \begin{Bmatrix} \Delta(t) \\ \dot{\Delta}(t) \end{Bmatrix} = \begin{Bmatrix} \mathbf{F}(t) \\ O \end{Bmatrix} \tag{2}$$

that can be rewritten as:

$$[\overline{M}]\,\dot{\mathbf{q}}(t) = [\overline{K}]\,\mathbf{q}(t) + [I]\,\overline{\mathbf{F}}(t). \tag{3}$$

This is the system's equation (order 2N) in the state space form, where q(t) is the state vector. The free vibration problem is:

$$[\overline{M}]\,\dot{\mathbf{q}}(t) = [\overline{K}]\,\mathbf{q}(t), \tag{4}$$

where $[\overline{M}]$ and $[\overline{K}]$ are non-symmetric matrices. Considering the solution in the form:

$$\mathbf{q}(t) = \mathbf{r}\,e^{\lambda t}, \tag{5}$$

we can obtain the associated eigenvalue problems:

$$(\lambda_j[\overline{M}] - [\overline{K}])\mathbf{r}_j = 0 \text{ , for: } j = 1, 2, \ldots, 2N \tag{6}$$

where: λ_j: system's eigenvalue of order "j"

$\mathbf{r}_j$: system's eigenvector of order "j"

The pairs $(\lambda_j, \mathbf{r}_j)$ are always complex:

$$\lambda_j = \xi_j + i\,w_j \text{ and } \mathbf{r}_j = \mathbf{s}_j + i\,\mathbf{t}_j \tag{7}$$

The corresponding motion of the entire system, in the mode "j", is described by Ref. (2):

$$\mathbf{q}_j(t) = A_j \cdot \mathbf{r}_j \cdot e^{\lambda_j \cdot t} + A_k \cdot \mathbf{r}_k \cdot e^{\lambda_k \cdot t} \tag{8}$$

where:

$$\lambda_k = \lambda_j^* = \xi_j - i.w_j \text{ and } \mathbf{r}_k = \mathbf{r}_j^* = \mathbf{s}_j - i.\mathbf{t}_j \tag{9}$$

Introducing Eq. (7) and Eq. (9) in Eq. (8), after some algebraic work, we have:

$$q_j(t)=e^{\xi_j \cdot t}\left[s_j\{(A_j+A_k)\cos(w_j t) + i(A_j-A_k)\sin(w_j t)\}+ \right.$$
$$\left. + t_j\{i(A_j-A_k)\cos(w_j t) - (A_j+A_k)\sin(w_j t)\}\right] \tag{10}$$

In order to have: $q_j(t) \in \mathbb{R}^{2N}$, its necessary:

$$A_j = C'_j - iD'_j = \frac{1}{2}(C_j - iD_j) = A_k^* \tag{11}$$

it follows:

$$q_j(t) = e^{\xi_j \cdot t}\left[s_j\{C_j \cdot \cos(w_j \cdot t)+D_j \cdot \sin(w_j \cdot t)\}+ \right.$$
$$\left. + t_j\{D_j \cdot \cos(w_j \cdot t) - C_j \cdot \sin(w_j \cdot t)\}\right] \tag{12}$$

The above equation can be developed in:

$$q_j(t) = e^{\xi_j \cdot t}\{(C_j s_j + D_j t_j)\cos(w_j t) + (D_j s_j - C_j t_j)\sin(w_j t)\} \tag{13}$$

Now we define:

$$\mu_j = C_j s_j + D_j t_j \quad \text{and} \quad \beta_j = D_j s_j - C_j t_j, \tag{14}$$

Developing the vector $q_j(t)$ in its components in order to clarify, we have:

$$q_j(t) = \begin{bmatrix} q_{1j}(t) \\ q_{2j}(t) \\ \vdots \\ q_{2N_j}(t) \end{bmatrix} = e^{\xi_j t}\left\{\begin{bmatrix} \mu_{1j} \\ \mu_{2j} \\ \vdots \\ \mu_{2N_j} \end{bmatrix}\cos(w_j \cdot t) + \begin{bmatrix} \beta_{1j} \\ \beta_{2j} \\ \vdots \\ \beta_{2N_j} \end{bmatrix}\sin(w_j \cdot t)\right\} \tag{15}$$

If we take two orthogonal degrees of freedom ("u" and "w"), from $q_j(t)$, corresponding to the displacements of certain node "m" of the system, it follows:

$$u_{mj}(t) = e^{\xi_j \cdot t}(P_{mj} \cdot \cos(w_j \cdot t) + Q_{mj} \cdot \sin(w_j t)) \tag{16}$$

$$w_{mj}(t) = e^{\xi_j \cdot t}(R_{mj} \cdot \cos(w_j t) + V_{mj} \cdot \sin(w_j t)) \tag{17}$$

Observing the trajectory described by the preceding equations, we can verify that the node "m" describe an elliptic-spiral orbit. This one may be unstable ($\xi_j > 0$), or stable ($\xi_j \leq 0$) in which the case $\xi_j = 0$ is pure elliptical (stationary). In order to better explore the characteristics of the elliptical term, we will neglect hereafter the efect of the exponential term "$e^{\xi_j t}$".

ELLIPTICAL TRAJECTORIES

From the Analytical Geometry, as described in Ref. (3), taking the equations of the planar centered ellipse, with its principal axes rotated in relation to the reference frame, we can write:

$$x(t) = P' \cos wt + Q' \sin wt \tag{18}$$

$$y(t) = R' \cos wt + V' \sin wt. \tag{19}$$

It is known that P', Q', R' and V' allow us to evaluate the four quantities to describe the elliptical orbit: magnitude of principal axes, inclination "α" of the maximum axis with respect to the "x" axis of the reference frame, and the direction of the velocity vector for increasing "wt".

It's important to observe that, without any loss of generality, any ellipse can be represented by means of its principal axes. This kind of approach has advantages with respect to the algebraic work, and enable a better geometric visualization of the orbit's properties.

For this kind of representation, the following equations are valid:

$$P' = P = A.\cos\alpha \qquad R' = R = A.\sin\alpha$$
$$\text{and} \qquad (20)$$
$$Q' = Q = -B.\sin\alpha \qquad V' = V = B.\cos\alpha$$

where: "2A" and "2B" are the lenght of the principal axes (maximum and minimum, respectively).

"α" is the angle of inclination of the maximum axis with respect to the "x" axis of the reference frame.

Thus, Eq. (18) and Eq. (19) can be rewritten:

$$x(t) = A\cos\alpha . \cos wt + (-B\sin\alpha).\sin wt \qquad (21)$$

$$y(t) = A\sin\alpha . \cos wt + B\cos\alpha . \sin wt \qquad (22)$$

Some relations easily arise:

$$A^2 = P^2 + R^2 , \qquad B^2 = Q^2 + V^2 \qquad \text{and} \qquad (23)$$

$$tg\alpha = \frac{R}{P} = -\frac{Q}{V} \qquad (24)$$

STUDY OF THE EIGENVECTOR'S PROPERTIES

We can introduce the above equations in order to quantify some values and properties of the trajectories described by complex eigenvectors.

For this purpose we will take Eq. (21) and Eq. (22) in the form presented by Eq. (16) and Eq. (17), without considering the exponential term, and keeping in mind that from the Eq. (14) we have:

$$P_{mj} = C_j . s_{u_{mj}} + D_j . t_{u_{mj}} \quad ; \quad Q_{mj} = D_j . s_{u_{mj}} - C_j . t_{u_{mj}} \qquad (25)$$
$$R_{mj} = C_j . s_{w_{mj}} + D_j . t_{w_{mj}} \quad ; \quad V_{mj} = D_j . s_{w_{mj}} - C_j . t_{w_{mj}} \qquad (26)$$

Linear trajectory (Elliptical trajectory degeneration).

For the node "m" in the mode "j", we have:

$$u_{mj} = K_{mj} . w_{mj} \qquad (27)$$

where "K_{mj}" is a real scalar. It follows from Eq. (16) and Eq. (17):

$$(P_{mj}-K_{mj}.R_{mj}).\cos w_j t + (Q_{mj}-K_{mj}.V_{mj}).\sin(w_j t)= 0. \qquad \text{or}$$

$$P_{mj} - K_{mj} \cdot R_{mj} = 0 \qquad \text{and} \qquad Q_{mj} - K_{mj} \cdot V_{mj} = 0 \tag{28}$$

Introducing in Eq. (28) the definitions of "P_{mj}", "Q_{mj}", "R_{mj}" and "V_{mj}", we obtain after some algebraic manipulation:

$$(s_{u_{mj}} - K_{mj} \cdot s_{w_{mj}})^2 = -(t_{u_{mj}} - K_{mj} \cdot t_{w_{mj}})^2 \tag{29}$$

and finally: $\quad s_{u_{mj}} = K_{mj} \cdot s_{w_{mj}} \quad$ and, $\quad t_{u_{mj}} = K_{mj} \cdot t_{w_{mj}} \tag{30}$

Equations (30) presents the general relations that must be verified simultaneously among the eigenvector's components in order to describe a linear trajectory (straight line).

Some information about these trajectories such as maximum amplitude "ψ_{mj}" and phase angle "ν_{mj}" (referred to t = 0, to reach the amplitude "ψ_{mj}") are:

$$\psi_{mj} = \sqrt{(1 + K_{mj}^2) \cdot (C_j^2 + D_j^2) \cdot (s_{w_{mj}}^2 + t_{w_{mj}}^2)} \tag{31}$$

$$\text{tg } \nu_{mj} = \frac{C_j \cdot S_{w_{mj}} + D_j \, t_{w_{mj}}}{D_j \cdot S_{w_{mj}} - C_j \, t_{w_{mj}}} \tag{32}$$

We can observe the dependency of these quantities on the coefficients "C_j" and "D_j" that are intrinsically related to the conditions at t = 0.

Circular trajectory

Circular trajectory or orbit is a particular case of an elliptical one, in which we have:

$$A_{mj} = B_{mj} \tag{33}$$

Taking Eq. (24) and substituting the correspondent terms:

$$R_{mj} \cdot V_{mj} = -P_{mj} \cdot Q_{mj} \tag{34}$$

It follows from this equation:

$$C_j \cdot D_j (s_{w_{mj}}^2 - t_{w_{mj}}^2 + s_{u_{mj}}^2 - t_{u_{mj}}^2) - (C_j^2 - D_j^2)(s_{w_{mj}} \cdot t_{w_{mj}} + s_{u_{mj}} \cdot t_{u_{mj}}) = 0 \tag{35}$$

From Eq. (23), by an analogous procedure, we have:

$$(C_j^2 - D_j^2)(s_{u_{mj}}^2 + s_{w_{mj}}^2 - t_{w_{mj}}^2 - t_{u_{mj}}^2) + 4C_j \cdot D_j (s_{u_{mj}} \cdot t_{u_{mj}} + s_{w_{mj}} \cdot t_{w_{mj}}) = 0. \tag{36}$$

Observing Eq. (35) and Eq. (36) it is possible to associate a system of equations in the following form:

$$\gamma . \overline{X} - \rho . \overline{Y} = 0 \quad \text{and} \quad \rho . \overline{X} + 4 . \gamma . \overline{Y} = 0 \tag{37}$$

Its solution is:

$$\overline{X} = 0 \quad \text{and} \quad \overline{Y} = 0 \tag{38}$$

Thus, the condition for the existence of circular trajectories are represented by the following relations, that must be simultaneously fulfilled:

$$s_{u_{mj}}^2 + s_{w_{mj}}^2 = t_{u_{mj}}^2 + t_{w_{mj}}^2 \quad \text{and} \quad s_{u_{mj}} . t_{u_{mj}} = - s_{w_{mj}} . t_{w_{mj}} \tag{39}$$

The radius of the circular trajectory is:

$$\zeta_{mj} = \sqrt{(C_j^2 + D_j^2)(t_{u_{mj}}^2 + t_{w_{mj}}^2)} \tag{40}$$

General properties of elliptical trajectories

In the general case of elliptical trajectories described by complex eigenvectors, some of its principal properties may be evaluated.

The maximum and minimum amplitudes "A_{mj}" and "B_{mj}" of the orbit are calculated directly from the Eq. (23) introducing the relations of Eq. (25) and Eq. (26). We can note their dependency on the terms "C_j" and "D_j".

In the study of inclination angle it is important to check its independency on the parameters "C_j" and "D_j" due to their characteristic of normalizing factors. The following development shows this fact. Taking:

$$P_{mj} = \frac{1}{tg\ \alpha_{mj}} . R_{mj} = \phi_{mj} . R_{mj} \tag{41}$$

$$V_{mj} = - \frac{1}{tg\ \alpha_{mj}} . Q_{mj} = - \phi_{mj} . Q_{mj} \tag{42}$$

Introducing Eq. (25) and Eq. (26) in Eq. (41) and Eq. (42) we have, after some algebraic work:

$$\phi_{mj}^2(t_{u_{mj}} . t_{w_{mj}} + s_{u_{mj}} . s_{w_{mj}}) + \phi_{mj}(t_{w_{mj}}^2 - t_{u_{mj}}^2 + s_{w_{mj}}^2 - s_{u_{mj}}^2) -$$

$$-(t_{u_{mj}} . t_{w_{mj}} + s_{u_{mj}} . s_{w_{mj}}) = 0 \tag{43}$$

If we introduce:

$$t_{u_{mj}} . t_{w_{mj}} + s_{u_{mj}} . s_{w_{mj}} = A_{mj} \qquad \text{and} \tag{44}$$

$$t_{w_{mj}}^2 - t_{u_{mj}}^2 + s_{w_{mj}}^2 - s_{u_{mj}}^2 = B_{mj} \tag{45}$$

We have, after introducing the definition in Eq. (41) and Eq. (42):

$$A_{mj} \, tg^2\alpha_{mj} - B_{mj} \, tg\alpha_{mj} - A_{mj} = 0 \qquad (46)$$

Its solution is immediate:

$$tg \; \alpha_{mj} = \frac{B_{mj} \pm \sqrt{B_{mj}^2 + 4 \, A_{mj}^2}}{2 \, A_{mj}} \qquad (47)$$

There are two distinct solutions for "$tg \, \alpha_{mj}$" in Eq. (47), and they are exactly "α_{mj}" and "$(\alpha_{mj} + \pi/2)$".

Thus, as we have mentioned before, it is verified the independency of "α_{mj}" on the parameters "C_j" and "D_j". It follows that the magnitude of the angle "α_{mj}" related to the node "m" in the mode "j" is completely defined by the eigenvector's component (real and imaginary).

One can note that despite the influence of the two terms "C_j" and "D_j" over all the orbits associated to the mode "j", they are completely determined by the position of a certain node at certain time (t = 0, for example). It means that these terms are the connecting elements among the nodal orbits associated to a fixed mode and they are important in the evaluation of the nodal trajectory amplitudes, ones related to the others.

GRAPHICAL RESULTS

Some simple examples are presented here in order to better demonstrate some properties of complex eigenvectors in Rotordynamics. The examples shown below were obtained, as yet mentioned, from Program ROTMEF developed at CEPEL - Electrical Energy Research Center. They refer to a simple rotor supported by two bearings. The rotor is modeled by 3 shaft elements with a disc not equidistant from the bearings, as can be seen in Figure 1. In the two cases shown the rotor is running and the bearings are non-symmetric. The first case refers to a conservative system while the second case refers to a non-conservative one (viscous-damped bearings). Figure 2 and Figure 3 are presented in two ways being the first a tri-dimensional visualization and the other an axial projected one.

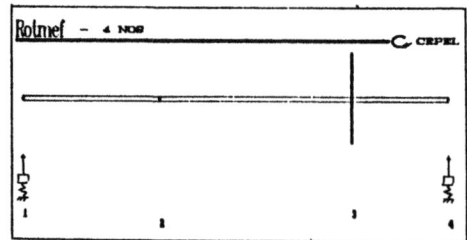

Fig. 1 - Rotating system's model.

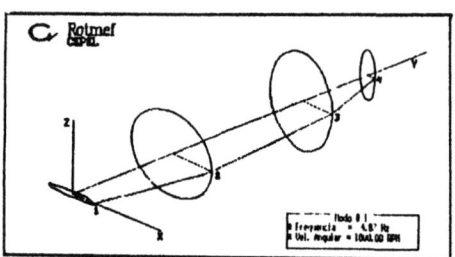 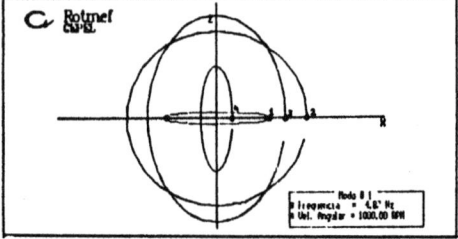

Fig. 2 - Conservative, non-symmetric bearings.

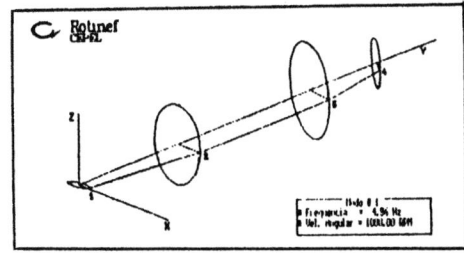 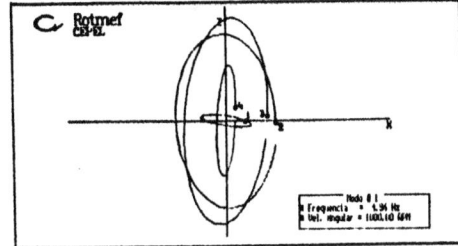

Fig. 3 - Dissipative, non-symmetric bearings.

It may be seen in these figures 4 points numbered 1, 2, 3 and 4. These points correspond to the position of the shaft center at t = 0.

It's possible to observe from these figures two kinds of "offset": a geometric "offset" represented by the different inclination angles relative to the ellipses principal axes, and a time "offset" (or time delay) which represents the non-planar, non-stationary mode shapes described in the case (shown by the points 1, 2, 3 and 4) of Figure 3.

CONCLUSIONS

It is intended, from the development presented in this paper, to obtain basis for a systematic treatment of the complex eigenvectors associated to Rotordynamics. These basis can be applyed to a better physical understanding of the dynamical behaviour of rotating machines or to help in the development of efficient graphical computational routines when dealing with this kind of modelling.

These equations were implemented in ROTMEF and are now being used as support in the research of model updating techniques applied to large hidrogenerating units.

REFERENCES

(1) Azevedo, H.R.T., et al., "On the Dynamical Modelling of Rotating Machinery by the Finite Element Method and Its Application to Hydroelectric Power Generators", **Proceedings of the 15th International Seminar on Modal Analysis**, Vol. III, pp. 1047 a 1062, Leuven, Belgium, 1990.
(2) Feltz, C., **"Calcul des Valeurs Propres et Vecteurs Propres d'un Rotor de Turbine"**, DEA Thesis on Vibration and Oscilators, Université de Besançon, Faculté des Sciences et des Techniques, Besançon, France, 1983.
(3) Leithold, L., **"The Calculus With Analytic Geometry"**, Edited by Harper & Row, Publishing. Inc, 1981.

The Evaluation of the Safety of the Shaft System of the Large Turbine Generator Set

Huang Ya Luo
Central China Electric Power Administration in Wuhan

Abstract

It is necessary to optimize the dynamic characteristics of the shaft system of the set, while optimizing the performance of the steam turbine and generator in order to ensure the safe operation of the set. In the paper, it is discussed how to evaluate the safety of the shaft system of the set from the dynamic characteristics of the shaft system, such as the traverse vibration, the torsional vibration and the stability etc. , and some calculation examples for 300MW set are analyzed.

Introduction

With the increase of the capacity of the set, the safety of the shaft system of the turbo-generator set (including the exciter) is threated for the following three reasons:

1). The length of the shaft system increases. The length of the shaft system is 15m for a 50MW set, 32—40m for a 300MW set and 90m for a 900MW nuclear power set. The increased number of the concentrated masses leads to the increase of both the number of the critical speeds and their dispersion as well as the smaller ratio of radial dimension to axial dimension, which yields more flexible shaft system and the drop of the first critical speed. Hence, the shaft system easily comes to operate at the edge of unsafety.

2). The longer shaft system leads to the lightened foundation of the set, which requires that the vibration of the set and the basis should be analyzed together.

3). The detriment of the oscillation of the electromechanical coupling and the relevant safety criteria are pointed out, owing to the operation of the large set in large electric—net work.

The safety of the shaft system should be evaluated in the initial stage of selecting the type of the set, which has been shown necessary for a series of accidents such as the fracture of the shaft system and the destruction of the set in the recent years.

Generally speaking, the manufacturing property of the set and the dynamic characteristics of the shaft system are two aspects in evaluating the safety of the shaft system, the latter of which is discussed in the paper.

Traverse vibration and the critical speed

1). The vibration standard.
The vibration standard of bearing is stipulated for middle and small scale set, but for a large set the vibration standard of the shaft is added. It is specified in many factory of

our courtry that at the rated speed of 3000r/min, the vibration standard of the bearing is 0. 025mm, that of the shaft is 0. 050mm and at the critical speed, that of the bearing is 1. 8mm/s(0. 0324mm) and the permissible vibration standard of the bearing in operation is 11. 2mm/s(0. 1mm). The satisfaction of the vibration standard mainly depends on the proper balance of the single rotor in the shaft system, which demands small residual unblance of thr rotor and the smallest internal torque after thr proper balance of the external end. So for the set with three—point support structure and the rotor with attached connecting short axis, the balance of exteral end should be solved. Hence after the dynamic balance at the high speed is carried out for the rotor, the short axis should be conbined with the rotor to take the dynamic balance at high speed. The unblance response characteristics at the rated speed need to be given out in order to judge the residual unbalance of the shaft system. For the factory of a high technology, tthe thermal stability and balance should be tested too. In addition, the intensity test of the rotor at super speed should be performed on the test rig.

2). The critical speed

The first and second order critical speeds should be calculated for the shaft system with both the rigid support and the elastic damping support. If the difference between the vertical and horizontal stiffnesses of the bearing is great, the vertical and horizontal critical speed should be given out respectively. For example, the first order critial speed in horizontal direction is 400r/min lower than that in vertical direction and the difference between the second order critical speeds is 1350r/min for the bearing of the generator of a 300MW set. For the rotor of the generator, the critical speed rigion should be given owing to the different values of bending stiffness between d—axis (big tooth) and g—axis (small tooth). Larger critical speed region causes not only the increase of the component of doubling frequency vibration, but also the appearance of resonance when the operaring speed is half of the critical speed. The distribution principle of the critical speed of the shaft system ought to be able to ensure the safe warming—up speed and the safe test of the danger protector, in addition to ensuring suitable anti—interference ability of the set and enough safely—operating margin. Hence, it is specified that every order crritical speed should be ±10%——±15% out of the rated speed, and in order to avoid the self oscillation and the instability of the shaft system it is required that the critical speed evade half or one third of the the rated speed. The critical speed of the adjacent rotor had better be different in order to enhance the anti—accident ability of the set under the condition of the unsatisfactory balance. Generally speaking, it is difficult to select the critical speed of the generator because the lower first order crritical speed may cause the oil film whip. So that in order to improve the stability of the shaft system, the first order critical speed of the generator is raised and the second order critical speed is greater than the rated speed in some manufacturing factory. Table 1 shows the optimized comparison of the selected plans of the critical speeds of the shaft system of a 300MW set, in which plan c is chosen.

The torsional stress and the torsional frequency

1). The checking operating condition of the torsional stress.

Table 1. the comparison of the plans of the critical
speed of the shaft system of a 300MW set

r/min

rotor / plan	high & middle pressure		low pressure		generator		exciter
	No. 1	No. 2	No. 1	No. 2	No. 1	No. 2	No. 1
A	1650	3900	1510	3800	950	2400	2530
B	1550		1575		1366	3675	1600
C	1700	3980	1540	3660	1480	3800	1420

NO. 1: the first order critical speed
NO. 2: the second order critical speed

In the past, it was believed that the short circuit of the triphase exit end of the generator is the most crucial mechancal stress condition faced by the turbogenerator. But in fact, it is not true. When the two—phase exit end of the generator is short, the electromechancal torque summed up by the foundermantal frequency torque and the doubling frequency torque will surpass that of the short triphase exit end. There is a definite value of the torsional stress for large scale set under the rated load, owing to its great current density and high magnetic line density, which causes bigger short current supplied by the set or the electric—net work when short circuit happens. Hence, the safety of the shaft system should be evaluated according to the torque and the largest torsional stress born by the shaft system when two—phase exit end of the generator is short. It is generally specified that the permissible torsional stress of the material is 1. 1 times over the larrgest torsional stress on the shaft sections under the condition of the short two—phase exit end. The calculated results of a 300MW set show that the torsional stress of the exciter shaft is up to 467. 4MPa. So the material and the structure ought to be changed and improved, which yields that the permissible shear stress of the exciter shaft is raised from 476MPa to 530MPa.

2). The torsional natural frequency.

Similar to the existance of the natural frequency spectrum of the bending vibration, there also exists the natural frequency spectrum of the torsional vibration in the shaft system, which should be avoided in operation. The cumulative effect of the fatigue failure due to the torsional vibration will greatly shorten the operating life of the shaft system.

In order to prevent the non—periodic component or the negative—phase sequence component of the current in failure, including the resonance of working and doubling frequency of the electrical and mechanical interaction excited by the unbalance load from damaging the set, it is specified in our country that the torsional natural frequency of each part of the shaft of the large scale set should not be within the ranges of 0. 9 to 1. 1 times ad 1. 9 to 2. 1 times over the working frequency. The turbine with long last—stage blades (851,869,900,1000mm) is sensitive to the torsional resonance of high frequency (about 100Hz). For example, as to a 350MW set the diameter of the middle part of the low pressure rotor is machined smaller about 254mm, because the torsional natural frequency of the shaft system could not evade the definite range of the doubling frequency.

Because of its big inertia of the rotor of the generator and the low pressure rotor of the turbine as well as its low damping to the low frequency oscillation, the shaft system of the large scale set is sensitive to the low frequency oscillation. In order to prevent the subsynchronous resonance from damaging the set, it is specified that the factory should supply the torsional natural frequency of each part of the shaft, so as to evade the subsynchronous torsional natural frequency when the primary equipment are used in the electric—net work such as the direct—current transmision, the series capacitance compensation, the electric power system stabilizer and the silicon control.

The stability of the shaft system

1). The index of the stability of the shaft system.
Since the first 200MW set of our country was put into operation in 1972, oil whip has taken place in fourteen identical sets. The low frequncy resonance happened in a 200MW set reported by Northeast Electric—net Work belongs to steam—exciting vibration. With the increase of the capacity of the set and the raise of the steam parameters, the self—exciting oscillation may happen within the range of working speed, leading to the instability of the shaft system mainly owing to the oil film force of the bearing and the aerodynamic force of the impeller of the turbine. There are two indexes, the instability speed n_{st} and the logarithmic decrement δ, to evaluate the stability of the shaft system comprehensively. Generally, it is required that $n_{st} \geqslant 1.25 n_H$, i. e. $n_{st} \geq 3750 r/min$, but there is not a united standard suitable for the logarithmic decrement, but only a permissible range. For example, since most of the instability of the shaft system caused by the steam—exciting force takes place on the high pressure rotor, it is required that $\delta \geq 0.05 - 0.10$ when loaded; on the other hand, most of the instability caused by the oil film force of the bearing happens on the low pressure rotor of the turbine and the rotor of the generator, so $\delta \geq 0.15 - 0.30$. The calculated results show that for a 200MW set in which oil film whip happens many times, the values of its n_{st} and δ are less than the range mentioned above. In fact, the plus δ corresponds to definite stability margin. In operation, the bench mark of the bearing seat will be changed by the variation of the lubricant temperature, the inhomogeneous sedimentation of the foundation, the vacuum of the coagulator and the hydrigon prssure of the generator as well as the differences of the heat expansion between the the cilinder and the foundation in the process of start and stop of the set, thus alterring the distribution of the load on the bearing and the eccentricity of the bearings, i. e., changing the stability of the shaft system. On the other hand, the bench mark of the bearing seat changes at great random in operation. The abnormal soaring of the speed, still within the range of the adjusting system and the rotor intensity, may also cause the drop of the logarithmic decrement at the definite slope, so while δ under rated condition is given, its changing regularity with every factors above should be studied to ensure $\delta > 0$ under every operating condition. As to the set having been put into operation, the best method to evaluate the stability of the shaft system is to measure live the components of the low frequency vibration of the set. When the temperature of the supplied oil

changes in possibly larger range, the vibration is live measured between the middle warming—up speed and the danger protector action speed of the set, especially at the speed of 3000r/min, to see whether the components of the low frequency vibration appear. When any ratios of the components of the low frequency vibration to those of the working frequency vibration are larger than 10% or the value of the low frequency vibration, over 0. 005mm, and some order critical speed which corresponds to the component of the low frequency vibration is in harmony with half and one third of the operating speed, the type of the pad should be changed and the structure should be modified, until the components of the low frequncy vibration are controlled at the permissible level. If possible, the live—measured data of many operating sets are very useful to evaluate the stability of the bearing of the identical sets to be put into operation.

2). The improvement of the stability of the shaft system.

The suitable assignment of the load on the bearing and the sensitivity analysis of the change of the bench mark of the bearing seat are the basic foundations to improve the stability of the shaft system. The reasonable load assignment is often decided by the mounting height curves, meanwhile, the bench mark of the shaft center at the support point, the opening of the couple and the error of the shaft center are also given to ensure that the curve of the shaft center under the rated operating condition satisfies the ideal height curve. After the type of the set and the structural parameters are decided, the stability of the shaft system mainly depends on the type, the structural parameters and the geometric parameters of the bearing. as to the selection of the bearing of the generator, three plans are compared, which shows that the logarithmic decrements differ from one another greatly, although the instability speeds are all over 3750r/min. When the cilinder pad is used, $\delta < 0. 10$; when the elliptic pad with groove in the upper part is used, $\delta = 0. 16$; and for the elliptic pad with no groove, $\delta = 0. 25$, with its temperature rising over 10°C. In analyzing the stability of the same set, the sensitivity of the bench mark of the bearing seats should also be considered. When the bench mark of every bearing rises or falls 0. 10mm, the maximum change of the load on every bearing is less than 20%. In selecting the type of the bearing, the oil flow supplied by the lubricating system should be verified, which ought to satisfy the safety of the oil supple in the bearing under different abnormal operating conditions. Meanwhile, the restriction orifice of the oil supple of the bearing should be carefully selected to ensure the proper assignment of the flow to every bearing in actual operation.

The foundation of large scale set is comparatively lightened with its comparatively —reduced radial dimensions and its comparatively flexiable shaft system. It is necessary to reanalyze the dynamic characteristics of the shaft system in the set—foundation system, on the basis of the designed dynamic characteristics of the shaft system. As to a 300MW set put into operation in 1970s, since the reanalysis of its dynamic characteristics was not carried out when designed, the worse stiffness of the bearing seat and the foundation frame became increasingly protrusive after operation for 10 years, which yielded that the vibration of the bearing was larger than that of the shaft and out of the vibration standard. By taking some temporary measures to reinforce the bearing seat, the set maintained the operation with difficulty.

Conclusions

1). It is required that the evaluation of the safety of the shaft system and the design optimization should be carried out during selecting the type of the large scale turbogenerator set because of the flexibility of the shaft system, the comparatively—lightened foundation and the electromechanical coupling action.

2). The reasonable distribution of the critical speed and the proper unbalance resonance are the basic requirement for the safety of the shaft system.

3). The two—phase short circuit condition at the exit end of the generator is chosen as the condition to verify the torsional stress, and the torsional natural frequency of the shaft system should evade the working frequency, the doubling frequency and the susynchronous harmonic components which may appear in the electric power system.

4). In actual engineering design, The instability speed n_{st} and the logarithmic decrement δ are the two indexes in evaluating the stability of the shaft system. As to the set in operation, it is evaluated by means of live measuring the ratio of the component of the low frequency vibration to that of the working frequency vibration. If the structure of the set has been determined, the stability of the shaft system is decided by the bearing design.

5). On the basis of the analysis of the dynamic characteristics of the shaft system, the combined reanalysis of the set—foundation dynamic characteristics and the evaluation of the safety of the shaft system should be carried out respectively.

References

1. Zhang Han Ying, He Yi "The Guide Rule of the Design for the Static and Dynamic Characteristics of the Rotor—bearing Foundation System of the Large—scale Turbogenerator Set" 1990. 8.
2. Zhang You Zhu & Shi Wei Xin "The Vibrarion of the Turbogenerator Set and the Finding Balance of the Rotor" 1986. 10.

SESSION 11   BLADED SYSTEMS

Vibration Reduction of Blading Using Interconnecting Elements

G.M. Chapman * , E. Swain # , X. Wang \ and M. Yang # .

* School of Engineering and Manufacture, Leicester Polytechnic, U.K.
# Engineering Design Institute, Loughborough University, U.K.
\ Mechanical Engineering Department, Loughborough University, U.K.

ABSTRACT

Bladed discs often require devices to reduce blade vibration to eliminate the risk
of fatigue failure. Different industries employ different approaches to the problem
with varying levels of success. The most common vibration reduction method is to
use an interconnecting element between each adjacent blade and create coupling
effects which introduce a form of damping. Small steam turbines commonly employ
riveted shrouds, large turbo-chargers invariably use lacing wires, some gas turbines
use blades with integral shroud elements and many turbine assemblies use connecting
elements which lock into place with increase in rotational speed.

This paper explores the vibration reduction opportunities offered by using
continuous connections between blades with lacing wires and compares the vibration
characteristics with the behaviour of integrally shrouded blades. The paper
describes the mathematical modelling work required to obtain a true representation
of a blade system and describes the multi-bladed response of both laced blading and
integrally shrouded blades.

HISTORICAL BACKGROUND

This work was started in an attempt to determine the most satisfactory method of
reducing vibration in blading in large turbocharger discs. The initial
investigations were concentrated on establishing the best location to place the
lacing wire hole to reduce the risk of fatigue failures as a consequence of locating
the lacing wire at a nodal position. (Connor et al, 1986). The investigation
stimulated a more thorough research programme into determining an understanding of
blade vibration modes and how high stress nodal lines could be located.
The turbocharger blade chosen was a complex tapered and twisted configuration.
Initial attempts at modelling this blade by finite elements became extremely costly
and because of a lack of understanding of the blade root clamping conditions the
results obtained were less than satisfactory. To improve upon this situation,
experimental methods were used to identify the fundamental frequencies of the blade.
Firstly, a root block fixing device was designed in which loading could be applied
through the base of the root to simulate the effect of centrifugal force pulling the
blade in the disc. (Chapman and Wang, 1988). Care was taken so that repeatable
loading conditions could be achieved for consecutive tests. The experimental
techniques used initially were based upon modal analysis techniques in which hammer
accelermometer readings were taken to determine the vibration characteristics of
many points across the blade surface. This method gave a good insight into blade
performance, but because of the mass of the accelermometer the results never truly
represented the real characteristics of the blade. Supporting tests were undertaken
using an inductive displacement transducer in place of the accelermometer and this
method proved highly satisfactory for the lower order frequencies, but because of
the small amplitudes associated with the high order frequencies the higher modes
were not measurable.

To support this analysis, electronic speckle pattern interferometry (ESPI) was used. This method is based upon an interference technique generating an image of a vibrating object by comparing the distance moved between the two extremes of vibration amplitude. The visual image can be displayed on a T.V. screen. The technique requires very careful tuning of the excitation frequency of the blade and much skill was required in interpreting the ESPI patterns produced on the screen. Together the two experimental methods were used as the basis for developing a simplified finite element model of the blade so that the mathematical model could be investigated later.

SIMULATED LOADING FOR LACED BLADING

Having developed confidence that the finite element model of the simple single blade could be truly representative of the more complex twisted blade for the first modes of vibration, the investigation then concentrated on the effects of the inter-relationship between the blade and the lacing wire passing through it. Parameters to simulate the characteristics of the lacing wire were selected to represent the element as a spring in extension mode with a limited amount of torsional resistance to add to the blade torsional characteristics.

A small test rig was built to introduce simulated centrifugal force. (Chapman and Wang, 1990). Pre-load could be applied to the lacing wire to simulate the pull on the blade root typically created by the rotational speed of the disc. The force induced in this way was set at a value determined by considering the mass of the wire element in the blade and the centripetal acceleration created by the rotational speed of the turbine disc.

The experimental tests employed excitation of the blade by an electro-magnetic vibrator in front of an ESPI facility. This method gave a fairly good insight into the true natural frequencies associated with this configuration and the finite element model was adjusted so that the performance was very similar to that of the physically tested system. The results obtained are shown in the following table.

Table 1. Comparison of natural frequencies for unlaced and laced simple
          rectangular blades.

Vibration mode	1F	1T	1E	2F
Unlaced	1290	4700	2960	5810
Laced	2575	4845	4181	7777

In the table the notation 'F' represents 'Flap mode' (that is a bending mode of the cantilever); 'T' represents 'Torsional mode' (that is a twisting mode along the length of the cantilever); 'E' represents 'Edge mode' (that is a bending mode in the stiffer plane of the cantilever).

The results from this initial investigation indicate that in the case of the torsion and edge modes there is little effect of the lacing wire on the blade. Whereas in the flap mode the lacing wire has a considerable effect, dependent upon the location of the lacing wire hole in the blade. In the case quoted the lacing wire was located 80% along the length of the blade towards the tip. By placing the lacing wire at this location effectively the blade is restrained from its normal free bending mode and is caused to vibrate nearer to the second flap mode as a consequence of the restraint caused by the lacing wire. The finite element model of the laced blade was modified to match the experimentally determined natural frequencies.

272

Multi Blade Model

Having achieved a reasonable comparison between the physical system and the finite element model the investigation then was directed to look at the vibration characteristics of multi-bladed systems. To achieve this, several blades were connected together via their lacing wires and the investigation first considered two blades connected, then three and eventually a ten blade system. For each configuration the 1F, 1T and 1E modes were investigated and for each configuration each of the vibration modes were documented. Whereas for a single blade there are unique 1F, 1T and 1E frequencies, when two blades are connected then two 1F, two 1T and two 1E modes exist. As the number of blades is increased to ten, there will be ten frequencies for each of the modes concerned and these families of frequencies are likely to overlap each other.

This lead to considerable difficulties in identifying the natural frequencies by both finite element analysis and by ESPI practical analysis. To classify these modes a system was used whereby the mode was described as 1Fn - the 'n' representing the number of nodes along the blade chain. Hence, for a two bladed system there would be a 1F0 mode where both blades were vibrating in unison, a 1F1 mode where the two blades vibrate in the first fundamental mode, but in antiphase.

Consequently, when investigating the ten bladed system there could be 1F9, 1T9 and 1E9 modes associated with the system.

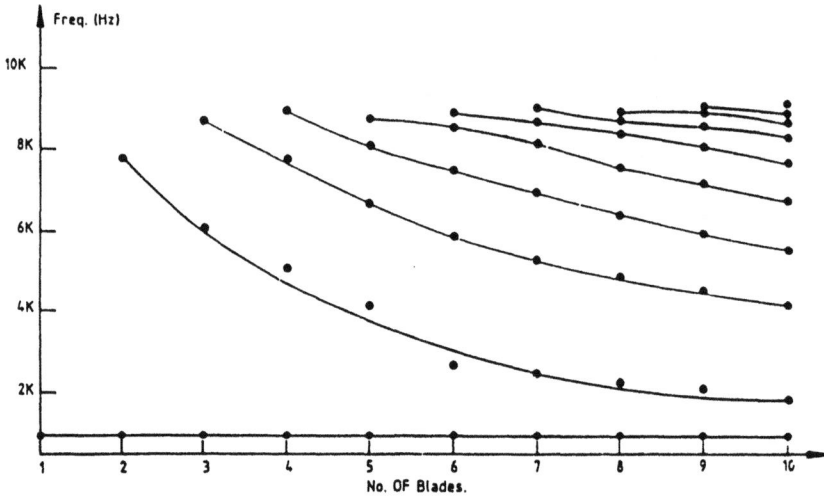

Fig. 1. 1F family of frequencies for laced blading.

Figure 1 shows the result of an FE analysis of a range of multi-bladed systems for the first flap modes for a system with up to ten blades. From the graph it can be seen that as the number of blades increase, the number of natural frequencies increase. The fundamental frequency, (1F0 mode) is the same for all blade configurations because all blades move together and the lacing wire has no effect. For all other modes the natural frequency value obtained with the lower number of blades is always higher than the same mode for a system of a greater number of blades. As the number of blades increase then the compaction of the number of 1F frequencies within a certain band-width becomes much more intense. These results have an identical pattern for the 1T and 1E modes, with families of frequencies overlapping the frequencies for the 1F modes. The consequence of this is that there is extreme difficulty in identifying these modes as consecutive frequencies may originate from differing fundamental modes.

Vibration Reduction Characteristics of Laced Blading

The investigation of the laced blading would imply that by the introduction of a lacing wire the number of possible natural frequencies increases as a consequence of the coupling of all the blades. Also implied by the FE analysis is that the overlapping of families of frequencies of the flap, torsion and edge modes causes a higher possibility of coupled vibrations between modes, particularly if damping exists within the blading when a flap mode could easily excite a torsion mode and vice-versa.

Experience shows that the introduction of lacing into bladed assemblies reduces the risk of blades failing as a consequence of fatigue caused by vibration. The commonly held view has always been that the blades slide on the lacing wire and the induced friction reduces the likelihood of excessive vibrations. However, little evidence is detected of fretting between the lacing wire and the blade hole. It, therefore, follows that this mode is unlikely to be the predominant mode when turbines are running at full speed. It is highly likely, however, that this mode is very relevant as the turbine runs up from zero speed to its operating range where loose blades can be controlled by the effects of the lacing wire inter-relating with the blades.

At higher operating speeds when the lacing wire locks into the blading as a consequence of centrifugal forces then the wire becomes an integral part of the blading system and the whole assembly locks as a multiple blade system with a fixed connecting wire between the blades. Each element of wire between each blade is subjected to centrifugal force and the wire takes on a deflected shape similar to a figure C (as a C spring). Such elements exhibit a non-linear characteristic in that in compression they are rather stiffer than in tension. It is this characteristic that contributes to the vibration reduction characteristic of laced system. Additionally there is the likelihood that there is an inter-reaction between the edge of the lacing wire hole and the lacing wire itself, sufficient to cause microslip which could introduce damping to reduce the vibration amplitude at higher frequencies. These possibilities need further investigation.

SIMULATED EXCITATION OF INTEGRATED SHROUDED BLADING

An alternative approach to the lacing wire in a blade is the integral shroud attached to the tip of a blade so that when several blades are pushed into the disc the shroud elements contact each other by rubbing face-to-face on the periphery of the disc assembly. These elements are subject to tolerance in size, and therefore it cannot be assumed that the geometry of the shrouds totally makes up the full circumference of the bladed disc assembly. The interference between each shroud element is likely to cause some form of vibration reduction by friction, damping and elasticity between the elements.

Finite Element Representation

With the confidence gained in the previous investigation where a physical system was represented by finite element modelling, the initial approach for this type of blading attempted to develop a finite element model to represent the prototype blade. The blade was tested and various parameters were identified such as the 1F, 1E and 1T mode and then the finite element model was built. The model represented the geometry of the blade and that of the shroud and introduced a mass distribution representative of the real blade system. Single blade parameters were set to be identical to the real model. The interface between blades was simulated by introducing sliding elements at the surface of the shroud, by adding stiffness and damping elements at each corner of the shroud contact point to ensure that flap, edge and torsion were truly simulated.

An attempt was made to couple two blades together with these elements. Results obtained were encouraging, but it was not felt that this model represented the true performance of the blade in a real turbine, particularly if the excitation was to be representative of stimulation by the mismatch between the number of blades and number of nozzles in the turbine stage. In this case each blade around the turbine would experience a force slightly phased in amplitude to its neighbour because of the difference between blade nozzle number.

Lumped Mass Parameter Model

To simulate transient excitation a lumped mass parameter method was used where the blade was considered to be represented firstly as a simple mass supported on a spring/damper system. One spring represented the stiffness of the blade system, with a second spring representing stiffness between consecutive blades. A gap between the end of the second spring and the next blade simulated the discontinuity between consecutive blades.

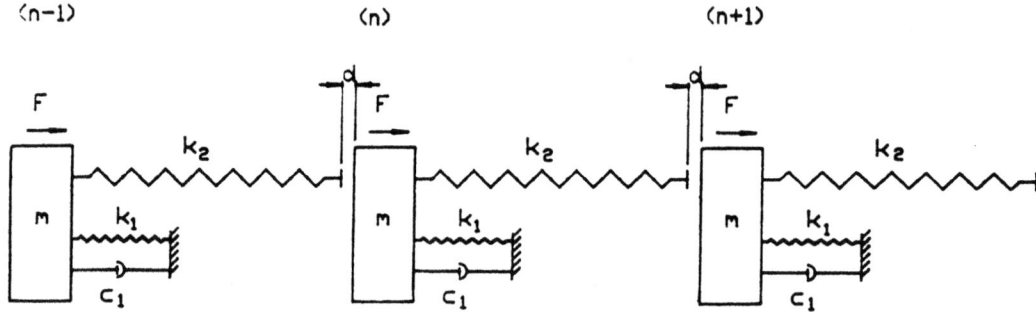

Fig. 2. Single mass representation of each blade.

Figure 2 shows a typical initial representation of the lump mass parameter system used to represent several blades connected to each other, the stiffness K3 represents stiffness of the blade interface at the shroud face. This investigation progressed satisfactorily by deriving the equations of motion for a multi-bladed system and then re-writing the differential equations in a finite different form so that the equations could be solved in a time stepping mode. To get a better representation in later analyses, however, it was decided that the blade under investigation should be modelled as a large mass concentrated within the shroud element and the remainder of the mass connected between the shroud and the base to represent the blade itself.

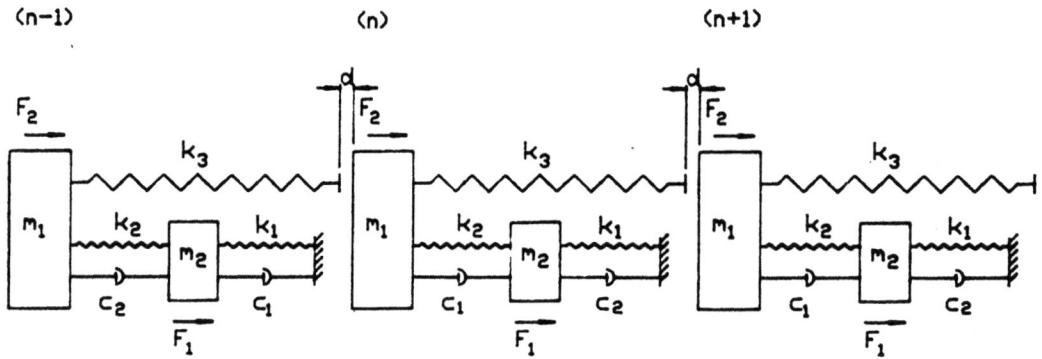

Fig. 3. Two mass representation of each blade.

The configuration shown in Figure 3 was used as the major form for investigation. The dual mass system gave a better representation of the shroud and the blade elements. Checks were made to ensure that this new configuration behaved in an identical way to the single blade system represented initially in the FE model, thus confidence in the lumped mass parameter system was established. The time step analysis required knowledge of the forces being applied to each blade element within the real system. To identify the characteristics of such a system a five bladed representation of the disc was investigated. The blades were considered to be a continuous ring around the circumference of a disc and so the first blade was connected via the spring representing the interface to the last blade, thus producing a continuous chain of blades.

The analysis also required that the force applied at all points within the system be known at all points in time so that the time step procedure could progress with all forces at all input points known at the same time. To understand this Figure 4 represents the effect of the force on a blade as it passes behind a nozzle blade and then becomes exposed to the gas pressure passing through the turbine.

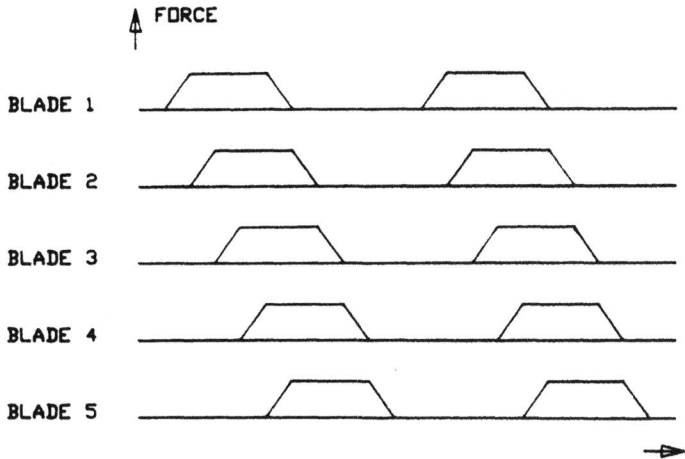

Fig. 4. Typical force excitation on each blade.

Whilst behind the nozzle the blade experiences virtually zero load and as it comes out from behind the nozzle and takes up the position between consecutive nozzles then the force becomes largest. For consecutive blades, because there is a mismatch between the nozzle and blade number each blade has a force applied to it slightly phased from the previous blade. Hence, at any one point in time the forces on each of the five blades are different around the disc and this can be derived by building up a look-up table for every incident in time as the turbine disc rotates. Hence the force pattern applied to the system can be input at all times.

It was considered essential that the force applied to the two masses in the lumped mass system representing each blade should be determined and these were calculated on the basis of the projected area exposed to the gas inlet pressure. The forces applied to each mass was proportioned to the projected area of the shroud and the blade. The damping values selected for the blading were purely arbitrary in order that a steady state situation could be determined from a transient excitation analysis to ensure that the mathematical analysis reached a stable solution.

Vibration Reduction Characteristics of Integral Shroud Blading

The analysis identified that the vibration characteristic of the blade assemblies was very dependent upon the forcing amplitude and the gaps between each blade. Two forms of vibration characteristic could occur. When the force was small, such that the blades were unable to travel across the gaps between consecutive blading, then vibration would occur as a single blade with fairly large stresses being generated within the blades. When the force was sufficient to cause the blades to vibrate in such a way that they travelled across the gap, then the natural frequencies of the

276

system changed completely to a system totally coupled as if the interface springs were permanently coupled to the blade elements.

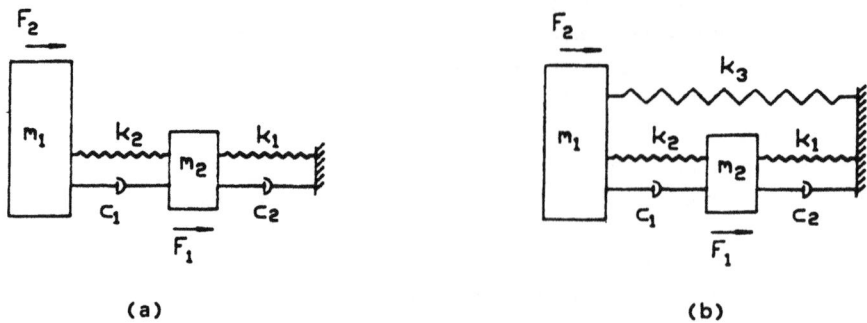

Fig. 5. (a) System parameters for small force excitation.
        (b) System parameters for large force excitation.

Investigations were conducted into the response of the bladed disc assembly and these results are presented in Swain, Chapman and Yang (1992). In that paper a thorough treatment of the maximum blade stress versus the blade gap for various force excitations is considered. Clearly, if all blades move in unison in the same mode, then the effect of the blade shroud has no consequence and this is similar to the situation associated with the lacing wire in the previous example. The mass of the shroud also has little consequence when the blade moves into the 2F mode where the nodal point on the blade becomes almost coincident with the shroud mass centre of gravity. Hence the blade will vibrate with the shroud having little effect and possibly not even passing through the gap between the blades. For the blade under investigation there was no twist along the blade length and hence the torsional mode was not considered of consequence. The edge mode was also insignificant because its natural frequency was so much higher than that of the flap mode.

CONCLUSION

The analysis of these two forms of blade inter-connection have similar conclusions. It would appear that the vibration reduction obtained in both systems is achieved by the introduction of non-linearity between consecutive blades. In the case of the laced blading the non-linearity is created by the distortion in the wire caused by centrifugal force, but in the case of the integral shrouded blade the non-linearity is introduced as a consequence of the irregular gaps between consecutive blades. Both these criterion contribute to the reduction in vibration amplitude. However, in both cases the inter-connecting element is incapable of reducing vibration amplitude if the first fundamental mode is excited where all blades move in unison. Fortunately this is quite difficult to stimulate as the force excitation upon consecutive blades is phased around the circumference of the turbine disc.

There is still a considerable amount of work to be done to tie together the steady state analysis used in the first application with the transient analysis used in the second application. Both systems require further mathematical analysis and would benefit from experimental testing of fully configured bladed discs subjected to real excitation force.

ACKNOWLEDGEMENTS

The authors wish to thank Napier Turbochargers Limited, W.H. Allen of Bedford, Loughborough University of Technology and Leicester Polytechnic for permission to present this paper and SERC for their assistance with instrumentation into the vibration analysis associated with rotor systems provided to support research contracts at the University.

REFERENCES

W. Connor, E. Swain, A. Bellamy and G.M. Chapman.
Excitation of Turbine Blade Vibrations in Large Turbochargers.
Third International Conference on Turbocharging and Turbochargers, IMechE, May 1986.

G.M. Chapman and X. Wang.
Interpretation of Experimental and Theoretical Data for Prediction of Mode Shapes of Vibrating Turbocharger Blades.
Journal of Vibration, Acoustics, Stress and Reliability in Design, ASME, Volume 110, No. 1, January 1988.

G.M. Chapman and X. Wang.
Vibration Analysis of Laced Blading.
Third International Conference on RotorDynamics, Lyon, September 1990.

E. Swain, G.M. Chapman and M. Yang.
Vibration Characteristics of Integrally Shrouded Blades.
Vibrations into Rotating Machinery, IMechE, September 1992.

1483.GC/JL
29.08.91

VIBRATION CHARACTERISTICS OF AN AXIAL BLADING WITH DIFFERENT TYPES OF COUPLING LINKS

J F Mayer

Institut für Thermische Strömungsmaschinen
und Maschinenlaboratorium
Universität Stuttgart
Germany

ABSTRACT

The vibrational behaviour of an axial blading with different types of couplings is investigated by numerical calculations and by experiments. The blades of simple geometry are coupled by lashing pins loosely inserted, pins soldered with the blade foil, a coverband (continuous shroud) or shroud segments, respectively. Natural frequencies and natural modes are calculated by means of the finite element method combined with the wave propagation technique. The computed frequencies are compared with measured resonant frequencies for the different configurations. The experiments were carried out making use of a special measurement set–up and a special processing technique. The blading was excited either by a shaker or, in the case of rotation, by a magnet fixed to the casing of a vacuum chamber.

1   INTRODUCTION

The availability of an axial turbomachine is essentially controlled by the blading. Considerations of structural integrity lead to a solid blade design that is contradictionary to the objective of achieving a high efficiency which leads to slim profiles and therefore to blades which are sensitive to vibration.

Mechanical coupling of the blades decreases in most cases the dynamical stresses, but increases the number of natural frequencies. However, for coupled structures the resonance conditions become more complicate. To avoid resonances, the vibratory behaviour of the blading, i.e. the natural frequencies and the natural modes, must be known.

In this paper the numerical and experimental investigation of a model blading is presented. The vibrational characteristics of the blading are determined for different types of blade couplings.

Blade coupling by lashing pins is a frequently used method in stages with long blades. These pins can be found in low pressure parts of steam turbines or axial compresssors. Two types are examined: pins loosely inserted and pins soldered with the blade foil.

Another configuration which was investigated is the coupling of the blades by a coverband (continuous shroud) welded to the tip of the blades.

Nowadays in high and medium pressure stages of steam turbines the blades are commonly coupled by shroud segments. This type of coupling is also used in gas turbines. Consequently coupling of blades by shrouds was included in this investigation, whereby the blades were pretwisted to achieve the connection of the shroud interfaces.

2   EXPERIMENT

2.1   Model for Axial Blading

An axial blading was manufactured to study the vibrational characteristics on the basis of a simple model (fig. 1). The most important design features are:

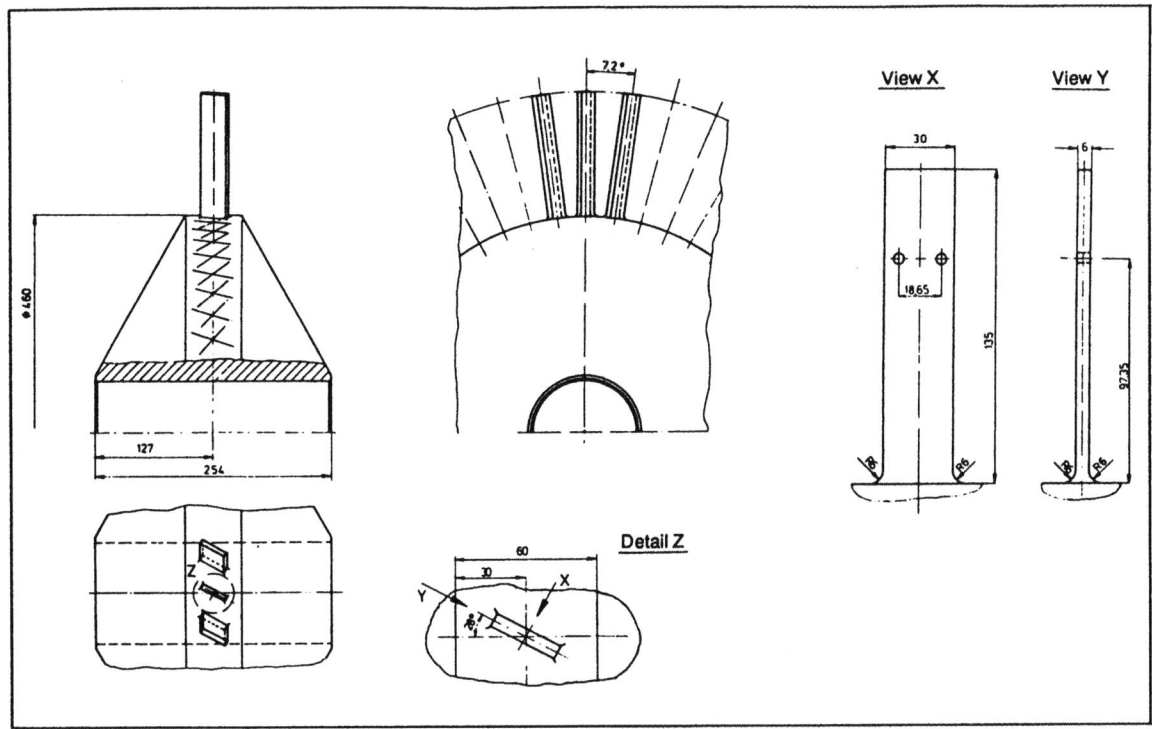

Figure 1: Model rotor

- No major coupling between blade vibration and disc vibration, therefore a very stiff disc design was chosen.

- Simple geometry simplifies interpretation of vibrational behaviour.

- Since the blades and the disc are manufactured out of one piece additional phenomena due to an incomplete root fixture need not to be considered.

- This model design makes it possible to mount the different coupling elements that have been investigated (fig. 2).

The rotor has 50 blades and is made of chromium–based steel.

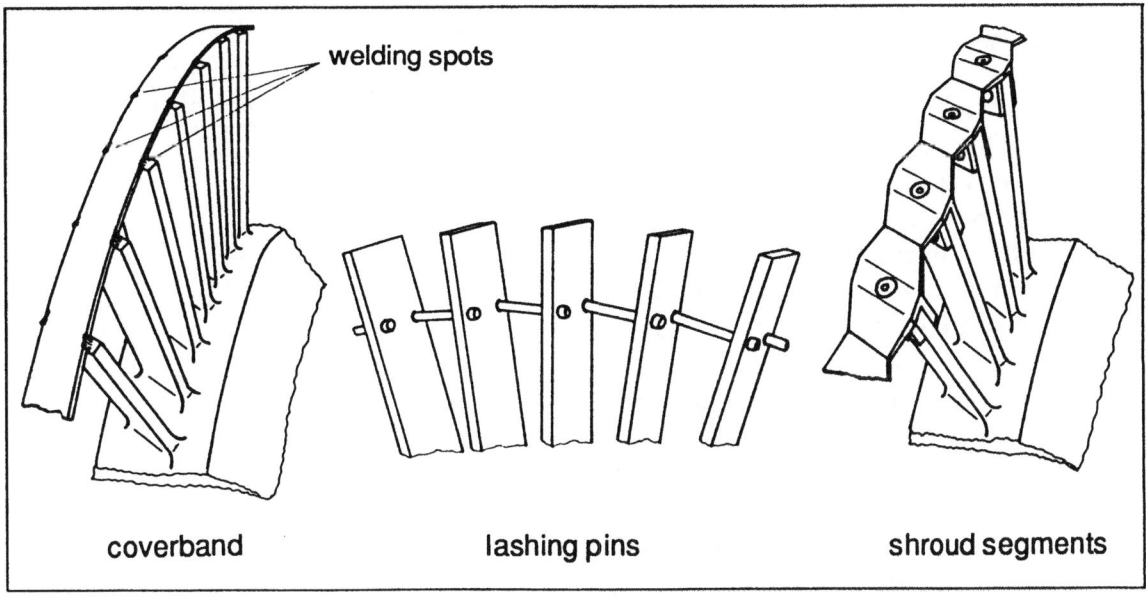

Figure 2: Performed coupling constructions

2.2 Experimental Set–up

In the non–rotating case the loosely inserted lashing pins were pulled in radial direction by means of a pneumatic installation to simulate the centrifugal force (fig. 3). Two tubes filled with pressurised air reacted on a plate which was connected to a tension wire fixed by hinges to the lashing pin. In this way it was possible to achive a homogenious radial force for the connection of pins and blades. The blading was excited by a shaker.

Some of the detected modes of the non–rotating rotor have been verified by measurements made under rotation in a vacuum chamber. In this case, the blades were excited by a single permanent magnet fixed to the vacuum chamber.

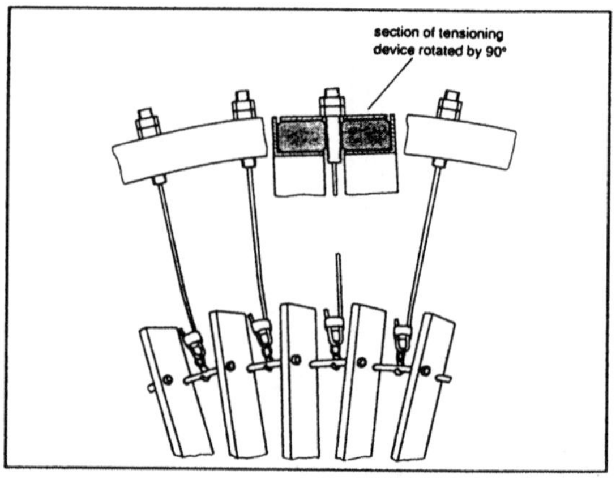

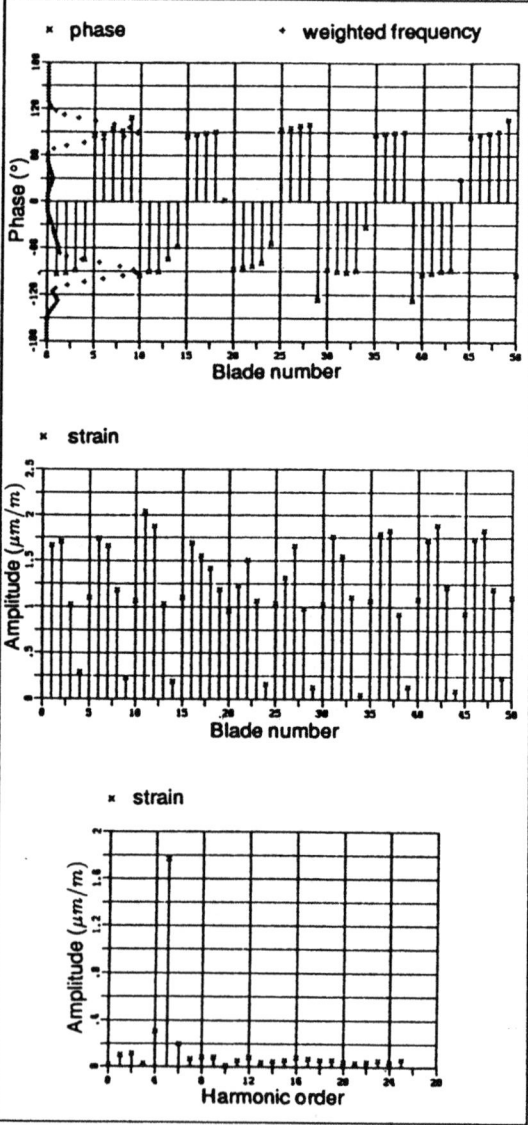

Figure 3: Pneumatic installation to achieve radial pressing of the lashing pins

2.3 Data Aquisition and Data Processing

Most of the measurements reported here have been carried out by Lange [1], some by Schaber [2].

Three semiconductive strain gauges were applicated on each blade to distinguish amplitude directions i.e. flapwise, edgewise and torsional components. These three strain gauges are arranged in series and powered by a constant current source. The voltaic potentials are hooked up to an electronic switching system fixed on the rotor which allows to choose the three signals belonging to one desired blade even in the rotating case [3]. One signal is given by the voltage drop at one strain gauge. A process computer controls the signal selection.

A mercury slipringless transmitter with eight channels connects the rotating and the non–rotating system. A Fast Fourier Transformer (FFT) computes for each amplified strain gauge signal the power spectrum and determines the phase angle regarding to a reference signal.

The analysis of the measured data is to be explained with the aid of an example, which is given in figure 4. First a smoothened frequency distribution of the phase values was computed. With the exception of the umbrella modes, a phase offset was introduced in such a way, that the two maxima have the same distance to the phase 0°. In most cases +90° and −90° were observed.

The above diagram in figure 4 shows the phase values and the corresponding statistical frequency distribution. The appertaining amplitude distribution to the 50 blades is shown below the phase distribution. Since the amplitudes represent the maximum deflection of each blade, this amplitude distri-

Figure 4: Measured mode of vibration

bution must not refer to one point of time. To analyse the mode shape of a measured amplitude distribution, it is necessary to observe one point of time of the vibrating system. This is done by multiplying the blade maximum amplitude by the sine of the respective blade phase (fig. 4 mid). After this a harmonic analysis can be carried out (fig. 4 below). The maximum harmonic amplitude defines the experimental observed nodal diameter of the considered vibration mode.

3 CALCULATION METHOD

Apart from manufacturing inaccuracies and material inhomogenities a blading with a circumferentially closed coupling can be regarded as a system of composed substructures, which itself consists of one blade and one coupling element and, if necessary, the respective disk sector. The vibration characteristics of this type of structures can be computed by means of the wave propagation technique [4],[5]. This technique reduces the calculation model to one substructure with the appropriate low number of degrees of freedom. In this way the required amount of computation time is decreased substantially. Moreover, using this method, in most cases it is not necessary to analyze the natural modes of the system, because the number of nodal diameters belonging to each eigenfrequency is known a priori.

A very flexible tool for computation is given by the combination of the wave propagation technique with the finite element method. The calculations presented in this paper are based on the Stuttgart finite element code PERMAS. The special algorithm of the wave propagation technique as well as routines for taking into account centrifugal force effects have been added to the given finite element code. Details of theory and implementation are given in [6].

For modelling the blade and the shrouds a very fine mesh of the well-known 4-node thick shell element QUAD4 was used. The radii at the root of the blade and the bores for the lashing pins have been taken into account (fig. 5). Coarser meshes have been studied as well. Beam elements were used for the lashing pins. The number of degrees of freedom was reduced by means of the Guyan's reduction.

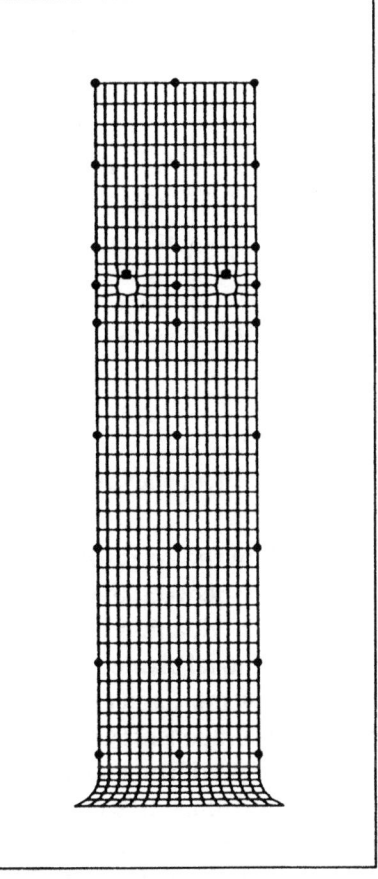

Figure 5: Finite element mesh of the blade (dots refer to nodal points with master degrees of freedom)

4 RESULTS

4.1 Loosely Inserted Lashing Pin

The calculated and measured frequencies of the model blading with loosely inserted lashing pins are shown in figure 6 as a function of the numbers of nodal diameters in the corresponding mode shapes. Adjacent frequency points in the diagram whose modes have similar portions of motion of the blades have been connected to form the vibration families of the system. On the right hand side in figure 6 the first natural frequencies of the cantilevered uncoupled blade are indicated.

In the calculation model the lashing pin is coupled excentrically to the blade by a hinge that is located at the bore edge which is orientated towards the center of gravity of the pin (fig. 7).

An analysis of the calculated mode shapes of the coupled assembly shows that the umbrella modes — in other words the modes with 0 nodal diameters — of the first three families are char-

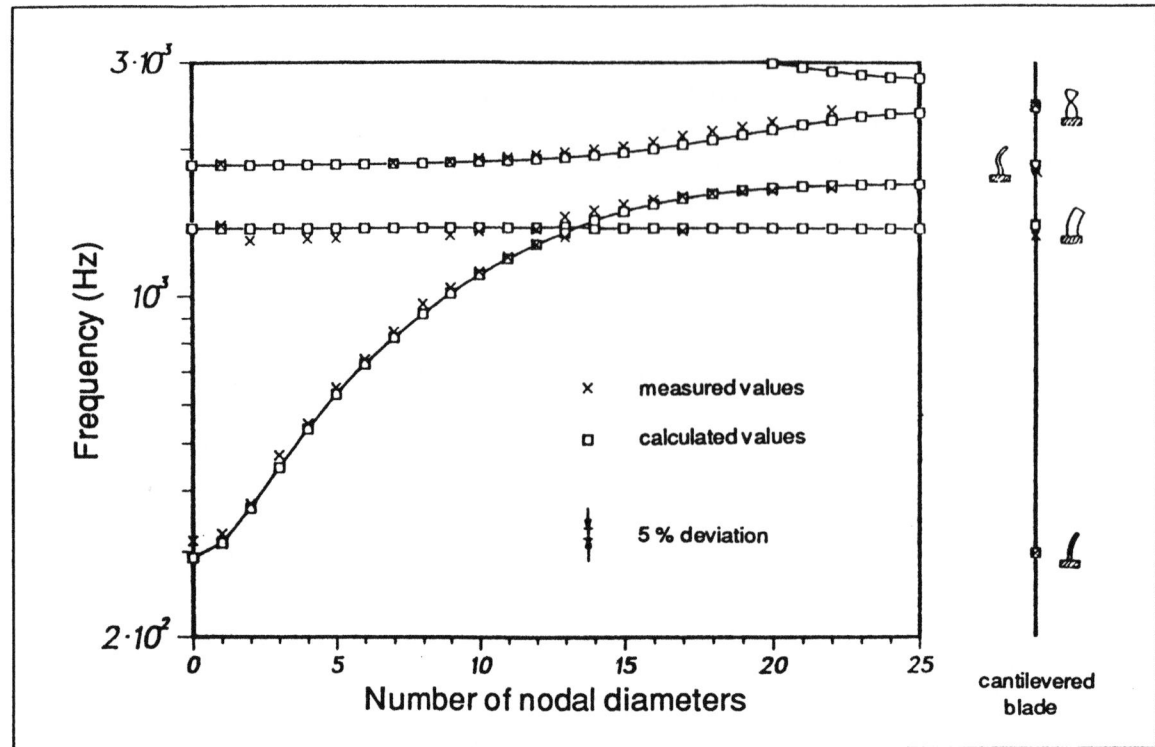

Figure 6: Frequency/nodal diameter diagram of the configuration "loosely inserted lashing pin"

acterized by the corresponding first three modes of the single blade. Therefore there is nearly a coincidence between the frequencies of the coupled system and of the cantilevered blade.

The first vibration family presents an increasing restrain of the blade motion with increasing number of nodal diameters. This effect leads to higher frequencies. Simultaneously the mode shapes at first show more and more torsional components. With further increasing number of nodal diameters these components are reduced and at $N/2 = 25$ nodal diameters there is a pure flapwise bending of the blades with one node over the blade length (N being the number of blades).

The horizontal line of frequencies in figure 6, which is defined as the second vibration family, is characterized by a blade motion in edgewise direction. The fact, that the pin is orientated nearly perpendicularly to the blade foil and therefore to the blade motion in this case, is reflected in the frequency vs. nodal diameter behaviour of this family, because there is no restrain for small blade motions by the hinged coupling pin.

The third family begins with an in–phase flapwise bending motion of the blades with one node over the blade length and ends in an anti–phase torsional motion.

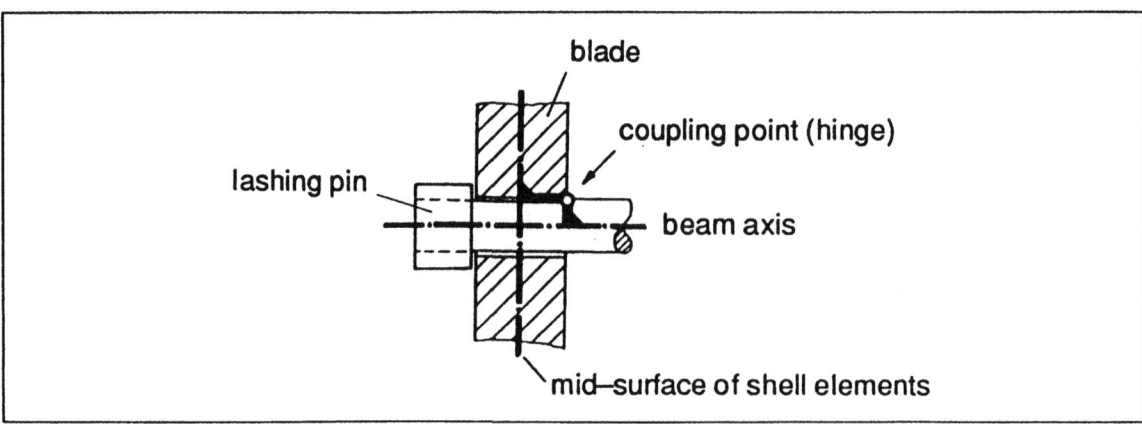

Figure 7: Model for coupling by loosely inserted lashing pins

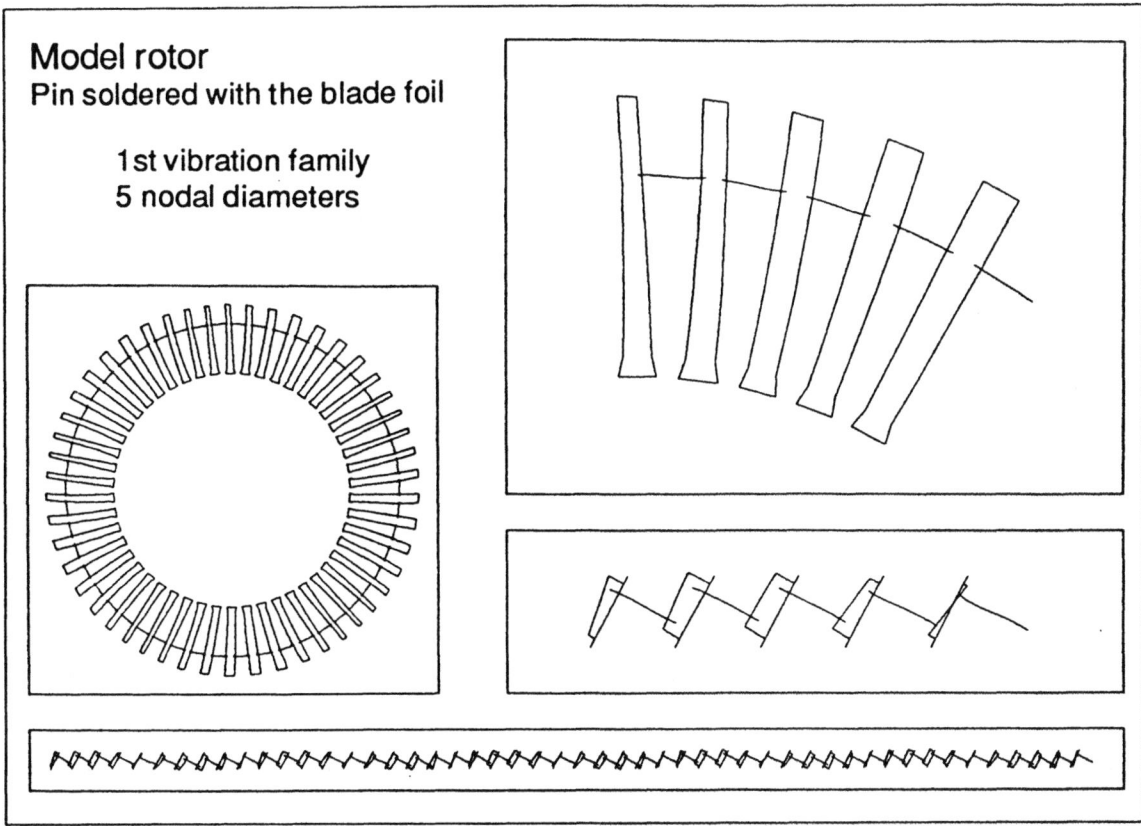

Figure 8: Calculated mode shape with 5 nodal diameters

4.2 Pin Soldered with the Blade Foil

If the pin is soldered with the blade foil, a new configuration with a stiffer coupling is obtained. Now moments can be transfered at the coupling point, which was placed in the point of intersection of the blade's middle plane and the pin axis. For taking into account in the calculation model that the free length of the pin is reduced by the blade thickness, the stiffness of the pin was doubled in the area of intersection.

Fig. 8 shows a calculated mode shape of this configuration. Side view and top view of the unwinding are plotted generally and in an enlarged opening. The different portions of motion that appear in this mode can be recognized. The corresponding measured resonant mode of the assembly with the soldered pin is the one presented in figure 4. Deviations of the phase angle from the 90° relationship can be explained by the superposition of double modes that may lead to travelling waves. The resulting frequency vs. nodal diameter diagram of the soldered pin configuration is shown in figure 9.

4.3 Coverband

If static stresses of mechanical origin — mainly due to centrifugal forces — in combination with thermal stresses are small enough, a blading can be linked by a coverband (continuous shroud). In most practical cases coverbands were riveted to the tip of the blades. For this investigation the coverband was welded to the model wheel at the tip of the blades along the blade thickness (fig. 2). In the finite element model the stiffness of the coverband was increased in the area of the welds, because the interface between blade and coverband cannot directly modelled by shell elements.

In figure 10 the frequency vs. nodal diameter diagram for the blade rim with coverband is presented. An analysis of the calculated mode shapes shows that in this case the blade motions in general differ distinctly from the vibration of the cantilevered blade — more than for the pin configurations. The blade motions now bear more resemblance to a cantilevered blade that is hinged at the tip.

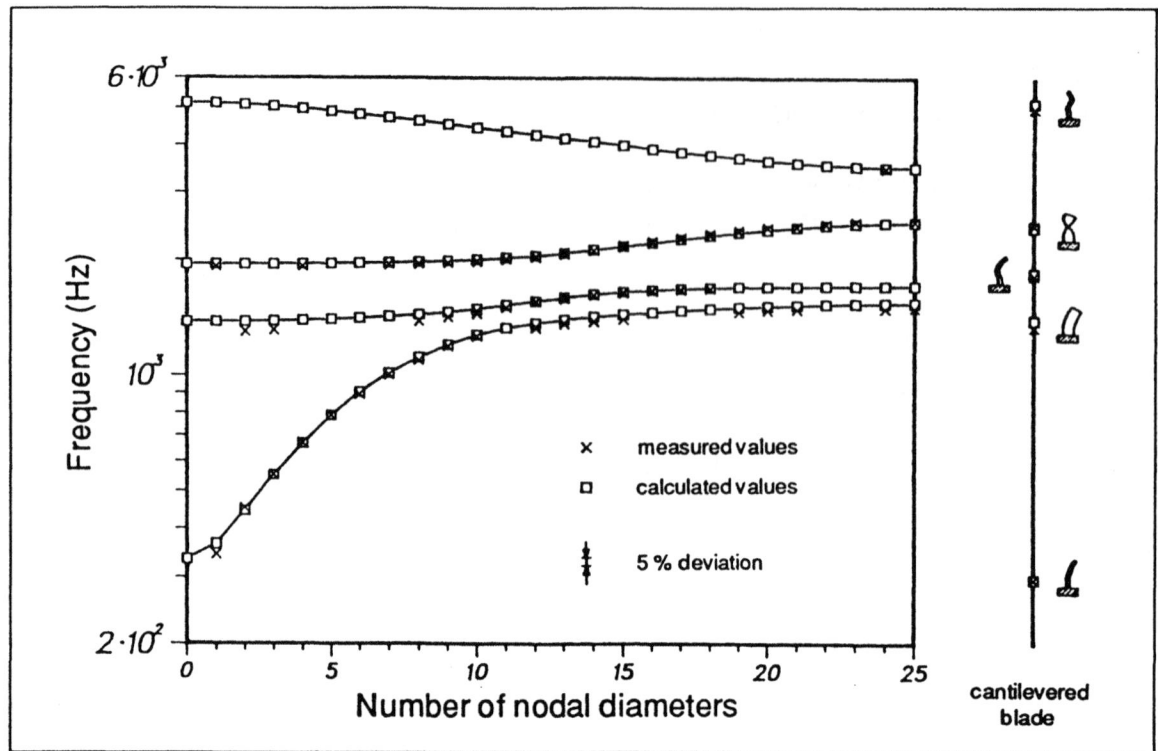

Figure 9: Frequency/nodal diameter diagram of the configuration "pin soldered with the blade foil"

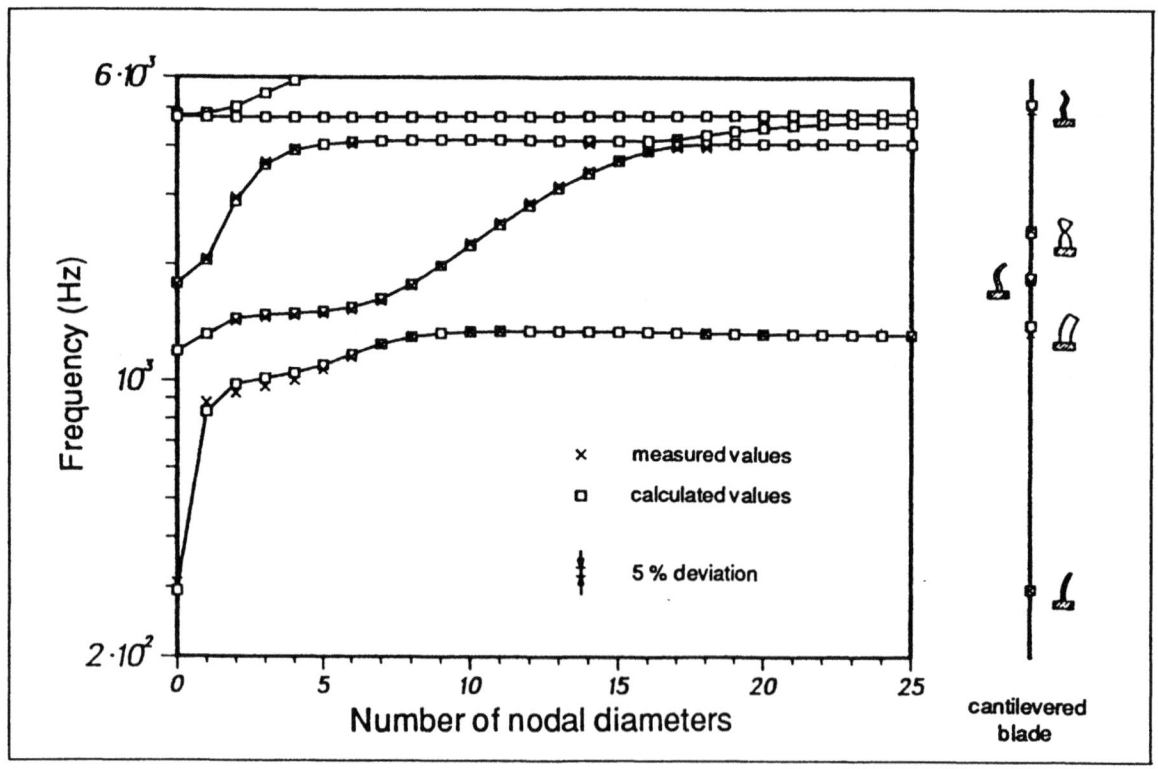

Figure 10: Frequency/nodal diameter diagramm for the blade rim with coverband

285

4.4 Shroud Segments

Blade coupling by shroud segments leads to a similar stiffening as a coverband linkage does. A shroud segment construction however reduces mechanical static stresses and thermal stresses by compensating displacements at the shroud segment interfaces. As an additional effect, the damping of the structure increases, when slip or microslip occurs at these interfaces.

In real bladings shroud and blade foil are integrated. For the model wheel the shroud segments were put on the blades and each was fixed by two press fitted cylindrical pins. The blades had been pretwisted by one degree to achieve a safe connection of the shroud interfaces. The shroud interface angle was chosen 5 degrees less than the blade angle at the root so that no self–locking occurs.

To simulate the stiffening effect of the integrated connecting piece of the shroud segment, the shell elements modelling this area have been given an augmented elastic modulus (multiplier 10), because the connecting piece was modelled by thickening the blade elements at the tip. Contrary to the other configurations, that had been stiffened additionally, this procedure had a major influence on the dynamic behaviour of the assembly, because in this case the stiffened area and therefore the reduction of the free length of the shroud is substantially greater.

The coupling conditions at the shroud–to–shroud interface were chosen so that a hinge band as connection was given. Fig. 11 shows a frequency vs. nodal diameter diagram for the shrouded blade assembly. The agreement between measured and calculated frequencies is not as good as in the other cases. This fact indicates that slip or microslip effects probably occur at the shroud interfaces. These effects cannot be simulated by a linear calculation model.

A solid connection of the interfaces with transfer of rotational degrees of freedom at the interface nodal points yielded no drastic frequency deviation. However, larger and very large deviations were given, when frictionless slip was allowed at the interfaces.

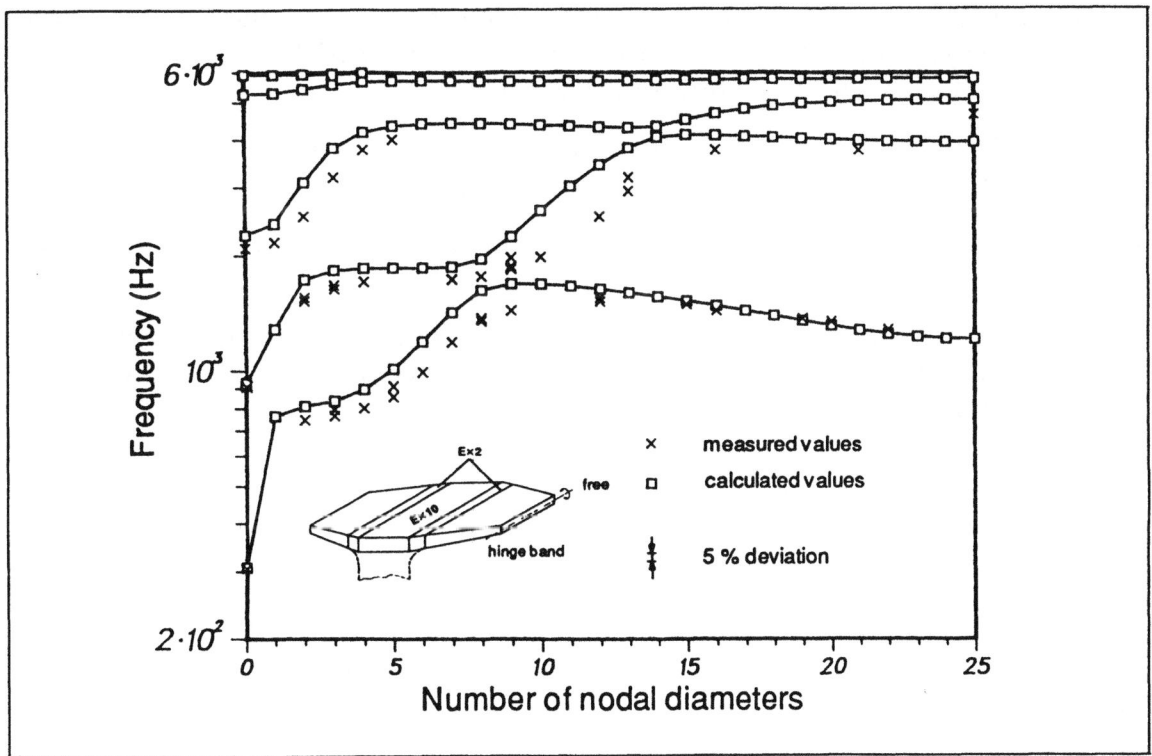

Figure 11: Frequency/nodal diameter diagram of the configuration with shroud segments

5 CONCLUSIONS

The objective of this investigations was to get a deeper knowledge about the vibratory behavior of complex blade systems. The calculations were carried out by means of a finite element code. By using the wave propagation technique it was possible to reduce the calculation time significantly.

For verification and to test the calculation parameters extensive experiments with different types of coupling elements have been done. In most cases a very good agreement between measured and calculated frequencies was achieved.

ACKNOLEDGEMENT

The studies this paper is based on were supported by the Arbeitsgemeinschaft Industrieller Forschungsvereinigungen e.V.(AIF) and the Forschungsvereinigung Verbrennungskraftmaschinen e.V. (FVV).

REFERENCES

[1] Lange W. Experimentelle Untersuchungen von Schwingungen einer Modellaxialbeschaufelung mit unterschiedlichen Rundumbindungen. Universität Stuttgart, 1987 (Mitteilungen des Instituts für Thermische Strömungsmaschinen Nr. 24)

[2] Mayer JF, Schaber U. Gekoppelte Systeme. Forschungsvereinigung Verbrennungskraftmaschinen e.V., Frankfurt/M., 1990 (Forschungsberichte Verbrennungskraftmaschinen, Heft 450)

[3] Diettrich H–P, Lange W. Entwicklung einer elektronischen Umschaltung zur Übertragung von Schwingungssignalen aus einem rotierendem System. Universität Stuttgart, 1985 (Interner Bericht, Institut für Thermische Strömungsmaschinen)

[4] Mead DJ. A General Theory of Harmonic Wave Propagation in Linear Periodic Systems with Multiple Coupling. J Sound Vibr 1973;27(2):235–260

[5] Thomas DL. Standing Waves in Rotationally Periodic Structures. J Sound Vibr 1974;37(2):288–290

[6] Mayer JF. Zur Berechnung des Eigenschwingungsverhaltens gekoppelter Beschaufelungen axialer Turbomaschinen. Universität Stuttgart, 1987 (Mitteilungen des Instituts für Thermische Strömungsmaschinen Nr. 25)

THE USE OF FINITE ELEMENT METHOD TO SOLVE VIBRATION PROBLEMS ON VERTICAL PUMPS

Massimo Scali
Giacomino Marenco
Pompe Gabbioneta S.p.A., Sesto S.Giovanni (MI), Italia

ABSTRACT

Vertical pumps are often a matter as far as regards vibrations. Their design is very simple but often is cause of high level vibrations.

The only way to prevent this problem is to know the dynamic behaviour of the machine in design phase, by finite element method calculations.

We made such a study on an our pump. Many approximations were made and the model is somehow poor, nevertheless the results were encouraging.

INTRODUCTION: THE GENERAL PROBLEM

In fluid pumping sometimes the available Net Positive Suction Head (NPSH) isn't enough to avoid cavitation. One way to solve the problem is to increase it by setting the impeller several meters under ground level. Therefore you must dig a hole in the ground and put in it a barrel which will contain the fluid and the impeller, maintaining the motor at ground level. Thus the power will be transmitted to the impeller by a long overhung shaft. The depth of the barrel depends by NPSH available at ground level and by NPSH required by 1st stage of the pump. The number of impellers is variable too depending by required head (fig.1).

All these conditions make it impossible to build a stiff shaft, because we should

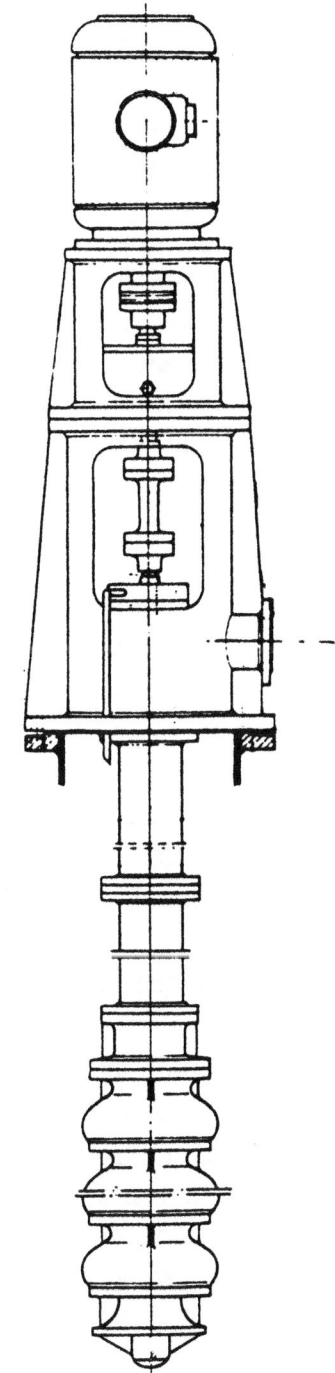

Fig.1 - A VB type pump.

design a very large diameter. Thus the pump is to run among its resonance speeds. It's easy to understand that it's necessary to know exactly where these resonances are.

There are two way to know it: the first is to build the machine and then dynamically test it. Of course this solution is very expansive and doesn't allow strong modifications on the original design, without implying high costs and long times.

The second is to calculate them using a finite element model of the entire pump considering every single component like motor, baseplate, as well as sleeve bearings and so on. This way allows the designer to change the pump before it's built. Besides, if you can use a precise method of calculation for seals and bearings, you can use the exact pumped fluid and not simply water, which can have a dynamic behaviour other than actual one.

A TROUBLE CASE

Our study was developed on a vertical pump which was installed in Venezuela two years ago. Its task is very important and without it the whole plant stops. The pumped fluid is isobutane. The operating conditions required a four stages pump, running at 3550 rpm, corresponding to nearly 60 Hz, (the actual speed turned out to be 3480 rpm, i.e. 58 Hz). The NPSH imposed to put the first stage at 5 meters under ground level. Along the shaft are set out four guide bushings at a distance of 750 mm each other.

We tested it in our test room and verified a perfect running, both for hydraulic performances and for dynamic behaviour. Of course the test was carried out at 2950 rpm (50 HZ) and with water as pumped fluid. Nevertheless when it was started on actual plant the result was very different: the vibration level was too high and for

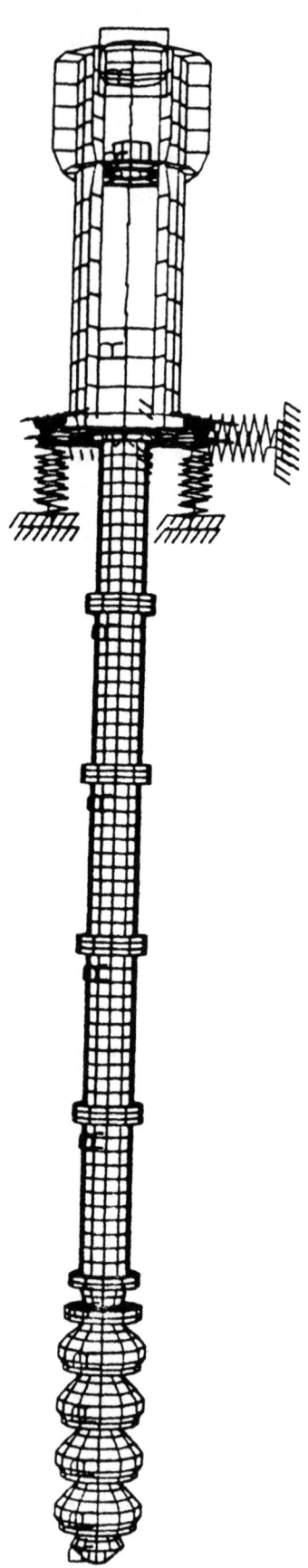

Fig.2 - A view of the mesh used for the calculation.

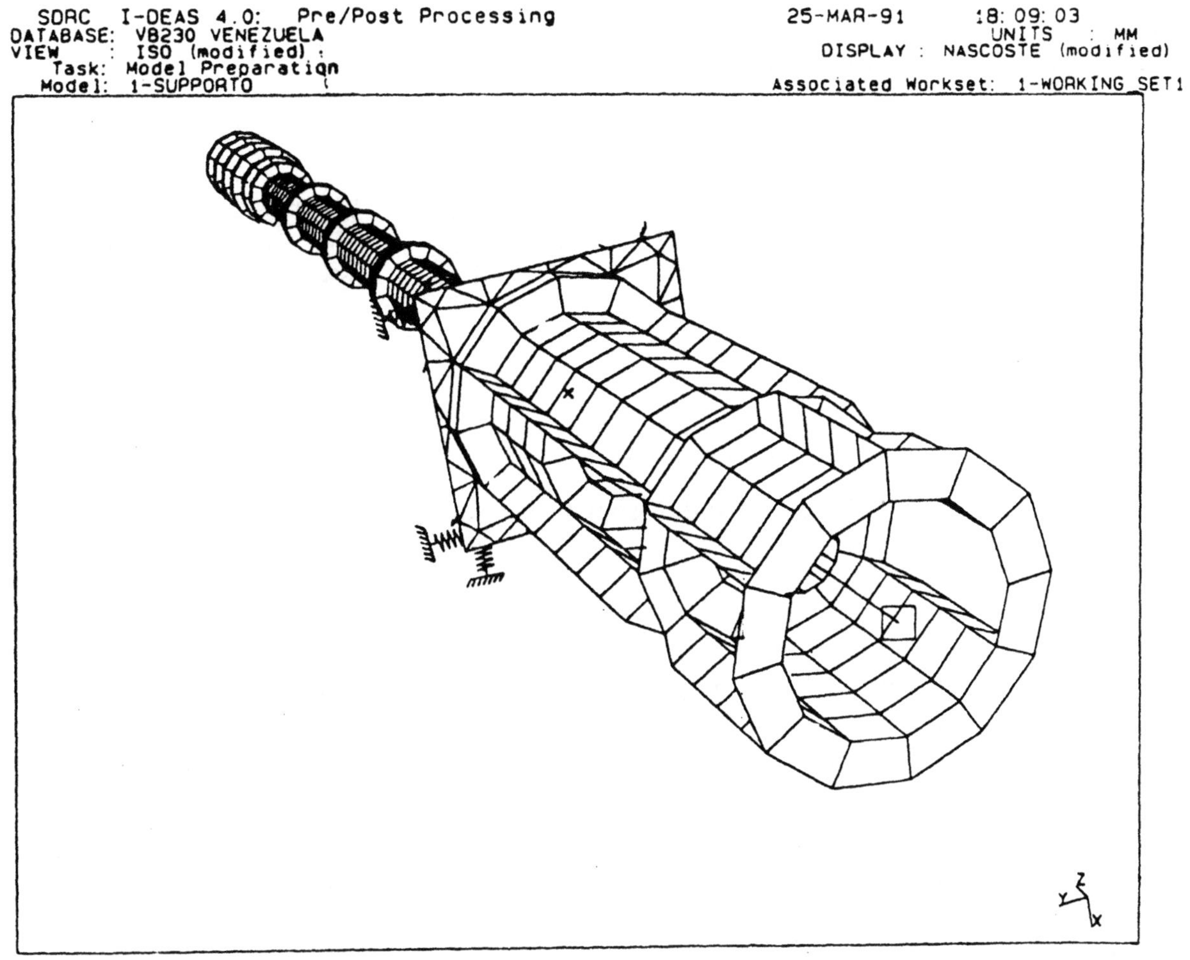

SDRC I-DEAS 4.0:   Pre/Post Processing   25-MAR-91   18:09:03
DATABASE: VB230 VENEZUELA   UNITS : MM
VIEW   : ISO (modified) :   DISPLAY : NASCOSTE (modified)
     Task: Model Preparation
     Model: 1-SUPPORTO   (Associated Workset: 1-WORKING_SET1

Fig.3 - A close up view of the support mesh.

this reason some bushings were damaged making the vibrations increase again (until 14 mm/s).

THE APPROACH TO THE PROBLEM.

Customer was of course very impatient to solve the problem. In spite of this we had poor informations from the plant. Thus we had to make a full simulation of the phisical phenomenon on our computer. For this task we decided to use the finite element method (FEM). In fact if you make a precise mesh and the boundary conditions are properly set, it's possible to reach the right solution without any test on real structure.

We had to make some hypothesis to begin to move in one precise direction. Our basic hypothesis was that the machine was running near a critical speed (we didn't know which one). We knew that some shaft guide bushings were worn and we knew the amount of the wear too: this information was helpful to detect the modal shape.

A finite elements model of the pump was made using the FEM program I-DEAS by SDRC: it's drawn in figg.2 and 3. It was composed by nearly 1200 elements and 1700 nodes. The elements used were bricks (linear, 8 nodes) and

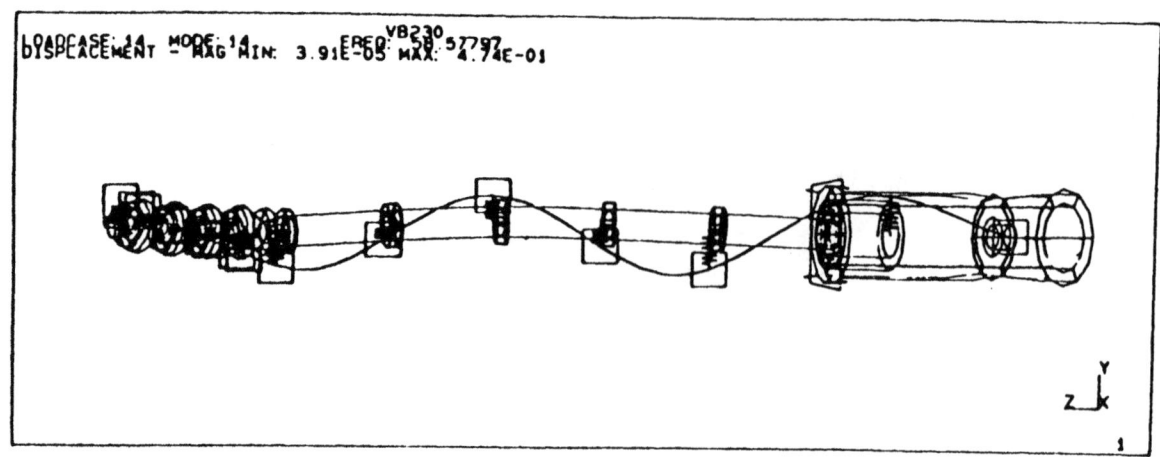

Fig.4 - Modal shape resonating at 58 Hz.

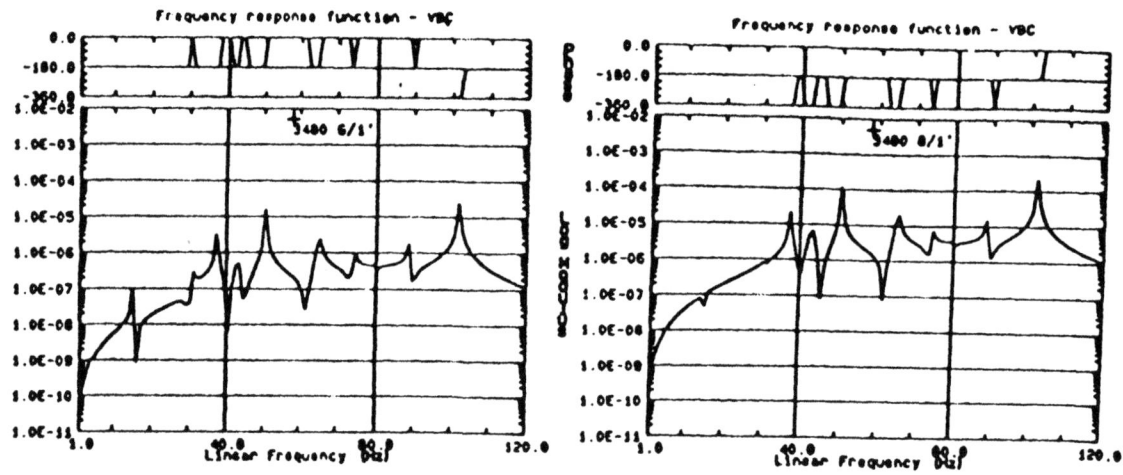

Fig.5 - Frequency response of shaft node on motor joint.

Fig.6 - Frequency response of the shaft node on the seal.

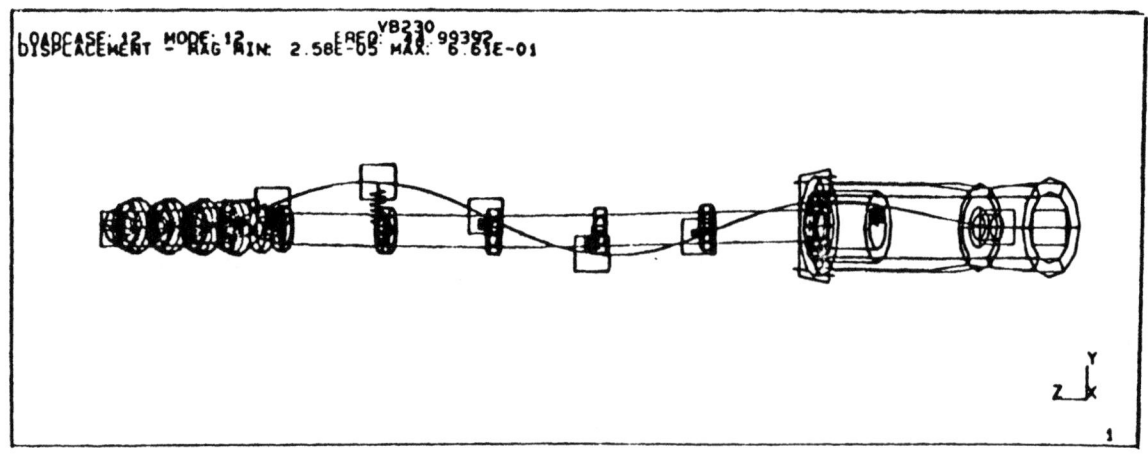

Fig.7 - Modal shape at 45 Hz.

quadrilateral shells (linear, 4 nodes) for the column and the casing, while linear beams were used for the shaft. Lumped masses simulated the motor, the impellers and bushings inertia. The shaft and the column were connected by spring element, whose stiffnesses were calculated by a self-made program for evaluating dynamic characteristics of seals and oil-film bearings.

The whole structure was restrained by four springs simulating terrain effect, whose stiffness is not infinitive. The stiffness used for these elements was read by literature and was assumed to be 9.806e9 N/m.

We had an uncertainty about the value of shaft guide bushings stiffness, because our method of evaluating oil-film dynamic behaviour is quite precise if there is a pressure gradient across the film, but is not if pressure is nearly constant, and that's our case.

We founded our process on the fact that all parameters were well known, but dynamic behaviour of shaft guide bushings. Therefore we fixed all known parameters and varied only the unknown one. Several calculations were made to see how the structure would react to the variations of bushings stiffness. We used at this pione the hypothesis we made: the pump was operating in resonance. The informations regarding the worn bushings too were useful in this process phase to detect the modal shape and by this to determine the right value of the stiffness for shaft guide bushings.

Here are the values used for every spring stiffness in the model:

1. seal on the casing: 5.394e5 N/m;
2. bushings along the shaft: 4.903e5 N/m;
3. impellers seals: 3.383e6 N/m
4. ball bearings: 9.807e8 N/m.

The next step was to modify the structure only for the parameters we were able to evaluate with good precision and then to analyse the results to understand whether the modifications made could reduce vibration levels enough.

THE CALCULATION RESULTS ANALYSIS AND THE MODIFICATION.

Following the previously described steps we were able to say that the pump resonance excited during normal running was in corrispondence of the 4th frequency of the shaft (the column and the support resonance frequency were far from operating one). In fig.4 is the mode corresponding to the resonating frequency, while in fig.5 you can see the frequency response for the shaft node on the motor joint and in fig.6 the frequency response for the shaft node on the seal.

Observing these diagrams we can see that under the operating frequency severals resonances exist and they belong to column, to support and, of course, to shaft. The lower modes of the shaft are in a little range near operating speed and for this reason we couldn't make drastical modifications on the structure because it was possible that one of these modes increased its frequency until operating one. In fig.7 it's shown the 3rd mode of the shaft, the nearer to running speed. These

292

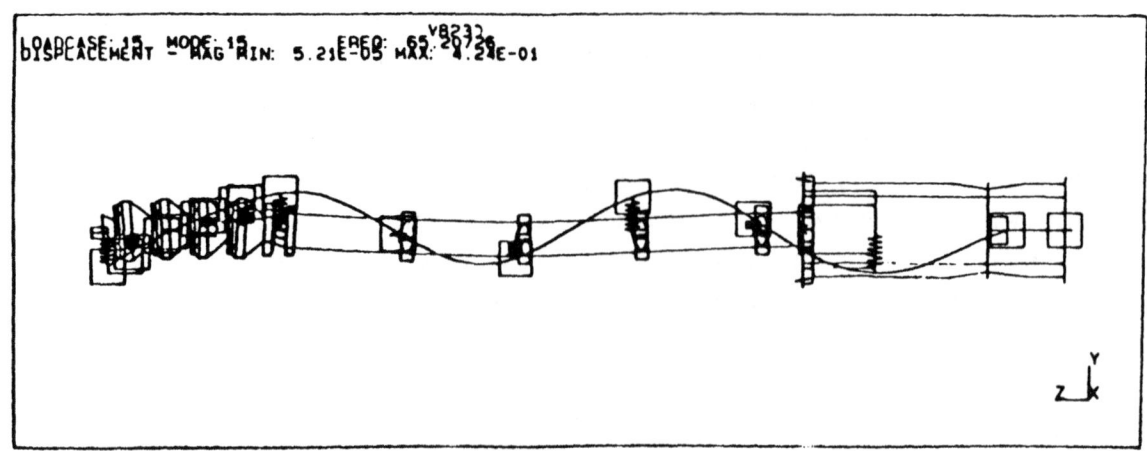

Fig.8 - Modal shape n.4 for modified pump (65 Hz).

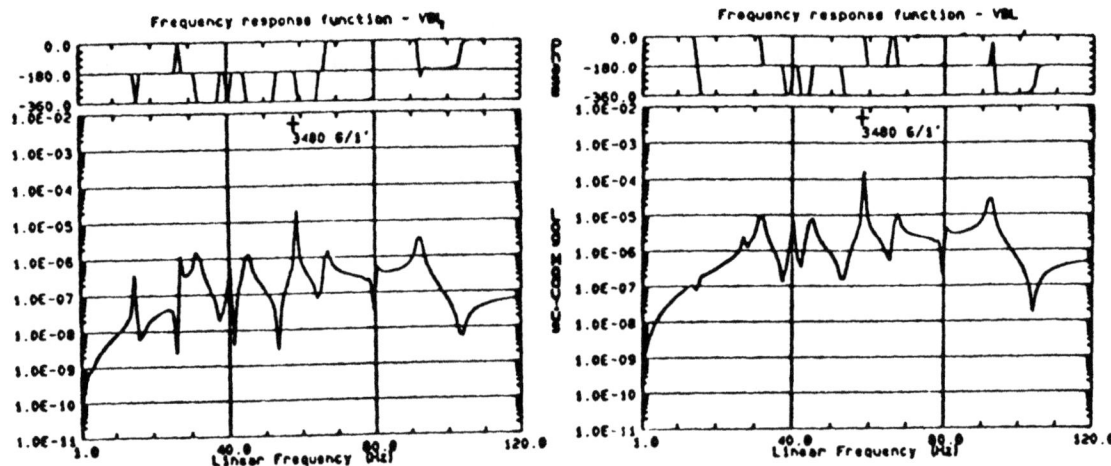

Fig.9 - Frequency response of shaft node on motor joint.

Fig.10 - Frequency response of the shaft node on the seal.

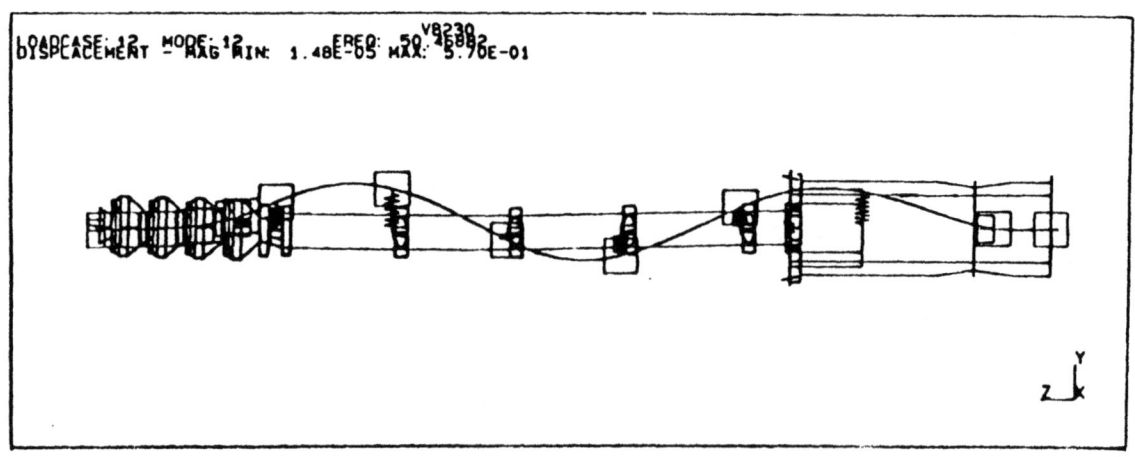

Fig.11 - Modal shape n.3 for modified pump (50 Hz).

considerations make it clear that we could neither make large errors in computation, because a wrong evaluation could bring to critical situations for the machine.

We thought that the best and fastest way to solve the problem was to shift the frequency that was found to coincide with running one by shortening the shaft and the column. The measures taken on the plant showed an available NPSH high enough, thus it was possible to make this modification without compromise the good hydraulic operation. This solution was even quite fast to implement, requiring few modifications on the original design. Thus we decided to shorten the pump by 500 mm.

The calculations results were encouraging: the 4th shaft mode was icreased to 65 Hz (fig.8), while the 3rd shifted to 50 Hz (fig.11); the frequency response diagrams for the same nodes we examined previously show that the pump is going to run in a minimum point as far as regard vibrations amplitude (fig.9 and 10).

Therefore this solution was implemented. The test on the plant confirmed the calculation, in fact the vibration levels decreased until 7 mm/s and mantained this value.

CONCLUSIONS AND FUTURE DEVELOPMENTS

We know that many uncertainties affected the whole study. On the other side we didn't get any informations from the plant, but the vibration level (no impact test, nor order tracking). Anyway the results were satisfactory and encouraging.

We're acquiring the modal analysis package for I-DEAS, and by this we think to develope a method to analyze systematically our vertical pumps before they are manufactured. Of course the results obtained will be compared with tests on actual pumps in our plant, by impact test and order tracking using a dynamic signal analyzer.

We intend to develope a more precise and wider application range program too for evaluating the dynamic behaviour of seals and bearings.

Anyway we think it's today possible to reach a very high computational precision by finite element analysis using a powerful workstation, and it's in our opinion that by these means you can fully predict the dynamic behaviour of a machine when it is still in designer's mind.

THE INFLUENCE OF PULSATING MECHANICAL LOADS ON ACOUSTIC NOISE GENERATED BY AC INDUCTION MOTORS

R. P. LISNER and R. J. ALFREDSON

Monash University,
Clayton, Victoria, Australia

Abstract

The trend towards the use of variable frequency inverter drives for AC induction motor control has resulted in the radiation of increased levels of acoustic noise from these motors. Thus, while aerodynamic noise, bearing noise and other mechanical noise sources are still relevant, noise generated as a result of the *electrical* operation of the motor is assuming an increased level of importance. The radiation of acoustic noise due to this factor is primarily attributable to the generation of radial force waves, which are rich in harmonics, in the rotor-stator air gap. Various factors influencing the generation of these forces are discussed.

This paper also presents the results of investigations into a previously unexplored area of study, concerning the effects of load torque pulsation upon the noise radiation behaviour of the motor. Previous analytical and experimental work in this area has generally been concerned with the investigation of motor noise under conditions of no-load or steady-load operation.

The investigation forming the basis for this paper includes an extensive experimental program. The experimental equipment supporting the investigation includes an induction motor (mounted inside an anechoic chamber), coupled to an electromechanical load which is external to the chamber. Means are provided for producing either a steady or pulsating torque characteristic within the load. A dynamic torque transducer is also incorporated, allowing the actual load torque pulsation behaviour to be monitored and correlated with motor vibration and noise. Results of the experimental investigations are presented, and the noise generating mechanisms are discussed.

1 Recent Developments in the Study of Acoustic Noise Radiated From Electric Motors

During the past two decades, a major change of emphasis has arisen in the study of electric machine noise. The widespread introduction of solid-state electronic speed control during this period has shifted the interest from the basic internal noise generating mechanisms, to include the effect of highly non-sinusoidal driving functions. In addition, advances in materials technology have allowed the production of smaller motors, running at substantially higher temperatures than motors of the past. The motors are almost invariably running at higher levels of magnetic flux density, and these motors have much thinner stator cores than earlier machines. These are factors which will make the motor far more likely to be a source of unwanted acoustic noise, especially when being driven from a solid-state speed controller.

There have also been significant developments in the analysis tools available for use in the study of electric motor noise. Finite element and finite difference techniques are becoming more widely used, and suitable computer packages are becoming more readily available. Modal analysis techniques offer the opportunity to extensively analyse the coupled structures within an induction motor, with sophisticated modal testing equipment becoming more readily available. The use of finite elements, finite differences and modal analysis has allowed more complex studies of motor stator natural frequencies to be carried out, and the vibration characteristics of stator teeth, end-windings, etc. can now be much more readily incorporated into the vibration model of the motor.

In order to address the motor noise and vibration emanating from AC induction motors when driven from electronically controlled power converters, much current research is concentrated on the study of motor drive modulation techniques, and the effects of different drive strategies on the noise and vibration characteristics of the motors [1]. The different modulation strategies that have been investigated include ultrasonic carrier techniques, low frequency optimal waveform techniques, and selective harmonic elimination [2], [3], [4].

2   Theoretical Background to the Generation of Induction Motor Acoustic Noise

2.1   Scope

The following treatment will focus primarly on the mathematical analysis of vibration and acoustic noise generated as a result of *electrical* excitation of three-phase electric asynchronous (induction) motors. The discussion will not be concerned with the generation of noise due to secondary mechanical aspects associated with the operation of the motors, such as bearing noise and aerodynamic noise.

2.2   Mathematical Analysis

In seeking to analyse the vibrational and acoustic behaviour of electrically-excited motor stators, it is useful to recognise that there are several aspects to be addressed:

- Electromechanical excitation forces within the motor stator

- Mechanical behaviour of the motor stator in response to the electromechanical excitation forces

- Acoustic performance of the motor stator in response to the mechanical behaviour induced by the abovementioned forces.

In the treatment that follows, a mathematical basis for this analysis is introduced.

2.2.1   Electromechanical Excitation Forces

Torque generation within a three-phase induction motor is dependent upon the existence of radial magnetic flux in the main air-gap of the motor, between rotor and stator. The direction of the resulting forces generated is primarily tangential, but some radial forces are also produced. It is the presence of these radial forces which is mainly responsible for the vibrations leading to generation of acoustic noise in the motor. In order to analyse the force waves produced in the machine air-gap, it is first necessary to study the air-gap flux distribution. This, in turn, is determined as a function of the magnetomotive force (*mmf*) distribution, and the air-gap permeance. The following discussion therefore commences with an examination of air-gap permeance and *mmf* distribution. Recognition that the product of these two quantities yields an expression for the magnetic flux distribution, then leads to a general expression for the stator radial force wave, which is of primary importance in determining the acoustic noise radiation characteristics of the motor.

2.2.2 Air-gap Permeance

It can be shown [5] that the air-gap permeance of an electric motor with slotted rotor and slotted stator, $\Lambda_{st,rt}(\phi,t)$, can be expressed as a function of angle ϕ and time t in the following form:

$$\Lambda_{st,rt}(\phi,t) = \sum_{i_{st}=0}^{\infty}\sum_{i_{rt}=0}^{\infty} \Lambda_{i_{st},i_{rt}} cos\left[(i_{rt}Z_{rt} \pm i_{st}Z_{st})\phi - i_{rt}Z_{rt}\omega_{rt}t\right] \tag{1}$$

where i_{rt} and i_{st} are integers, rt and st denote rotor and stator respectively, Z_{rt} and Z_{st} are the rotor and stator slot numbers, ω_{rt} is the rotor angular speed.

This expression gives the air-gap permeance for the ideal case, where the motor and stator exhibit perfect symmetry and no eccentricity, and saturation effects are absent. When these effects *are* taken into account, the expression for air-gap permeance due specifically to the presence of static eccentricity, dynamic eccentricity and saturation is:

$$\Lambda_{ec,sa}(\phi,t) = \sum_{i_{ec,rts}=0}^{\infty}\sum_{i_{ec,rtd}=0}^{\infty}\sum_{i_{ec,st}=0}^{\infty}\sum_{i_{sa}=0}^{\infty}\Lambda_{i_{ec,rts},i_{ec,rtd},i_{ec,st},i_{sa}}$$
$$\times cos\left[(i_{ec,rtd} \pm i_{ec,st} \pm i_{ec,rts} \pm 2i_{sa}p)\phi - (i_{ec,rtd}\omega_{ec} \pm 2i_{sa}\omega_1)t\right] \tag{2}$$

where $i_{ec,rts}$, $i_{ec,rtd}$, $i_{ec,st}$ and i_{sa} are integers, subscript *ec,rts* denotes "due to rotor static eccentricity", subscript *ec,rtd* denotes "due to rotor dynamic eccentricity", subscript *sa* denotes "due to saturation", p is the number of motor pole pairs, ω_{ec} is the dynamic rotor angular speed, ω_1 is the fundamental exciting angular frequency.

By combining equations (1) and (2), an expression for the total permeance $\Lambda_{tot}(\phi,t)$ is obtained, where the effects of rotor slotting, stator slotting, static eccentricity, dynamic eccentricity and saturation are all included:

$$\Lambda_{tot}(\phi,t) = \sum_{i_{st}=0}^{\infty}\sum_{i_{rt}=0}^{\infty}\sum_{i_{ec,rts}=0}^{\infty}\sum_{i_{ec,rtd}=0}^{\infty}\sum_{i_{ec,st}=0}^{\infty}\sum_{i_{sa}=0}^{\infty}\Lambda_{i_{st},i_{rt},i_{ec,rts},i_{ec,rtd},i_{ec,st},i_{sa}}$$
$$\times cos\left[(i_{rt}Z_{rt} \pm i_{st}Z_{st} \pm i_{ec,rtd} \pm i_{ec,st} \pm i_{ec,rts} \pm 2i_{sa}p)\phi\right]$$
$$- \left[(i_{rt}Z_{rt}\omega_{rt} \pm i_{ec,rtd}\omega_{ec} \pm 2i_{sa}\omega_1)t\right] \tag{3}$$

2.2.3 Magnetomotive Force Distribution

It has been established [5] that the space harmonic mmf wave produced by the flow of current in an AC machine can be expressed as:

$$F = \sum_{k_{st}=1}^{\infty}\sum_{q_{st}=-\infty}^{\infty} F_{k_{st},q_{st}}cos\left(k_{st}p\phi - q_{st}\omega_1 t - \frac{k_{st}\alpha_{st}zp}{l_c} - \psi_{k_{st},q_{st}}\right)$$
$$+ \sum_{k_{rt}=1}^{\infty}\sum_{q_{rt}=-\infty}^{\infty} F_{k_{rt},q_{rt}}cos\left(k_{rt}p\phi - q_{rt}s\omega_1 t - k_{rt}p\omega_{rt}t - \frac{k_{rt}\alpha_{rt}zp}{l_c} - \psi_{k_{rt},q_{rt}}\right) \tag{4}$$

where the k subscripts represent the order of the space harmonics, the q subscripts represent the order of the current-time harmonics, st and rt refer to stator and rotor respectively, α_{st} and α_{rt} are the slot skew angles for stator and rotor, z is the *axial* distance from the centre of the machine, l_c is the rotor/stator core length, $\psi_{k_{st},q_{st}}$ and $\psi_{k_{rt},q_{rt}}$ are phase angles.

2.2.4 Radial Force Wave

Multiplying the above expressions for Λ_{tot}, and F together will yield the magnetic flux density B as a function of ϕ, t and z. Recognition of the fact that the magnitude of the radial force wave, $\sigma_{rad}(\phi, t)$, is proportional to the square of the magnetic flux density wave [6] leads to a simplified result of the form [5]:

$$\sigma_{rad}(\phi, t) = \sum_{m_i, \omega_i} \sigma_{m_i, \omega_i} \cos(m_i \phi - \omega_i t) \tag{5}$$

where the m_i are functions of the rotor and stator slot numbers, the pole-pair number and a set of integers, and where the ω_i are functions of rotor slot number, the angular frequencies ω_{rt}, ω_1 and ω_{ec}, the orders of the current-time harmonics, and a set of integers.

The result in equation (5) above refers to a *set* of rotating force waves, not simply a single force wave. These rotating radial force waves correspond to sets of different mode numbers and frequencies.

In terms of acoustic noise generation, the most significant of the above set of force waves will be those having low mode numbers and those with large amplitudes.

At the most ideal level, where the effects of all possible imperfections are neglected (eccentricity, asymmetries, slotting and saturation), the above expression reduces to the form:

$$\sigma_{rad}(\phi, t) = \sigma_{m_i = 2p, \omega_i = 2\omega_1} \cos(2p\phi - 2\omega_1 t) \tag{6}$$

which represents a force wave having a frequency of twice the fundamental supply frequency, and a mode number of twice the pole-pair number.

If however the stator/rotor slotting is also taken into account the resulting force wave expression is of the form:

$$\sigma_{rad}(\phi, t) \propto \cos[(Z_{rt} \pm Z_{st} \pm 2p)\phi - (Z_{rt}\omega_{rt} \pm 2\omega_1)t] \tag{7}$$

It can be seen that the resulting force waves have mode numbers which are functions of stator slot number, rotor slot number, and pole pair number, while the force wave frequencies are functions of rotor slot number, rotor angular speed, and fundamental supply frequency.

For an induction motor, operating at slip s, the frequency term $(Z_{rt}\omega_{rt} \pm 2\omega_1)$ in equation (7) becomes $\left(Z_{rt}\frac{\omega_1(1-s)}{p} \pm 2\omega_1\right)$ since the rotor angular speed is equal to $\left(\frac{\omega_1(1-s)}{p}\right)$.

2.2.5 Stator Vibration Behaviour and Acoustic Response

For a motor stator of generally cylindrical form, the amplitude of radial vibration, δ_{rad}, can be expressed as

$$\delta_{rad} = \sum_{i=0}^{\infty} \delta_{rad, m_i, \omega_i} \cos(m_i \phi + \omega_i t + \psi_{m_i, \omega_i}) \tag{8}$$

where the m_i are circumferential vibration modes, ϕ represents angular position, ω represents angular frequency, and ψ represents phase angle.

Alternatively, for a particular mode number m_k, frequency ω_q, and neglecting phase angle, this can be expressed as

$$\delta_{rad}(\phi, t) = \delta_{rad_{k,q}} e^{j(m_k \phi - \omega_q t)} \tag{9}$$

where the m_k and ω_q values are from equation (7) as discussed above.

Application of the three-dimensional wave equation, in polar form, for pressure p results in a solution for the sound pressure around a cylindrical machine, of the form

$$p_i(r, \phi, z, t) = \frac{\delta_{rad_{k,q}} \omega_q^2 L}{2\sqrt{2\pi}} e^{j(m_k \phi - \omega_q t)} f(r, z) \tag{10}$$

where L is the machine length, $f(r, z)$ is a function of r and z incorporating the system constants and based on Bessel functions, while r and z have their usual meanings in polar co-ordinate notation.

2.3 Summary

The foregoing discussion provides an introduction to the mathematical basis for analysis of radiated acoustic noise from electric motors. The general expression obtained for the radial force wave $\sigma_{rad}(\phi, t)$ acting on the motor stator allows determination of the force wave frequencies and mode numbers. The results obtained from the force wave equation can then be applied to the expression for radial vibration $\delta_{rad}(\phi, t)$ at the stator surface. This determination of the radial vibration of the stator surface provides a basis for further study of the acoustic noise radiation behaviour of the motor.

3 Investigations into the Effects of Load Torque Pulsation on Induction Motor Noise

3.1 Introduction

This paper is based on a research program which involves the study of acoustic noise generation in converter-fed three-phase induction motors. Of particular concern within this program is the investigation of acoustic noise in such machines when they are coupled to mechanical loads possessing a pulsating torque characteristic, such as hydraulic pumps and gearbox-driven loads.

Most studies of electric motor acoustic noise are based on the condition of no-load running or, at best, running with a steady mechanical load applied. The research program upon which this paper is based involves extending the theoretical basis presented above, to include the effects of torque pulsations emanating from the mechanical load connected to the shaft of the induction motor. A number of postulations have been made, arising from this research program, relating to ways in which load torque pulsation may influence the vibrational and acoustic behaviour of the motor. Some of the suggested mechanisms for such interaction are:

- Influence of load-coupled circumferential rotor vibration upon the stator/rotor currents, thereby resulting in modification of the *radial* force waves in the motor.

- Generation of motor vibrations as a result of load torque pulsations being coupled into the motor rotor, particularly where the load torque pulsation frequency is close to electrical excitation or mechanical resonance frequencies. Alternatively, vibrations may result from this form of coupling as a result of beat-frequency phenomena.

- The interaction of higher frequency load torque pulsations with electromechanical force components arising from the presence of harmonic components in the electrical driving source. This is of particular relevance where the motor is to be driven by a semiconductor switching converter.

- The establishment of axial bending waves in the motor rotor, in rigidly-coupled motor/load configurations, where radial force components exist within the load. In such a case, the rotor bending wave will cause cyclic motor permeance variations, resulting in the generation of radial motor/stator force wave components.

3.2 Experimental Investigations

The investigation described in this paper is supported by an extensive experimental program. A three-phase, 7.5 kW four-pole induction motor has been installed inside an anechoic chamber, and is coupled to an electrically-controlled load; this load is mounted *outside* the anechoic chamber, with the means of connection being a drive shaft which passes through the 0.3 metre thick concrete wall of the chamber. The load is capable of being operated either with a steady torque characteristic, or as a load exhibiting a controlled pulsating torque characteristic with variable pulsation frequency.

Experiments have been carried out, in order to relate motor acoustic noise to the electrical excitation and load characteristics. Results of these experiments are presented here.

In figures 1 to 6, load torque and acoustic noise spectra are shown for three different types of mechanical loading:

- Steady torque (figures 1 & 2)
- 100 Hz torque pulsation (figures 3 & 4)
- Severe 50 Hz torque pulsation (figures 5 & 6)

A common set of test conditions applies for all spectra:

- Fundamental induction motor drive frequency: 13.9 Hz
- Switching frequency for the motor drive converter: 942 Hz
- Mean torque level: 25 Nm

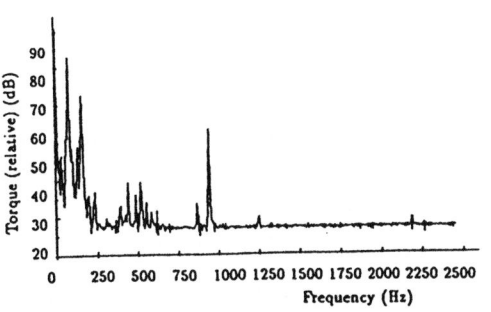

Fig. 1 Torque Spectrum,
Steady Torque

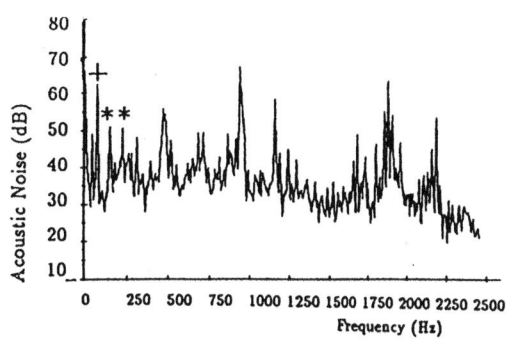

Fig. 2 Acoustic Noise Spectrum
Steady Torque

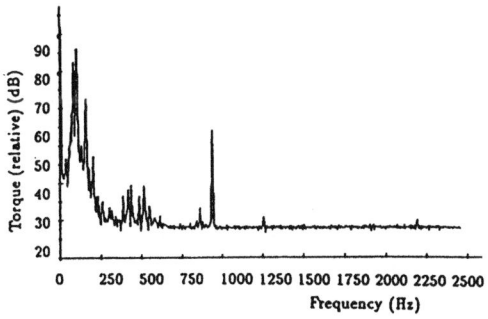

Fig. 3 Torque Spectrum,
100 Hz Load Torque Pulsation

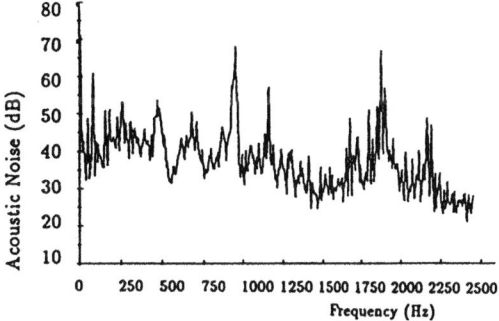

Fig. 4 Acoustic Noise Spectrum
100 Hz Load Torque Pulsation

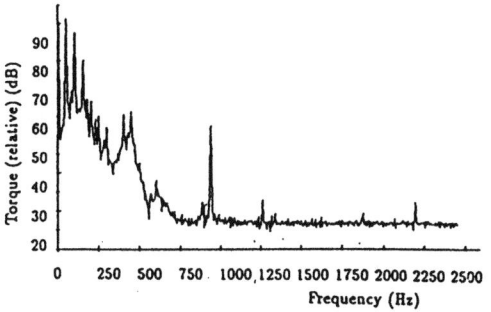

Fig. 5 Torque Spectrum,
Severe 50 Hz Load
Torque Pulsation

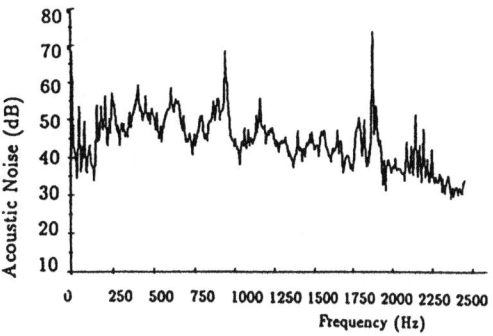

Fig. 6 Acoustic Noise Spectrum
Severe 50 Hz Load
Torque Pulsation

3.2.1 Behaviour With Steady Load

The theoretical background developed in section 2.2 was based on purely sinusoidal electrical excitation, at a single frequency, ω_1. However, the experimental results in figures 1 to 6 relate to a highly complex electrical excitation, generated by an electronically-controlled power switching converter. The power delivered from this converter to the electric motor is highly non-sinusoidal, and contains a broad range of spectral components, in addition to the underlying fundamental excitation frequency ω_1.

In view of this, the task of relating the experimental results to the theory of section 2.2 is far less straightforward than for purely sinusoidal excitation. However, a basis for the relationship between theory and experimental results can be developed with reference to the case depicted in figure 2. For this case, the fundamental component of the complex excitation waveform has a frequency of 13.9 Hz. As derived earlier (from equation (7)), the force wave frequency term is given as $\left(Z_{rt}\frac{\omega_1(1-s)}{p} \pm 2\omega_1\right)$. For the motor under test, considering only the fundamental excitation frequency, the numerical values to be substituted into the above expression are: $Z_{rt} = 28$, $p = 2$, $\omega_1 = 2\pi \times 13.9$ rad/sec, $s = 0.04$. Performing this substitution, and converting the result from rad/sec into Hz, gives calculated force wave frequencies of 159 Hz and 215 Hz. These two calculated frequencies correspond closely with the two peaks identified with the * symbol in figure 2: the measured frequencies relating to these two peaks are 156 Hz and 219 Hz. The adjacent peak, identified by the + symbol, corresponds to six times the fundamental exciting frequency ($6 \times 13.9 = 83$ Hz), as expected for three-phase AC excitation.

There are several other spectral peaks in evidence in figure 2; although the investigation leading to a full understanding of this behaviour is still in progress, there is another correlation that can readily be commented upon: the two major peaks, at 942 Hz and 1885 Hz, correspond to integer multiples of the power converter switching frequency.

3.2.2 Behaviour With Pulsating Load

The dominant components in the torque spectrum for steady load torque, which is the reference case, are related to the fundamental driving frequency and the power converter switching frequency.

Comparison of the torque spectra in figures 1 and 3 very clearly indicates the presence of the 100 Hz load torque pulsation. The 100 Hz component dominates all other components in figure 3.

In the acoustic noise spectra, the effect due to load torque pulsation is not so pronounced. Examination of figure 4 reveals the presence of added spectral components at 100 Hz and multiples there-of, but there is also an overall level increase of approximately 5 dB in the acoustic noise over the range 100 to 500 Hz.

The result of applying a different type of load torque pulsation is shown in figures 5 and 6. In this case, the load torque pulsation is severe, and is of the periodically-unloading type. Although the effect on the acoustic noise spectrum is very pronounced, it is considered that some of the acoustic noise measured in this case is due to noise generation in the drive shaft and coupling.

Overall, the results presented here indicate that, although a substantial degree of load torque pulsation has been generated, the effect on radiated motor noise is of somewhat lower order.

Part of the explanation for this lies in the fact that the rotor inertia of the induction motor under test is substantial. Although considerable load torque pulsation is coupled back to the motor, the high rotor inertia reduces the ability of the load torque pulsation to cause modulation of the rotational motion of the rotor. Dynamic modulation of the rotor angular velocity would provide a basis for modification of the permeance and magnetomotive force relationships (equations (3) and (4) respectively); both these expressions will be affected by modulation of the angular velocity term, ω_{rt}. This in turn will have an impact on the radial force wave expression (equation (7)), leading to modification of the expected surface vibration behaviour of the motor stator, and thereby affecting the acoustic noise radiation from the motor. These aspects are the subject of further theoretical investigation under the current research program, and the experimental investigation has been extended to include induction motors having low values of rotor inertia.

4 Conclusion

The radiation of acoustic noise from induction motors is due to the interaction of a number of complex mechanisms. Substantial levels of noise can be attributed to operation of the motors from electronically-controlled solid state inverter drives, and these effects have been demonstrated in this paper.

The effects of load torque pulsation have also been examined, and the results of investigations into the acoustic noise generated with steady and pulsating loads have been presented. Some correlation between load torque pulsation and radiated acoustic noise has been observed, and further investigations, based on low rotor inertia machines, are being pursued.

Acknowledgement

This project has been supported by funding from the Australian Electricity Supply Industry Research Board. This support is gratefully acknowledged.

References

[1] Belmans, R., D'Hondt, L., Vandenput, A. and Geysen, W., "Analysis of the Audible Noise of Three Phase Squirrel Cage Induction Motors Supplied by Inverters", *Conf. Record of the 1986 IEEE Ind. App. Soc. Annual Meeting*, 28 Sept - 3 Oct 1986, vol. 1, pp. 870-875.

[2] Kato, T., "Precise PWM Waveform Analysis of Inverter for Selected Harmonic Elimination", *ibid.*, vol. 1, pp. 611-616.

[3] Murphy, J. M. D. and Egan, M. G., "An Analysis of Induction Motor Performance With Optimum PWM Waveforms", *Proc. Int. Conf. on Electrical Machines*, Athens, Sept. 15-17, 1980, vol. 2, pp. 642-655.

[4] Takahashi, I. and Mochikawa, H., "Optimum PWM Waveforms of an Inverter for Decreasing Acoustic Noise of an Induction Motor", *IEEE Transactions on Industry Applications*, vol. IA-22, No. 5, Sept/Oct 1986, pp. 828-834.

[5] Yang, S. J., "Low-Noise Electrical Motors", *Oxford University Press*, 1981.

[6] Yang, S. J., "Acoustic Noise from Small 2-Pole Single-Phase Induction Machines", *Proc.IEE*, vol. 122, No. 12, Dec 1975, pp. 1391-1396.

Dynamics of Electric Submersible Pumps

R. D. Neilson
Department of Engineering, University of Aberdeen, U.K.

ABSTRACT

This paper deals with the dynamics of electric submersible pumps. Initially the linear torsional oscillations occurring under both transient start-up conditions and steady state excitation by harmonic torques generated from a variable speed drive are described with the aid of a realistic model. Subsequently, a nonlinear model which couples torsion to lateral vibration is presented, as an aid to understanding the possible interactions between the two types of motion. A simple analysis is given to show the spectral properties of the lateral vibration during torsional oscillation. This is confirmed by numerical simulations. The results may be used as an aid to diagnosing the presence of torsional vibration in the system.

INTRODUCTION

In the oil industry, electric submersible pumps (ESPs) are used for the recovery of oil in wells where there is insufficient pressure to produce a self generated flow. The pumps are generally suspended vertically in position in the oil by a hanger assembly. An umbilical cable feeds power to the motor(s) to drive the pump. Because of the need for the entire assembly to be placed in a hole possibly as small as 150 mm diameter, the pumps used are multi-stage axial flow units, and the motor units consist of multiple induction motor rotors placed in series on a common shaft. As a consequence of this long thin geometry, which may exceed 30m, there may be several torsional natural frequencies of the system within the running speed range of the assembly. In addition to the possibility of transients induced during start up, the use of Pulse Width Modulated (PWM) and Quasi Square Wave (QSW) variable speed drive systems which have a rich harmonic content in their current outputs introduces the possibility of exciting the torsional modes because of the harmonic torques induced in the motors [1].
   Although work has been undertaken on the prediction of the torsional natural frequencies and starting characteristics of ESPs [2,3] and other work reported on the monitoring of lateral vibration [4], little appears to have

been presented on the problem of the mechanism of exciting torsional oscillations in such systems. This paper attempts to examine the problem of torsional oscillations in ESPs. Both the effects of starting transients and steady state oscillations caused by non-sinusoidal fluxes in the motors are considered.

STARTING TRANSIENTS

There are several sources of torsional excitation during starting. There is the possibility of wind up and release of the long flexible pump assembly, and also the generation of unsteady torques by the induction motors until steady state running conditions are achieved. In addition disturbances of the electrical system may induce transients [5]. Only the former source will be discussed here.

To assess the effects of starting transients on a pump system, a model which is representative of a large tandem motor ESP assembly, was used. The model (shown in Fig. 1) is a lumped mass model comprising 6 inertias coupled by 5 elastic shafts. This was validated against a larger scale finite element model of the system prior to use in the dynamic analysis. The inertias I_1 and I_2 represent the motor units while I_3 to I_6 model the axial flow pump sections. The shafts in the system are model by the torsional springs k_1 to k_5. The data used for the model is included in Table 1. The value of k_2 is less than the others because of the longer length of shaft between the motor and pump assemblies. This is because of the seal and protector assemblies mounted there.

The equations of motion for the system are simply given by the matrix problem

$$[I]\{\ddot{\theta}\} + [C]\{\dot{\theta}\} + [K]\{\theta\} = \{T(t)\}$$

This was solved numerically for a variety of starting configurations which assumed that one of the rotors was locked at start-up and then suddenly released.

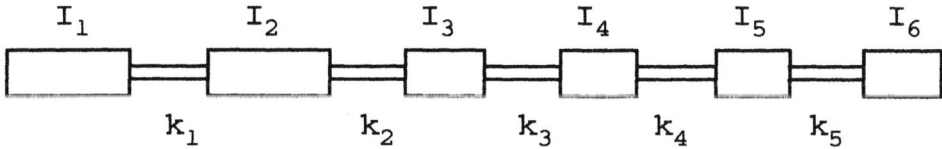

Fig. 1. Torsional model of a tandem motor ESP with four pump units.

Inertial Data

$I_1 = I_2 = 0.1514$ kgm^2
$I_3 = I_4 = 0.0268$ kgm^2
$I_5 = I_6 = 0.0212$ kgm^2

Stiffness Data

$k_1 = 1288$ Nm/rad
$k_2 = 747$ Nm/rad
$k_3 = 1443$ Nm/rad
$k_4 = 1591$ Nm/rad
$k_5 = 1782$ Nm/rad

Table 1. Data for the linear torsional model.

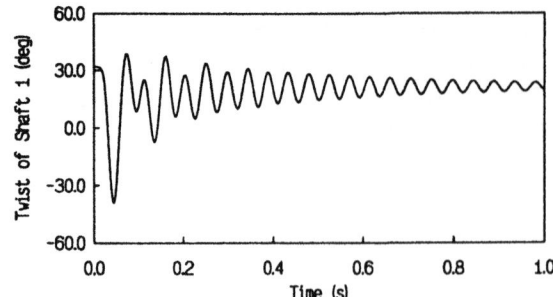

Fig. 2. Twist in the shaft k_1 for rotor I_6 initially locked.

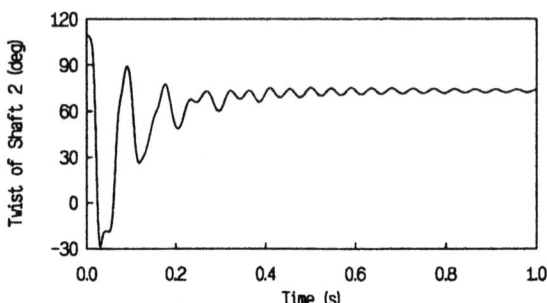

Fig. 3. Twist in the shaft k_2 for rotor I_6 initially locked.

The initial conditions of the rotors were calculated by applying one and a half times full rated torque of the motors to the system with the appropriate rotor clamped. Damping in the system was included in the form of proportional damping in the shafts and large damping terms on rotors I_3 to I_6 to simulate the pumping load. Figures 2 and 3 show the resulting twist in the shafts k_1 and k_2 for the last pump unit (I_6) being locked

As can be seen from these results the twist generated in shaft k_1 by the starting transient can be as large in the reverse direction as the steady state running torque. This gives the possibility of failure at start up in the reverse direction to that of the normal torque.

STEADY STATE TORSIONAL OSCILLATIONS

As well as the possibility of transient vibration due to start-up, it may also be possible to excite torsional oscillations by the presence of harmonic torques on the motor units caused by harmonics and inter harmonics present in the current waveform driving the motor. These occur because the waveform is produced by switching thyristors on and off to generate a waveform of suitable frequency from a previously rectified A.C supply. The consequence of this is that the waveforms applied to the motor are not sinusoidal and therefore small oscillatory torques are induced in addition to an average D.C. torque.

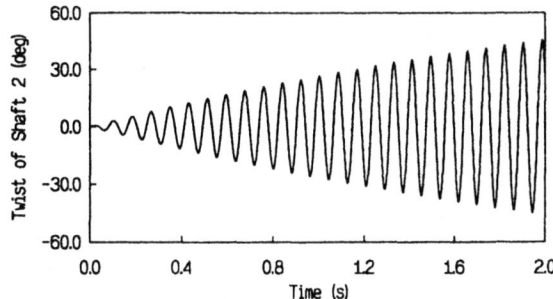

Fig. 4. Twist in the shaft k_2 during the build up of torsional oscillation with light damping.

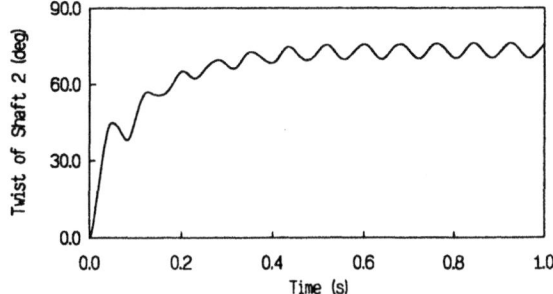

Fig. 5. Twist in the shaft k_2 during build up of torsional oscillation with heavy damping to model the pumping load.

Using the previous model with D.C. and oscillating torque components applied to the motors it is possible to look at effect of these fluctuating torques. Fig 4 shows the build up in torsional oscillation for in-phase fluctuating torques of 23.85 Nm being applied to both I_1 and I_2 at 12.2 Hz. when there is only light damping in the system. This frequency is close to the first torsional natural frequency of the system. In this case, although the applied torque is small, only 5% of full-rated torque, because of the dynamic effects, the twist and consequent stress in the shaft k_2 is appreciable. This may be a cause for concern because of the possibility of fatigue and premature failure.

Fig 5 shows the twist in shafts k_2 for the steady full rated torque of 477 Nm with a superimposed fluctuating torque of 23.85 Nm being applied to each motor again at a frequency of 12.2 Hz. Obviously in this case the amplitude of the torsional oscillation is reduced by the higher damping in the system. In reality it is likely that the system will possess damping levels between the two cases and that an intermediate level of oscillatory motion may occur. It is possible that the steady state motion may be less damaging than the starting effects, unless large amplitudes of oscillation are maintained for long periods.

NONLINEAR MODEL

To assess the problem of coupled lateral and torsional motion in a pump system, the nonlinear equations of motion of the simplified system shown in Fig. 6 have been derived using Lagrange's Equations.

306

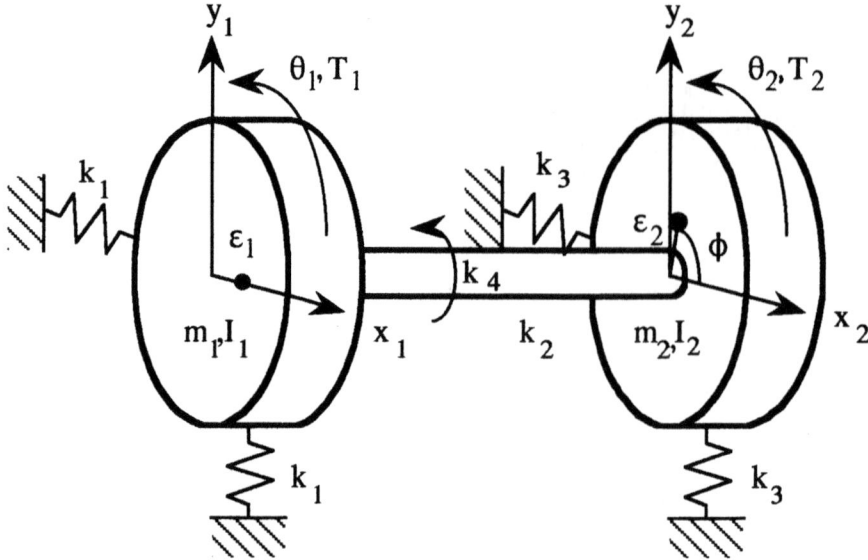

Fig 6.   Coupled nonlinear model of a simplified pump system.

The resulting equations of motion are as follows

$$m_1\ddot{x}_1 + (k_1 + k_2)x_1 - k_2x_2 = m_1\varepsilon_1(\dot{\theta}_1^2\cos\theta_1 + \ddot{\theta}_1\sin\theta_1)$$

$$m_1\ddot{y}_1 + (k_1 + k_2)y_1 - k_2y_2 = m_1\varepsilon_1(\dot{\theta}_1^2\sin\theta_1 - \ddot{\theta}_1\cos\theta_1)$$

$$(I_1 + m_1\varepsilon_1^2)\ddot{\theta}_1 + k_4(\theta_1 - \theta_2) = m_1\varepsilon_1(\ddot{x}_1\sin\theta_1 - \ddot{y}_1\cos\theta_1) + T_1(t)$$

$$m_2\ddot{x}_2 + (k_2 + k_3)x_2 - k_2x_1 = m_2\varepsilon_2(\dot{\theta}_2^2\cos(\theta_2 + \phi) + \ddot{\theta}_2\sin(\theta_2 + \phi))$$

$$m_2\ddot{y}_2 + (k_2 + k_3)y_2 - k_2y_1 = m_2\varepsilon_2(\dot{\theta}_2^2\sin(\theta_2 + \phi) - \ddot{\theta}_2\cos(\theta_2 + \phi))$$

$$(I_2 + m_2\varepsilon_2^2)\ddot{\theta}_2 + k_4(\theta_2 - \theta_1) = m_2\varepsilon_2(\ddot{x}_2\sin(\theta_2 + \phi) - \ddot{y}_2\cos(\theta_2 + \phi)) + T_2(t)$$

These equations are similar in essence to a model presented by Tondl [6] for a turbo-generator set, but do not include the gravitational terms as the pump would normally work in a vertical direction. In addition, the torques $T_1(t)$ and $T_2(t)$ will not be assumed steady in this analysis but will include fluctuating terms which model the torques generated by a variable speed drive.

As an aid to understanding the effect of a steady state torsional oscillation on the vibration in the lateral direction a simple model can be used. Assuming that the torsional oscillation of the inertia I_1 can be written as

$$\theta_1 = \Omega_0 t + \psi\sin\Omega_T t + \theta_{1,t=0}$$

$$\dot{\theta}_1 = \Omega_0 + \Omega_T \psi \cos\Omega_T t$$

$$\ddot{\theta}_1 = -\Omega_T^2 \psi \sin\Omega_T t$$

where Ω_0 is the mean shaft speed, Ω_T is the torsional excitation frequency and ψ is the amplitude of the torsional oscillation, then substituting this into the equation for x_1 gives

$$m_1\ddot{x}_1 + (k_1 + k_2)x_1 - k_2 x_2$$
$$= m_1\varepsilon_1(\Omega_0 + \Omega_T\psi\cos\Omega_T t)^2\cos(\Omega_0 t + \psi\sin\Omega_T t + \theta_{1,t=0})$$
$$- m_1\varepsilon_1\Omega_T^2\psi\sin\Omega_T t \, \sin(\Omega_0 t + \psi\sin\Omega_T t + \theta_{1,t=0})$$

This may be expanded in terms of trigonometric identities and Bessel functions. By doing this and retaining only the largest terms containing the Bessel functions, namely those involving $J_0(\psi)$, this gives

$$m_1\ddot{x}_1 + (k_1 + k_2)x_1 - k_2 x_2$$
$$= -\frac{m_1\varepsilon_1\Omega_T^2\psi}{2}[J_0(\psi)\{\cos(\Omega_T t - \Omega_0 t - \theta_{1,t=0}) - \cos(\Omega_T t + \Omega_0 t + \theta_{1,t=0})\}]$$

$$+ m_1\varepsilon_1 J_0(\psi)[\{\Omega_0^2 + \frac{\Omega_T^2\psi^2}{2}\}\cos(\Omega_0 t + \theta_{1,t=0})$$

$$+ \Omega_0\Omega_T\psi\{\cos(\Omega_T t - \Omega_0 t - \theta_{1,t=0}) + \cos(\Omega_T t + \Omega_0 t + \theta_{1,t=0})\}$$

$$+ \frac{\Omega_T^2\psi^2}{4}\{\cos(2\Omega_T t - \Omega_0 t - \theta_{1,t=0}) + \cos(2\Omega_T t + \Omega_0 t + \theta_{1,t=0})\}]$$

This shows that when a steady state torsional oscillation is present, there will be a series of forcing functions at sums and differences of the torsional and rotational frequencies which will excite lateral vibration of the system. These result from the fluctuations in speed caused by the torsional oscillation. The amplitude of many of these terms will be small because of the diminishing size of the higher Bessel functions. It is likely however that the larger of these components will be present in the lateral vibration spectrum.

This may be readily verified by numerical simulation of the complete equations of motion of the system. Fig. 7 shows the vibration spectrum for the case where a steady state torsional oscillation is superimposed on the mean rotational speed. This was achieved by exciting the rotor system with a sinusoidally varying torque close to the

308

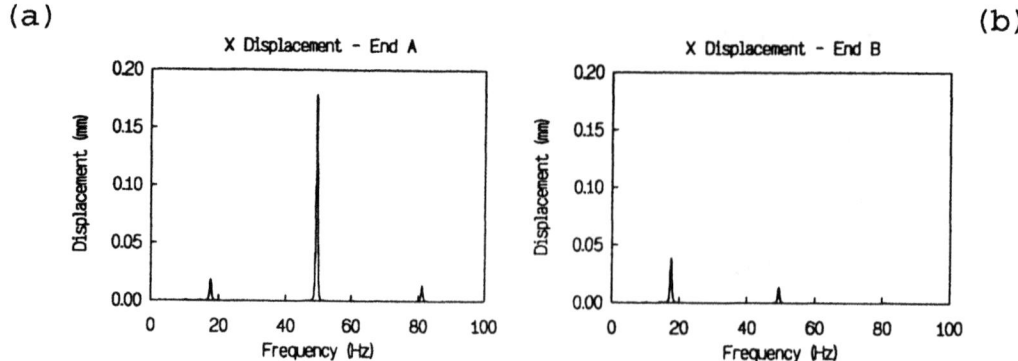

Fig 7 : Spectra of the lateral vibration of the inertias I_1 (a) and I_2 (b) during torsional oscillation.

torsional natural frequency of 31.7 Hz. The model used consisted of inertias of $I_1/2$ and I_3 coupled by a shaft of stiffness k_2. As can be seen, there are sidebands spaced at $\pm \Omega_T$ about the shaft speed in the spectrum as was predicted by the analysis. A particular case of interest would be the case where one of these lateral excitation frequencies corresponds closely to a natural frequency of lateral vibration. In this case severe lateral vibration at a frequency other than shaft speed could occur.

In interpreting the presence of sidebands in a vibration signal, caution is required as sidebands may also indicate the presence of bearing clearance effects in a rotor system rather than a torsional oscillation. This effect has been noted by Thomas [4] and a more full analysis presented by Neilson and Barr [7]. In general however torsional oscillation is only like to occur when there is an excitation frequency close to a torsional mode. In this case the sidebands will be spaced at $\pm \omega_T$ around the shaft speed. In the case of clearance this is unlikely to be the case generally. Consequently it should be possible to differentiate between the two effects if information about the torsional characteristics of the machine are available.

CONCLUSIONS

From the simulations and the literature it is apparent that torsional oscillations may be a cause for concern in electric submersible pump systems. It appears that perhaps the most severe oscillations occur as a result of starting transients with the possibility of the large reverse torques generated causing fatigue failure in the reverse direction to the normal steady torque. Depending on the damping there would also appear to be a possibility of steady state torsional oscillations being excited by small amplitude fluctuating torques generated by variable speed drives if the excitation frequency is close to a natural torsional frequency. If this does occur it may be possible to detect the phenomenon by close examination of the spectrum of the lateral vibration. Components at $\Omega_0 \pm n\Omega_T$ in the spectrum will generally indicate the presence of a torsional oscillation. These spectral characteristics need to be confirmed by testing of experimental systems.

ACKNOWLEDGEMENTS

The author would like to acknowledge the support of B.P. Exploration for some of the work reported and the contribution made by his colleague Mr. Yacamini, of the Department of Engineering, University of Aberdeen.

REFERENCES

[1] Yacamini R., "How HVDC schemes can excite torsional oscillations in turbo-alternator shafts", Proc. I.E.E., 1986, vol. 133, part C, No. 6, p301-307.

[2] Brinner T. R., Traylor F. T. and Stewart R. E., "Causes and prevention of vibration induced failures in submergible oilwell pumping equipment", SPE Paper 11043,1982.

[3] Hyde R. L. and Brinner T.R., "Starting characteristics of electric submergible oil well pumps", I.E.E.E. Trans. on Industry Applications, 1986, vol. IA-22, No 1, p133-144.

[4] Thomas, T. J. M., Condition monitoring of borehole submersible pumps. Proceedings of the I. Mech. E. Seminar on Machine Condition Monitoring, I. Mech. E. London, 1990, p. 17-30.

[5] Tsao T. P., Smith J. R. and Cudworth C. J., "Simulation of turbine generator blade and shaft vibrations as caused by electrical network disturbance", National Symposium on Electrical Power Engineering, Taiwan, Chinese Inst. of Engr., R.O.C. 1988, part D, p707-720.

[6] Tondl, A., Some Problem of Rotordynamics, Chapman and Hall, London, 1965.

[7] Neilson, R. D. and Barr, A. D. S., Dynamics of a rigid rotor mounted on discontinuously nonlinear elastic supports. Proceedings of the I. Mech. E., 1988, 202, C5, 369-376.

High Vibration on a 90MW Gas Turbine-Generator due to a Supporting Structure Resonance

G. L. Lapini* - M. Zippo* - A. Vallini** - G. Diana*** - A. Collina***
(*) CISE SpA, Segrate
(**) ENEL DPT-VDT-STE, Pisa
(***) Dipartimento di Meccanica, Politecnico di Milano

ABSTRACT

A 90MW gas turbine and a generator connected by means of a synchro-self-shifting joint, showed high level of vibration at the intermediate shaft, caused by a supporting structure resonance. The measurements performed and the analytical simulation of the machine are reported. The proposed modification of the supporting structure has been optimized and tested on the analytical model.

1. INTRODUCTION

The design of the 90 MW gas turbine generator described in the paper was modified before plant construction, from the original basic configuration, so that the generator could also be used as a synchronous power factor corrector. For this purpose it was necessary to re-design the intermediate shaft between turbine and generator, and its supporting structure. A heavy synchro-self-shifting (SSS) joint and clutch was introduced between the two machines, so that it was possible to operate them separately; to do so it was necessary to support the intermediate shaft by two bearings and by a quite complex box-type metallic structure.Fig.1 shows a sketch of the machine shaft line and reports the bearing numbering that will be used in this paper. Fig.2 shows a sketch of the intermediate shaft and of its supporting structure.

Since machine commissioning, vibrations were very high, particularly at the intermediate shaft bearings.

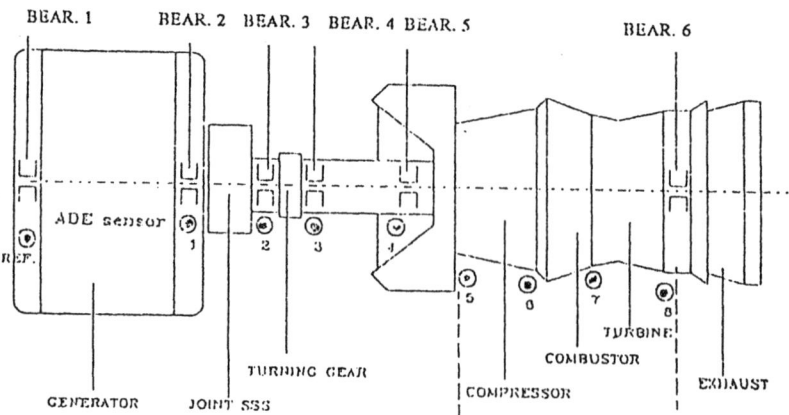

Fig. 1. Scheme of turbine-joint-generator and bearings denomination.

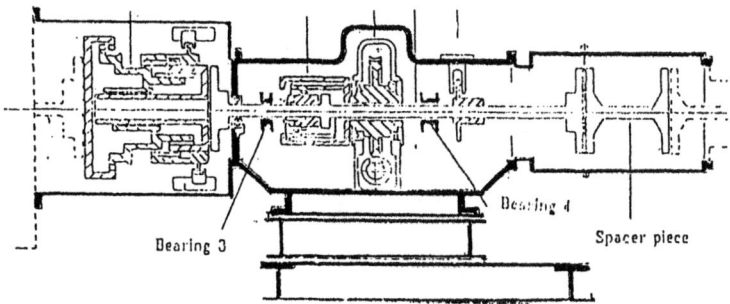

Fig. 2. Schematic drawing of the intermediate shaft, bearings and joint assembly.

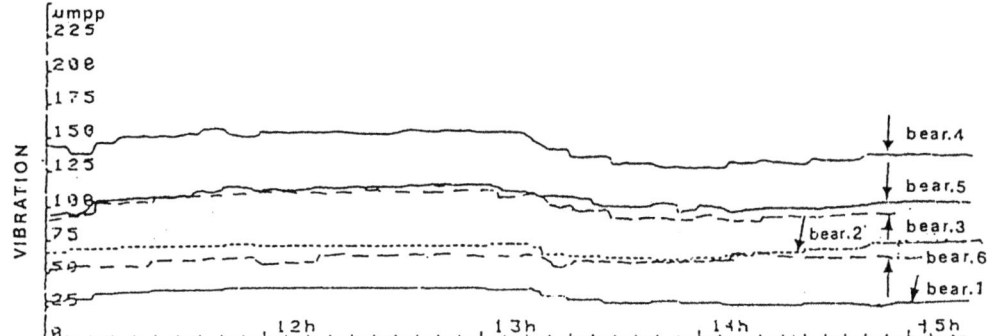

Fig. 3. Machine vibration at full power, measured as the maximum of the vibration orbit by a couple of proximity sensors.

The initial attempts to balance the shafts were not successful; balancing was also difficult because every time turbine and generator rotors were disconnected and connected again (by the SSS joint), the angular position of the two rotors changed and the residual unbalances combined in a different way, causing also very different vibrations.Going on in its attempts to solve the problem the machine manufacturer tried to change the dynamic characteristics of the intermediate bearings. Different types of bearings were mounted on the machine and at the end, adopting four-grooves bearings, vibration levels were reduced to more reasonable, but not yet completely satisfactorly values. In the meantime the machine was equipped with new vibration sensors and monitors. A couple of proximity sensors, installed at ±45 degrees from the vertical, was mounted on bearings number one to five and a single sensor, in vertical direction, on bearing number six. Total vibration levels, measured by the installated monitors as the maximum of the relative vibration orbit, are reported for each bearing on Fig.3 for a typical working period at full power. As it can be seen vibration levels on the intermediate bearings (3 and 4) are still quite high. For this reason ENEL, the power plant owner, decided to go on independently in analyzing the machine behaviour, deepening the analysis of the alignment and of the vibration.

2. MEASUREMENTS OF ALIGNMENT VARIATIONS

During the numerous disassemblies of the machine performed in an attempt to solve vibration problems it had been noted that the alignment condition of the machine changed significantly with time and with ambient temperature. In order to analyse the possible influence of the alignment variations on the machine vibratory behaviour it was decided to install on the machine an ADE system, developed by CISE for continuous measurements of the vertical alignment variations of large rotating machinery (for a complete description of the ADE system characteristics see ref. [1]). Fig.1 shows a simplified plan view of the turbine-generator where the positions of the bearings and turbine pedestals are indicated along with the position of eight ADE sensors installed on the machine. Each of these sensors measures the vertical movement of the point of the machine to which it is fixed (see for instance on Fig.4 the one fixed near bearing 4) relatively to the sensors mounted near bearing number one, which was assumed as a reference.

Fig. 4. ADE sensors placed to the intermediate shaft casing.

Fig.5 reports a typical example of the machine vertical movements during a warm-up.The straight horizontal line represents on this figure the condition of the machine, stopped and cold, at the date and time assumed as a reference, (12/09/89, h.8:40).When the turbine began the start-up, on 20/09/89, and then accelerated to operating speed, generating power along the day, the progressive warm-up caused the changes in vertical position of the ADE sensors that can be seen. As it can be clearly observed particularly the turbine movements are noteworthy, while changes in alignment in the joint section are less important. The comparison among different alignment situation with measured vibration showed no determinant influence of alignment on the dynamic behaviour of the machine.

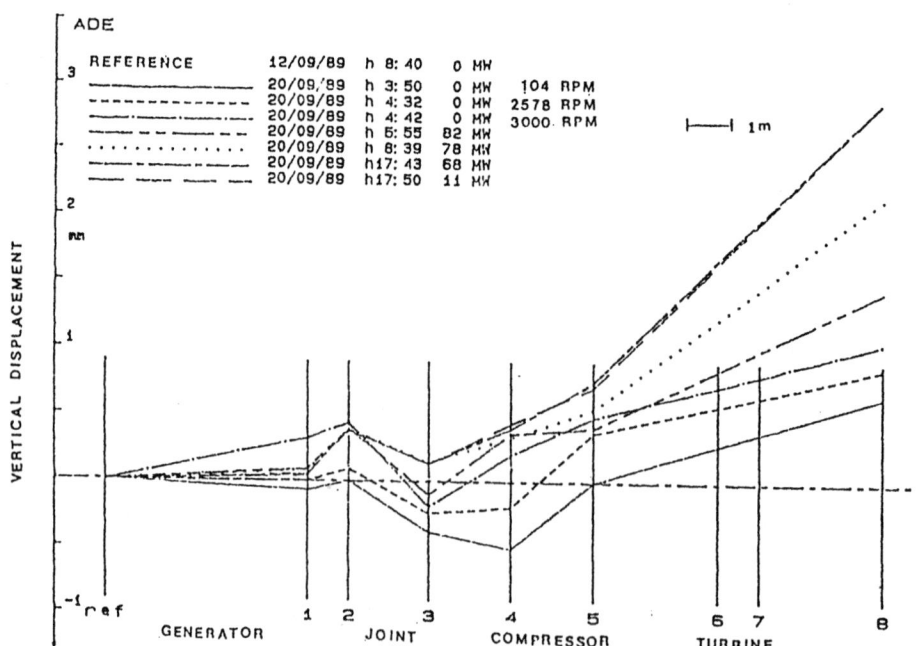

Fig. 5. Typical evolution of the machine vertical alignment during a warm-up (after elaboration of ADE system measurements).

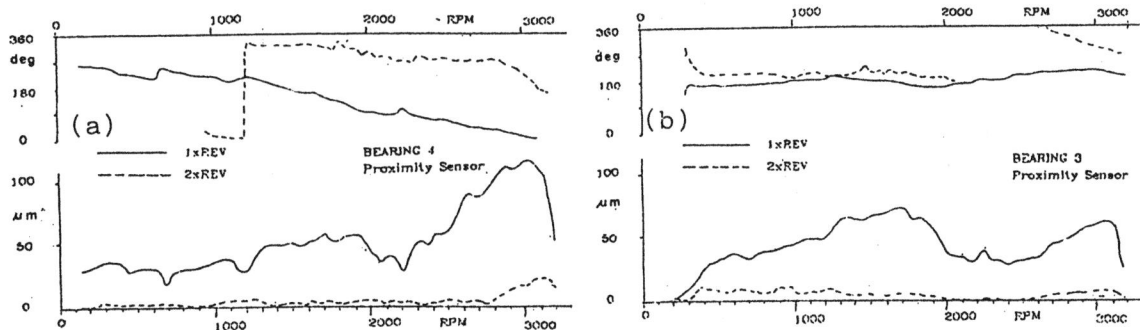

Fig. 6 Amplitudes and phases of shaft relative vibrations measured by one of the proximity sensors on bearing 3 (a) and 4 (b), along a run-down.

3. VIBRATION MEASUREMENTS AND ANALYSIS

During the same period in which alignment measurement were performed, tests were also conducted in order to deepen the analysis of the machine vibratory behaviour.After adding several accelerometers and a key-phasor to the already existing proximity sensors, in order to measure the support vibration too, vibration signals were recorded on magnetic tape during several machine run-ups and run-downs, and analysed with a multi-channel computerized system that calculated amplitude and phase for one per revolution and two per revolution vibration components.

The most significant results were obtained along run-down following overspeeds up to 3200 rpm. As it can be observed in the examples reported relative vibrations (Fig.6a,b), measured by proximity sensors mounted on the intermediate bearings three and four, and absolute support vibration, measured by accelerometers in horizontal direction (Fig.7a,b), show a rapid increase of the 1 x rev. vibration component between about 2500 and 3000 rpm and a rapid drop above 3000 rpm. Moreover support absolute vibrations have amplitudes comparable to relative vibrations and the phases of both relative and absolute vibrations change continuously in the same speed range. These facts seemed to indicate that the intermediate shaft was working, at 3000 rpm operating speed, very near to a shaft critical speed or to a structural resonance.

For this reason an accurate vibration mapping of the structure supporting the intermediate shaft and bearings was performed, with the machine working at running speed. A schematic drawing of this structure is reported on Fig.8, where also reported are the positions of the measuring points. From the measured vibration amplitudes (μm peak to peak) in horizontal directions along the shaft axis, reported in Fig.9a, and from the horizontal vibration amplitudes along a vertical axis at transverse sections A-A and B-B (Fig.9b), which are about in the same positions as the intermediate bearings 3 and 4, it can be clearly observed that this structure is mainly vibrating, at bearings 3 and 4, in horizontal direction and with amplitudes increasing from foundation to mid span.

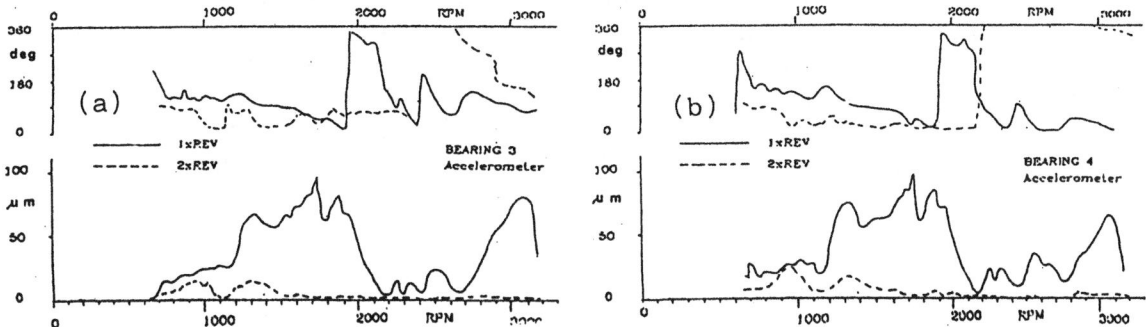

Fig. 7 Amplitudes and phases of support absolute vibrations measured by an accelerometer in horizontal direction on bearing 3 (a) and 4 (b), along a run-down.

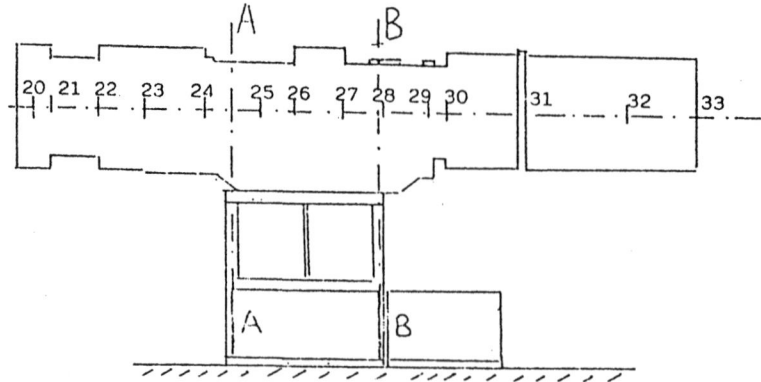

Fig.8 Schematic drawing of the structure supporting joint and the intermediate bearings; the reported numbers indicate the points used for vibration mapping.

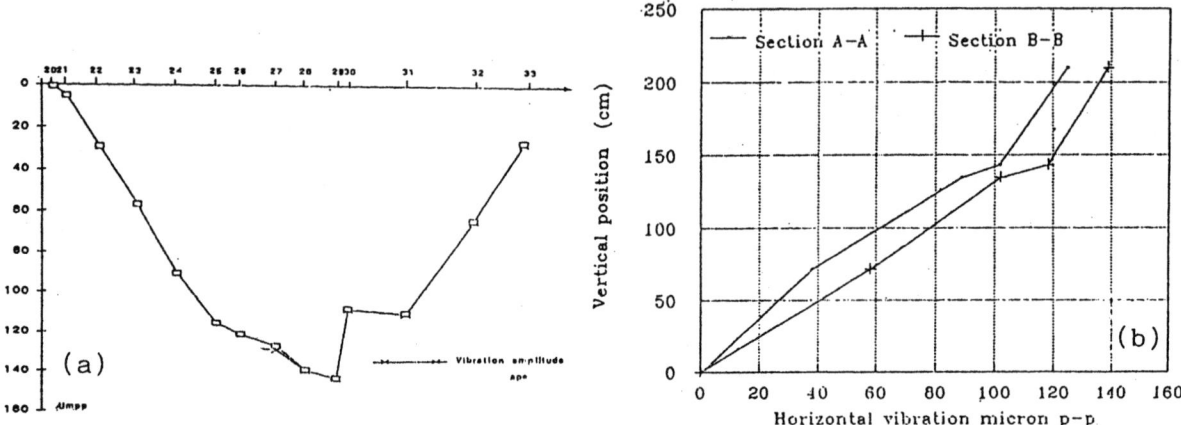

Fig. 9 Horizontal vibration amplitudes (a) along the longitudinal axis C-C of the structure of Fig.8. Horizontal vibration amplitudes (b) along the vertical section A-A and B-B of the structure of Fig.8

That looks roughly like a first bending cantilever mode of vibration of the complete structure. These results confirmed the hypothesis of a structural resonance and suggested the action described in the following paragraph to correct the problem.

4. ANALITYCAL SIMULATION OF MACHINE DYNAMIC BEHAVIOUR

Subsequently to the measurements, the analytical study of machine dynamic behaviour was performed. The first step of the analytical study of the machine was the simulation of the dynamic behaviour of the gas-turbine generator, with the original joint support, as a second step the modification of the intermediate support was introduced into the analytical model and tested, in order to reduce the vibration amplitudes at operating speed.

A finite element model of the rotating machine and its supports has been used: in Fig.10 the f.e. schematization is shown for the gas-turbine (A), the joint (B) and the generator (C). The characteristics of the support of the joint (Tab.I) were calculated separately by means of a f.e.m. adopting beam element, and introduced into the machine+foundation analytical model; the effect of the lubricated bearings was taken into account too.

Tab.I

stiffness [N/m] of support 3 and 4

	before modification	after modification
vert.	$35.0 \cdot 10^8$	$7.0 \cdot 10^8$
hor.	$6.5 \cdot 10^8$	$0.7 \cdot 10^8$

Tab.II

unbalance distibution [kgm]

sect.	3	4	9	13	14
unb.	0.65	0.1	0.15	0.1	0.4

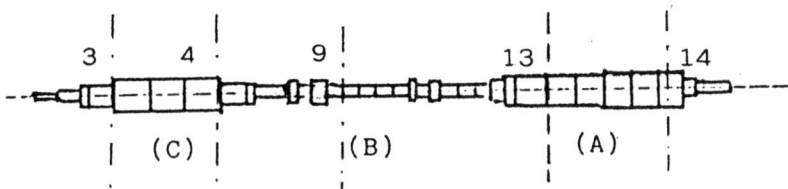

Fig.10 Finite element schematization of the group gas turbine-intermediate shaft-generator.

Several hypotheses were made on the unbalance distribution on the shaft, in order to simulate a condition similar to the real one: Tab.II reports the adopted unbalance distribution. A first comparison between analytical and experimental results has been made, in order to validate the machine schematization. In particular the amplitude and phase of the response of the shaft and the support are reported, as a function of rotating speed.

Comparing the analytical results with the experimental one, it must be pointed out that the calculated curves refer to the horizontal and vertical component of vibration amplitudes, while the experimental sensors are positioned at ±45 degree from the vertical direction, moreover the experimental diagram show the peak to peak amplitude. Observing the analytical frequency response for the first support (Fig.11, "a" horizontal, "b" vertical) it was possible to find the first two critical speeds, respectively equal to 1280 rpm and 2450 rpm for the gas-turbine , while the calculated critical speeds for the generator are equal to 1200 rpm and 2600 rpm. These results are in satisfactory agreement with the measured ones. The behaviour of the intermediate shaft is well reproduced: observing Fig.12a the amplitude of the horizontal shaft oscillation at support N.4 increase approaching the speed of 3000 rpm, (reaching the value of 35μm) while the level of the vertical oscillation is less (4μm, Fig.12b), in agreement with the experimental behaviour. The trend of support horizontal vibration (Fig.13a) is also similar to the experimental one.

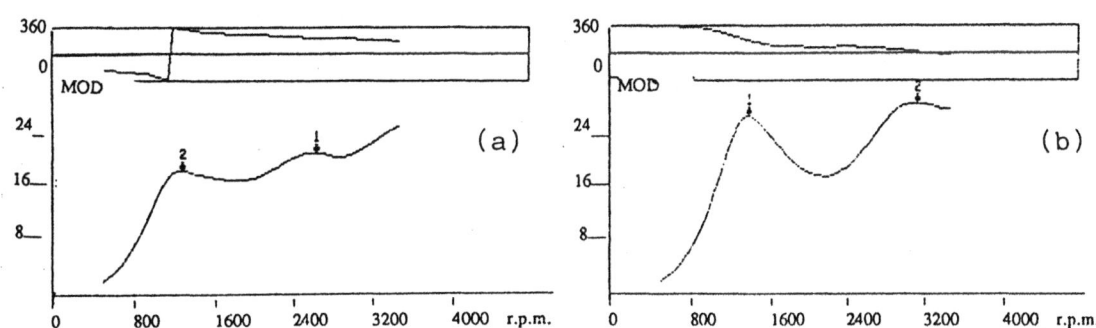

Fig.11 Horizontal (a) and vertical (b) amplitudes (μm) of shaft vibration at bearing 6, before support modification.

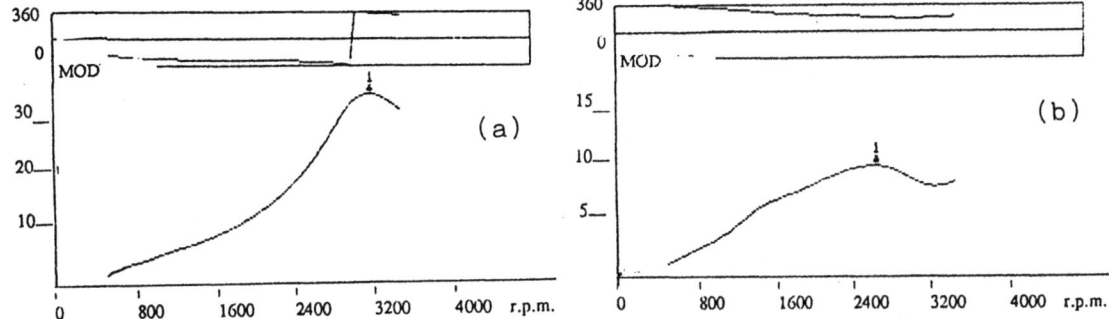

Fig.12 Horizontal (a) and vertical (b) amplitudes (μm) of shaft vibration at bearing 3, before support modification.

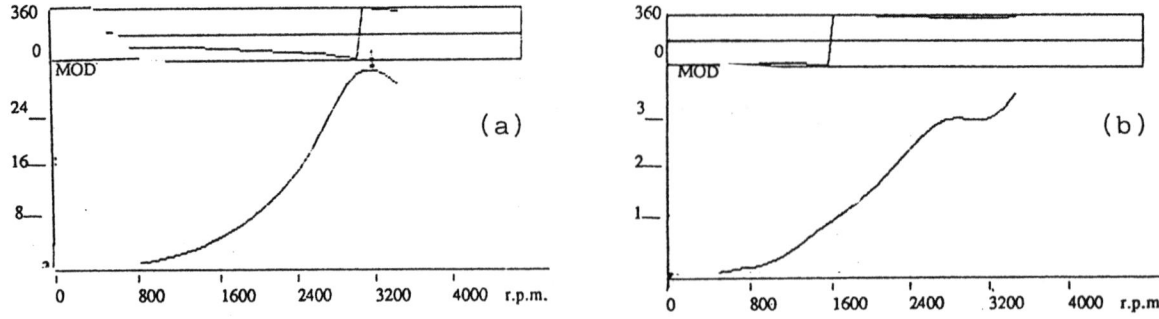

Fig.13 Horizontal (a) and vertical (b) amplitudes (μm) vibration of support 3, before support modification.

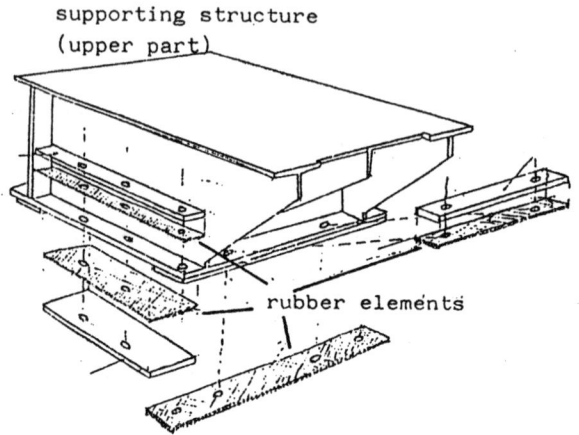

Fig.14 Insertion of rubber element for supporting structure modification.

5. THE DESIGN OF THE MODIFIED JOINT SUPPORT

Once the mathematical model of the system has been assessed, comparing the analytical and the experimental results, it is now possible to predict the effect of a modification in the supporting structure of the joint. It was decided to modify the support inserting rubber "carpets" between the various element of the supporting structure of the joint, in order to modify its horizontal frequency, leaving as much as possible inalterate its vertical one. In particular the rubber elements were inserted between the upper and the lower part of the supporting structure (Fig.14) The stiffness of the modified structure has been calculated (Tab.I) and its characteristics introduced into the analytical model of the system machine+foundation.

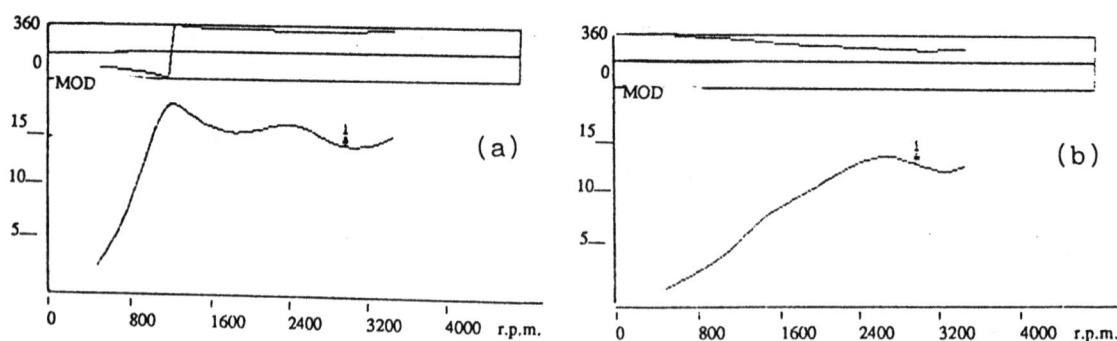

Fig.15 Horizontal (a) and vertical (b) amplitudes (μm) of shaft vibration at bearing 3, after support modification.

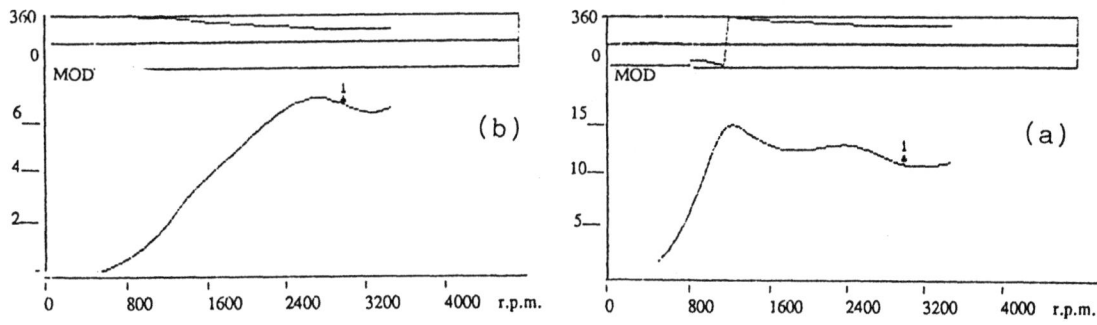

Fig. 16 Horizontal (a) and vertical (b) amplitudes (μm) of vibration of support 3, after support modification.

6. ANALYTICAL SIMULATION OF THE MODIFIED MACHINE

Several simulations were performed in order to optimize the characteristics of the rubber elements: only the final configuration is here shown. While the behaviour of the gas-turbine is almost the same, the amplitude of the joint oscillation is considerably less. Comparing the horizontal oscillations of the joint at the fourth support, after the modification (Fig.15a) and before (Fig.12a), the amplitude at the operating speed is now of 14μm, against 35μm. The vertical oscillations is now of 13μm (Fig.15b), with a moderate increment with respect to the unmodified situation. Looking at the support vibrations, the horizontal amplitude (Fig.16a) is now of 11μm at the operating speed, against 30μm before the modification (Fig.13a); the presence of a horizontal resonance much below the operating speed, can also be noticed. The vertical amplitude shows (Fig.16b) a maximum just below the operating speed, which is due to the generator's second critical speed: its small value does not affect the validity of the proposed solution.

7. CONCLUSIONS

An accurate experimental investigation has been made on a gas turbine generator, whose intermediate shaft bearings were affected by a very high vibration level. The experimental test results showed that the problem was due to a horizontal resonance of the supporting structure. An analytical f.e.m permitted to calculate the dynamic behaviour of the system in the described situation and to simulate different solutions for lowering the horizontal resonance. The optimal solution was choosen, the validity of the obtained results (such as the natural frequency of the supporting structure in its actual situation and the natural frequencies of the modified structure) will be checked by some experimental tests which will be performed by means of an electrodynamic shaker at the end of September '91. After the restart of the group the experimental results in normal operating conditions will be available allowing to draw the final conclusions on the adopted solution.

ACKNOWLEDGMENTS

This study received a grant by MURST and by CRN.

318

REFERENCES

[1] A.Clapis, G. Possa, T. Rossini. Mesure de Variation D'alignment Vertical des Paliers de Turbo-alternaeurs en Fonctionnement. Journees d'Etudes sul la "Survellance des Machines en Fonctionnament et Seuils d'Intervention", Paris, 13-14 Janvier 1982

[2] G. Diana, B. Pizzigoni, A. Curami, "Computer analysis of a rotor-supporting structure system', Rotordynamics 2. Problems in Turbomachinery. 1988. Springer Verlag pp.191-260

STUDY ON TORSIONAL SHAFT VIBRATION RESPONSE OF TURBINE-GENERATOR UNIT

Huang Xiuzhu, Zhang Xueyan, Sun Daixia
Thermal Power Research Institute Ministry of Energy, PRC

ABSTRACT

This paper presents a study of a turbine-generator unit with its connected electrical network, the coupling bolts of which experienced failures after a number of fast valve closing tests 13 months later. Through comparative calculations of the torsional vibration response of shaft system with fast valve closing initiated under normal operation of electrical network and with or without fast valve closing in case of the network faults, it is concluded that fast valve closing is independent of the coupling bolt failure.

Two calculation methods have been adopted to define the mechanical torque of shaft system. The first one simulates the shaft system by using a lump-mass model with a consideration to electrical-mechanical coupling and the second takes the shaft system as a rigid system to define first the electromagnetic torque with different network faults and switching operations and then add the defined electromagnetic touque as an external moment to the lump-mass model. It is illustrated by the obtained curves of the mechanical torque of shaft system that results of the two methods have no much difference.

1. INTRODUCTION

It has been well recognized in the past ten years that the electrical faults of generator (such as 2-phase and 3-phase terminal short circuits), electrical network faults and switching operations (such as 2-phase and 3-phase short circuits , reclosing, etc.) have led to an unbalance between the mechanical torque and the electromagnetic torque, resulting in a torsional vibration of shaft system and a torsional stress on rotor.

Fast closing the IP governing valve of reheat turbine to immediately reduce the output of turbine-generator in case of the network faults is usually applied to avoid the acceleration or deacceleration of rotor and the existence of swing angle caused by the unbalance between the mechanical input and the electrical output, so as to maitain a stable operation of the network. However, a 200MW unit experienced some bolt failures on the HP/IP coupling after a number of fast valve closing test runs. As there are divided opinions towards the cause of bolt failure, a study has been made into the effect of fast valve closing on the torsional vibration

response and into the calculation methods for the torsional
vibration response as well.

2. NOTATION

S_n rated output of generator U_n rated voltage of generator
I_n stator current f_n frequency
X_d d-axis synchronous reactance X_d' d-axis transient reactance
X_d'' d-axis subtransient reactance X_q q-axis synchronous reactance
T_{HI} HP/IP coupling torque T_{IL} IP/LP coupling torque
T_{LG} LP/GEN coupling torque T_{GE} GEN/EXC coupling torque
X_q'' q-axis subtransient reactance T_d'' d-axis subtransient short-
T_d' d-axis transient short-circuit   circuit time constant
      time constant

3. STUDY OBJECT AND PARAMETERS

The system for simulation is shown in Fig. 1, the
structure of the shaft system, in Fig. 2 and the turbine
governing system with fast valve closing, in Fig. 3. The
generator's electrical parameters are as the follows:

S_n = 235 MVA, U_n = 15750 V, I_n = 8605 A, f_n = 50 Hz, X_d = 0.834 P.U.
X_d' = 0.118 P.U., X_d'' = 0.094 P.U., X_q = 0.834 P.U., X_q'' = 0.094 P.U.
T_d' = 0.863 s, T_d'' = 0.108 s, T_q'' = 0.108 s ,   1 P.U. = 100 MVA

The faults discussed in this paper are at point ① in
Fig. 1. The process of the fault is in the time order that
the network fault occurs at 0.04 second, switching-off at
0.15 second, reclosing at 0.7 second and switching-off again
at 0.8 second in case of unsuccessful reclosing . The fast
valve closing discussed in this paper only considers closing
the IP governing valve. The operation sequence of which is
as shown in Fig. 4.

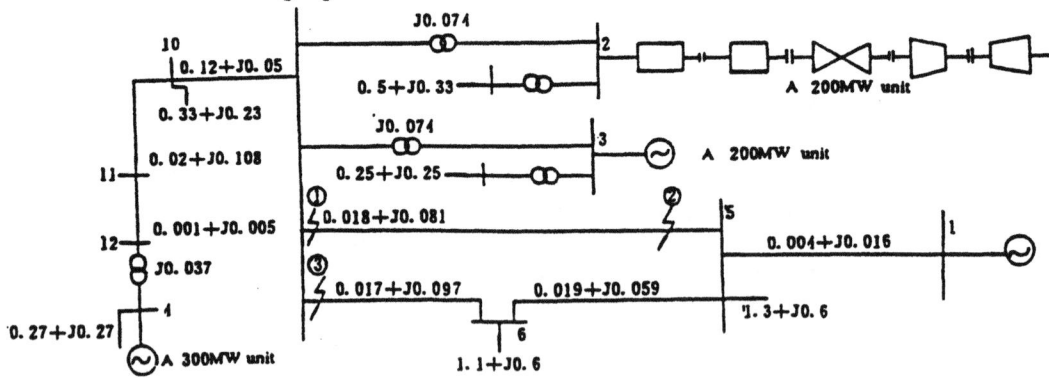

Fig. 1 Schematic diagram of mechanical and electrical system

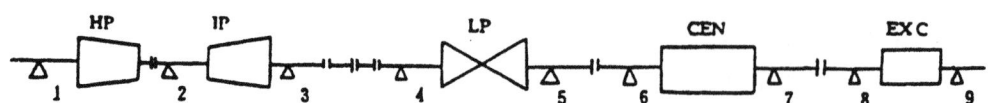

Fig. 2 Schematic diagram of a shaftline of a 200MW unit

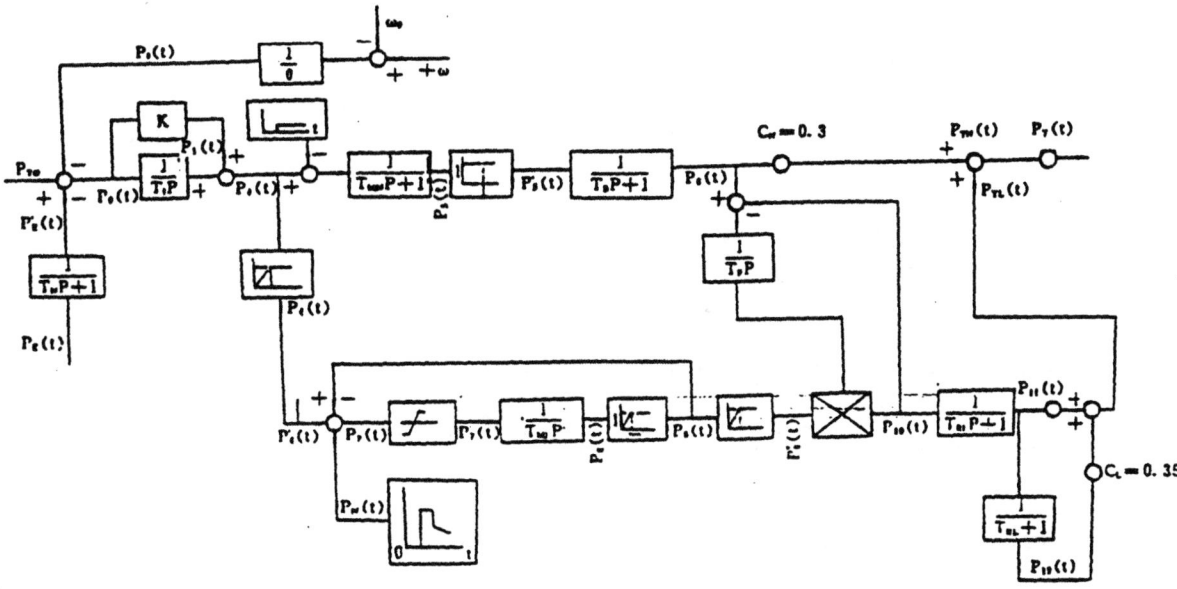

Fig. 3 Block diagram for transmission function of turbine

Fig. 4 Action time order of IP regulator

Fig. 5 15 Lumped Parameter Model

4. EFFECT OF FAST VALVE CLOSING ON THE TORSIONAL VIBRATION RESPONSE OF THE SHAFT SYSTEM

To study the effect of fast valve closing on the torsional vibration response of the shaft system in a comprehensive way, calculations have been made for the torsional vibration response with fast valve closing initiated under normal operation of the network and with or without fast valve closing in case of successful or unsuccessful reclosing after single-phase, 2-phase and 3-phase short circuits of the network. The results of the study is summarized in the sections here after.

4.1 Torsional Vibration Response with Fast Valve Closing in Normal Operation of the Network

A simulation calculation has been given to the torsional vibration response of shaft system with fast valve closing in normal network operation. Table 1 shows the maximum values of the mechanical torques in each of shaft sections. It can be noted in Table 1 that the effect of fast valve closing on the shaft torque in normal operation of the network is very small, which has no large torque impact to the shaft system

Table 1. Maximum mechanical torques with fast valve closing under normal operation of the network (10^6 N.m)

Coupling	T_{HI}	T_{IL}	T_{LG}	T_{GE}
Transient value	0.2191	0.4525	0.6981	0.0006
Stable value	0.1856	0.3464	0.4825	0.6186

4.2 Effect of Fast Valve Closing on the Torsional Vibration Response with the Network Faults

Table 2 shows the calculation results of the effect of fast valve closing on the torsional vibration response of shaft system with different network faults. It can be seen from Table 2 that, with different network faults and switching operations, the mechanical torque of shaft system have been reduced to variable degrees after fast closing the IP governing valve. But, the reducing of the mechanical torque is dependent on the network faults, that is to say, some are significant and some are not. However the mechanical torques have been actually found slightly increased in some shaft sections after fast valve closing and the value is < 3%.

The curves of the electromagnetic torque with or without fast valve closing following an unsuccessful reclosing of 3-phase short-circuit in the network are shown separately in Fig.6 and 7. The curves of the mechanical torque in the shaft section between LP rotor and generator are illustrated in Fig.8 and 9. It can be seen from Fig.6~7 that, no matter whether fast valve closing is initiated, the peak electromagnetic torque of generator appears always soon after a network fault taking place. So, the high-frequency component of the electromagnetic torque magnitude remains unchanged with or without fast valve closing. Nevertheless, the low-frequency component of the electromagnetic torque magnitude becomes small and attenuates very soon after fast valve closing, which helps not only to improve the stable operation of the electrical system, but also to reduce the magnitude of the electromagnetic torque.

It can be seen from Fig.8~9 that, with fast valve closing, the peak mechanical torque in shaft section between LP rotor and generator is reduced from 5.4265 P.U. to 4.523 P.U., which is equal to 14%.

The above study has come to light that fast valve closing is not the cause of coupling bolt failure and this is confirmed by application of some new coupling bolts with a better smoothness and improved corner radius, which experience no more breaking in operation.

Table 2. Max. mechanical torque in each of shaft sections (10^6 N.m)

Type of fault and switching mode	T_{HI}	T_{IL}	T_{LG}	T_{GE}
Unsuccessful HSR of SL-G fault	0.771	1.736	2.406	0.616
Successful HSR of SL-G fault	0.694	1.531	1.850	0.396
Unsuccessful HSR of SL-G fault with fast valving	0.760	1.517	2.217	0.650
Successful HSR of SL-G fault with fast valving	0.690	1.429	1.954	0.390
Unsuccessful HSR of 2φ short-circuit and 3φ switching-off	0.924	2.506	3.305	0.717
Successful HSR of 2φ short-circuit and 3φ switching-off	0.799	1.983	2.435	0.626
Unsuccessful HSR of 2φ short-circuit and 3φ switching-off with fast valving	0.826	1.702	2.478	0.629
Successful HSR of 2φ short-circuit and 3φ switching-off with fast valving	0.826	1.702	2.478	0.626
Unsuccessful HSR of 3φ short-circuit	0.658	2.25	3.342	0.489
Successful HSR of 3φ short-circuit	0.706	1.948	2.880	0.498

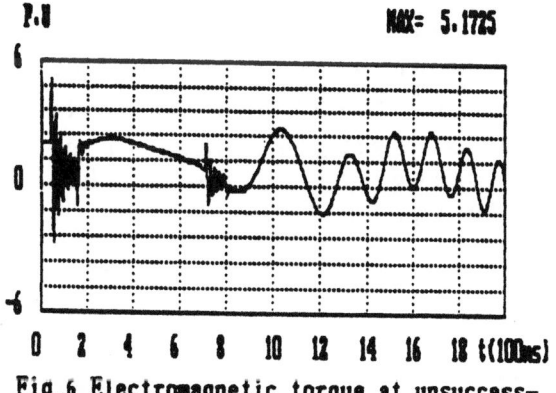

Fig.6 Electromagnetic torque at unsuccessful HSR of a 3-phase short circuit

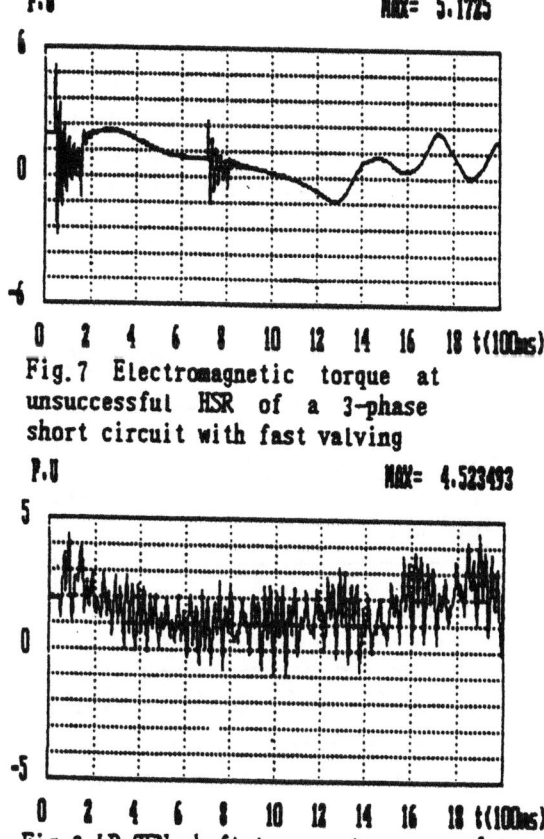

Fig.7 Electromagnetic torque at unsuccessful HSR of a 3-phase short circuit with fast valving

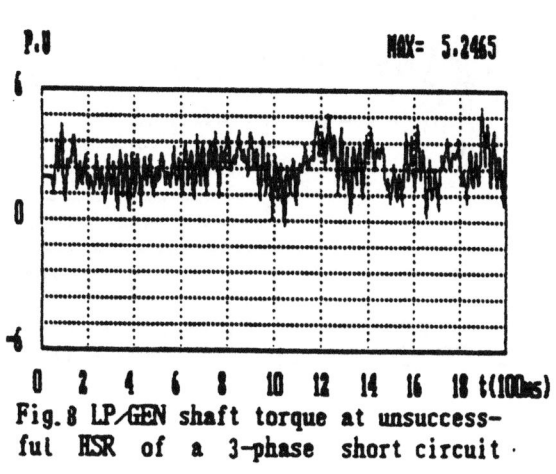

Fig.8 LP/GEN shaft torque at unsuccessful HSR of a 3-phase short circuit·

Fig.9 LP/GEN shaft torque at unsuccessful HSR of a 3-phase short circuit with fast valving

5. CALCULATION METHODS FOR THE TORSIONAL VIBRATION RESPONSE OF SHAFT SYSTEM FOLLOWING THE NETWORK FAULTS

Two calculation methods have been adopted for the torsional vibration response of shaft system following network faults. One is the calculation which uses a lump-mass model to find out the electromagnetic torque of generator and the mechanical torques in each of shaft sections by giving a consideration to the electrical-mechanical coupling; and the other is the calculation which takes the shaft system as a rigid system to define the curves of the electromagnetic torque of generator following different network faults and switching operations first and then find out the mechanical torques in each of shaft sections by appling the defined electromagnetic torques, as an external moment, onto the lump-mass model. The results of these two calculation methods are presented in this paper.

A 15-lump-mass model (Fig.5) and a rigid system have been used to simulate the shaft system (Fig.2). The inertia value of the rigid system is equal to the total one of the 15-lump-mass model.

First, the 15-lump-mass model is used to define the electromagnetic torque of generator and the mechanical torques in each of shaft sections following the network faults with a consideration to the electrical-mechanical coupling. Second, a rigid system is used to define the curves of the electromagnetic torque of generator following the network faults and find out the mechanical torque in each of shaft sections by appling the defined electromagnetic torques of generator, as an external moment, onto the 15-lump-mass model.

This paper presents only the curves of the electromagnetic torques of generator and the mechanical torques in the shaft section between LP rotor and generator following a successful reclosing of single-phase short circuit and an unsuccessful reclosing of 3-phase short circuit of the network, see Fig.10 to 17.

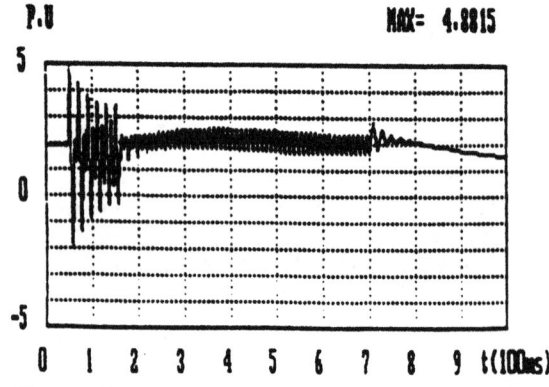

Fig.10 Electromagnetic torque at successful HSR of a SL-G (for 15-lump-mass model)

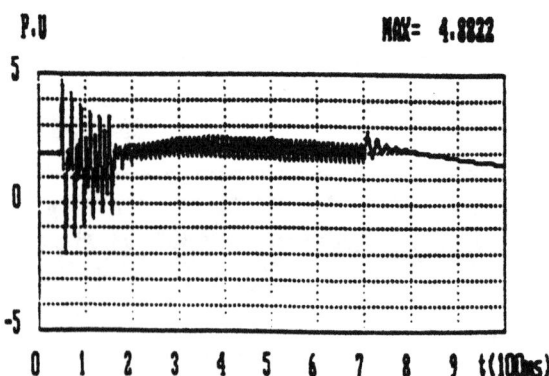

Fig.11 Electromagnetic torque at successful HSR of a SL-G (for rigid system model)

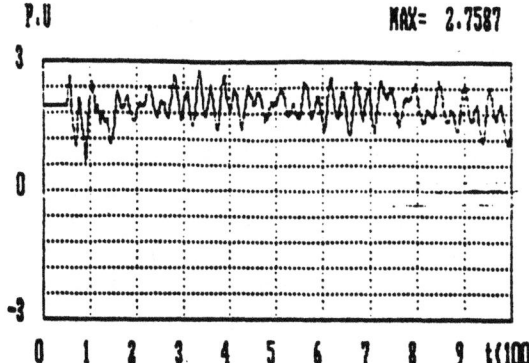

Fig.12 LP/GEN shaft torque at successful HSR of a SL-G (mechanical-electrical coupling)

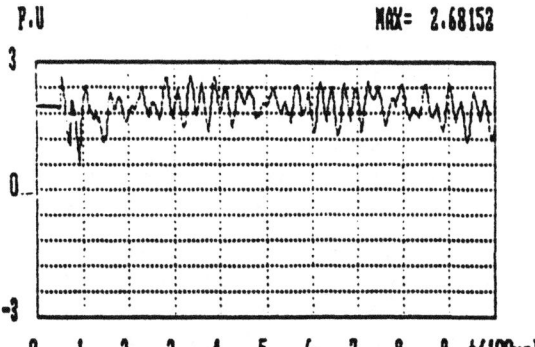

Fig.13 LP/GEN shaft torque at successful HSR of a SL-G (only for mechanical system)

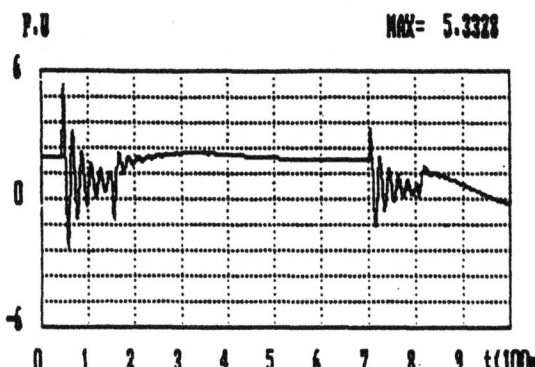

Fig.14 Electromagnetic torque at unsuccessful HSR of a 3-phase short circuit (for 15-lump-mass model)

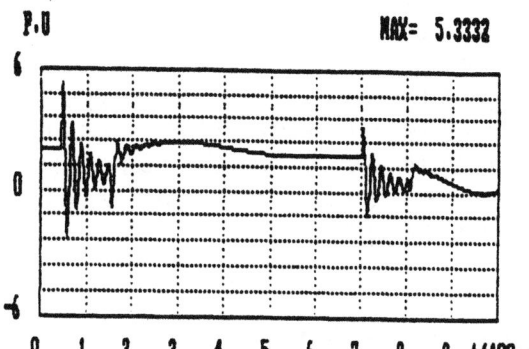

Fig.15 Electromagnetic torque at unsuccessful HSR of a 3-phase short circuit (for rigid system model)

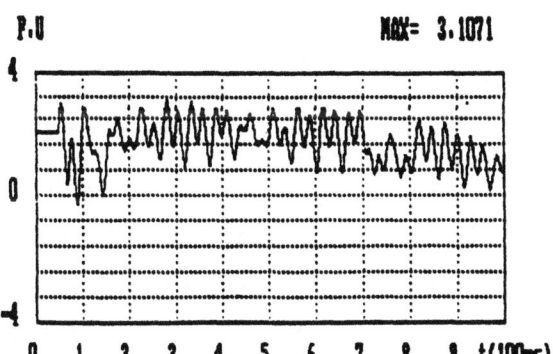

Fig.16 LP/GEN shaft torque at unsuccessful HSR of a 3-phase short circuit (mechanical-electrical coupling)

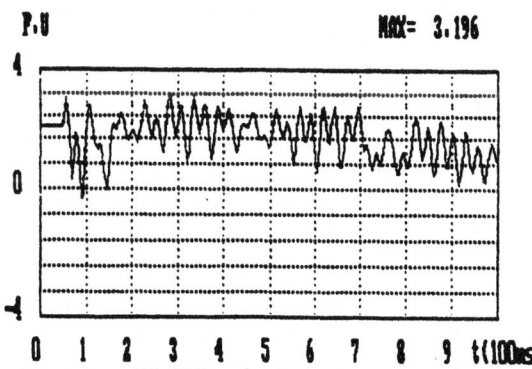

Fig.17 LP/GEN shaft torque at unsuccessful HSR of a 3-phase short circuit (only for mechanical system)

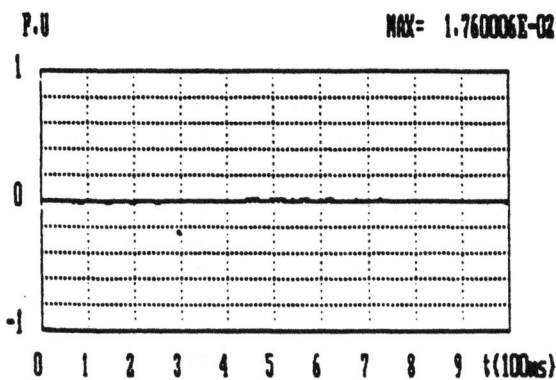

Fig.18 Difference of electromagnetic torque between Fig.10 and 11

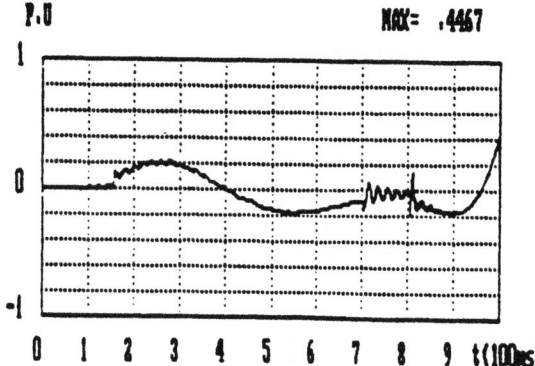

Fig.19 Difference of electromagnetic torque between Fig.14 and 15

Fig. 18 and 19 show separately the difference of electromagnetic torques illustrated in Fig. 10~11 and Fig. 14~15. It can be noted from Fig. 18~19 that, in the case of equal inertia value for the simulated model, there is little difference in the electromagnetic torque of generator under the network faults, no matter how calculations are made either by taking the shaft system as a rigid system or by using a multiple-lump-mass model with a consideration to the electrical-mechanical coupling. The reason for this is that the time constant of the electrical system is much smaller than that of the mechanical system.

Based on a same simplified model of shaft system, the mechanical torques in each of shaft sections are of no much difference (within the range of 5%) according to the calculations, one of which simulates the shaft system by using a lump-mass model with a consideration to the electrical-mechanical coupling and the other takes the shaft system as a rigid system to define the electromagnetic torque of generator and then add the defined electromagnetic torque as an external moment onto the lump-mass model. This can be seen from the analytical study of Fig. 12~13 and Fig. 16~17.

6. CONCLUSIONS

a. Fast closing the IP governing valve has little effect to the torsional vibration response of shaft system under normal operation of the network. The mechanical torques in each of shaft sections have been reduced to variable degrees in most of the cases of network faults and switching operations after fast valve closing. The reducing of the mechanical torques are dependent on the network faults, some are very small and some are significant. However, the mechanical torques have been found slightly increased only following a number of the network faults and switching operations, the value of such increase is about < 3%.

b. For the same simplified model of shaft system, there are no much difference in the mechanical torques in each of shaft sections with the two calculation methods for the torsional vibration response. These two methods have the same accuracy in fact.

Refference

1. T.J.Hammons, "Effect of Three Phase System Faults and Faulty Synchronizations on the Mechanical Stressing of Large Turbine-Generators", Rev. Gen. Elect., Vol.86, No.7/8, 1977, pp. 558-580
2. T.J.Hammons, "Accumulative Fatigue Life Expenditure of Turbine-Generator Shafts Following Worst-Case System Disturbances", IEEE Trans. (PAS), 1982, Vol.101, No.7, pp. 2364-2374
3. J.S.Jovce, T.Kuling and D. Lambrecht, "The Impact of High-Speed Reclosure of Single and Multi-Phase System Faults on Turbine-Generator Shaft Torsional Fatigue", IEEE Tuans. (PAS), 1980, Vol.99, No.I pp.279-291

Wide Band Active Vibration Controller for an Out-of-Balance Rotor

A.V. Metcalfe and J.S. Burdess

Departments of Engineering Mathematics and Mechanical Engineering, The University of Newcastle upon Tyne, Newcastle upon Tyne, England.

ABSTRACT

A method for reducing the forced vibration of an out-of-balance rotor-bearing system by the application of external control forces is presented. Design of the controller only requires very approximate estimates of the rotor mass and structural stiffness. It does not require estimates of the damping or mass unbalance distribution. The proposed controller is independent of the frequency of vibration and therefore has potential for rotors which have rapidly varying speeds. It could also be used on rotors, such as vehicle axles, which are subject to non-harmonic forced vibration. Results from a computer simulation are given.

INTRODUCTION

There are several approaches to the design of active vibration controllers when the disturbance is at a known frequency (e.g. Burdess and Metcalfe 1983). These are applicable to an out-of-balance rotor if its speed of rotation is a known constant. This strategy is less restrictive than it may at first sound, because speed of rotation is relatively easy to monitor and a control algorithm can be tuned to this frequency if it remains constant for reasonable lengths of time (Metcalfe and Burdess 1986). There are nevertheless many applications in which speed of rotation varies rapidly, and a controller which is effective against any frequency of vibration would have practical advantages. A different application is reducing the vibration of the rear axle of a road vehicle which is subject to forced non-harmonic vibrations, due to road irregularities, which will vary with the surface and the vehicle speed.

Burrows and Sahinkaya (1983) presented a method for obtaining an open-loop control strategy which will minimise the vibration of a rotor-bearing system by the application of external control forces. The optimisation was performed for a simulated rotor over a frequency range from 10 radian/s up to the instability threshold (critical) frequency of 360 radian/s. The optimal control forces varied considerably, including changes in sign, with frequency making the open–loop strategy impractical if the speed of rotation varies rapidly throughout the range. Their method also requires estimates of the system parameters and out–of–balance response, and these quantities may be difficult to obtain for some practical systems. In contrast to this, the closed loop feedback control strategy proposed here only requires very approximate estimates of the rotor mass and structural stiffness and does not change with the speed of rotation. The drawback is that it, in common with all closed–loop feedback control, has the potential to be unstable. It will be designed to be stable for a mathematical model of the rotor, sensors and actuators but this unfortunately does not guarantee stability in practice (Balas 1982). The external forces required for the control could be provided by electromagnetic actuators (Nikolajsen *et al.* 1979).

ROTOR MODEL

The rotor used for the simulation is shown in Figure 1. It is based on that used by Burrows and Sahinkaya (op. cit.) with a, less extreme, mass unbalance distribution:

[(4.0,0 deg) (3.3,73 deg) (9.6,196 deg) (2.9,118 deg) (11.9, 7 deg) (2.3,266 deg) (9.3, 303 deg) (1.6, 116 deg) (2.7,160 deg)]

where the pairs represent mass (g) at periphery and angle for the nine stations. The displacements in the x and y directions at station i are denoted by x_i and y_i respectively for i running from 1 up to 9. The rotor bearing system is modelled (Burrows and Sahinkaya op. cit.) by:

$$M\ddot{q} + C_e \dot{q} + (K_s + K_e)q = Bf \qquad (1)$$

where $q^T = (x_1, ..., x_9, y_1, ..., y_9)$ and $f^T = (\cos \omega t, \sin \omega t)$ with ω representing the speed of rotation in radian/s. M and K_s are the mass and structural stiffness matrices of dimensions 18 × 18. The bearing forces are accounted for by introducing damping and stiffness matrices C_e and K_e respectively of dimensions 18 × 18. The out–of–balance forces

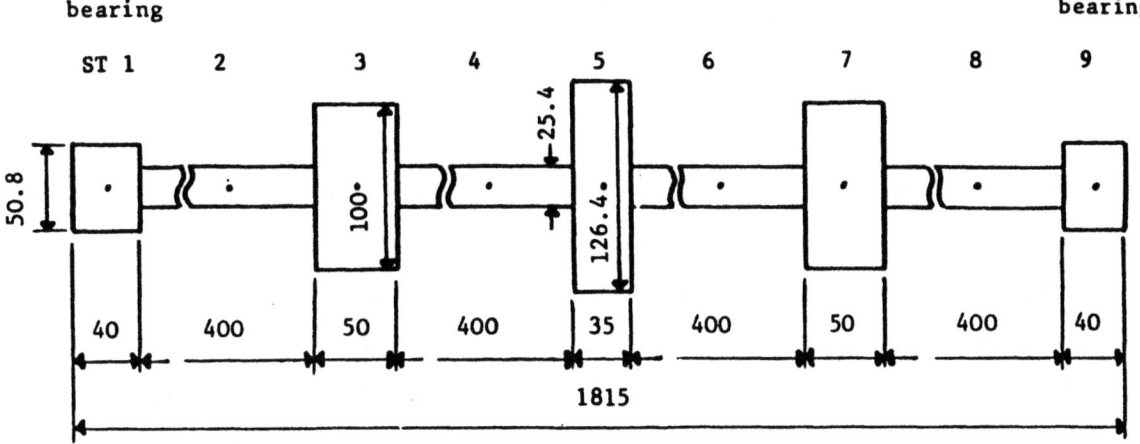

Figure 1   Rotor bearing model

are obtained by introducing the matrix B of dimensions 18 × 2.   The
structures of these matrices are given below as they have implications for
the design of the controller.

$$M = \begin{bmatrix} \text{diag } (m_1, \ ..., \ m_9) & 0 \\ 0 & \text{diag } (m_1, \ ..., \ m_9) \end{bmatrix} \quad (2)$$

$$K_s = \begin{bmatrix} K'_s & 0 \\ 0 & K'_s \end{bmatrix} \quad (3)$$

and the only non zero elements of the 9 × 9 'banded' matrix K'_s are of
the form $k_{i\,i-1}$ or $k_{i\,i}$ or $k_{i\,i+1}$.   The only non zero elements of C_e and
K_e are in positions (1, 1), (1, 10), (9, 9), (9, 18), (10, 1), (10, 10),
(18, 9) and (18, 18).   Their values depend on the speed of rotation
(Holmes 1960).

DESIGN OF VIBRATION CONTROLLER

Theory for Single Input Single Output Structure

The theory (Metcalfe and Burdess 1990) is summarised in Figure 2. The reasoning behind the controller is to estimate the net input to the system (ε), that is disturbance (w) plus control (u), using an electronic system referred to henceforth as a 'disturbance observer', subtract the known control to obtain an estimate of the disturbance ($\hat{\varepsilon}$) and finally set the control equal to the negative of this estimate. This ideal is not physically realisable but it can be approximated by using a large gain (k) and setting the control,

$$u = -k\hat{\varepsilon} \quad . \tag{4}$$

It is straightforward to verify that the Laplace transform of disturbance to output is given by

$$Y(s) = \frac{(1 + KG_e(S))G(s)}{1 + KG_e(s) + kKG(s)} W(s) \quad . \tag{5}$$

Provided the closed loop system remains stable, large values of k and K will lead to a reduction in the magnitude of Y(s) from its uncontrolled value of G(s) W(s). For example, if the product kK G(s) is large relative to 1 over the frequency range of interest then Y(s) is reduced by a factor of 1/k from its uncontrolled value.

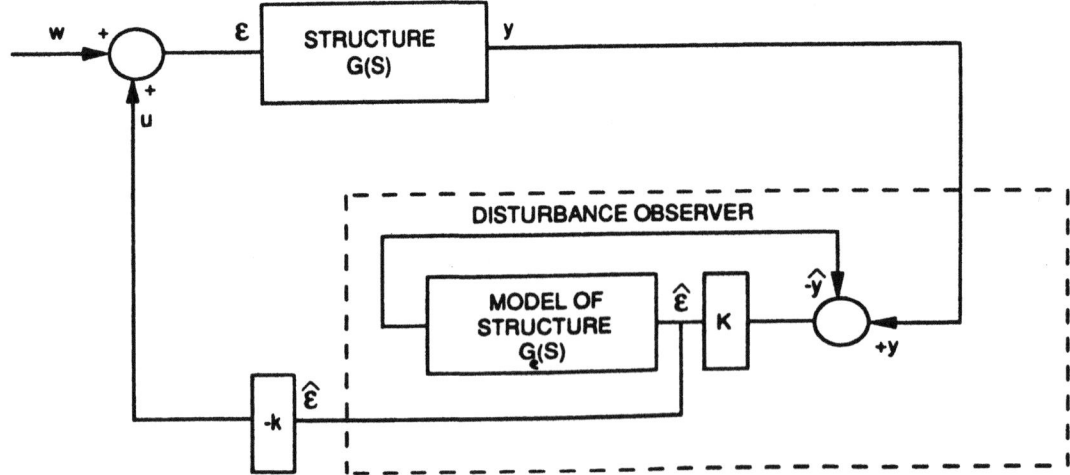

Figure 2   Disturbance observer and closed loop control applied to a single input single output structure

Application to Rotor

An electromagnetic actuator that can provide control forces in both the x and y directions is positioned at Station 6. A transducer is also positioned at Station 6 to measure displacements in the x and y directions. The relevant subset of the equations of motion is

$$\ddot{x}_6 = 3961626\, x_5 - 7917414\, x_6 + 3955788\, x_7 + b_{61}\cos\omega t + b_{62}\sin\omega t - u \quad . \tag{6}$$

A disturbance observer, consisting for example of a second order analogue electronic circuit, described by:

$$\ddot{\hat{x}}_6 = -5628\, \dot{\hat{x}}_6 - 7917414\, \hat{x}_6 + K(x_6 - \hat{x}_6) \tag{7}$$

was set up to model "$\ddot{x}_6 = -7917414\, x_6 + \varepsilon$". Fictitious critical damping is included as Metcalfe and Burdess (op. cit.) found that this tends to improve the closed loop stability — at least for single mode structures. The forces associated with the displacements at Stations 5 and 7 are included with the out–of– balance forces and the control. Modelling error, purposefully introduced by overestimating the damping, can also be thought of as incorporated with the other disturbances. The actual control u satisfies the differential equation

$$\dot{u} = -au + akK(x_6 - \hat{x}_6) \tag{8}$$

which includes a first–order system model for the actuator dynamics. The control for the y direction was set up in an identical manner. This simple control law might, intuitively, be expected to reduce vibrations at Station 6 provided the system remains stable. However, as the higher modes of the shaft are hardly affected by out of balance forces at frequencies below its critical frequency (about 240 radian/s) reducing vibrations at Station 6 might also have the desirable consequences of reducing vibrations at the other stations. This was investigated by computer simulation.

332

SIMULATION RESULTS

Results of simulations at frequencies from 10 radian/s up to 240 radian/s in 10 radian/s intervals, with the parameter a set at 50000 and both the parameters k and K set at 1000 are shown as a solid line in Figure 3. The "sum of squared displacements" is the sum of squares of the x and y displacements at the nine stations sampled over a two second interval, which is long enough for the transient response to become negligible. The vertical scale is, 20 times logarithm base ten of, the ratio of the controlled to the uncontrolled response. Thus negative values reflect an attenuation and any values above zero would correspond to a detrimental effect. At a speed of 80 radian/s the vibrations are reduced by a factor of nearly 100. The controller also stabilised the rotor up to at least 400 radian/s. The coefficient of $\hat{x}_6$ in Equation (7) is equal to k_{66}/m_6. The simulation exercise was repeated with this overestimated by 50% and underestimated by 50%. The results were similar to those obtained with 'exact' estimates, and they suggest that very approximate estimates of the mass and structural stiffness will suffice.

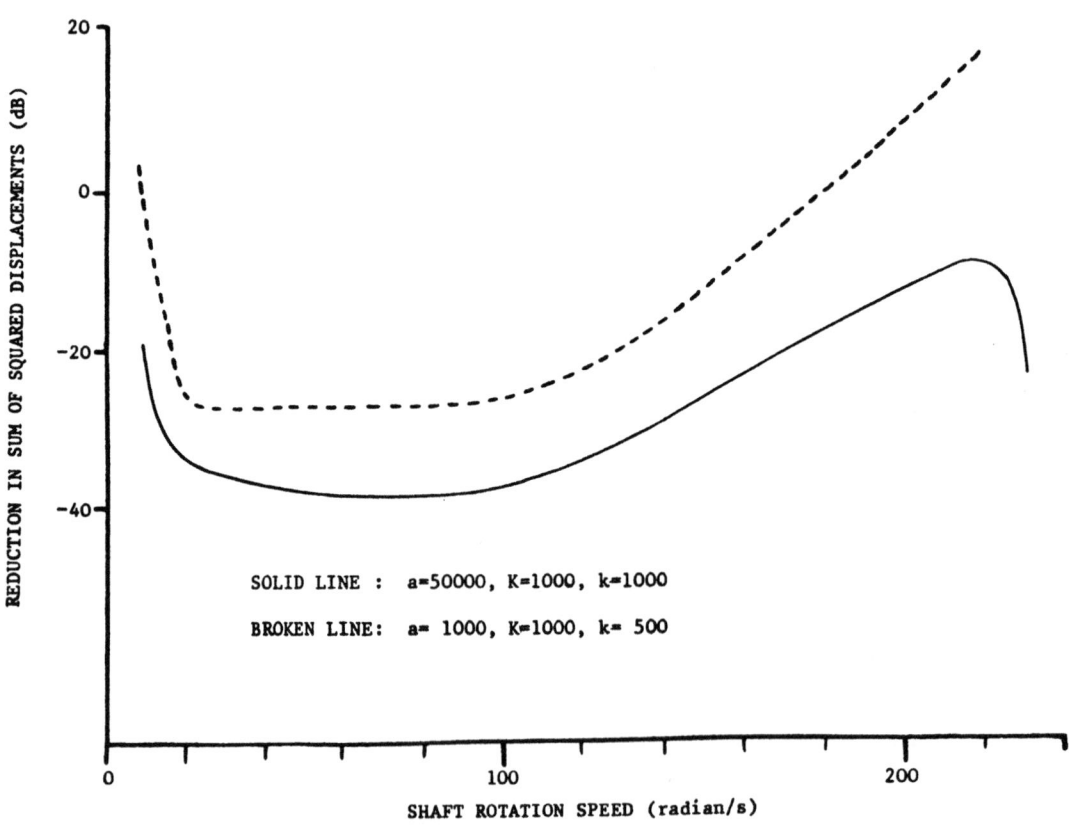

Figure 3   Controller performance

A third set of simulations were carried out with the parameter a reduced to 1000, corresponding to a slower actuator. The gain parameter k was reduced to 500 in order to maintain stability at low rotation speeds. The performance is shown by the broken line in Figure 3. The controller stabilised the rotor up to 350 radian/s but the increase in vibration between 190 and 240 radian/s limits the usefulness of this.

DISCUSSION

The simulation results are sufficiently encouraging for the proposed vibration controller to be considered for practical applications. The actuator design will clearly be crucial, although inherent damping in a real rotor, which was not modelled in the simulations, may make the system stability less sensitive to its response time. However the large gains used in the simulations may promote instability in practice, partly due to unmodelled high frequency modes. It should also be noted that only the first mode of the rotor was excited by the vibrations, making vibration control relatively easy. Additional actuators would be needed if higher modes were excited. Whilst further theoretical and simulation studies could be made it is the authors' opinion that these are not a substitute for practical experimentation, and that testing the algorithm on a real rotor would be more valuable.

REFERENCES

Balas MJ (1982) Trends in large space structure control theory: fondest hopes, wildest dreams. IEEE Transactions on Automatic Control Vol AC27:522–535.

Burdess JS and Metcalfe AV (1983) Active control of forced harmonic vibration in finite degree of freedom structures with negligible natural damping. Journal of Sound and Vibration 91:447–459.

Burrows CR and Sahinkaya MN (1983) Vibration control of multi–mode rotor–bearing systems. Proceedings of The Royal Society of London A 386:77–94.

Holmes R (1960) The vibration of a rigid shaft on short sleeve bearings. Journal of Mechanical Engineering Science 2:337–341.

Metcalfe AV and Burdess JS (1986) Active vibration control of multi–mode rotor–bearing system using an adaptive algorithm. ASME Journal of Vibration, Acoustics, Stress and Reliability in Design 108(2):230–231.

Metcalfe AV and Burdess JS (1990) Experimental evaluation of wide band active vibration controllers. ASME Journal of Vibration and Acoustics 112(4):535–541.

Nikolajsen JL, Holmes R and Gondhalekar V (1979) Investigation of an electromagnetic damper for vibration control of a transmission shaft. Proceedings of the Institution of Mechanical Engineers 193:331–336.

Vibration Control of a Flexible Rotor Suspended by Active Electromagnetic Bearings

Y.Suzuki, S.Michimura and A.Tamura

Department of Mechanical Systems Engineering, Takushoku University, Tokyo, Japan

ABSTRACT

The PID controller with an analog circuit is widely used in the active magnetic bearing system. Although this system is simple and non-expensive, the ability of vibration control for a flexible rotor is not sufficient. To reduce resonance vibrations of a flexible rotor suspended by active electromagnetic bearings, the state feedback control whose design is based on an approximate reduced-order model is effective. For a realization of this control, it is necessary to estimate unbalanced forces. One method which is often used for estimating the state is so called a state observer and this method is valid only for external forces being periodic. The other method is an inverse system in which external forces have not to be periodic in comparison with a state observer. In this paper using these estimator we design the state feedback controller which enables to improve modal damping of the mechanical system independently. In the experimental system applied above concept, it is verified that the feedback control which relates a modal velocity has a capability to improve a damping property of the mechanical system on each mode.

1. INTRODUCTION

An electromagnetic bearing has been used for a turbomolecular pump and a high speed spindle of manufacturing machine because of having the advantage of supporting a rotor without any contact and without lubrication. Mainly for economic reasons the PID controller with an analog circuit is often used in the active magnetic bearing system. Although this system is simple and non-expensive, the ability to improve the dynamic behavior of a flexible rotor is not sufficient.

It is well known that the state feedback control using state-space methods is more effective than the PID control. And a modal control design based on an approximate reduced-order model to a flexible rotor system has been investigated [1,2,3]. For realization of the state feedback control to a flexible rotor system, it is necessary to estimate external forces due to residual unbalances. One method which is often used for estimating the state is so called a state observer and this method is valid only for external forces being periodic. The other one is the method by using an inverse system in which external forces have not to be periodic in comparison with a state observer. In this paper, using these estimators we design the state feedback controller which enables to improve modal damping of the mechanical system independently.

First the state-space description of a flexible rotor suspended by active electromagnetic bearings will be given in modal form. Then the estimator design methods to obtain an estimate of the entire state and external forces is discussed. Also, in order to demonstrate the damping capability of the active electromagnetic bearing system, the experiments on the dynamic characteristics of a flexible rotor is carried out.

2. MODELING OF ELECTROMAGNETIC BEARING-ROTOR SYSTEM

Fig. 1 shows a system consisting of active electromagnetic bearings and rotors where the flexible rotor is supported by n_c electromagnetic bearings. The discretized equation of motion for a flexible rotor in x-direction is given by

$$M\ddot{x}+Kx=f \qquad \text{(1)}$$

with the vectors $x \in R^{n_p}$ as the generalized position coordinates of the rotor and $f \in R^{n_p}$ as the external force vector. The matrices $M \in R^{n_p \times n_p}$ and $K \in R^{n_p \times n_p}$ denote mass and stiffness matrices respectively. Without the gyroscopic terms, an analogous equation holds in y-direction. In the remainder of this section a description of the mechanical system will be made only in x-direction. The external force vector f consist of the actuator force vector $f_c \in R^{n_c}$ resulting from the n_c locally discrete active bearings and unbalanced force vector $f_p \in R^{n_p}$.

$$f=\Gamma f_c + f_p \qquad \text{(2)}$$

where the matrix $\Gamma \in R^{n_p \times n_c}$ describes the influence of the actuator forces on the different coordinates. The adequate arrangement of the coils for the bearing leads to a linear input/output characteristic between the magnetic force vector f_c, the displacement vector $x_c (=\Gamma^T x) \in R^{n_c}$ at discrete active bearings, and the control current vector $i_c \in R^{n_c}$

$$f_c = D x_c + E i_c \qquad \text{(3)}$$

with the force-displacement-factor matrix $D \in R^{n_c \times n_c}$ and the force-current-factor matrix $E \in R^{n_c \times n_c}$ both of which are diagonal.

After modal transformation and subsequent truncation

$$x = \Phi \xi \qquad \text{(4)}$$

where the matrix $\Phi \in R^{n_p \times m}$ contains the eigenvectors normalized with the mass matrix M, EQ.(1) becomes to the familiar form

$$\ddot{\xi} + \Omega^2 \xi = \Phi^T f \qquad \text{(5)}$$

where $\xi \in R^m$ denotes the vector of modeled modal coordinates, $\Omega \in R^{m \times m}$ the diagonal modal stiffness matrix whose elements are angular eigenfrequencies. Considering EQ.(2),(3) and (4), the right hand side of EQ.(5) can be written as

$$\Phi^T f = \Phi_c^T D \Phi_c \xi + \Phi_c^T E i_c + \Phi^T f_p \qquad \text{(6)}$$

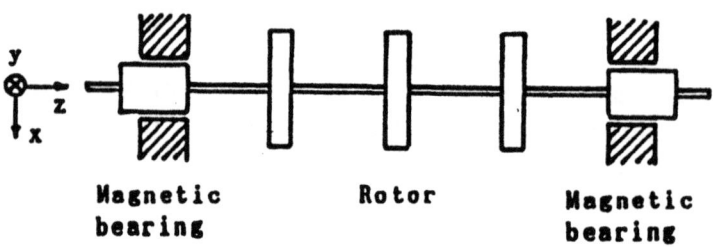

Fig.1. Analytical model

where the matrix $\Phi_c(=\Gamma^T\Phi)\in R^{n_c\times m}$ is submatrix of the matrix Φ of eigenvectors. Substitute EQ.(6) into EQ.(5), we have following equation.

$$\ddot{\xi}+(\Omega^2-\Phi_c^T D\Phi_c)\xi=\Phi_c^T E\, i_c+\Phi^T f_p \qquad \text{......(7)}$$

The voltage equation

$$e_c=L\,\dot{i}_c+R\,i_c \qquad \text{......(8)}$$

for the magnets of the bearing, characterized by a diagonal matrix $L\in R^{n_c\times n_c}$ whose elements are inductivities and a diagonal matrix $R\in R^{n_c\times n_c}$ whose elements are ohmic resistances is obtained. The equation of motion (7) and the voltage equation (8) can rewritten as state equation.

$$\dot{\hat{x}}=A_c\,\hat{x}+B_c u \qquad \text{......(9)}$$

where the state vector $\hat{x}\in R^n$ ($n=2m+n_c$) consists of displacements and velocities of a flexible shaft in modal coordinate and control currents and the input vector $u\in R^{m+n_c}$ is containing modal forces and control voltages.

$$\hat{x}^T=\{\xi^T,\dot{\xi}^T,i_c^T\}$$
$$u^T=\{f_m^T,e_c^T\}$$
$$f_m=\Phi^T f_p$$
$$A_c=\begin{bmatrix} 0 & I & 0 \\ -\Omega^2+\Phi_c^T D\Phi_c & 0 & \Phi_c^T E \\ 0 & 0 & -L^{-1}R \end{bmatrix}$$
$$B_c=\begin{bmatrix} 0 & 0 \\ I & 0 \\ 0 & L^{-1} \end{bmatrix}$$

where I denotes identity matrix and the modal force vector $f_m\in R^m$ characterizes the disturbing influences in modal coordinate caused by unbalance.

If the control is applied from the computer by a zero-order hold, EQ.(9) has an exact discrete representation as

$$\hat{x}(k+1)=A\,\hat{x}(k)+B u(k) \qquad \text{......(10)}$$

$$A=\exp[A_c T]$$

$$B=\int_0^T \exp[A_c t]d t\, B_c$$

with the sampling period T. On the one hand the measurement vector v(k) can be expressed as

$$v(k)=C\,\hat{x}(k) \qquad \text{......(11)}$$

3. ESTIMATOR DESIGN

3-1. Observer

The modal forces caused by unbalance are periodic. Then the x-directional modal force vector $f_{mx}\in R^m$ and y-directional modal force vector $f_{my}\in R^m$ can be written as

$$f_{mx}(k+1)=\cos(\omega T)f_{mx}(k)-\sin(\omega T)f_{my}(k)$$
$$\qquad\qquad\qquad \text{......(12)}$$
$$f_{my}(k+1)=\sin(\omega T)f_{mx}(k)+\cos(\omega T)f_{my}(k)$$

Considering EQ.(10) and EQ.(12), the state equation in x-y plane is assembled as

$$z(k+1) = A_{xy} z(k) + B_{xy} e(k) \quad \text{-----(13)}$$

where

$$z^T = \{ \hat{x}^T, f_{mx}^T, \hat{y}^T, f_{my}^T \}$$

$$e^T = \{ e_{cx}^T, e_{cy}^T \}$$

$$A_{xy} = \begin{bmatrix} A & B_1 & 0 & 0 \\ 0 & C_s & 0 & -S_n \\ 0 & 0 & A & B_1 \\ 0 & S_n & 0 & C_s \end{bmatrix}, \quad B_{xy} = \begin{bmatrix} B_2 & 0 \\ 0 & 0 \\ 0 & B_2 \\ 0 & 0 \end{bmatrix}, \quad B = [B_1 \, B_2]$$

$$C_s = \cos(\omega T) \cdot I, \quad S_n = \sin(\omega T) \cdot I$$

The measurement vector $v(k)$ is given as

$$v(k) = C_{xy} z(k) \quad \text{----- (14)}$$

Therefore the description of state observer leads to

$$z(k+1) = A_{xy} z(k) + B_{xy} e(k) + L\{v(k) - C_{xy} z(k)\} \quad \text{-----(15)}$$

where the feedback gain matrix L is chosen so that the roots of $A_{xy}-LC_{xy}$ fulfill the stability condition.

3-2. Inverse System

Using 1-delay inverse system, the estimates of $u(k)$ and $x(k)$ in EQ.(9) are obtained as

$$\overline{u}(k) = K_0 v(k) + K_1 v(k-1) - G \overline{\hat{x}}(k) \quad \text{----- (16)}$$

$$\overline{\hat{x}}(k+1) = A \overline{\hat{x}}(k) + B \overline{u}(k) \quad \text{----- (17)}$$

where "$\overline{}$" denotes the estimated value and the matrix K_0 and K_1 are the solution of the following linear equations.

$$\begin{bmatrix} 0 & B^T C^T \\ C^T & A^T C^T \end{bmatrix} \begin{bmatrix} K_1^T \\ K_0^T \end{bmatrix} = \begin{bmatrix} I \\ G^T \end{bmatrix} \quad \text{----- (18)}$$

The estimation errors can be defined as

$$\varepsilon_u(k) = \overline{u}(k) - u(k-1) \quad \text{----- (19)}$$

$$\varepsilon_x(k) = \overline{\hat{x}}(k) - \hat{x}(k-1) \quad \text{----- (20)}$$

Using EQ.(10),(11),(16),(17) and (18) this estimation errors leads to

$$\varepsilon_u(k) = -G \varepsilon_x(k) \quad \text{----- (21)}$$

$$\varepsilon_x(k) = (A - BG) \varepsilon_x(k-1) \quad \text{----- (22)}$$

If the feedback gain matrix $G \in R^{(m+n_c) \times n}$ is chosen so that the roots of $A-BG$ fulfill the stability condition, the estimation errors converge to zero as time is going by. Therefore from EQ.(19) and (20) the estimates converge to 1-sample delayed values.

Choosing displacements and velocities in m-locations of the flexible shaft and control currents of n_c magnetic circuits as elements of measurement vector $v(k)$ in EQ.(11), the sufficient condition to find the matrix G, K_0 and K_1 is fulfilled.

4. CONTROL DESIGN

In order to improve modal damping of the mechanical system independently, the control input depend linearly on the measurements of displacement, velocity and current of magnetic circuits and estimated modal velocity. A control law is expressed as

$$e_c(k) = e_0(k) + e_m(k) \qquad\qquad \text{........ (23)}$$

$$e_0(k) = -\left\{ g_P x_c(k) + g_I \sum_{j=0}^{k} x_c(j) + g_d \dot{x}_c(k) + g_c i_c(k) \right\}$$

$$= -\left\{ [g_P \Phi_c \ g_d \Phi_c \ g_c \Phi_c] \hat{x}(k) + g_I \sum_{j=0}^{k} x_c(j) \right\}$$

$$e_m(k) = -G_m \hat{x}(k)$$

if the collocation-requirement is fulfilled and time-lag for calculation can be ignored. In EQ.(23), $e_0(k) \in R^{n_c}$ is the feedback voltage based on direct output with scalar gains g_P, g_I, g_d and g_c. And $e_m(k) \in R^{n_c}$ is the feedback voltage based on estimated modal velocity with gain matrix $G_m \in R^{n_c \times n}$. The reason why EQ.(23) is used to a control law is that the direct use of measured values which contain less error than estimates as far as possible will result in good control. Considering the coefficient matrix of $\hat{x}(k)$ in EQ.(23), the feedback control relating modal velocities has a capability to improve damping properties of the mechanical system on each mode compared with the direct feedback control using only measured values.

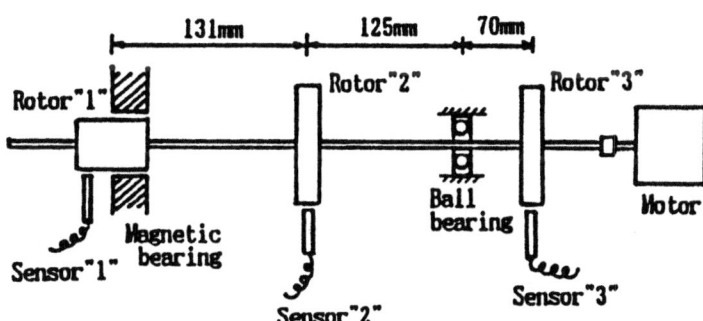

Fig.2. Experimental apparatus

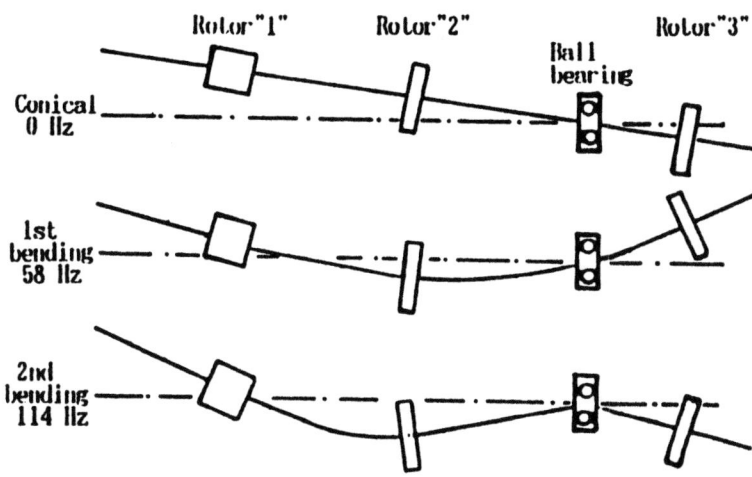

Fig.3. Eigenmodes and eigenfrequencies

5. EXPERIMENT (MODAL CONTROL WITH OBSERVER)

A flexible rotor is supported by an active electromagnetic bearing and a ball bearing (Fig.2). The rotor consists of a flexible shaft whose diameter and length are 6mm and 470mm respectively and three rigid mounted discs whose mass are 0.15Kg respectively. Using flexible coupling the rotor is joined to a motor. The calculated eigenfrequencies and eigen modes of the flexible rotor are shown in Fig.3. Considering the time required in calculation of the state vector with observer, the sampling period to attain with this implementation is about 625 μs. Then the dimension of the truncated design model is chosen to be m=3. Eddy current sensors are used for measuring displacement of rotors and this displacement signal is transformed to the velocity through an analog circuit for derivative action.

In our experiments, choosing the measurement vector in EQ.(14) as the following

$$\mathbf{v}^T = \{x_1, \dot{x}_1, \ddot{x}_1, i_{cx}, x_3, y_1, \dot{y}_1, \ddot{y}_1, i_{cy}, y_3\} \quad \text{------} \quad (24)$$

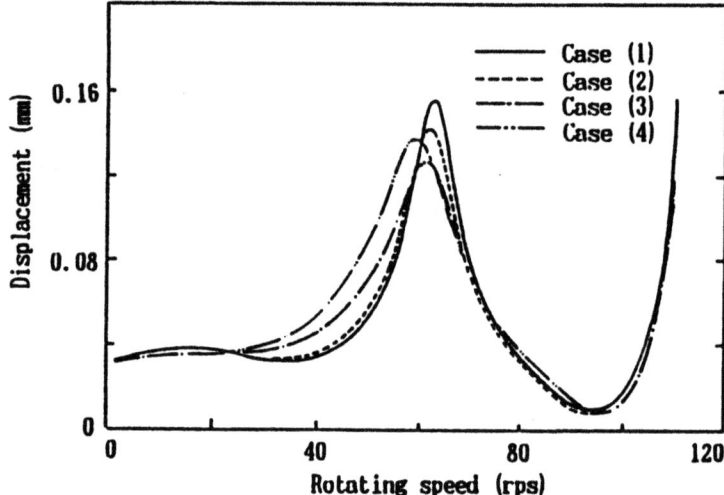

Fig.4. Unbalance response with a direct output feedback

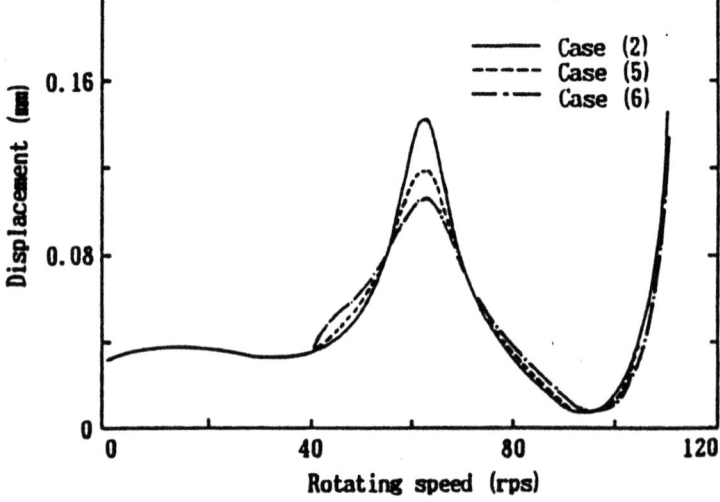

Fig.5. Unbalance response with a feedback of the modal velocity

it is found out that the system is observable. So only sensor "1" and sensor "3" are used for measurements. Feedback gains used in experiments are listed in table 1.

First, unbalance response with a direct output feedback is measured (Fig.4). While the eigenfrequency of conical mode is well damped, the peak amplitude at the critical speed of the first bending mode has a minimum value for a certain optimal value of the derivative feedback gain. Second, unbalance response with a feedback of the modal velocity estimated by observer is measured (Fig.5). The peak amplitude of the first bending mode is reduced to an extent below the direct output feedback being used. The modal damping ratio at this critical speed is calculated (Fig.6). In case of the direct output feedback the modal damping ratio for the first bending mode has a optimum value against derivative feedback gain. On the one hand, in case of the modal velocity feedback, it is found out that the damping ratio for the first mode increases monotonously over the level achieved with the direct output feedback. Both experimental and calculated results demonstrate the performance of this approach.

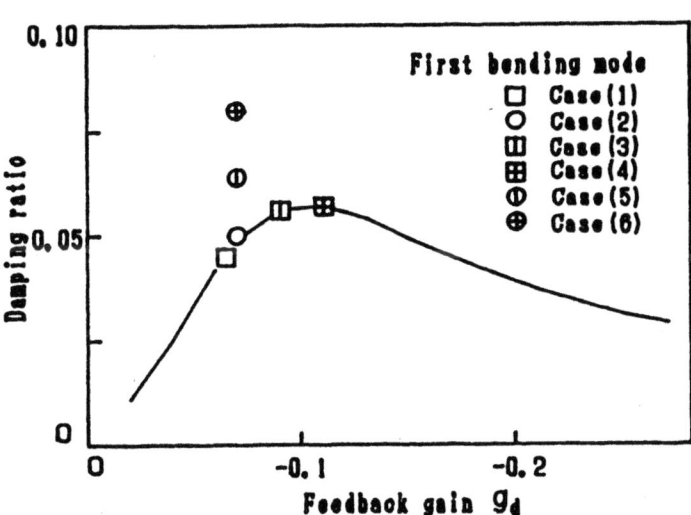

Fig.6. Modal damping ratio (calculated)

Table 1. Gains of designed controller

| | g_p | g_i | g_d | g_c | G_m | | |
					$\dot{\xi}_1$	$\dot{\xi}_2$	$\dot{\xi}_3$
Case (1)	-25	-0.01	-0.065	5	0	0	0
Case (2)	-25	-0.01	-0.07	5	0	0	0
Case (3)	-25	-0.01	-0.09	5	0	0	0
Case (4)	-25	-0.01	-0.11	5	0	0	0
Case (5)	-25	-0.01	-0.07	5	0	-0.02	0
Case (6)	-25	-0.01	-0.07	5	0	-0.04	0

342

6.CONCLUSIONS

A modal velocity feedback control system based on an approximate reduced-order model has been designed to improve damping properties of a flexible rotor suspended by active electromagnetic bearings. In this investigation a state observer and an inverse system are employed to estimate the state and external forces due to residual unbalances. To demonstrate the damping capability of the designed controller, the experiments on the unbalance response of a flexible rotor is carried out and it is verified that the feedback control which relates modal velocities has a capability to improve modal damping of the mechanical system on each mode.

REFERENCES

[1] Balas, M.J., "Feedback control of flexible systems", IEEE Trans. Automat. Contr., vol.AC-23, pp.673-679, 1978.
[2] Salm, J. and Schweitzer, G., "Modelling and control of a flexible rotor with magnetic bearings", IMechE C277/84, pp.553-561,1984.
[3] Salm, J., "Active electromagnetic suspension of an elastic rotors: modelling, control, and experimental results", ASME 11th Vib. Conf. Rotating Mathinery Dynamics., vol.1, pp.141-149, 1987.

Reduction of Periodic Rotating Vibration Using the Repetitive Control in the Rotating Servo System

Zhao-Wei Zhong

Materials Fabrication Laboratory, The Institute of Physical and Chemical Research
Hirosawa 2-1, Wako-shi, Saitama, 351-01, Japan

ABSTRACT

In rotating servo systems, disturbances which cause rotor vibrations are often periodic. Repetitive control is useful in reducing such periodic rotating vibrations. However, the typical repetitive control has a peculiar weak point. The author tried to solve this problem by means of a new iterative learning compensator. To realise this proposed compensator, several methods including the use of Butterworth band-pass filters, infinite impulse response digital filters, and the fast Fourier transform technique were investigated. A numerical example was used, and numerical simulations were also carried out to confirm the effectiveness of the proposed method. The results proved that it is practical and useful to apply this repetitive control without deteriorating the controlled output of the rotating servo system at any frequency.

1. INTRODUCTION

In rotating servo systems, disturbances which cause rotor vibrations are often periodic. For instance, the disturbance causing rotational speed fluctuation in motors usually synchronises with the rotation of the armature. The torque fluctuation of load causing rotating vibration of the worktable in a NC gear cutting machine mainly synchronises with both one rotation and one knife edge of the cutting tool. Repetitive control is useful in reducing such periodic rotating vibrations.

Repetitive control was originally proposed by Inoue, Nakano, Kubo et al. (1980) to construct a servo system, whose reference commands to be tracked are periodic signals of a fixed period L, using the error of the preceding periods. After that, many studies were carried out (Hara 1986, Hara, Yamamoto, Omata et al. 1988).

This method is also useful in rejecting disturbances whose frequencies are the same or integral times higher than ω_0, the basic frequency of the iterative learning compensator, where $\omega_0 = 2\pi/L$, L is the period of disturbances and/or the periodic signals.

However, it is known that for disturbances whose frequencies are not the same or integral times higher than ω_0, the rejecting effect is small. Moreover, at certain frequencies, the influence of disturbances on a repetitive control system which has a basic servo system and the typical iterative learning compensator, on the contrary, is larger than that on the same basic servo system without the iterative learning compensator (Gotou, Matsubayashi, Miyazaki et al. 1987).

The author tried to solve this problem by means of a new iterative learning compensator. To realise this proposed compensator, the use of Butterworth band-pass filters, infinite impulse response digital filters, and a fast Fourier transform technique was investigated. A numerical example was used and numerical simulations were also carried out to confirm the effectiveness of the proposed method. The results proved that it is practical and useful to apply this repetitive control without deteriorating the controlled output of the servo system at any frequency.

In this paper, the principle of the proposed method is explained theoretically. The results of a numerical example that proved the effectiveness of the rejection method proposed are reported.

2. REPETITIVE CONTROL SERVO SYSTEM

In this paper, the repetitive control is discussed based on a basic SISO servo system, as shown in Fig. 1. In Fig. 1, the symbols represent the following:

$Y(s)$ = Output of the system
$R(s)$ = Reference input
$U(s)$ = Repetitive control
$E(s)$ = Error
$P(s)$ = Control object (Rotating system)
$T_L(s)$ = Disturbance
$C_1(s)$ = Dynamic compensator 1
$C_2(s)$ = Dynamic compensator 2

It is assumed that in the basic servo system, the system is asymptotically stable and the output y(t) tracks the reference input r(t).

From Fig. 1, $Y(s)$ is expressed as a linear combination as follows:

$$Y(s)=G_r(s)R(s)+G_L(s)T_L(s), \tag{1}$$

where

$$G_r(s)=C_1(s)G_L(s) \tag{2}$$

$$G_L(s)=P(s)/\Delta \tag{3}$$

$$\Delta = 1+P(s)[C_1(s)+C_2(s)]. \tag{4}$$

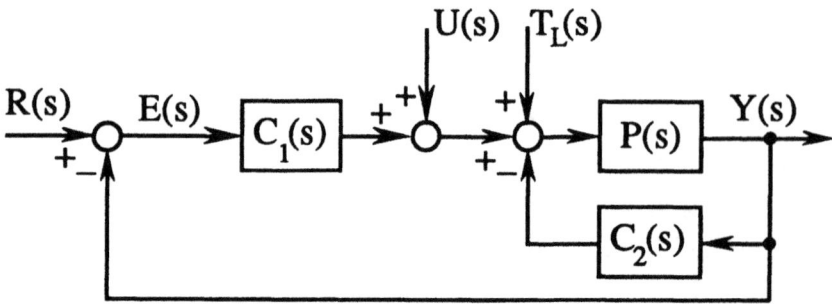

Fig. 1.   Scheme of the servo system.

For case I of the repetitive control, let

$$U(s)=H_r(s)C_1(s)E(s), \tag{5}$$

and for case II, let

$$U(s)=H_r(s)E(s), \tag{6}$$

where $H_r(s)$ is an iterative learning compensator given by

$$H_r(s)=H_u(s)H_0(s)/[1-H_0(s)]. \tag{7}$$

$H_u(s)$ is for compensating the gain and phase of the iterative learning compensator to guarantee the stability of the repetitive control system and convergency of the error. H_0 is the central component of the iterative learning compensator that we mainly discuss in this paper.

In the closed loop of the repetitive control system, the following equations are obtained:
for case I,

$$E(s) = \frac{[\Delta^{-1}+C_2(s)G_L(s)]R(s)-G_L(s)T_L(s)}{1+H_0(s)[G_r(s)H_u(s)-1]}[1-H_0(s)] \tag{8}$$

and for case II,

$$E(s) = \frac{[1-G_r(s)]R(s)-G_L(s)T_L(s)}{1+H_0(s)[G_L(s)H_u(s)-1]}[1-H_0(s)] . \tag{9}$$

From Eqs. (8) and (9), it is obvious that under the assumption mentioned above, if $H_0(s)$ satisfies Eq. (10), the influences, on the error, of $r(t)$ and $T_L(t)$ whose frequencies are the same or integral times higher than ω_0, become 0 in the steady state:

$$|1-H_0(j\omega_n)|=0, \tag{10}$$

where $\omega_n=n\omega_0$, $n=1,2,3,\dots$.

From the assumption, a sufficient stable condition of this repetitive control system is

$$\|H_0(s)[1-G_i(s)H_u(s)]\|_\infty<1, \tag{11}$$

where

$$G_i(s) = \begin{cases} G_r(s) & \text{(In case I)} \\ G_L(s) & \text{(In case II)}. \end{cases} \tag{12}$$

Here we have two choices, $G_L(s)$ and $G_r(s)$, to satisfy the sufficient stable condition and to obtain a better characteristic of disturbance rejection.

The typical $H_0(s)$ is

$$H_0(s)=\exp(-Ls) . \tag{13}$$

For the frequencies $\omega=\omega_n$, from Eqs. (8) or (9) and (10), the error becomes 0 in the steady state. However, for the other frequencies,

$$|1-H_0(j\omega)|\leq 2. \tag{14}$$

Particularly, when $\omega=\omega_0(2n-1)/2$, $n=1,2,3,\ldots$,

$$1-H_0(j\omega)=2 . \tag{15}$$

As shown by Eqs. (8) and (9), since the error $E(s)$ is in proportion to $[1-H_0(s)]$, $|1-H_0(j\omega)|$ can be used to approximately estimate the characteristic of disturbance rejection. Equations (14) and (15) mean, for disturbances $\omega\neq\omega_n$, the effect of disturbance rejection is small. Moreover, at certain frequencies, the influence of disturbances on the repetitive control system which has a basic servo system and the typical iterative learning compensator, on the contrary, is larger than that on the same basic servo system without the iterative learning compensator. This is a big problem (Gotou, Matsubayashi, Miyazaki et al. 1987).

3. PROPOSED METHOD

To solve the problem mentioned in the previous chapter, instead of Eq. (13), let $H_0(s)$ be

$$H_0(s) = a\,F\,\exp\,(-Ls) + b \tag{16}$$

and

$$F = \begin{cases} 1 & (\omega=\omega_n) \\ 0 & (\omega\neq\omega_n) \end{cases} \tag{17}$$

$$a +b =1. \tag{18}$$

The following equation can be obtained:

$$|1-H_0(j\omega)| = \begin{cases} 0 & (\omega=\omega_n) \\ |\,a\,| & (\omega\neq\omega_n). \end{cases} \tag{19}$$

Equation (10) is satisfied and the control error becomes 0 in the steady state. Choose a as

$$0<|\,a\,|\leq 1. \tag{20}$$

Then for any frequency,

$$|1-H_0(j\omega)|\leq|\,a\,|\leq 1. \tag{21}$$

Thus, a repetitive control without deteriorating the controlled variable at any frequency can be realised.

Also, instead of Eq.(17), let F be

$$F(j\omega)=1 \quad (\omega=\omega_n) \atop |F(j\omega)|\leq1 \quad (\omega\neq\omega_n).} \tag{22}$$

Equation (10) is also satisfied so that the steady-state error becomes 0. Since

$$|1-F(j\omega)\exp(-jL\omega)|\leq2 \tag{23}$$

$$|1-H_0(j\omega)|=|a||1-F(j\omega)\exp(-jL\omega)|, \tag{24}$$

choose a as

$$0<|a|\leq0.5 ; \tag{25}$$

then for any frequency,

$$|1-H_0(j\omega)|\leq1. \tag{26}$$

4. NUMERICAL EXAMPLE

The basic system represented by Eqs. (27), (28) and (29) was used as a numerical example for confirming the effectiveness of the proposed method.

$$P(s)=68.4/(s+22.1) \tag{27}$$

$$C_1(s)=100(s+22.1)/(68.4s) \tag{28}$$

$$C_2(s)=0. \tag{29}$$

Let K_d, K_u be constants and

$$H_u(s)=K_d s+K_u. \tag{30}$$

To realise the filter shown by Eq.(22), Butterworth band-pass filters given as follows were used:

$$F(s)=F_1(s)+F_2(s)+...+F_n(s) \atop F_n(s)=1/(\lambda^3+2\lambda^2+2\lambda+1) \atop \lambda=(s^2+\Omega_{n1}\Omega_{n2})/[s(\Omega_{n2}-\Omega_{n1})] ,} \tag{31}$$

where Ω_{n1} and Ω_{n2} are, respectively, the lower frequency and upper frequency that define the band-pass width. The central frequencies of the band-pass filters are ω_1, $\omega_2,...,\omega_n$.

In real systems, we often encounter the case in which it is sufficient for practical purposes to reject the influences of some special frequencies of T_L. For example, we can choose $n=1, 2, 8$ according to practical needs. Therefore, from now on, the computation is carried out assuming $n=1, 2$ for simplification.

Let the transfer function from the disturbance T_L to the output y be $G_c(s)$. $G_c(s)$ is given by

$$G_c(s) = \frac{Y(s)}{T_L(s)} = \frac{G_L(s)}{1+H_0(s)[G_i(s)H_u(s)-1]}[1-H_0(s)] . \qquad (32)$$

If the repetitive control is not used, $G_c(s)=G_L(s)$.

In chapters 2 and 3, $|1-H_0(j\omega)|$ is used to approximately estimate the characteristic of disturbance rejection. It is necessary for practical applications to compute $|G_c(j\omega)|$, to exactly evaluate the characteristic of disturbance rejection.

Figures 2, 3 and 4 show examples of the computation of $|G_c(j\omega)|$, where Eq. (6) is used to calculate the repetitive control u, $K_d=0,005$, $K_u=1$ and $L=0.1$.

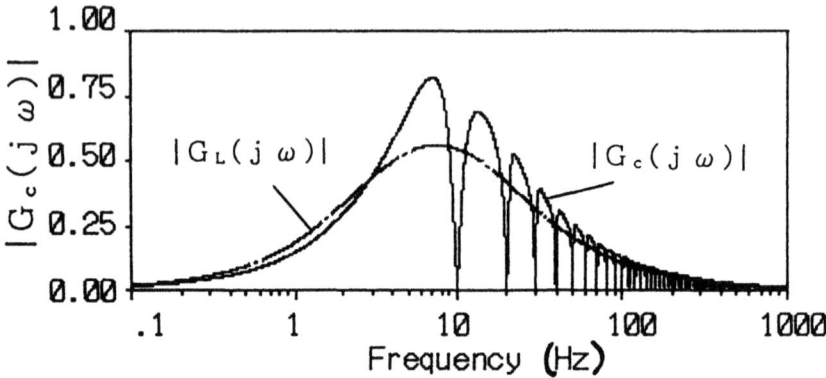

Fig. 2. Computation of $|G_c(j\omega)|$ with the typical compensator (using Eq. (6)).

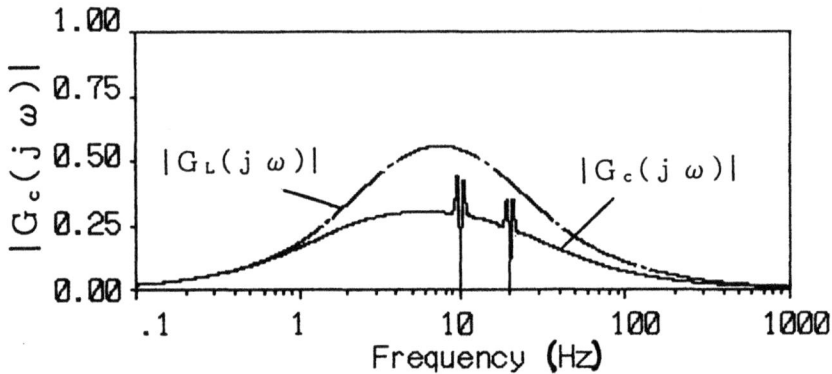

Fig. 3. Computation of $|G_c(j\omega)|$ with the proposed compensator (a=0.4, using Eqs. (6) and (31)).

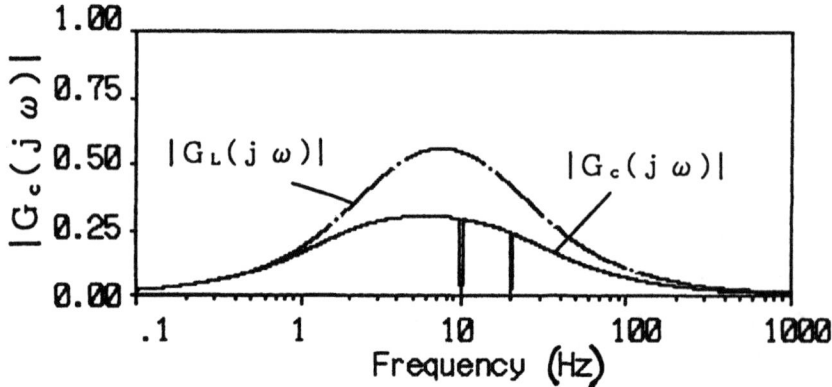

Fig. 4. Computation of $|G_c(j\omega)|$ with the proposed compensator (a=0.4, using Eqs. (6) and (17)).

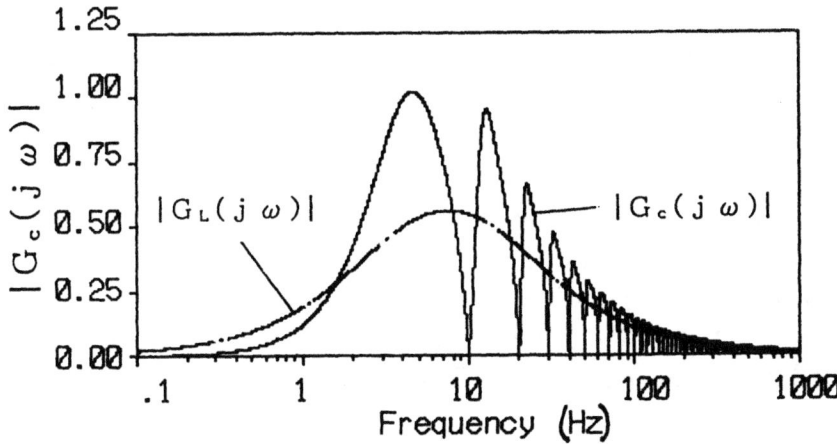

Fig. 5. Computation of $|G_c(j\omega)|$ with the typical compensator (using Eq. (5)).

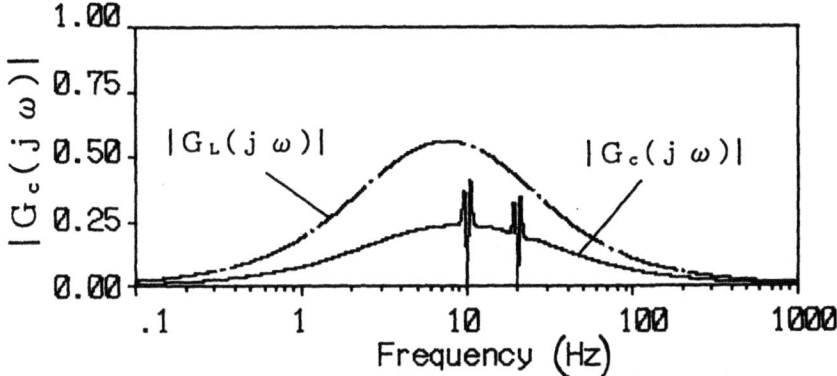

Fig. 6. Computation of $|G_c(j\omega)|$ with the proposed compensator (a=0.4, using Eqs. (5) and (31)).

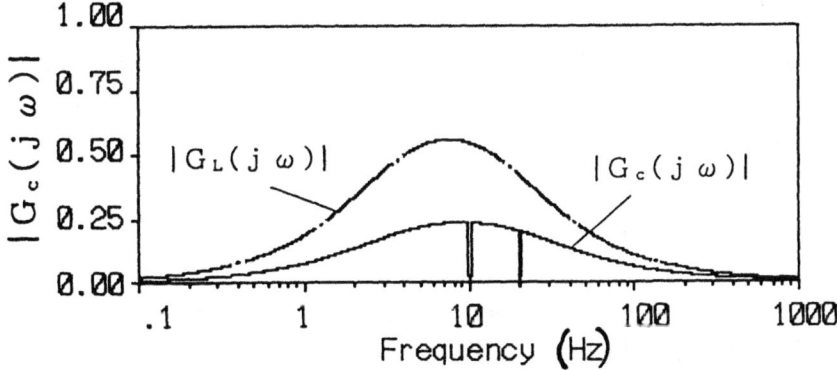

Fig. 7. Computation of $|G_c(j\omega)|$ with the proposed compensator (a=0.4, using Eqs. (5) and (17)).

We can see, when the typical $H_0(s) = \exp(-Ls)$ is used for repetitive control, $|G_c(j\omega)|=0$ at $\omega=\omega_n$; however, there are some frequencies $\omega\neq\omega_n$ where $|G_c(j\omega)|>|G_L(j\omega)|$. That is, at these frequencies, the characteristic of disturbance rejection becomes worse due to the repetitive control. When the filter of Eq. (31) or Eq. (17) is used for repetitive control, as shown in Figs. 3 and 4, $|G_c(j\omega)|=0$ at $\omega=\omega_n$; moreover, $|G_c(j\omega)|\leq|G_L(j\omega)|$ at any frequency over the whole frequency range. Thus, a repetitive control without deteriorating the controlled variable at any frequency can be realised.

Figures 5, 6 and 7 show other examples of the computation of $|G_c(j\omega)|$, where also $K_d=0,005$, $K_u=1$ and $L=0.1$ but Eq. (5) is used to calculate the repetitive control u.

350

Comparison with Figs. 2, 3 and 4 shows that the statement mentioned above for Figs. 2, 3 and 4 is also true for Figs. 5, 6 and 7. However, the characteristics of disturbance rejection are different. This means that we can choose Eq. (5) or (6) to calculate the repetitive control u to obtain a better characteristic of disturbance rejection. In this example, using Eq. (5) is better than using Eq. (6).

The filter of Eq. (17) was realised by using fast Fourier transform (FFT) technique, and simulation experiments were carried out. It usually takes time to complete the computation of FFT. However, it is possible for practical applications to realise this filter using FFT by taking advantage of the characteristic of the repetitive control in which the control u is computed using the error information of the preceding periods, or by using hardware and so on. Also, infinite impulse response digital filters were constructed by transforming the filters of Eq. (31) to the z domain, and simulation experiments were performed. Since the order of the digital filters is six, the computing time is very short and it is sufficient for practical applications. The results of the simulations also confirmed the effectiveness of the rejection method proposed in this paper.

5. CONCLUSIONS

Repetitive control is useful to reject periodic disturbances. However, it is known that at certain frequencies, the influence of disturbances on a repetitive control system having a basic servo system and the typical iterative learning compensator is larger than that on the same basic servo system without the iterative learning compensator. The author tried to solve this problem by means of a new iterative learning compensator. To realise this proposed compensator, the use of Butterworth band-pass filters, infinite impulse response digital filters, and a fast Fourier transform technique was investigated. A numerical example was used, and numerical simulations were also carried out to confirm the effectiveness of the proposed method. The results proved that it is practical and useful to apply this repetitive control without deteriorating the controlled output of the servo system at any frequency.

6. ACKNOWLEDGMENT

The author is grateful that this study was supported by the Basic Science Special Study Funds from the Institute of Physical and Chemical Research.

7. REFERENCES

Gotou M, Matsubayashi S, Miyazaki F et al. (1987) A robust servo system with an iterative learning compensator and a proposal of multi-period learning compensator. Systems and Control, Vol.31, No.5:367-374

Hara S (1986) Repetitive control. Jour. of SICE, Vol.25, No.12:1111-1119

Hara S, Yamamoto Y, Omata T et al. (1988) Repetitive control system: a new type servo system for periodic exogenous signals. IEEE Trans. Automat. Contr., Vol.AC-26:659-667

Inoue T, Nakano M, Kubo T et al. (1980) High accuracy control for magnet power supply of proton synchrotron in recurrent operation. Trans. IEE Japan, Vol.100, No.7:16-22

Steady State Analysis of a Novel Variable Impedance Oil Film Bearing

Y. Fang,   M. J. Goodwin,   P. J. Ogrodnik and M. P. Roach

Department of Mechanical and Computer-Aided Engineering, Staffordshire Polytechnic, Beaconside, Stafford, ST18 0AD United Kingdom

Abstract

This paper describes a theoretical investigation of the steady state characteristics of a novel variable impedance hydrodynamic oil film bearing, which it is proposed would result in lower machine vibration than is obtained with conventional bearings.   The novel bearing design is similar to a conventional bearing, but has two recesses machined into the bearing surface, each of which is connected to a hydraulic accumulator. The bearing dynamic characteristics can be controlled by opening and closing remote controlled valves between the bearing surface and the accumulators.   In this paper, a theoretical model of the new bearing is introduced and theoretical results of its steady state characteristics are presented. These are compared with those of a conventional bearing.

Introduction

Many modern large rotating machines are designed to operate at speeds in excess of one or more critical speeds. This can create problems of high levels of vibration associated with operation at or close to a critical speed, high levels of noise and corresponding high levels of dynamic bending stresses in the shaft. The designer has to predict the machine's critical speeds with considerable accuracy in order to avoid their coinciding with the normal operating speed of the machine. Unfortunately there is often inadequate knowledge of the dynamic characteristics of the foundations on which the machine is to be installed at the design stage. The result of this situation is that from time to time some machines have been constructed and found to have excessive critical speed related vibration during normal operation.

In order to solve the aforementioned problems,   it would be desirable for the bearing dynamic characteristics to be easily controlled during the commissioning of a machine so that the machine critical speeds can be tuned away from the operating speed.   It is also desirable that the bearings have a low value of dynamic stiffness and a moderate amount of support damping which will give rise to a near optimum unbalance response over the entire speed range of the machine [1,2] so that the machine has low levels of vibration when

352

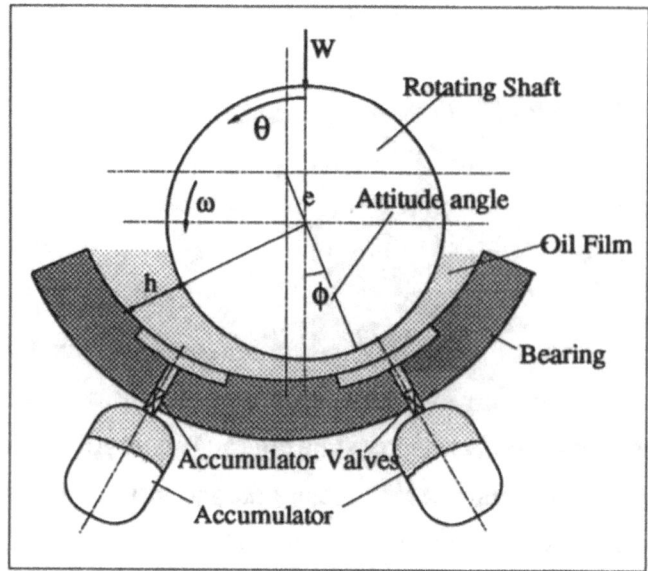

Figure 1 New Bearing Arrangement

it passes through its critical speeds.

However, for conventional bearings, it is extremely difficult to obtain low dynamic stiffness values and is impossible to change the dynamic characteristics during operation. One alternative is to use electrically controlled magnetic bearings where the magnetic flux is controlled by manipulating the magnitude and phase of the electrical supply to the electromagnets so as to ensure that the stiffness and damping of the bearings takes on suitable values [3,4]. However these bearings are expensive and so have not yet found widespread application. An inexpensive alternative is the variable impedance hydrodynamic oil film bearing developed recently, the dynamic characteristics of which are controlled by means of an accumulator connected to the oil film. This kind of bearing has a low value of oil film stiffness and a moderate value of oil film damping, such that machine unbalance response is lower than that obtained with conventional bearings.

This paper presents investigations concerned with the characteristics of a new form of variable impedance hydrodynamic oil film bearing which has two accumulators connected to recesses in the bearing surface, instead of just one investigated earlier [5]. A diagram of the new arrangement is shown in Figure 1. The purpose of the new proposed design is to provide more control of bearing stiffness and damping by a combination of opening and closing of two accumulator valves. The main concern of this paper is the steady load carrying capacity of the new bearing. The accumulators connected to the bearing recesses are for bearing dynamic control. Whether they are connected to the bearing does not affect the bearing's steady state performance. A theoretical model of the novel bearing is presented in this paper, and the pressure distribution and load capacity of the new bearing are then evaluated and compared with those of a conventional bearing and of the bearing with one recess .

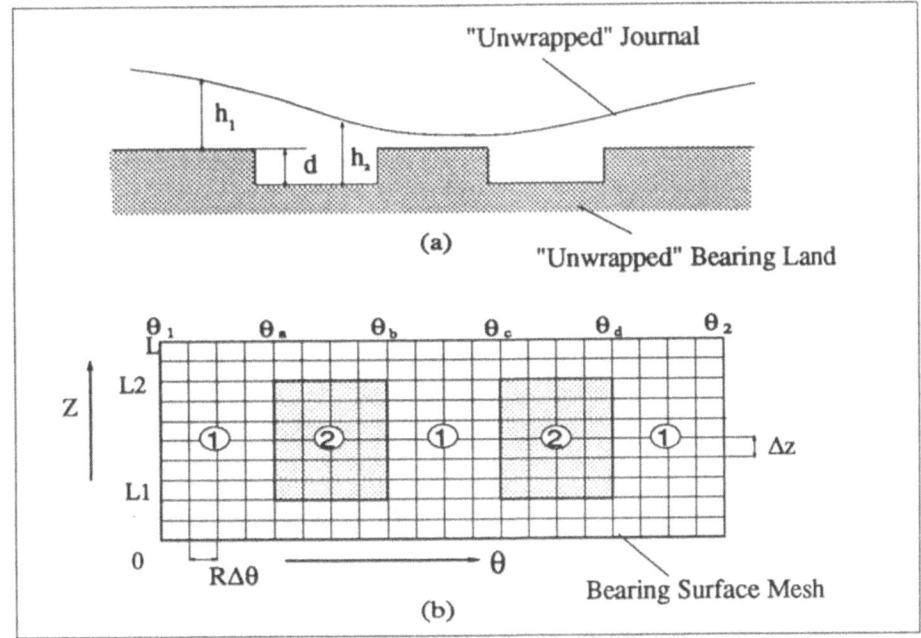

Figure 2 New Bearing Surface

Theoretical Model and Solution of the New Bearing

For hydrodynamic lubrication the oil film pressure distribution can be obtained by solving Reynolds' equation which may be written,

$$\frac{\partial}{R^2\partial\theta}\left(h^3\frac{\partial p}{\partial\theta}\right)+\frac{\partial}{\partial z}\left(h^3\frac{\partial p}{\partial z}\right)=6\mu U\frac{\partial h}{R\partial\theta}+12\mu\frac{\partial h}{\partial t}$$

(1)

where p is the oil film pressure, θ and z are circumferential and axial coordinates respectively, U is the tangential velocity of the journal surface, R is the journal radius, μ is the lubricant dynamic viscosity and h is the film thickness   given by

$$h=\begin{cases}c+x\cos\theta+y\sin\theta & (\text{region }1)\\ c+d+x\cos\theta+y\sin\theta & (\text{region }2)\end{cases}$$

(2)

where the region 1 is the bearing land area and region 2 is the recess area. The bearing surface was divided into a finite difference mesh as shown in Figure 2 where θ_1 and θ_2 are the bearing land starting and ending coordinates. θ_a and θ_c are recess starting coordinates, θ_b and θ_d are recess ending coordinates respectively. L_1 and L_2 indicate the starting and ending positions of the recesses in the z direction. $R\Delta\theta$ and Δz are the grid increments in the θ and z direction respectively.   The finite difference solution of Reynolds equation at the nth node of the bearing surface mesh is ,

$$p_n = \frac{\frac{3}{2}\Delta\theta\, h^2\frac{\partial h}{\partial\theta} + h^3}{2h^3\left[1+\left(\frac{R\Delta\theta}{\Delta z}\right)^2\right]}\, p_{n+1} + \frac{-\frac{3}{2}\Delta\theta\, h^2\frac{\partial h}{\partial\theta} + h^3}{2h^3\left[1+\left(\frac{R\Delta\theta}{\Delta z}\right)^2\right]}\, p_{n-1}$$

$$+ \frac{h^3 R^2\left(\frac{\Delta\theta}{\Delta z}\right)^2}{2h^3\left[1+\left(\frac{R\Delta\theta}{\Delta z}\right)^2\right]}(p_{n+m} + p_{n-m}) - \frac{6(R\Delta\theta)^2\left(\frac{\mu U}{R}\frac{\partial h}{\partial\theta} + 2\mu\frac{\partial h}{\partial t}\right)}{2h^3\left[1+\left(\frac{R\Delta\theta}{\Delta z}\right)^2\right]} \qquad (3)$$

where $\quad \frac{\partial h}{\partial\theta} = -x\sin\theta + y\cos\theta \qquad\qquad \frac{\partial h}{\partial t} = \dot{x}\cos\theta + \dot{y}\sin\theta$

For a $N \times M$ node bearing surface mesh, equation (3) represents $N \times M$ simultaneous equations. These were solved by an iterative numerical method. The bearing surface was divided into a finite difference mesh such that all the recess boundary nodes were on the mesh grid. The value of the journal attitude angle under steady load conditions for various values of eccentricity ratio was obtained by noting when the net horizontal force generated in the lubricant film was zero. The steady state analysis was carried out for a range of eccentricity ratios.

In order to investigate and compare the affects of recess dimensions, and their positions in the bearing surface, on the bearing load carrying capacity, a

Bearing Number	θ_a	θ_b	θ_c	θ_d	L_1 (mm)	L_2 (mm)
Bearing 1	24°	48°	72°	96°	14	36
Bearing 2	16°	52°	68°	104°	10	40
Bearing 3	30°	60°	75°	105°	14	36
Bearing 4	14°	44°	60°	90°	14	36
Bearing 5	36°	84°	—	—	14	36
Bearing 6	24°	96°	—	—	10	40
Bearing 7	46°	106°	—	—	14	36
Bearing 8	14°	74°	—	—	14	36

Table 1 Bearing Parameters

120^0 partial bearing with different recess dimensions and positions in the bearing surface was analyzed. The detailed geometry of these bearings is listed in Table 1. The recess area of bearing 1 is about 17.6% of the total bearing effective surface, and its two recesses were arranged symmetrically to the bearing axial central line. The recesses of bearing 2 were arranged on the same position on the bearing surface as those of bearing 1, however its recess area was 36% of the bearing surface. Comparison of bearing 1 and bearing 2 will indicate the affects of recess dimensions on the bearing load carrying ability. Bearing 3 and bearing 4 had the same recess dimensions, but their two recesses were arranged on different positions on the bearing surface. For the purpose of comparison, another four bearings with only one recess were also analyzed and they are refered to as bearing 5, bearing 6, bearing 7 and bearing 8. Their geometry is also indicated in table 1.

Results and Discussion

In order to facilitate the comparison of the load carrying capacity of the novel design bearing with that of the similar one recess bearing and conventional bearing, the theoretical results of attitude angle, Sommerfeld number and the pressure profile are presented. Figure 3 and Figure 4 are the bearing central plane pressure profile for bearing 1 with $\varepsilon = 0.5$ and

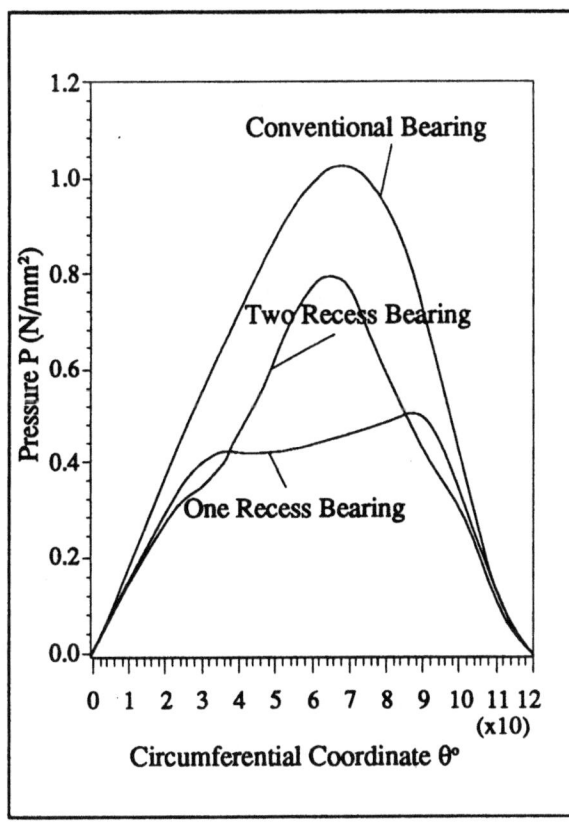
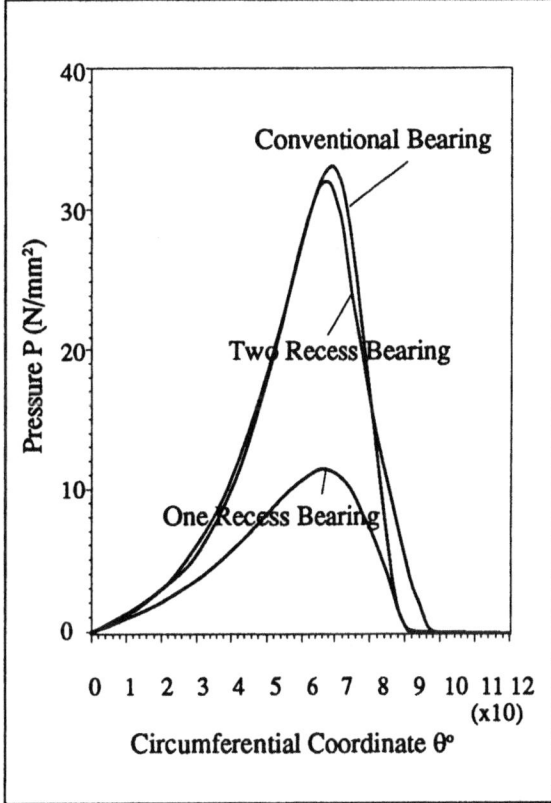

Figure 3 Bearing Central Plane
Pressure Profile at $\varepsilon = 0.5$

Figure 4 Bearing Central Plane
Pressure Profile at $\varepsilon = 0.9$

356

ε=0.9 respectively. It shows that the pressure distribution of the bearing is dictated by the recess numbers, dimensions and its position on the bearing surface. It can be seen that the maximum pressure of the recessed bearings is lower than that of the conventional bearing. This can be explained by the fact that the 'wedge' action in the recess area is eliminated because the recess depth value is much bigger than the bearing land film thickness. This means that the recess on the bearing surface reduces the effective bearing land 'wedge' area, and hence reduces the peak oil film pressure. The pressure value in the recess area is not zero because it is an enclosed chamber, its pressure is determined by the pressure values on the recess boundaries. It is interesting to note that the recess in the bearing trailing area can prevent cavitation happening when the shaft is running at high eccentricity ratios as shown in Figure 4.

Figures 5 and 6 illustrate the variation of attitude angle and Sommerfeld number with eccentricity ratio for bearings 1 and 2 . It can be seen that the one recess bearing has a very similar attitude angle to the two recess bearing when the bearings operate at low eccentricity ratios (ε<0.3) and that recessed bearings always operate at higher attitude angles than conventional bearings. It can also been seen that bearing 2 is operating at a higher attitude angle than that of bearing 1 at the same eccentricity ratio. As a consequence of the reduced lubricant pressure there is also a corresponding reduction in bearing load-carrying capacity. This can be seen in Figure 6 where for a particular

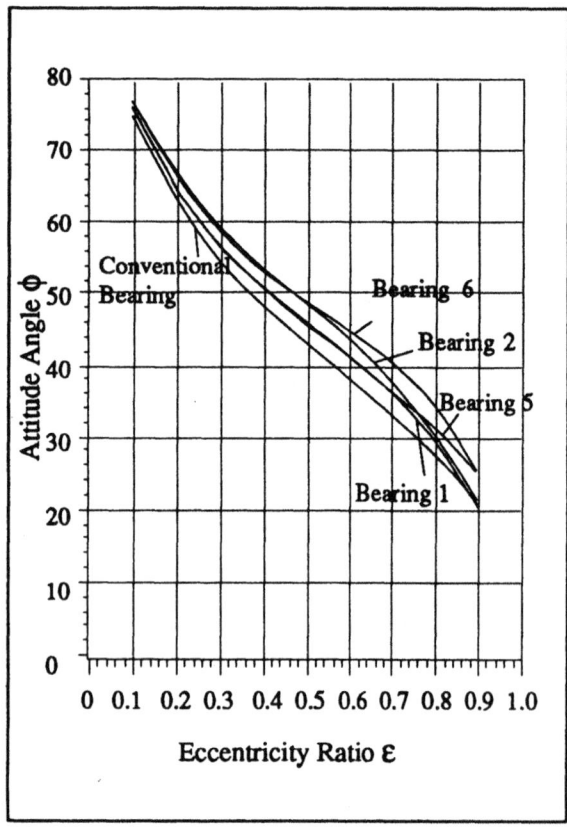

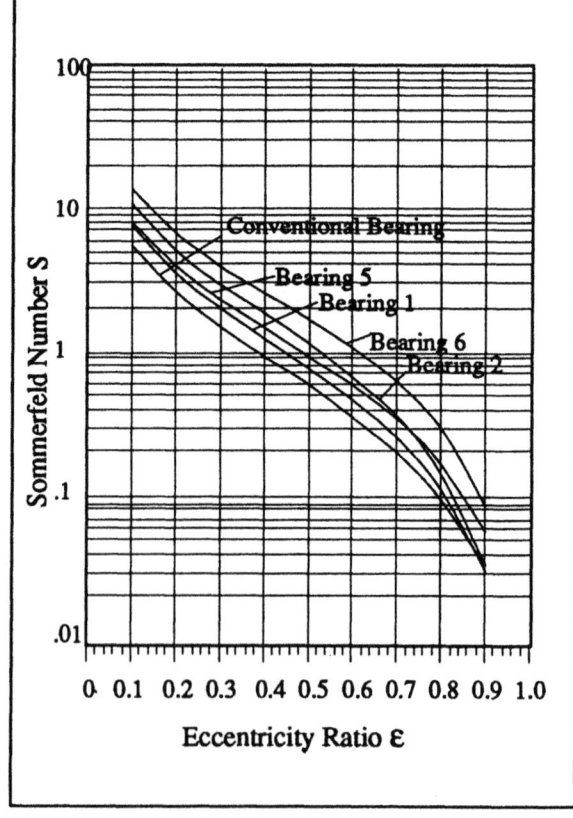

Figure 5 Attitude Angle of
Bearing 1 and 2

Figure 6 Sommerfeld Number of
Bearing 1 and 2

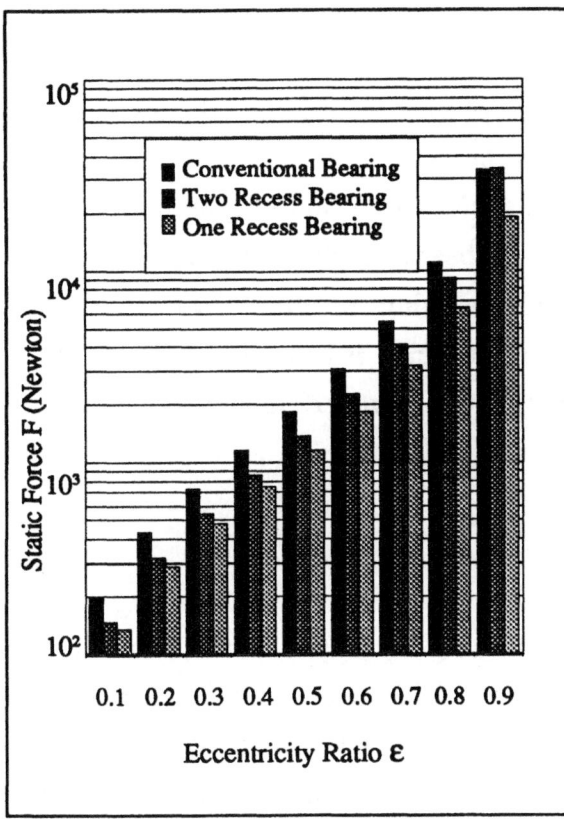

Figure 7 Static Load Chart

Sommerfeld number the bearing with one recess operates at a higher eccentricity than the bearing with two recesses, and the bearing with two recesses operates at a higher eccentricity than the conventional bearing. It can also be seen that the bearing with two larger recesses operates at a higher eccentricity than the bearing with two smaller recesses, which indicates that the larger recess tends to reduce load-carrying capacity slightly. For easy visualization and comparison, static loads for bearing 1 and 5 are presented in the form of a chart, in Figure 7, for bearings with $L/D = 0.5$, $L = 50$ mm and running at a rotating speed 1000 rve/min. In this particular case, the average static load carried by the two recess bearing is 7.7% lower than that of the conventional bearing over the whole

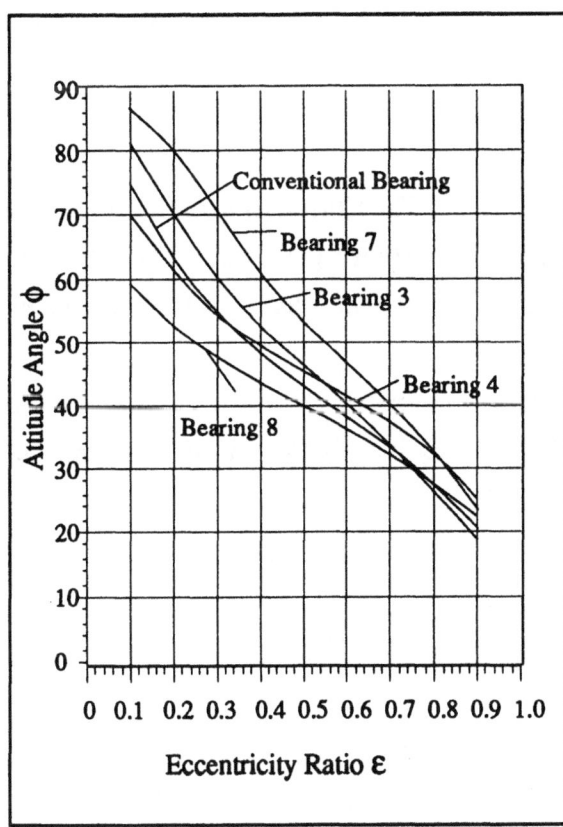

Figure 8 Attitude Angle of Bearing
3 and 4

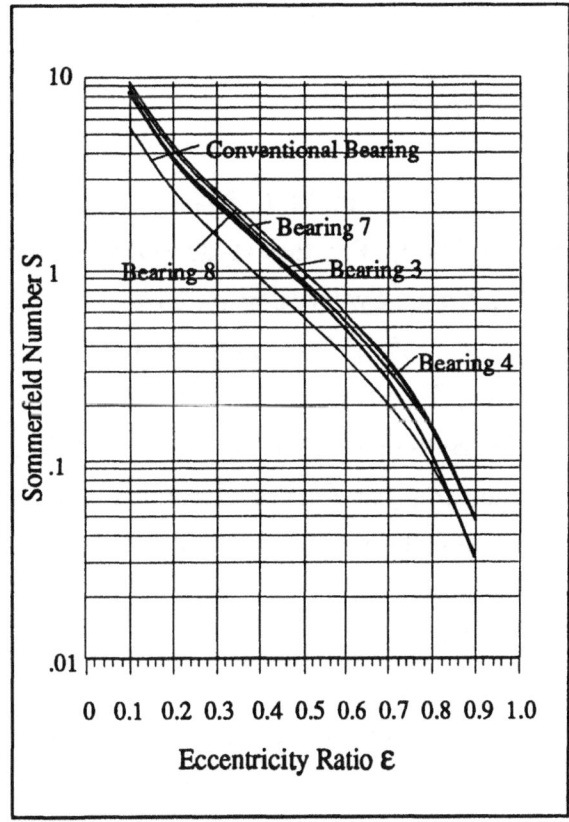

Figure 9 Sommerfeld Number of
Bearing 3 and 4

358

range of eccentricity ratios. It is interesting to note that when the bearing runs at high eccentricity ratios (around 0.9), the two recess bearing has almost the same load carrying ability as that of the conventional bearing. For the one recess bearing, the average static load is 32% lower than that of the conventional bearing.

Figure 8 illustrates the variation of attitude angle with eccentricity ratio for bearings 3 and 4. It can be seen that the recess position on the bearing surface has a very significant influence on the bearing stable equilibrium operating position especially at lower eccentricity ratios. However, the load carrying capacity is less sensitive to the recess position, this can be seen in Figure 9 where for a given Sommerfeld number there is little difference in operating eccentricity ratio.

Conclusion

The two recess bearing has a closer load carrying capacity to that of a conventional bearing than does a one recess bearing, and bearings with several smaller recesses have a higher load carrying capacity than that of a bearing with one larger recess if the effective recess areas are the same. The size of the recess has a significant effect on load carrying capacity too — a bearing with two large recesses tends to have a smaller load-carrying capacity than one with two small recesses. The position of the recesses has relatively little effect on load carrying capacity however, although it does have an effect on the attitude angle adopted by the journal. Attitude angle is less sensitive to recess size and number.

References

[1] Goodwin, M. J., Roach, M. P. and Penny, J. E. T. "The Elimination of Shaft Critical Speeds Using Parallel Hydrostatic And Squeeze Film Bearings" *Proc. The Second International Symposium on Transport Phenomena, Dynamics and Design of Rotating Machinery, Vol.2 Dynamics* 1988 pp 226-240

[2] Kirk, R.G. and Gunter, E.J. "The Effect of Support Flexibility and Damping on the Synchronous Response of a Single-Mass Flexible Rotor" *ASME Journal of Engineering for Industry* February 1972 pp221-232

[3] Habermann, H. and Liard, G. " An Active Magnetic Bearing System" *Tribology International* April 1980 pp. 85-89

[4] Schweitzer, G. " Magnetic Bearings" *CISM Lecture Course on Rotordynamics* Udine 1985

[5] Goodwin, M. J., Boroomand, T. and Hooke, C. J."Variable Impedance Hydrodynamic Journal Bearings for Controlling Flexible Rotor Vibrations" *Rotating Machinery Dynamics* Proc. the 1989 ASME Design Conf. on Mechanical Vibration and Noise Sept 1989 Canada

INFLUENCE OF ELASTICITY AND GEOMETRICAL PROFILE ON DYNAMIC TRANSMISSIBILITY OF CONNECTING-ROD BEARING

B. Fantino[*] and B. Bou-Saïd[*]

* Laboratoire de Mécanique des Contacts .URA 856 I.N.S.A Bât. 113
20 Avenue Albert Einstein 69621 Villeurbanne Cedex FRANCE

SUMMARY

When a bearing is under dynamic loads, the fluid transmits the forces generated by the lubricant to the frame. The amplitude of the transmitted forces may be greater than that of the exciting force, and can be expressed as a fundamental response and harmonics. Risks of resonance may appear and lead to mechanism deterioration.

In this paper, through discrete Fourier's transforms, a frequency analysis of hydrodynamic forces corresponding to a dynamically loaded connecting-rod bearing with an actual load diagram of a petrol engine running at 3000 rev/mn is performed. The influence of the fundamental frequency as well as harmonics on the forces transmitted to the frame is also evaluated.

The comparisons of the results obtained for a rigid and a deformable housing initially circular to a two lobed bearing and a three lobed bearing, for a given operating conditions, are made.

Results show that :
- the consideration of the deformability the housing modifies the shaft center trajectory and the forces transmitted to the frame,
- lobed bearings reduce the transmitted forces to the frame.

1. INTRODUCTION

The dynamic characteristics of bearings have been largely investigated for a rigid housing either by linear analysis in which the bearing stiffness and damping coefficients are calculated or by non linear analysis [1, 5]. For highly loaded journal bearings such as those encountered in automobile motors or in modern aircraft engines, the influence of bearing deformation is not negligible on its static characteristics such as the pressure field, the minimum oil film thickness etc... The influence of the bearing deformation has been studied exhaustively [6, 11] using elastohydrodynamic lubrication analysis. When a bearing is dynamically loaded, the oil film transmits the forces generated by hydrodynamic pressure to the frame. The amplitude of the transmitted forces may be greater than that of the exciting force and are expressed by a fundamental response and harmonics. Risks of resonance may appear and lead to mechanism deterioration.

An investigation on the dynamic response will indicate whether the oil film amplifies or attenuates the transmitted forces. In fact, the fluid acts as a system of forces transfer. To this end, a discrete Fourier transform (D F T) [12] is carried out on the hydrodynamic forces generated in one load cycle. Thus, we obtain a spectrum which characterises the different hydrodynamic forces frequencies as well as their amplitudes. A complete study of this problem requires the simultaneous solution of the following three equations :

- equation of the shaft motion,
- Reynolds equation governing the pressure generated in the oil film,
- housing deformation.

2. DYNAMIC AND REYNOLDS EQUATION

Figure 1 gives a schematic description of the shaft position and that of bearing housing, (O, X, Y) represents the fixed reference coordinates. O, O_a and G are respectively the bearing housing center, the shaft geometrical center, the shaft inertial center. Application of the fundamental principle of the shaft dynamic movement yields the following equilibrium equations :

$$M \, \overset{\circ\circ}{x} = Fx + M \, e_b \, \omega^2 \, \sin(\omega t) + W_x(t)$$

$$M \, \overset{\circ\circ}{y} = W_0 + Fy + M \, e_b \, \omega^2 \, \cos(\omega t) + W_y(t)$$

$$(1)$$

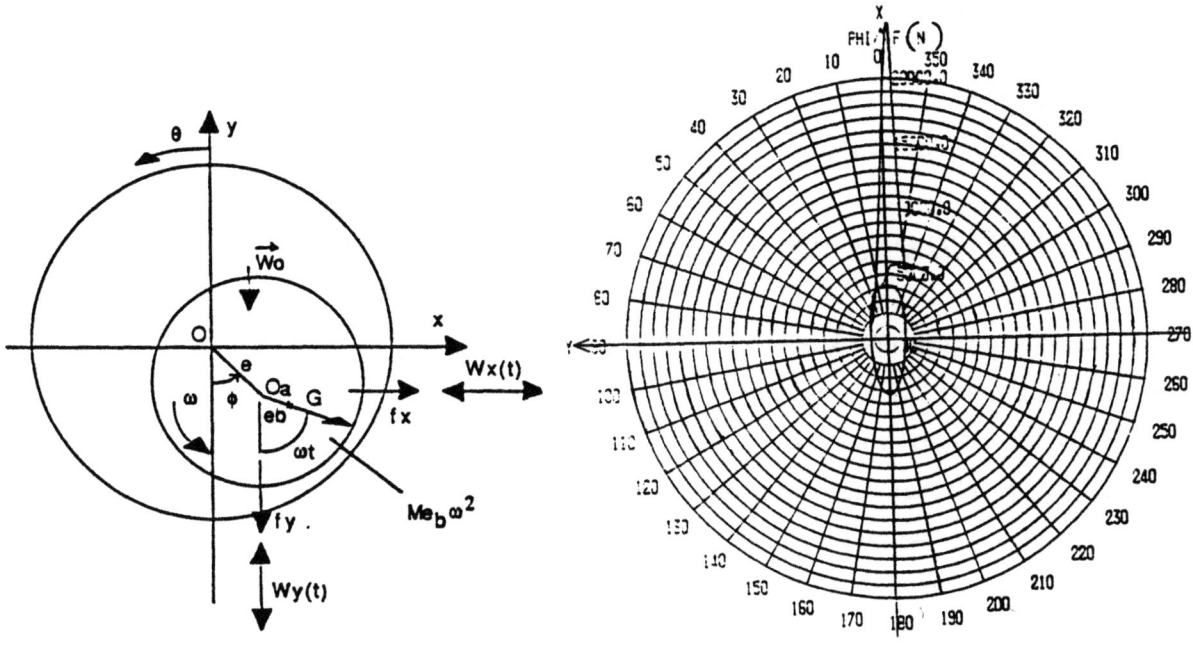

Fig. 1 : Bearing schematic

Fig. 2 : Petrol engine load diagram (ω_a = 3000 rv/mn)

where F_x and F_y represent the oil film forces on the shaft. F_x and F_y are non linear functions of the co-ordinates (x, y) of shaft geometrical center O_a as well as its velocities ($\overset{\circ}{x}$, $\overset{\circ}{y}$). F_x and F_y are determined by integrating the pressure field since the shear stress is negligible when compared to the pressure forces [7].

In dynamic regime, taking into account the housing deformation, the Reynolds equation for infinitely short journal bearing which governs the pressure field generation is written as [7] :

$$\frac{\partial}{\partial z} \left(\frac{H^3}{\mu} \frac{\partial p}{\partial z} \right) = 6 \ \omega \ \frac{\partial h}{\partial \theta} - 6\omega \ \frac{\partial \delta_{rb}}{\partial \theta} + 12 \ (\ C \ \dot{\varepsilon} \ \cos \theta \qquad (2)$$
$$+ \ C \ \varepsilon \ \dot{\varphi} \ \sin \theta + \dot{\delta}_{rb})$$

In this equation p is the pressure, μ the viscosity, $H=h+\delta_{rb}$ is the actual film thickness, where $h=C(1 + \varepsilon \cos \theta)$ is the classical film thickness, δ_{rb} is the radial bearing displacement. The dot indicates derivation with respect to time. ω represents the rotational speed

Notice that equation (2) is based on the assumptions that the oil flow is laminar and isothermal, fluid is newtonian and adheres to contact surface. The coupled differential system (1) and (2) is solved by modified Euler method [2].

3. STRUCTURE CALCULATION AND FLUID/STRUCTURE COUPLING

Considering the deformability of the bearing housing and its effect on structure/fluid coupling leads to treat an elastohydrodynamic lubrication problem [7]. The structure calculation is assumed to be static with linear elastic material property.

In static structural analysis which is used in elastohydrodynamic lubrication the equation which governs the housing deformation is written as :

$$\int_{\Omega} \sigma \delta \varepsilon \, d\Omega = \int_{\Gamma} T \, \delta u \, d\Gamma \qquad (3)$$

where δu are the virtual displacements, $\delta \varepsilon$ are the associated virtual strains, T is the external force, σ are the stresses. The domain of interest Ω is bounded by Γ

The finite element method is used to solve equation (3). The classical displacement formulation yields a system of algebric equations which is expressed in matrix form as :

$$[K] \{u\} = \{F\} \qquad (4)$$

in which [K] is the global stiffness matrice { u} represents the nodal displacements vector at the instant t and F the external forces. The fluid/structure coupling is realised by means of an iterative process. The direct method is retained in the solution of the resulting system of non-linear equations

4. DISCRETE FOURIER TRANSFORM

The solution of the equation (1), (2) yields the forces F_x, F_y at each time step of a load cycle. Note that F_x and F_y represent the oil film forces on the housing. A frequency analysis on F_x and F_y is necessary for examining the bearing dynamic transmissibility. By means of a discrete Fourier transform [12], the obtained Fourier transform spectrum provides the information relative to the force transmitted to the frame.

5. RESULTS AND DISCUSSIONS

5.1 - Data of the problem

We study a dynamically loaded connecting rod bearing (figure 2). The short bearing hypothesis is retained in the hydrodynamic pressure calculation, whilst the housing deformation is computed with plane stress assumption. The principal bearing characteristics are :

$L/D = 0.34$; $C/R = 1.3\ 10^{-3}$ shaft rotation $\omega_a = 3000$ rpm, lubricant viscosity $\mu = 7$ m Pa.s and housing thickness $B = 1.5\ 10^{-2}$ m.

During the calculation of the structure deformation, two cases are considered :

    a) a rigid housing,

    b) an elastic housing with $E = 1.05\ 10^{10}$ N/m^2 = E steel/20

We examine in this paper the influence of the bearing geometrical profile on the bearing dynamic transmissibility by means of discrete Fourier transform.

5.2 Results

The figure 3 represents the orbit of the shaft center obtained with a rigid and deformable bearing for one cycle of load. We note that the deformability of the bearing increases the amplitude of the shaft center trajectory. In the rigid case the hydrodynamic forces are overvalued.

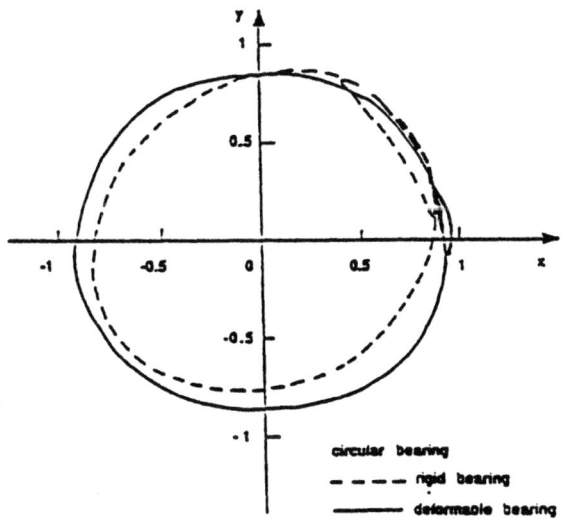

Fig. 3 : Orbit of the shaft center obtained with a rigid and deformable bearing, for the petrol engine load diagram given fig. 2.

ω_a = 3000 rv/mn ; M = 90 kg = rotor mass ; E = 1.05 x 10^{10} = E steel / 20 ; μ = 7 m.Pa.s ; L/D = 0.34 ; C/R = 1.3 x 10^{-3} ; Thickness of the shell bearing B = 1.5 x 10^{-2} m

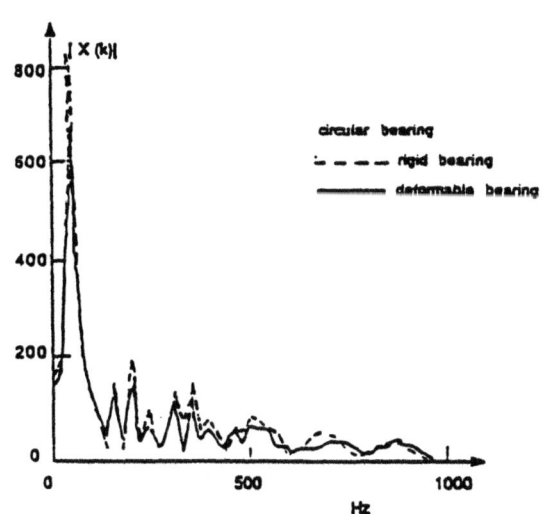

Fig. 4 : Comparison of discrete Fourier's Transforms of Fy obtained with a rigid and deformable bearing, for the petrol engine load diagram given fig. 2.

ω_a = 3000 rv/mn ; M = 90 kg = rotor mass ; E = 1.05 x 10^{10} = E steel / 20 ; μ = 7 m.Pa.s ; L/D = 0.34 ; C/R = 1.3 x 10^{-3} ; Thickness of the shell bearing B = 1.5 x 10^{-2} m

This is confirmed in figure 4 through discrete Fourier transform where we observe a greater value for F_y in the case of rigid bearing.

Static and dynamic properties of a large number of fixed geometry bearings have been studied by several authors [1-2]. The majority are concerned with the determination of the effect of parameters believed or known to affect these characteristics. These include preload and offset factors [1-2]. It has been shown that the geometry modification usually existing in turbine bearings can have a stabilizing or destabilizing effect. Preloaded lobes with offset have a stabilizing effect with respect to the circular bearing for all load conditions [2].

Figure 5 represents a general view of a lobed bearing. We take here m = 0.5 and α = 1. Figure 6 gives the orbit of the shaft center for a two lobed bearing. We note that the eccentricity ratio can be greater than 1 i.e. the hydrodynamic forces are more important compared to those obtained for a journal bearing.

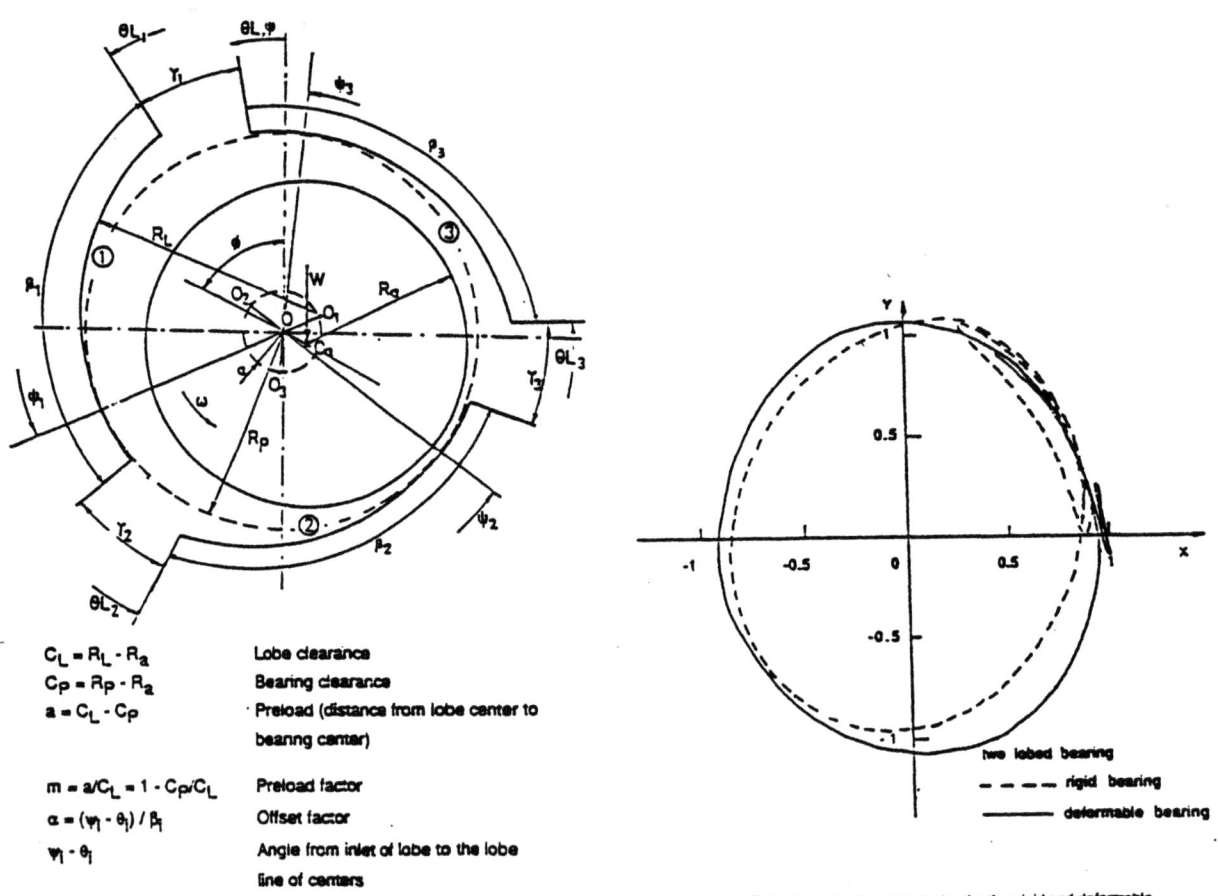

$C_L = R_L - R_a$	Lobe clearance
$C_P = R_P - R_a$	Bearing clearance
$a = C_L - C_P$	Preload (distance from lobe center to bearing center)
$m = a/C_L = 1 - C_P/C_L$	Preload factor
$\alpha = (\psi_1 - \theta_1)/\beta_1$	Offset factor
$\psi_1 - \theta_1$	Angle from inlet of lobe to the lobe line of centers
β_1	Lobe angular amplitude

Fig. 5 : Nomenclature of the multilobe bearing

Fig.6 : Orbit of the shaft center obtained with a rigid and deformable bearing, for the petrol engine load diagram given fig. 2.

ω_a = 3000 rwmn ; M = 90 kg = rotor mass ; E = 1.05 x 10^{16} = E steel / 20 ; μ = 7 m.Pa.s ; L/D = 0.34 ; C/R = 1.3 x 10^{-3} ; Thickness of the shell bearing B = 1.5 x 10^{-2} m

No significant change is noted for the D.F.T. of F_x but in figure 7 we note a decrease in the magnitude of F_y due to the greater deformation obtained for this specific geometric profile. Figure 8 presents the orbit of the shaft center for a three lobed bearing. The bearing geometry gives a specific trajectory. As for the two lobed bearing no significant change is noted for the D.F.T. of F_x. Figure 9 shows a decrease in the magnitude of F_y.

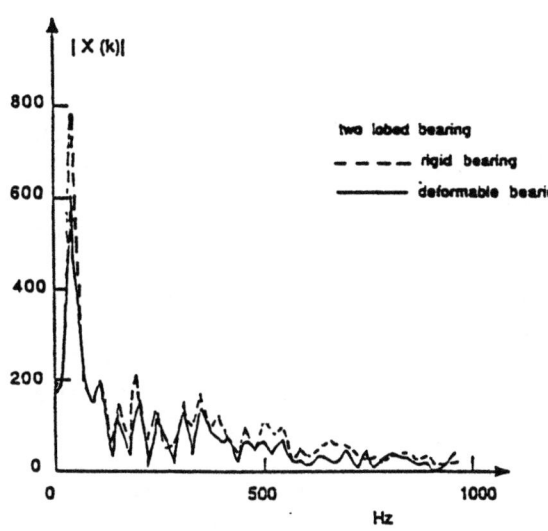

Fig. 7 : Comparison of discrete Fourier's Transforms of Fy obtained with a rigid and deformable bearing, for the petrol engine load diagram given fig. 2.

ω_a = 3000 rv/mn ; M = 90 kg = rotor mass ; E = 1.05 x 10^{10} = E steel / 20 ; μ = 7 m.Pa.s ; L/D = 0.34 ; C/R = 1.3 x 10^{-3} ; Thickness of the shell bearing B = 1.5 x 10^{-2} m

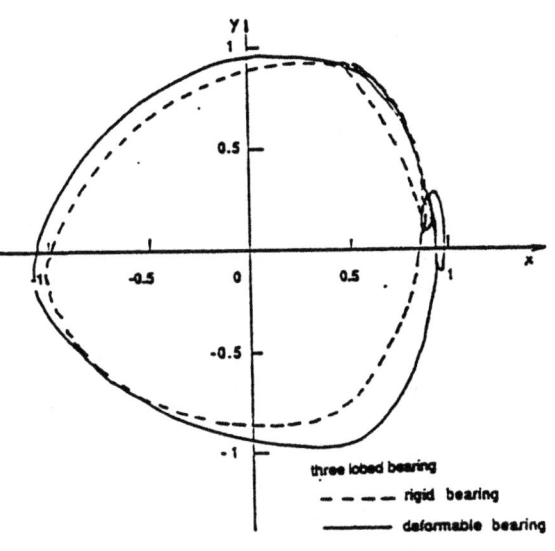

Fig. 8 : Orbit of the shaft center obtained with a rigid and deformable bearing, for the petrol engine load diagram given fig. 2.

ω_a = 3000 rv/mn ; M = 90 kg = rotor mass ; E = 1.05 x 10^{10} = E steel / 20 ; μ = 7 m.Pa.s ; L/D = 0.34 ; C/R = 1.3 x 10^{-3} ; Thickness of the shell bearing B = 1.5 x 10^{-2} m

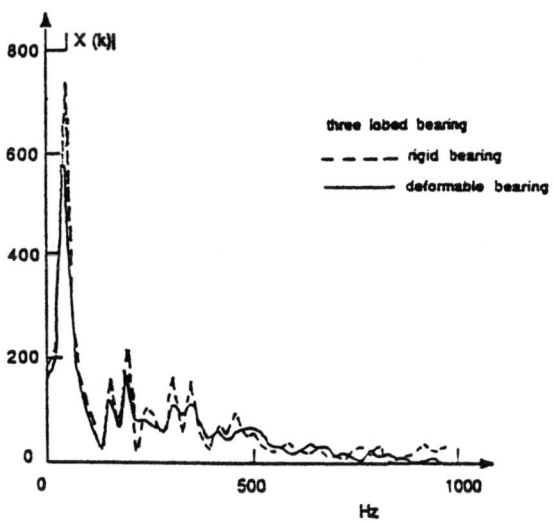

Fig. 9 : Comparison of discrete Fourier's Transforms of Fy obtained with a rigid and deformable bearing, for the petrol engine load diagram given fig. 2.

ω_a = 3000 rv/mn ; M = 90 kg = rotor mass ; E = 1.05 x 10^{10} = E steel / 20 ; μ = 7 m.Pa.s ; L/D = 0.34 ; C/R = 1.3 x 10^{-3} ; Thickness of the shell bearing B = 1.5 x 10^{-2} m

6. CONCLUSION

By accounting for the deformability of the housing it is found that for a dynamically loaded bearing the shaft center trajectory and the forces transmitted to the frame are notably modified. We have noted that the lobed bearings reduce the transmitted forces to the frame. This is due to the greater deformation of the housing obtained for these specific geometrical profiles.

7. REFERENCES

[1] Abdul-Wahed N "Comportement dynamique des paliers fluides, étude linéaire et non linéaire". Thèse Dr.-ès-Sciences, INSA/UCB, 1982.

[2] Abdul-Wahed N., Nicolas D. and Pascal M.T. "Stability and unbalance response of large trubine bearings". ASME Journal of Lubrication Technology, 1982, vol. 104, p. 66-75.

[3] Cusano C. and Funk P.E. "Transmissibility study of a flexibly mounted rolling element bearing in a porous bearing squeeze-film damper". ASME Journal of Lubrication Technology, 1977, p. 50-56.

[4] Gunter E.J., Barett L.E. and Allaire P.E. "Design of nonlinear squeeze-film dampers for aircraft engines". ASME Journal of Lubrication Technology, 1977, p. 57-64.

[5] Mohan S. and Hahn E.J. "Design of squeeze film damper supports for rigid rotors". ASME Journal of Engineering for Industry, 1974, p. 976-982.

[6] Fantino B., Frêne J. "Comparison of dynamic behavior of elastic connecting rod bearing in both petrol and diesel engines". ASME Journal of Tribology, 1985, vol. 107, p. 87-91.

[7] Fantino B., "Influence des défauts de forme et des déformations élastiques des surfaces en lubrification hydrodynamique sous charge statiques et dynamiques". Thèse Dr-ès-Sciences, INSA/UCB, Lyon, 1981.

[8] Fantino B., Godet M. and Frêne J. "Dynamic behavior of an elastic connecting rod bearing. Theoretical study". Published by SAE : Society of Automotive Engineers in "Studies of Engine Bearings and Lubrication", SP 539, p. 23-32, 1983.

[9] Labouff G.A. and Booker J.F. "Dynamically loaded journal bearings : a finite element treatment for rigid and elastic surface". ASME Journal of Tribology, 1985, vol. 107, p. 505-515.

[10] Goenka P.K., and Oh K.P. "An optimum connecting rod design - a lubrication viewpoint". ASME Journal of Tribology, 1986, vol. 108, p. 487-496.

[11] Bates T.W., Fantino B., Launay L. and Frêne J. "Oil film thickness in an elastic connecting-rod bearing : comparison between theory and experiment. STLE Tribology Transactions, 1990, Vol 33 n°2, p. 254-266.

[12] Brigham E.O. " The fast Fourier transform ". Prentice-Hill, 1974

Vibration Control in Rotating Machinery by the use of Accumulators or Aerated Lubricants

M P Roach and M J Goodwin

Department of Mechanical and Computer-Aided Engineering, Staffordshire Polytechnic, Beaconside, Stafford ST18 0AD, United Kingdom

Abstract

Recent work in the area of variable impedance supports for rotating machinery (1,2) has shown both theoretically and experimentally that in the majority of rotating machines the supports/bearings are over-stiff and that a reduction of the support stiffness can significantly reduce the vibration amplitude of the shaft, reduce the support force transmission and increase the system's stability reserve, provided that sufficient support damping is present. Such work (which is summerised in this paper) was primarily concerned with the reduction of squeeze film support stiffness by the use of accumulator modified hydrostatic squeeze-film supports surrounding the rolling element bearings on which the shaft runs. This work has recently been extended to investigate the effect of accumulator modified journal bearings (3). From the above theoretical analyses it is apparent that the support stiffness is dominated by a fluid compressibility parameter. As a result it should also be possible to attain the same improvements to a journal bearing supported machine dynamic characteristics by reducing the overall lubricant bulk modulus by aerating the lubricant. The following paper extends the above work by investigating the effect of lubricant aeration on the dynamic behaviour of a horizontal rigid shaft running on journal bearings. Preliminary experimental test results are presented which show that significant increase in machine stability may be achieved..

1. Introduction

Squeeze film dampers are frequently used for the stabilization and /or vibration control of rotating machinery. The operation of such dampers frequently assumes and in practice attempt to achieve the use of incompressible lubricant . Following the accidental aeration of the lubricant and resultant violent rotor vibrations during the operation of a vertical, centralised squeeze film supported rotating shaft running on rolling element bearings (4), Feng and Hahn (5) theoretically investigated the effect of gas entrainment on the operation of squeeze film dampers. The results showed that the use of such a 'spongy' lubricant does significantly effect the performance of the squeeze film. Unfortunately whether the effect was beneficial or detrimental was dependant upon the film model used.

A theoretical and experimental investigation was carried out into the effect of nitrogen bag accumulator attachment to hydrostatically centralised squeeze film supports (1). The squeeze film supports surrounded the rolling element bearings on which the shaft ran, with one bearing at each end of a horizontal flexible rotor (Fig 1). The squeeze film inner surface (rolling element bearing outer sleeve) was prevented from rotating by dogs in the usual manner.The results showed that the effect of the accumulators was to significantly increase the effective compressibility of the lubricant within the hydrostatic pad volume. The overall effect of this was to significantly reduce the dynamic stiffness of the support (Fig 2) whilst maintaining the static stiffness (and hence load centralisation within the film clearance). The hydrostatic damping component of the support damping reduced in a similar manner to the hydrostatic dynamic stiffness (Fig 3), due to the fact that, under the influence of a vibrating load, the fluid is now allowed to flow into the accumulators (attached to the high pressure oil feed recesses) as opposed to being force back down the high pressure supply line restrictor or over the squeeze film clearance. Despite these hydrostatic

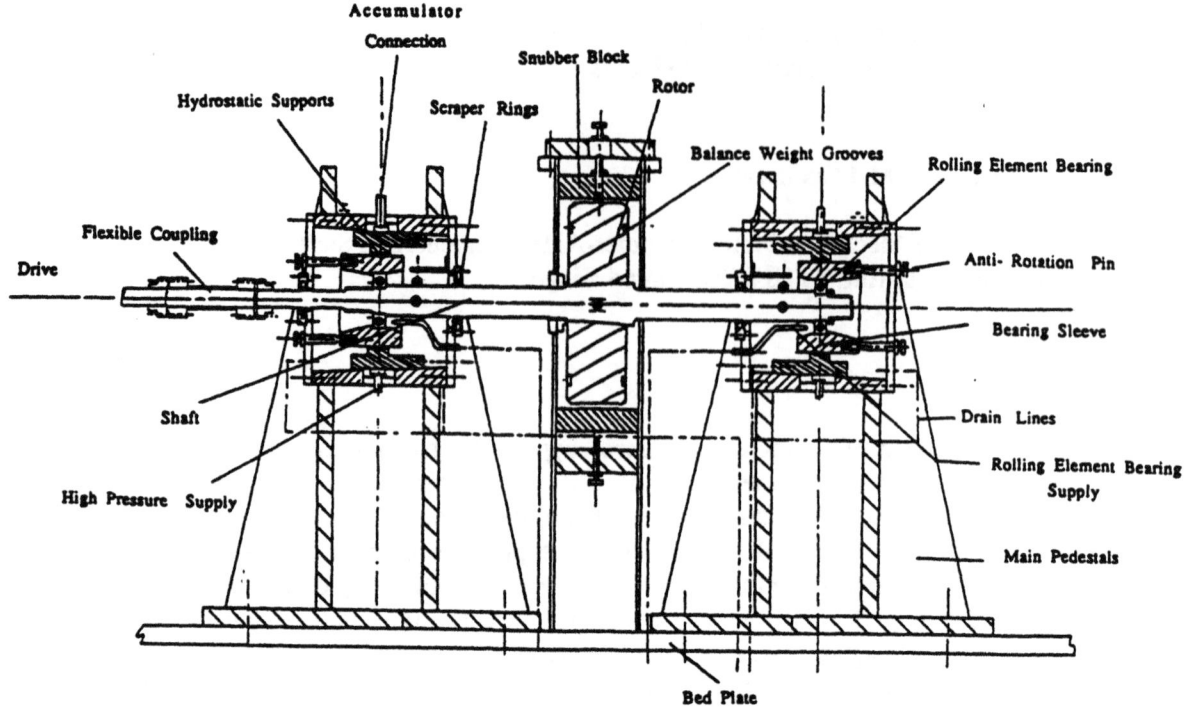

Fig. 1 Accumulator Modified Hydrostatic Squeeze-film Experimental Test Rig.

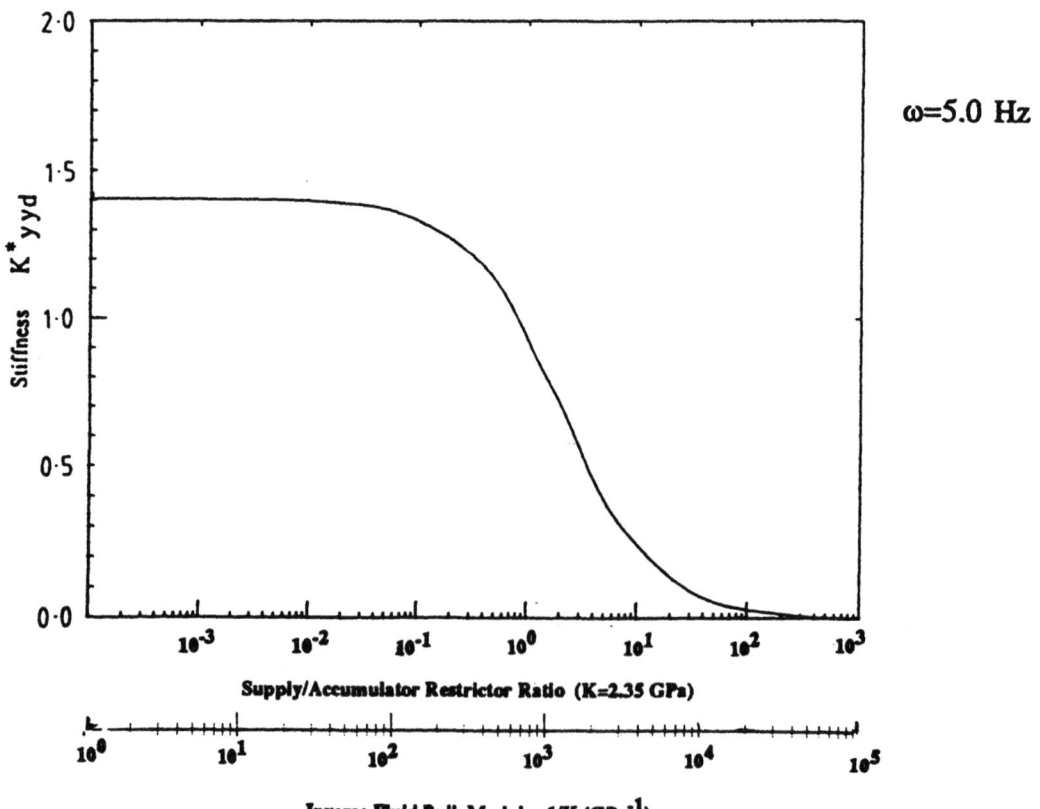

**Fig. 2 Effect of Supply/Accumulator Restrictor Ratio and Fluid Bulk Modulus
on Dynamic Load Hydrostatic Support Stiffness.**

stiffness and damping reductions the squeeze-film damping component was unaffected due to the lack of any reductions to the fluid bulk modulus over the land area.

The theoretical analysis presented in (1) showed that, the support dynamic stiffness reduction caused the optimum support damping value to be reduced, for machines with low shaft end/shaft midspan mass ratios Md (fig 4) Due to the reduced value for optimum support damping, the squeeze-film damping value was shown to be sufficient to provide optimum damping despite the severe reduction in hydrostatic damping levels. The theoretical results indicated that provided the optimum support damping value was present, the effect of reduced support dynamic stiffness would in many cases reduce the shaft midspan vibration due to unbalance, with the system stability showing similar improvements (fig 6). These results were experimentally confirmed (figs 5,7 and 8). It should be noted however that the system stability falls drastically if the support stiffness/shaft stiffness ratio Kd falls below a value determined by the shaft mass distibution irrespective of the level of support damping present.

It is therefore apparent that lubricant aeration is not desirable with spring centralised squeeze-film supports as the support stiffness does not reduce with the squeeze-film damping and therefore the optimum support damping remains unaffected, this results in an underdamped system exhibited in (4).

2. Reduction of Journal Bearing Stiffness by Lubricant Aeration.

Following the work on accumulator modified hydrostatic squeeze-films it is therefore proposed that the effect of accumulator attachment to hydrostatic pads may be attained with journal bearings by aeration of the lubricant. As a consequence, the work has been extended to experimentally determine the effects of lubricant aeration on the unbalance response and stability of a rigid shaft running horizontally in journal bearings at each end of the shaft.

3. Experimental Journal Bearing Test Rig

The experimental journal bearing test rig is shown in Fig 9. with the shaft, journal bearing and lubricant data being presented in table 1. The lubricant was aerated just before entry into the bearings by a standard commercial fish tank aeration pump with aeration stones attached to the end of each aeration line within the lubricant feed lines. The oil aeration stones were used as this was found to produce a finer and more even distribution of air bubbles within the lubricant which it was assumed would be preferential to single large bubbles during operation. The finer bubbles also tended to remain in suspension for a longer period than a few large bubbles. The problem of ensuring that the lubricant remained aerated right up to entry into the bearings also resulted in the need to aerate the lubricant as close as possible to the bearing lubricant inlet. This problem was exacerbated by the low viscosity of the lubricant used (see table 1).

It was possible that the fine air bubbles within the lubricant would be dissolved in the oil particularly under the high lubricant pressures experienced at the minimum clearance point under the shaft during shaft rotation. As such dissolution requires a finite time, it was hoped that the speed of the oil flow through this minimum clearance point would be high enough to allow the bubbles to remain in suspension.

The shaft vibration during operation was measured using two inductive probe transducers positioned one in the vertical plane and one in the horizontal plane at approximatly one third the shaft span. Due to the rigid nature of the shaft, these measurments were assumed to be representitive of the general shaft vibration. The inductive probe transducers were conected to a standard CED 1401 data logging unit which was controlled by an IMB PC. The data loggging unit allowed the rotational speed of the shaft, the shaft horizontal and the shaft vertical displacement to be measured and recorded similtaneously, thus the vibration response of the system could be

370

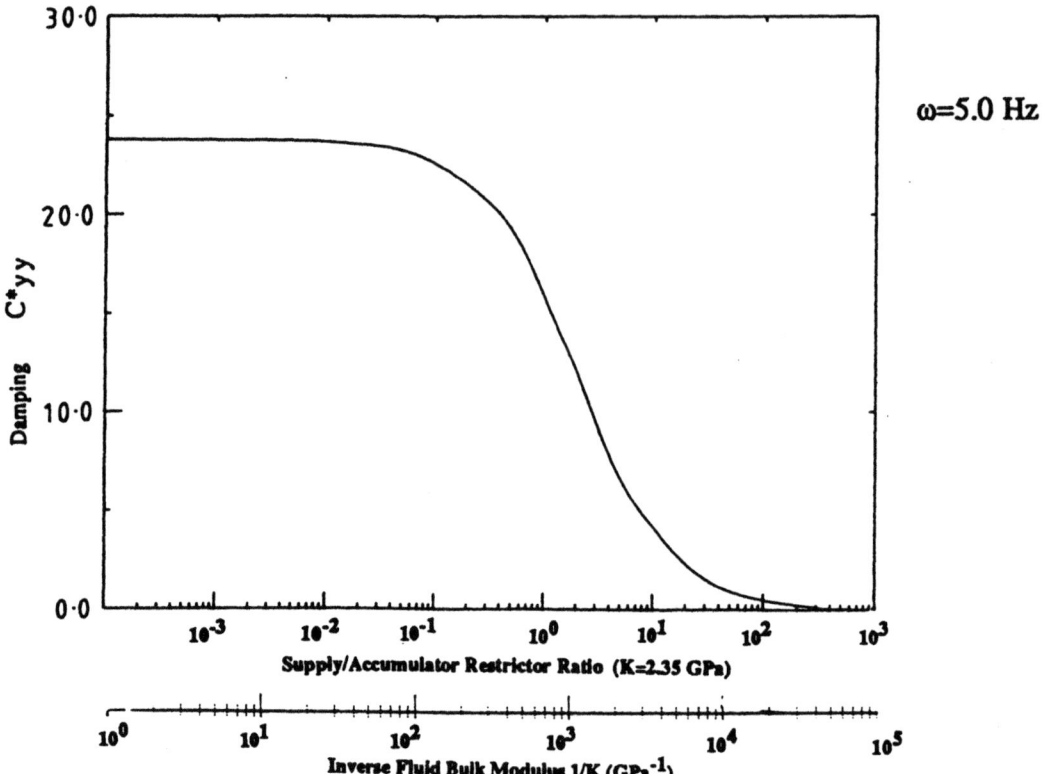

ω=5.0 Hz

Fig. 3 Effect of Supply/Accumulator Restrictor Ratio and Fluid Bulk Modulus on Hydrostatic Support Damping.

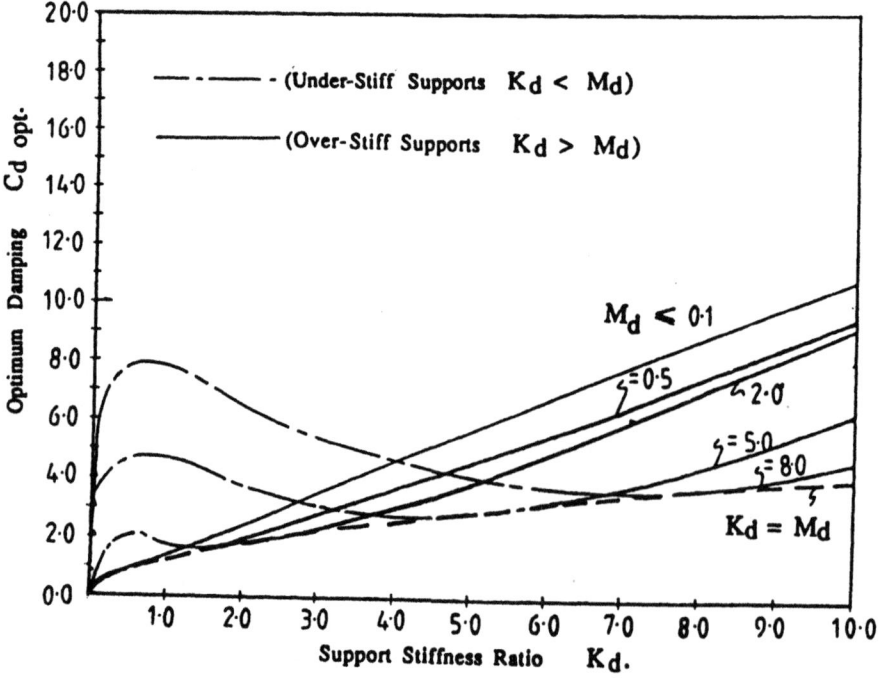

Fig. 4 Variation of Optimum Support Damping With Support/Shaft Stiffness Ratio (Kd)

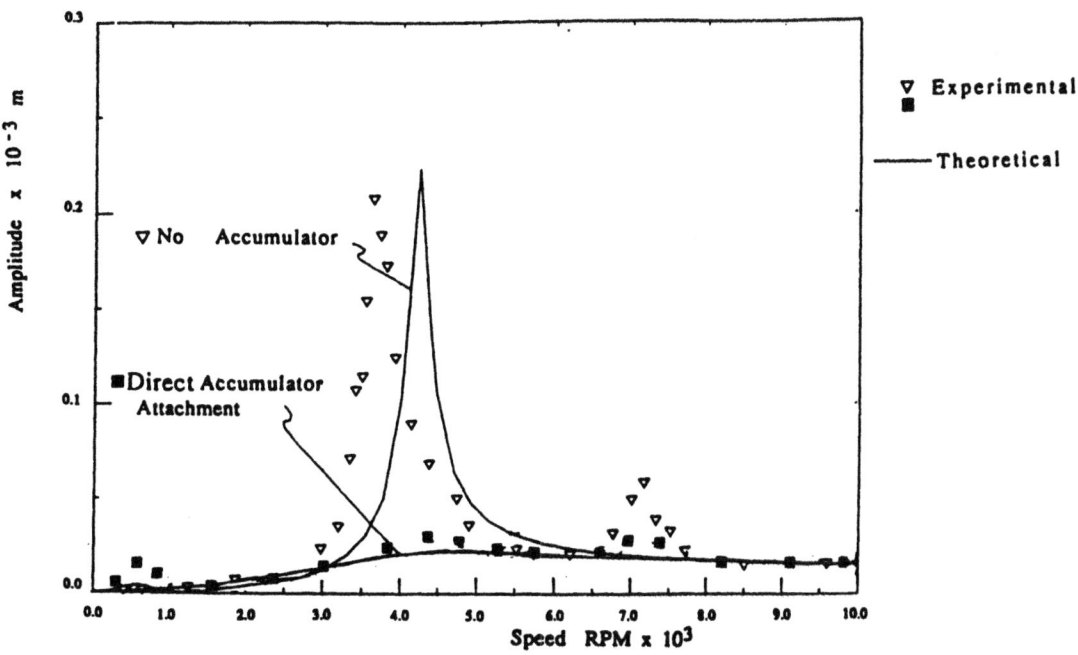

Fig. 5 **Experimental Shaft Midspan Response to Unbalance (Hydrostatic Test Rig)**

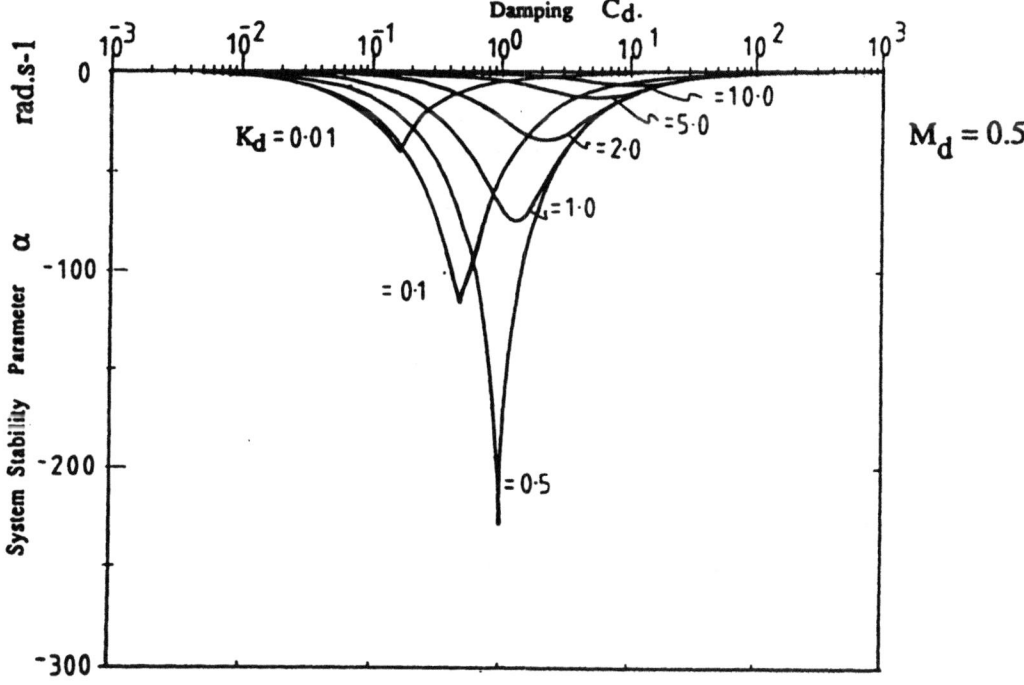

Fig. 6 **Theoretical Variation of System Stability with Support Damping
and Support/Shaft Stiffness Ratio
(Shaft end/midspan Mass Ratio Md=0.5)**

372

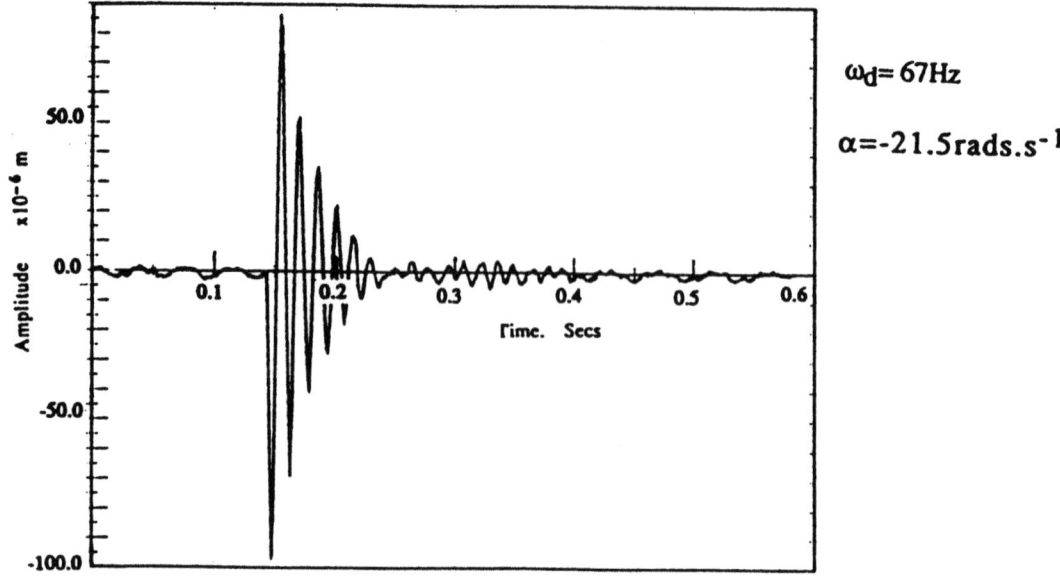

Fig. 7 Experimental Transient Response at Shaft Midspan (Hydrostatic Rig)
(No Accumulators)

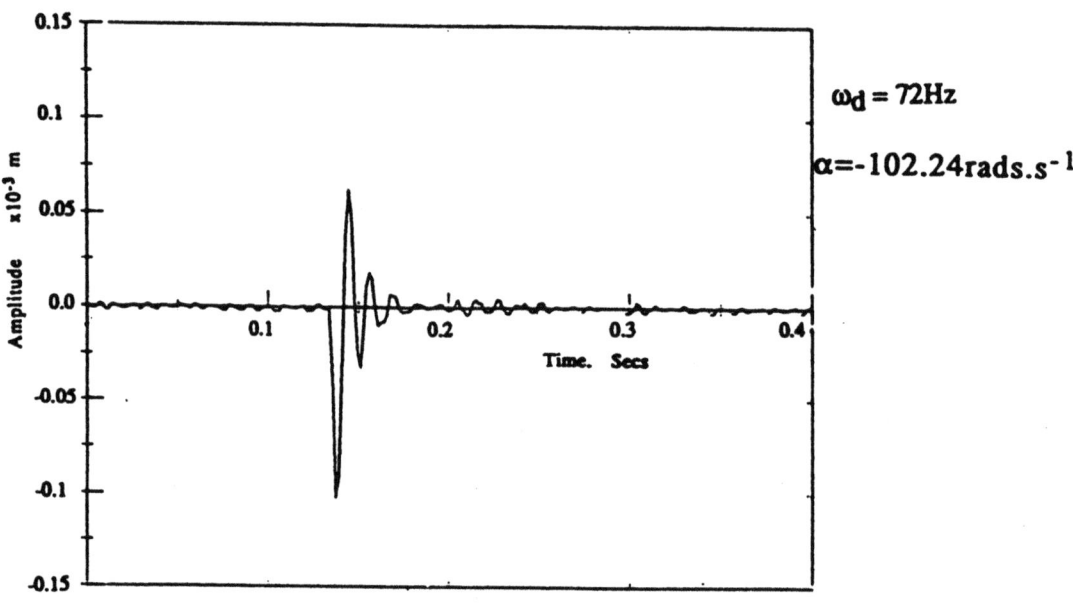

Fig. 8 Experimental Transient Response at Shaft Midspan (Hydrostatic Rig)
(Accumulators Attached)

recorded. The test rig was then run with no oil aeration (standard 'incompressible' lubricant) upto the oil whirl onset speed. This process was then repeated with the oil aeration unit switched on.

4. Journal Bearing Experimental Results

The experimental results are presented in Fig 10. As can be seen from this figure the stability limit (defined by the onset of oil whirl) has increased from 2700 rev/min to 3800 rev/min due to the aeration of the lubricant. It can however be seen that certain amplitude peaks occur at approximately 3000 rev/min and 3300 rev/min. These are not resonant peaks and were coinsident with the rotor jumping during operation. During operation with the lubricat aerated, such jumping was to a degree apparent at all running speeds above 2900rev/min from observation of the oscilloscope trace. It is suggested that such jumping was associated with 'large' air bubbles forming within the lubricant before the lubricant entered the bearing. When the lubricant was 'filled' with only fine air bubbles such jumping did not occur. It may be noted that the aeration of the lubricant has had little noticable effect on the steady-state response to unbalance of the system.

5. Discussion.

The theoretical results presented in figs 4 and 6 suggest that with rotating machines, the stability is maximised and steady-state response to unbalance minimised with certain support stiffness and damping values which are dependant upon the systems shaft mass distribution and shaft stiffness. This implies that the majority of rotating machines (which have shaft end/shaft midspan mass ratios below Md=0.5) are operating with supports or bearings which are over stiff.

It is apparent from the preliminary experimental results that such stability and unbalance response improvements may be achieved with systems running on rolling element bearings by the use of accumulator modified squeeze-film supports surrounding the bearings. The experimental results exhibit steady-state response reductions of 80% and stability improvements of nearly 500% over conventional hydrostatic supports.

Further it has been experimentally shown that the controlled aeration of the journal bearing lubricant resulted in a significant increase in the stability limit of the system. This is in agreement with the results from (1). Any stability reduction associated with a reduction of bearing damping due to the oil aeration (although not experimentally determined) has apparently been negated by the overall improvement in system stability due to reduced bearing stiffness and a consequent reduction in the optimum value of bearing damping. Similarly, although the bearing dynamic stiffness has been reduced by the lubricant aeration it does not follow that the bearings load capacity will be reduced, this has yet to be experimentally confirmed.

Although the effect of oil aeration on the system steady-state response to unbalance was not examined in detail, no significant effect was shown. It should be noted however that the rig was designed such that the first critical speed of the system occurred above the system stability threshold when non aerated lubricant was used. The increase of effective lubricant compressibility has been shown previously (1) to reduce the support stiffness and consequently reduce the speed at which this resonance occurs. As no significant vibration response increase was displayed below the stability threshold, either the reduced critical speed did not fall below the stability threshold or it was damped out by the available bearing damping. Further experimental work is required to investigate this case.

374

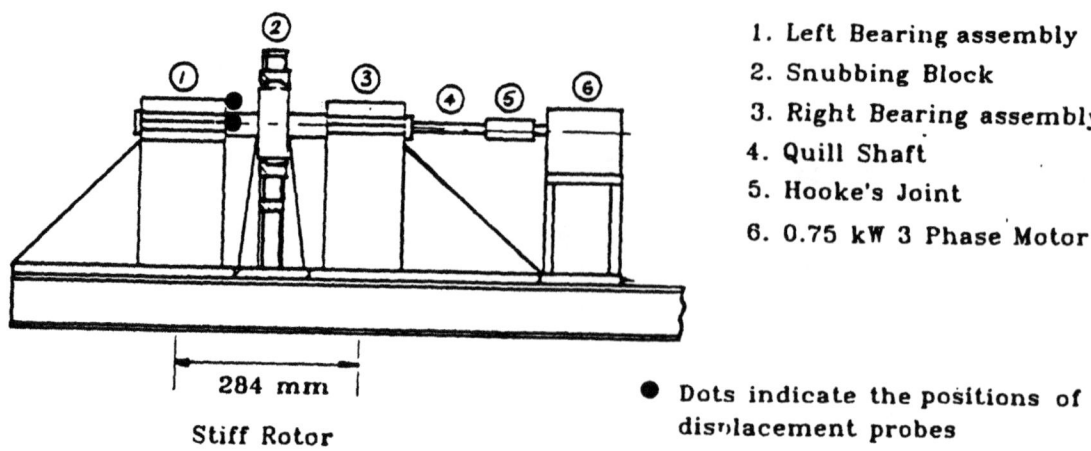

1. Left Bearing assembly
2. Snubbing Block
3. Right Bearing assembly
4. Quill Shaft
5. Hooke's Joint
6. 0.75 kW 3 Phase Motor

284 mm

Stiff Rotor

● Dots indicate the positions of displacement probes

Fig. 9 Journal Bearing Experimental Test Rig.

Journal Radius	38.1 mm
Journal Clearance	0.127 mm
Static Bearing Load	116 N
Rotor Mass	12 kg
Oil Viscosity	0.046 Ns/m²

Table 1 Journal Bearing Test Rig Data

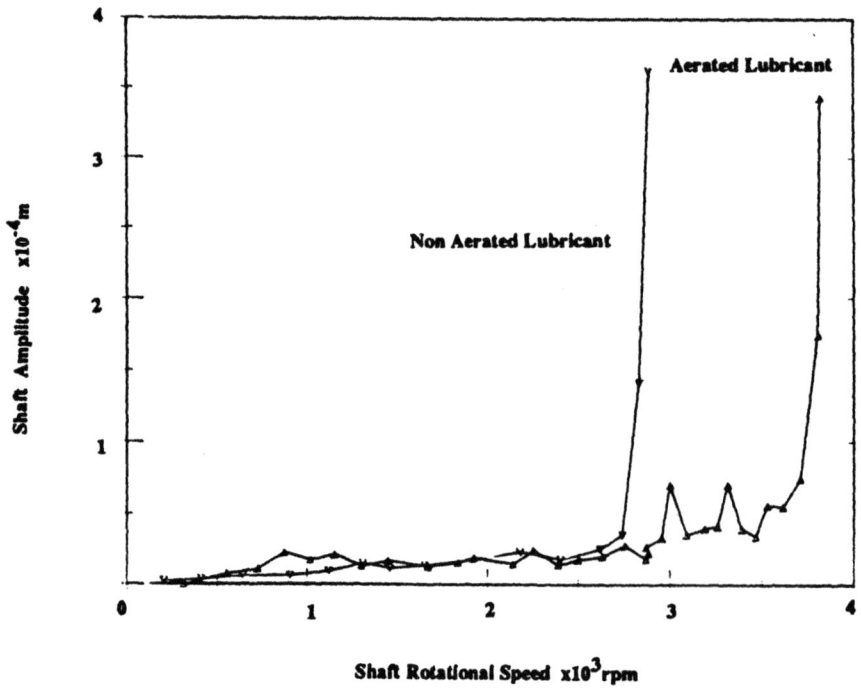

Fig. 10 Experimental Shaft Response Amplitude (Journal Bearing Test Rig)

6. Conclusions

With the majority of rotating machines which have a shaft end/midspan mass ratio of 0.5 or less it is likely that conventional bearings or support systems provide over stiff supports. Provided that optimum support damping is provided, a reduction of support or bearing stiffness may significantly inprove the system's stability and steady-state response to unbalance.

With rolling element bearing systems, this may be achieved with the use of accumulator modified hydrostatic supports around the rolling element bearings.

With shafts running on journal bearings oil aeration may be of value in improving the stability of the system. although it appears that it is important to control the degree and nature of the aeration in order to optimise any benefits in the system response.

The experimental work presented is of a preliminary nature and further work is required to confirm these findings.

References

1   Roach M P   " Vibration Control in Rotating Machinery using Variable Stiffness
                    Hydrostatic Squeeze-films "
                    PhD Thesis 1990

2   Goodwin M J   " Stability of Rotating Machines Running in Variable Stiffness Bearings "
     Roach M P   Proc NATO/ASI Conf. Vibration and Wear Damage in High Speed
     Penny J   Rotating Machinery, Troia Beach nr Lisbon
                    April 10-22 1988

3   Goodwin M J   " Variable Impedance Hydrodynamic Journal Bearings for Controlling
     Boroomand T   Flexible Rotor Vibrations "
     Hooke C J   Proc ASME Conf Mechanical Vibration and Noise
                    Montreal Sept 17-20 1989

4   Simandiri, S   " A Study of Squeeze-film Bearings "
                    PhD Thesis, University of New South Wales 1978.

5   Feng N S   " Effects of Gas Entrainment on the Squeeze-film Damper Performance "
     Hahn E J   Journal of Tribology Jan 1987 Vol 109 pp 149-154

SESSION 15   FLUID EFFECTS

MODELING OF RECIRCULATION IN AXIAL FLOW PUMPS
R. NOGUERA, F. MASSOUH, R. REY, S. KOUIDRI

R. NOGUERA : Maitre de Conférences   ENSAM-Paris.

F. MASSOUH : Maitre de Conférences   ENSAM-Paris.

R. REY : Professeur ENSAM-Paris.

S. KOUIDRI : Doctorant ENSAM-Paris.

1- ABSTRACT :

The object of this paper is to qualitatively point out the critical flow rate corresponding to recirculation in the axial pumps. This situation is reached when the global flow rate of the machine decreases from the nominal value Q_m to a first critical value Q_k corresponding to the cancellation of the flow rate at the hub. If the global flow rate decreases again, the flow becomes negative near the hub whereas it remains positive at the external part of the rotor. The existence of these two zones of flow, can be pointed out even if the radial equilibrium is supposed to be respected.

2- DIMENSIONING AND PERFORMANCE ANALYSIS PROGRAM :

The program for dimensioning axial pumps and fans (inverse calculation) is based on the definition of performance at the mean radius of the machine [ref. 1 and 2]. This program results from the correlations of deflexion and of losses estalished for the NACA 65 profiles [ref 3 and 4]. Apart from the usual degrees of freedom degrees concerning the load of rotor and stator blades (diffusion factor), the calculation is restricted the choice of two independent parameters $\bar{\bar{\alpha}}m$ and $\bar{\bar{\beta}}m$ defined at the mean radius (figure 1). These parameters are the angles of the mean velocity respectively at the stator and the rotor.

As it is shown [ref 5 and 6] the choice of the couple $(\bar{\bar{\alpha}}m, \bar{\bar{\beta}}m)$ makes it possible to define the geometric properties of the machine i.e external diameter, hub/tip ratio, and also the hydraulic properties such as : efficiency, NPSH, noise etc.

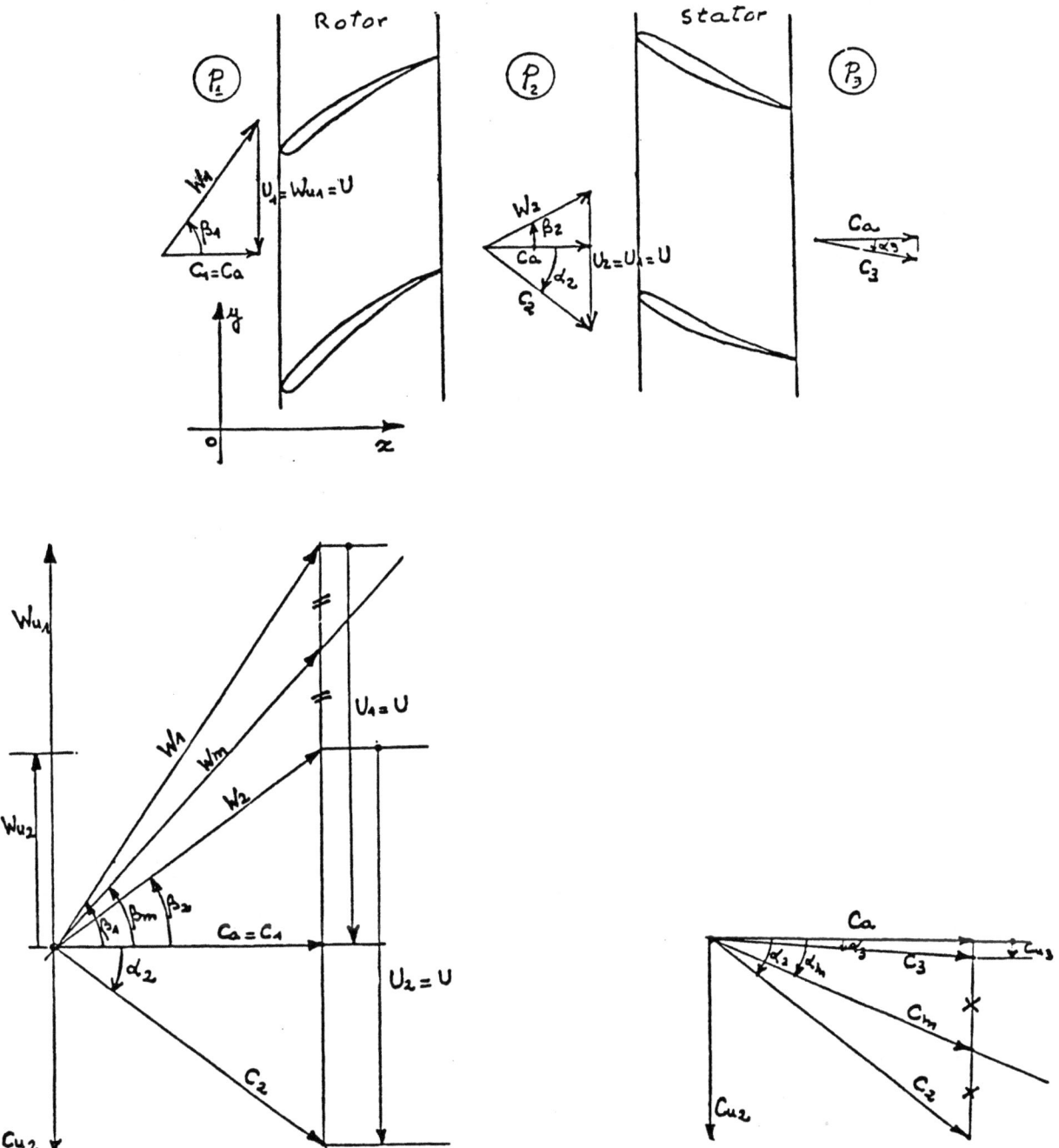

Figure 1 : Velocity diagram of a compression stage.

This dimensioning program is completed by an analysis programme (direct calculation) in which the boundary layers are taken into account but the radial desequilibrium and the prerotation phenomenon at the rotor upstream are neglected.

On the other hand, whatever the specific velocity of the
machine, the program shows that the angle $\bar{\bar{\beta}}m$, which is directly in
relation with the external diameter, determines the existence or the
non existence of the instable area of flow. [ref. 7]

3- DECOMPOSITION OF THE MACHINE INTO TWO ELEMENTARY MACHINES :

We have dealt with the case of a helical pump with a
specific velocity equal to 6, dimensioned with a mean rotoric angle
$\bar{\bar{\beta}}m=60°$ and furnishing the head 14 m and the flow rate 0.41 m^3/s.

The pump is designed using the free vortex hypothesis and is
divided into two elementary machines operating in parallel, with each
half of the total flow rate (fig 2). Radius $\bar{R}$ which constitutes the
limit between the two machines can be defined as :

$$\bar{R} = \sqrt{0.5(Re^2+Ri^2)}$$

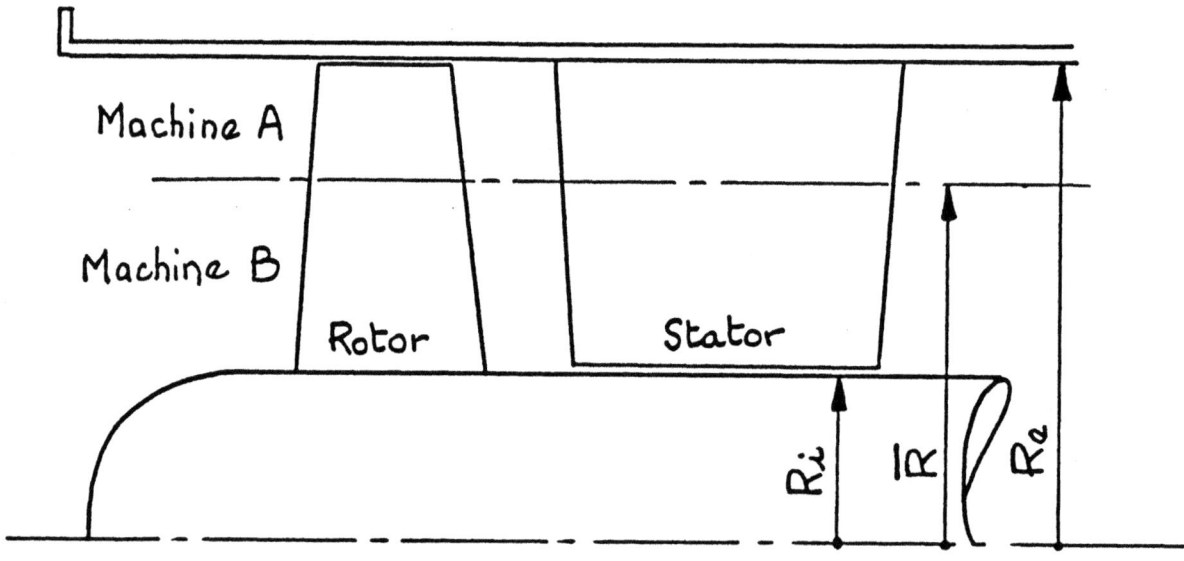

figure 2 : Definition of the radius $\bar{R}$

The different characteristics of these pumps are given in the following table :

	Pumps A	Pumps B	together
flow rate Qm (m^3/s)	0.205	0.205	0.41
head Hm (m)	14	14	14
specific speed Ω	4.24	4.24	4.24
$\bar{\beta}$ (deg)	67.4	54.7	60
Re (mm)	93.3	69.7	93.3
Ri (mm)	69.7	31.7	31.7

The program of performance analysis is applied to both the two elementary pumps A and B supposed to be independent. The results show two distinct characteristics and we can note that pump A which has a small angle $\bar{\beta}$m presents a great instable zone (fig. 3).

4- DEFINITION OF RECIRCULATION CRITICAL FLOW RATE :

As pumps A and B are supposed to operate in parallel, then the characteristic curve of the whole machine can be obtained by adding the flow rates under the same head. Inversely each point of the global characteristic allows to know the flow rate in each of the elementary pumps simply by horizontal translation.

In our modeling the recirculation critical flow rate Qk appears when the flow rate in the elementary central pump [machine B, fig(3)] disappears. Below this critical value the flow rate becomes negative in the center whereas the external flow rate remains positive.On the same figure we can observe that the decreasing of the global flow rate influences slightly the external flow rate, whereas the central flow rate is highly decreased and even reversed. We finally notice that the recirculation critical flow rate corresponds precisely to the head at which the flow rate in the internal pump desappears.

380

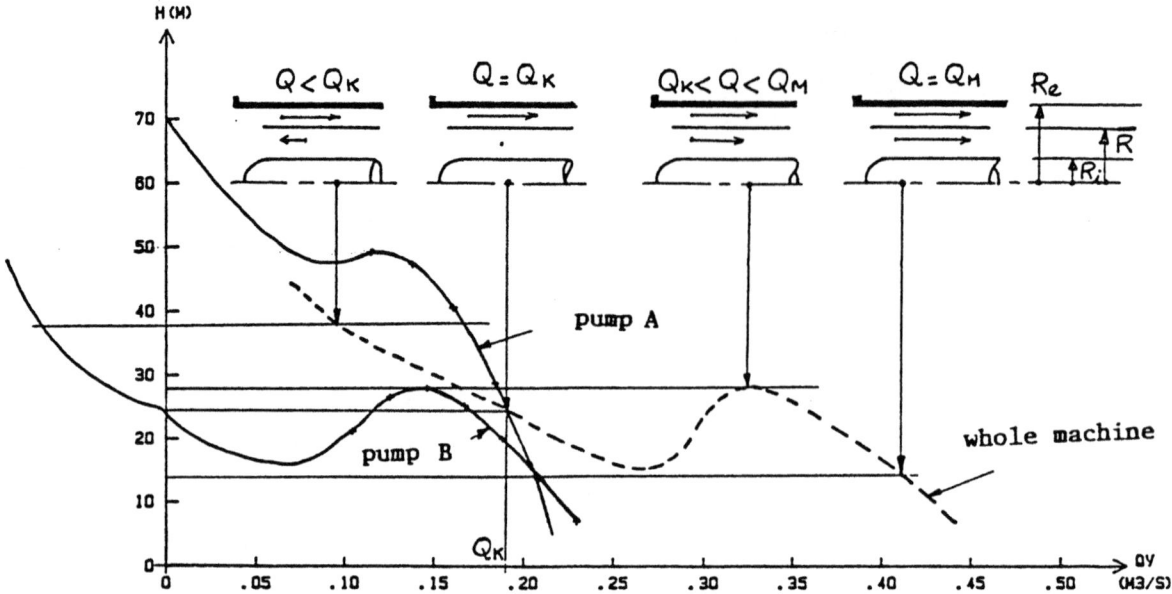

Figure 3 : characteristic curve

5- CONCLUSION :

This study has permitted us to qualitatively show the recirculation phenomenon in the helical pumps. If we follow the previous observations we can imagine to divide the reference pump into an infinite number of elementary pumps. Then the recirculation flow rate corresponds to the head at which the flow rate vanishes at the radius of hub Ri.

We can then imagine more stable machines at partial flow rate operating when the hub to tip ratio is higher.

REFERENCES :

1- REY R. "Méthode générale de détermination d'un étage de turbomachine axiale de compression", Thèse de Doctorat d'Etat, Université de PARIS VI, 1981.

2- REY R, NOGUERA R, "Optimisation of the axial flow pumps" 7th Conference on fluid machinery, Budapest, 1983 pp 706-718.

3- REY R "Représentation fonctiuonnelle des propriètés cinématiques des grilles d'aubes planes, ENTROPIE 1976 N° 70 pp 25-30

4- HERRIG L.J, EMERY J.C, ERWIN J.R, " Systematic two-dimensional cascade testes of NACA 65 - Series compressor blades at low speeds, NACA TN 3916 1957.

5- NOGUERA R "Contribution à la maîtrise du dimensionnement et du fonctionnement des machines axiales de compression . Etude des débits partiels et de la cavitation", Thèse de Doctorat d'Etat. Université de Paris VI 1986.

6- REY R, NOGUERA R, BRACHEMI B, MASSOUH F "Generalized represetation of hydraulic and geometric performances of axial pumps and fans" International Conference 35 years of Turboinstitut, LJUBLJANA, 1984, pp 499-512.

7- Rey R, NOGUERA R "Influence des critères de dimensionnement sur la stabilité des courbes caracteristiques des pompes hélices" 7èmeCongrès Français de Mécanique, Bordeaux 1985.

8- REY R, GUITON P, KERMAREC Y, VULLIOUD G, "Etude statistique sur les caracteristiques à débit partiel des pompes centrifuges et sur la détermination approximative du débit critique" La Houille Blanche, 1982, N° 2/3, pp 107-120.

9-COMOLET R "Mécanique des grilles d'aubes planes" Bulletin des Etudes et Recherches, E.D.F 1974, N° 2.

NOTATIONS :

C : Absolute velocity

Ca : Axial velocity

Cu : Tangential component of C

W : Relative velocity

U : Tangential velocity

ω : Angular velocity

H : Stage head

qv : flow rate

Re : Blade tip radius

Ri : Hub radius

g : Gravitational acceleration

ρ : Static pressure

α : Absolute flow angle

β : Relative flow angle

Statoric : αm mean angle defined as $\mathrm{tg}\ \overline{\overline{\alpha}}m = \dfrac{\mathrm{tg}\alpha_2 + \mathrm{tg}\alpha_3}{2}$

Rotoric : βm mean angle defined as $\mathrm{tg}\ \overline{\overline{\beta}}m = \dfrac{\mathrm{tg}\beta_2 + \mathrm{tg}\beta_3}{2}$

Ω : Specific speed defined as $\Omega = \dfrac{\omega\sqrt{qv}}{(gH)^{3/4}}$

FINITE ELEMENTS INTEGRATION OF THE BULK-FLOW EQUATIONS

B. Pizzigoni (*), E. Tanzi (**)

(*)University of Pavia, Italy
(**) Politecnico di Milano, Italy

ABSTRACT

In this paper a method to calculate dynamic coefficients for plain fluid film seals is shown. Bulk-flow equations, used as a mathematical model of the turbulent flow regime which takes place in the seal, are integrated by means of a finite element procedure.

INTRODUCTION

The type of seal considered in this paper (see Fig.1) is the plain one, used e.g. as a pump interstage seal in centrifugal pumps. Similar to bearings but with a larger clearance/diameter ratio (approx. 0.5%), these seals are generally also affected by a significant axial flow, due to the axial pressure drop between two successive stages. Due to these conditions, fully-developed turbulent flow conditions are normally present in these seals.

As well known, the presence of fluid film bearings and/or seals strongly affects the dynamic behaviour of a rotor. With reference to Fig.1, forces and moments exerted by the seal fluid film on the shaft can be written in the form

$$
\begin{Bmatrix} F_x \\ F_y \\ M_\alpha \\ M_\beta \end{Bmatrix} =
\begin{bmatrix} m_{xx} & m_{xy} & m_{x\alpha} & m_{x\beta} \\ m_{yx} & m_{yy} & m_{y\alpha} & m_{y\beta} \\ m_{\alpha x} & m_{\alpha y} & m_{\alpha\alpha} & m_{\alpha\beta} \\ m_{\beta x} & m_{\beta y} & m_{\beta\alpha} & m_{\beta\beta} \end{bmatrix}
\begin{Bmatrix} \ddot{x} \\ \ddot{y} \\ \ddot{\alpha} \\ \ddot{\beta} \end{Bmatrix} +
\begin{bmatrix} r_{xx} & r_{xy} & r_{x\alpha} & r_{x\beta} \\ r_{yx} & r_{yy} & r_{y\alpha} & r_{y\beta} \\ r_{\alpha x} & r_{\alpha y} & r_{\alpha\alpha} & r_{\alpha\beta} \\ r_{\beta x} & r_{\beta y} & r_{\beta\alpha} & r_{\beta\beta} \end{bmatrix}
\begin{Bmatrix} \dot{x} \\ \dot{y} \\ \dot{\alpha} \\ \dot{\beta} \end{Bmatrix} +
\begin{bmatrix} k_{xx} & k_{xy} & k_{x\alpha} & k_{x\beta} \\ k_{yx} & k_{yy} & k_{y\alpha} & k_{y\beta} \\ k_{\alpha x} & k_{\alpha y} & k_{\alpha\alpha} & k_{\alpha\beta} \\ k_{\beta x} & k_{\beta y} & k_{\beta\alpha} & k_{\beta\beta} \end{bmatrix}
\begin{Bmatrix} x \\ y \\ \alpha \\ \beta \end{Bmatrix} \quad (1)
$$

in which the so-called mass, stiffness and damping coefficients are evidenced. In a finite element model of the rotating system these coefficients become part of the overall mass, stiffness and damping matrices of the system itself. To calculate these coefficients bulk-flow equations ([1],[2]) have already been used (see e.g.[3]) to represent the turbulent flow regime for this kind of seal, showing a good agreement with experimental findings. In order to obtain the above mentioned coefficients, integration of the bulk-flow equation was performed by means of a perturbation method in paper [3]. More recently these coefficients have been obtained through the finite difference integration of the Navier-Stokes equations [4], which allows the analysis of any kind of seal. The computation method shown in this paper does not have such extended possibilities but, as far as plain seals are concerned, runs efficiently even on a personal computer with very short computation times.

SOLUTION PROCEDURE

Bulk-flow equations are fully described in [1], [2]. Geometry and coordinate system for plain seals considered in this paper are shown in Fig.1.

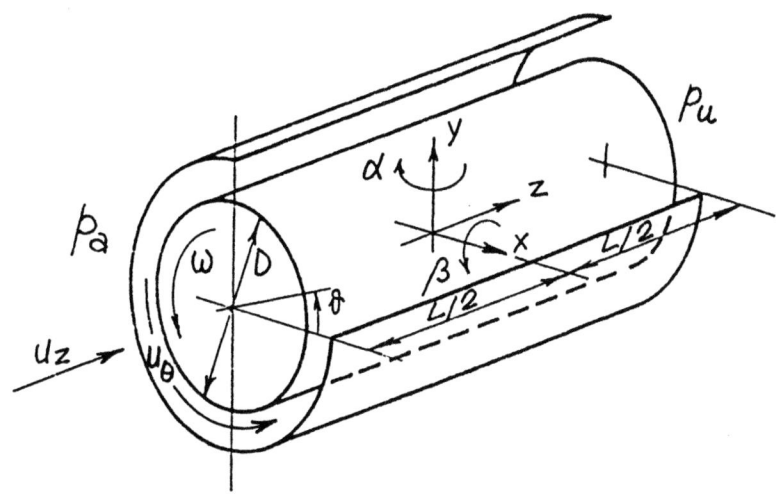

Fig.1 - Plain seal geometry and coordinate system

Fluid film flow is characterized by pressure distribution $p(z,\theta,t)$ as well as velocity distributions $u_z(z,\theta,t)$, $u_\theta(z,\theta,t)$ in axial and circumferential directions. These functions represent the unknowns of the problem. Dependence on time is accounted for by assuming that the journal oscillates steadily with a given pulsation Ω in the neighbourhood of a position x_c, y_c, α_c, β_c, i.e.

$$x = x_c + Re[\varepsilon_x e^{i\Omega t}] \qquad\qquad y = y_c + Re[\varepsilon_y e^{i\Omega t}] \qquad (2)$$

$$\alpha = \alpha_c + Re[\varepsilon_\alpha e^{i\Omega t}] \qquad\qquad \beta = \beta_c + Re[\varepsilon_\beta e^{i\Omega t}]$$

where complex quantities in square brackets are indicated with a boldface character. Pressure and velocity fields are then given by

$$p = p_0 + Re[\Delta p e^{i\Omega t}]; \quad u_z = u_{zo} + Re[\Delta u_z e^{i\Omega t}]; \quad u_\theta = u_{\theta o} + Re[\Delta u_\theta e^{i\Omega t}] \qquad (3)$$

where p_0, u_{zo} and $u_{\theta o}$ represent the pressure and velocity distributions when the journal center is located in x_c, y_c, α_c, β_c without oscillating. Boundary conditions for the problem are

$$p(-\frac{L}{2},\theta) = p_a - 0.5\rho V^2 u_z^2 (1+\xi) \qquad (4)$$

$$u_\theta(-\frac{L}{2},\theta) = \bar{u}_\theta \qquad (5)$$

$$p(\frac{L}{2},\theta) = p_u \qquad (6)$$

In eq.(4) the boundary pressure distribution is expressed in terms of an inlet pressure p_a outside the seal and kinetic energy head loss, being $V = \omega r$ the journal peripheral speed and ξ the local loss coefficient. Eq.(5) allows to keep account of inlet "swirl" i.e. prerotation of the fluid flow. Bulk-flow and continuity equations form a system of three partial differential equations of the first order which can be written as

$$A_n(.) = f_n \qquad\qquad n = 1,2,3 \qquad\qquad\qquad (7)$$

where $A_n(.)$ is a first order differential operator. Equation residuals (identically zero on the integration domain when the exact solutions p, u_z and u_θ of eqs.(7) are used) are defined as

$$R_n(.) = A_n(.) - f_n \qquad\qquad n = 1,2,3 \qquad\qquad\qquad (8)$$

Given a set of N linearly independent (shape) functions $\Phi_i(z,\theta)$ defined in the same domain, exact solutions are approximated by expressions:

$$p \approx \sum_1^N p_i \Phi_i; \qquad u_z \approx \sum_1^N u_{zi} \Phi_i; \qquad u_\theta \approx \sum_1^N u_{\theta i} \Phi_i \qquad\qquad (9)$$

When these approximate solutions are used, residuals (8) are no longer identically zero on the integration domain but become functions of the coordinates z,θ and of the 3N parameters p_i, u_{zi}, $u_{\theta i}$. By using Galerkin's method the following algebraic system

$$\int_D R_n(z,\theta, p_i, u_{zi}, u_{\theta i}) \Phi_j dA = 0 \qquad n=1,2,3; \quad i,j=1,2,...N \qquad (10)$$

of 3N equations in the 3N unknowns $p_i, u_{zi}, u_{\theta i}$ is obtained. From the operative point of view the integration domain is divided into rectangular elements. Let N be the number of nodal points of the resulting mesh.

The N functions F_j are chosen so as to satisfy the conditions $\Phi_j(z_k,\theta_k)=1$ if k=j or $\Phi_j(z_k,\theta_k)=0$ elsewhere and to be linear in each element. Eqs.(7) are non linear and so consequently are eqs.(8),(10). In order to deal with a linear system, operators $A_n(.)$ are linearized with respect to p, u_z, u_θ by developing $A_n(.)$ in Taylor' series in the neighbourhood of a known solution $p_o, u_{zo}, u_{\theta o}$. For this purpose the "centered journal" position is the most convenient choice, since it leads to a form of eq.(7) easy to solve. From this starting point, the general journal position x_c, y_c, α_c, β_c (with corresponding solutions p, u_z, u_θ) is reached by sufficiently small steps $\Delta x, \Delta y, \Delta\alpha, \Delta\beta$. Dynamic coefficients appearing in eq.(1) are computed as ratios between increments ΔF_x, ΔF_y, ΔM_α, ΔM_β and respective perturbations $\text{Re}[\varepsilon_x e^{i\Omega t}]$, ..etc. Force and moment increments can be found after computing the variation $\Delta p(z,\theta)$ with Galerkin's method. In matrix form we have

$$\{\Delta F\} = -\Omega^2[M]\{\varepsilon\} + i\Omega[R]\{\varepsilon\} + [K]\{\varepsilon\} \qquad\qquad (11)$$

By computing $\{\Delta F\}$ in correspondence to a series of values of Ω, the elements m_{ij}, r_{ij}, k_{ij} of the different matrices can be determined by least squares procedure.

NUMERICAL RESULTS

As a first application dynamic coefficients have been calculated, for different L/D ratios, for a seal having the following characteristics: D=152.4 mm, C=.254 mm, fluid density ρ=996.5 kg/m^3, viscosity μ=.8614x10^{-3} Poise, pressure drop across the seal ΔP=3.446 MPa, angular speed ω=3600 rpm, Hirs' turbulence coeffs. m_0=-0.25, n_0=-0.066. The stiffness and damping coefficients (divided by 1E6) are shown in Table 1 (FEM column) in comparison with those obtained with the perturbation method [3] (PM column) for different L/D ratios, v_0 being the "swirl coefficient".

Table 1. Stiffness and damping coefficients compared to those obtained in [3]

	L/D = 0.16 v_0 = 0		L/D = 0.5 v_0 = 0		L/D = 1.0 v_0 = 0		L/D = 0.5 v_0 = -0.5	
	PM	FEM	PM	FEM	PM	FEM	PM	FEM
K_{xx}	13.6	13.47	25.7	23.35	16.6	14.42	26.2	23.74
K_{xy}	2.59	2.34	18.8	17.04	61.3	55.6	2.37	1.76
$K_{\beta x}$	-0.88	-0.78	-13.5	-10.66	-60.9	-46.4	1.46	1.76
$K_{\alpha x}$	-0.57	-0.58	-3.75	-3.41	-8.10	-7.2	-3.7	-3.35
R_{xx}	1.37	1.24	10.0	9.0	32.5	29.6	9.97	9.07
R_{xy}	.55	.26	14.2	12.4	85.7	66.4	11.5	9.86
$R_{\beta x}$	-.46	-0.42	-7.1	-6.5	-32.4	-43.7	-7.12	-6.4
$R_{\alpha x}$	-.30	-0.28	-29.5	-20.7	-154.	-214.	-27.	-18.

Calculation of the coefficients appearing in eq.(1) has been also performed as a function of e_x/C ratio for another seal having the following characteristics: D=160 mm, C=.36 mm, fluid density ρ=1000 kg/m^3, viscosity μ=10^{-3} Poise, pressure drop across the seal ΔP=1 MPa, angular speed ω=4000 rpm, Hirs' turbulence coeffs. m_0=-0.25, n_0=-0.066, swirl ratio v_0=0.3. The results can be directly compared with those obtained by other authors [4], with a different method, and with some of the experimental results given in [5] for the same seal.

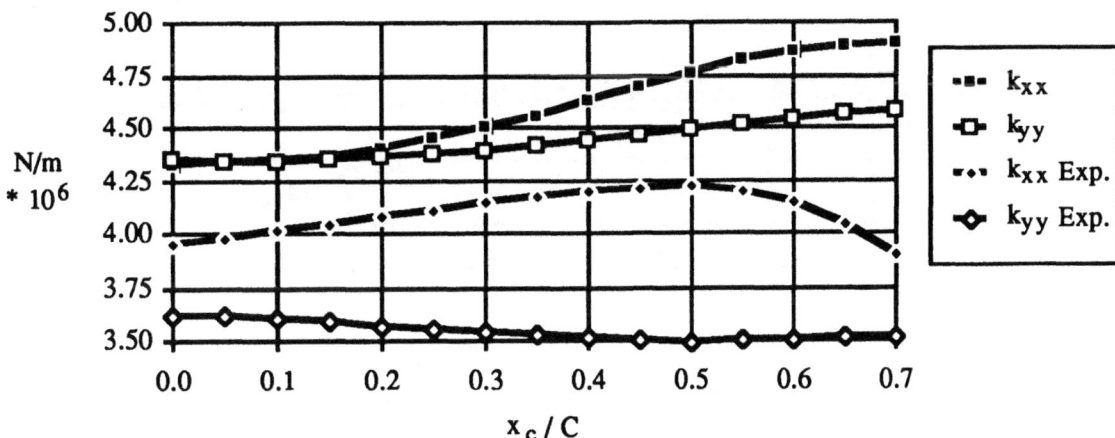

Fig.2 - Direct stiffness coefficients vs. eccentricity ratio, Exp. refers to experimental findings given in [5]

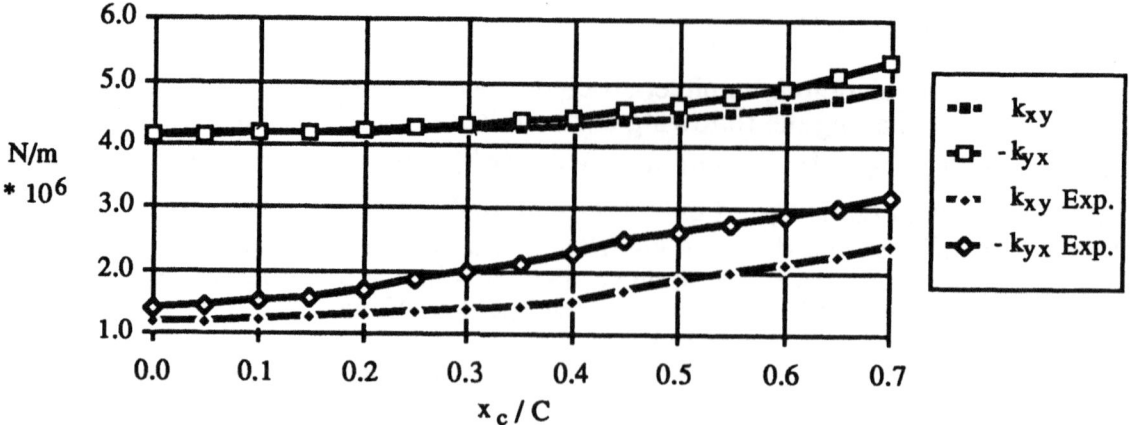

Fig.3 - Cross coupling stiffness coefficients vs. eccentricity ratio, Exp. refers to experimental findings given in [5]

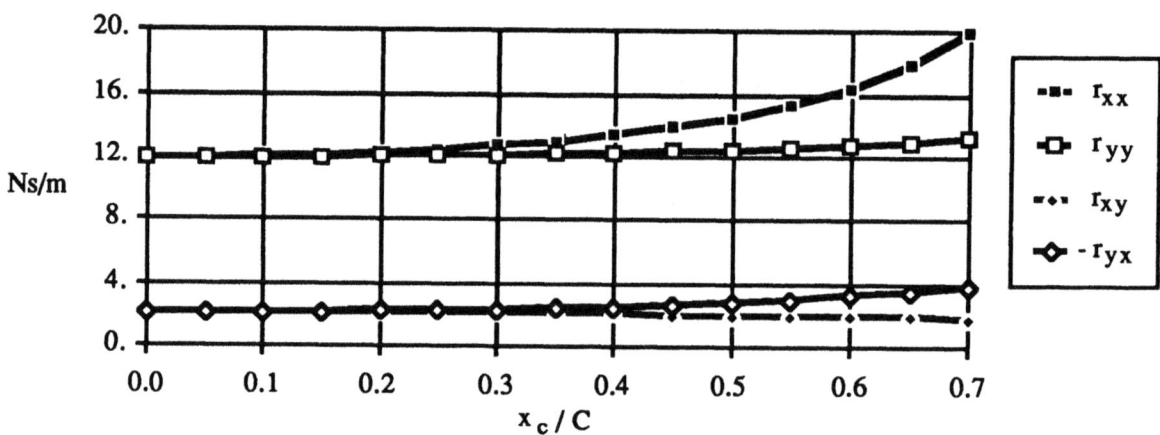

Fig.4 - Direct and cross-coupling damping coefficients vs. eccentricity ratio

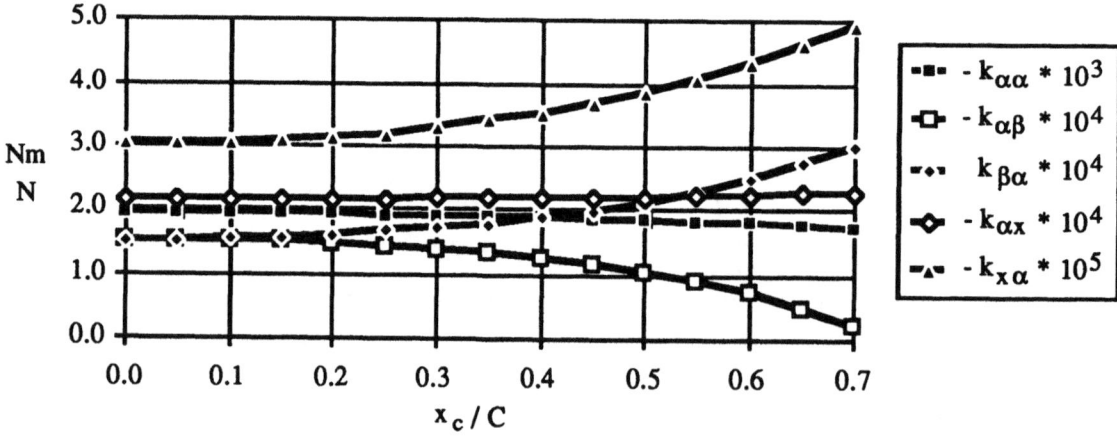

Fig.5 - Direct and cross-coupling moment and force stiffness coefficients vs. eccentricity ratio

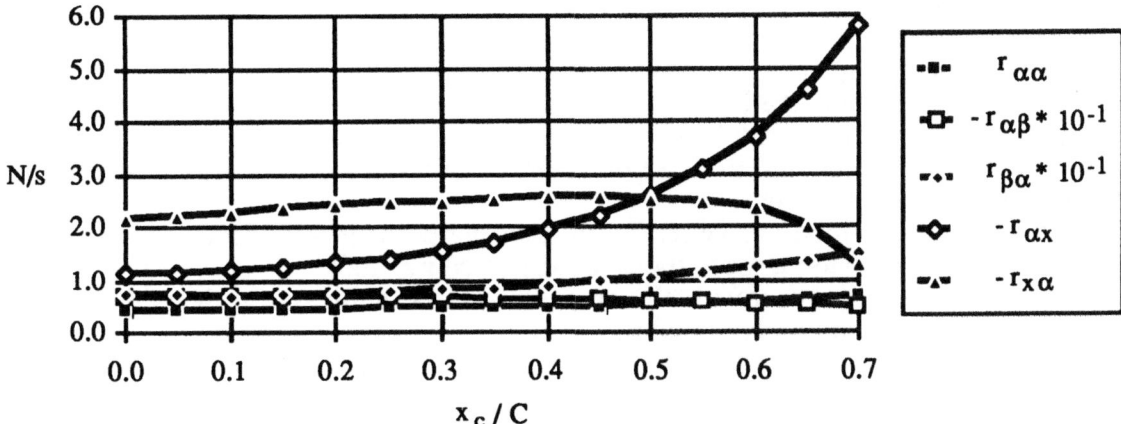

Fig.6- Direct and cross-coupling moment and force damping coefficients vs. eccentricity ratio

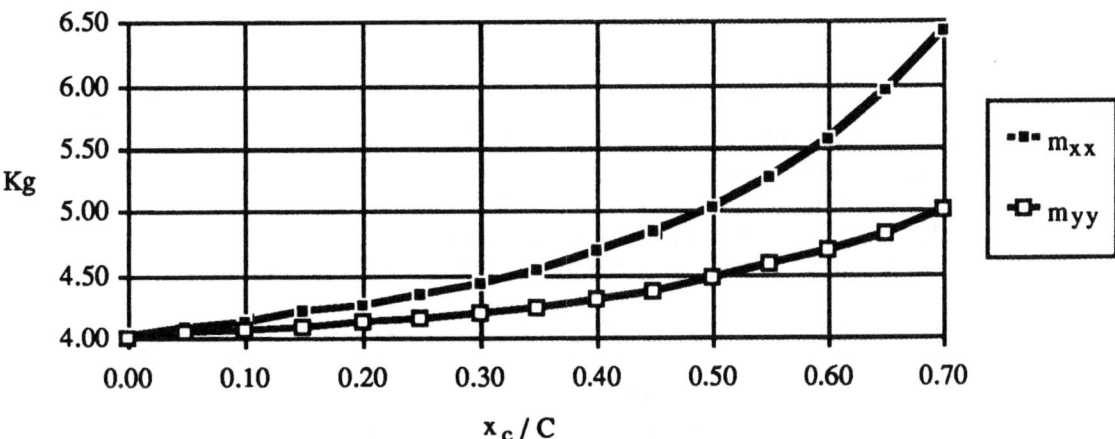

Fig.7- Direct force-added mass coefficients vs. eccentricity ratio

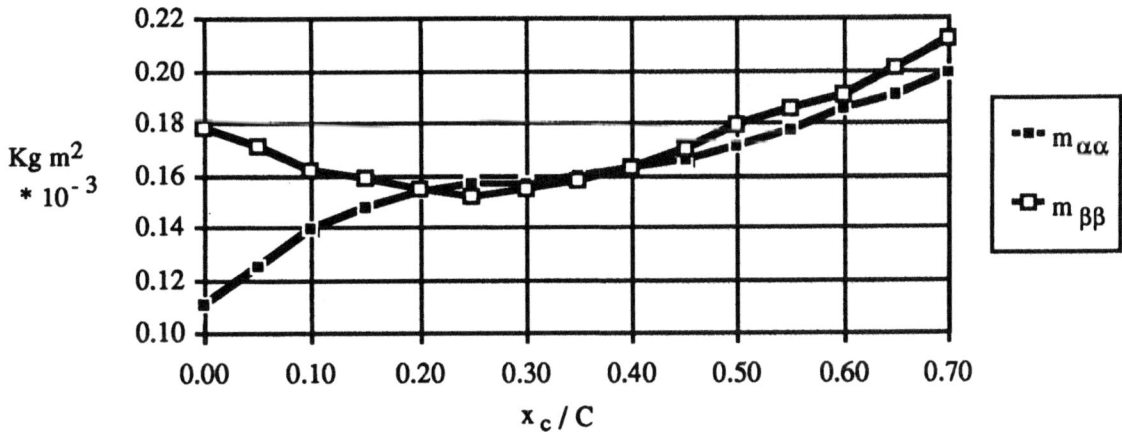

Fig.8- Direct moment -added mass coefficients vs. eccentricity ratio

388

CONCLUSIONS

Stiffness, damping and inertia coefficients for a plain fluid film seals are obtained in this paper by integrating the bulk-flow equations with a finite elements approach. With a computer program based on this method, these coefficients have been computed and the results have been compared to those obtained by other authors. It should be noted that with the method shown here:
1) the results obtained are generally in good agreement with those found by other authors [4], who use a much more sophisticated method;
2) accuracy can be improved (the same cannot be achieved with method [3]) by increasing the number N of nodes;
3) by a suitable iterative procedure, any journal-seal relative position can be analyzed
4) calculation of stiffness and damping coefficients can be performed in a reasonable computation time on a normal PC.

ACKNOWLEDGMENTS

Support of M.P.I.(Ministero Pubblica Istruzione) and C.N.R. (Consiglio Nazionale delle Ricerche) is gratefully acknowledged.

REFERENCES

[1] Hirs G.G.:"A Bulk-Flow Theory for Turbulence in Lubricant Films". Journal of Lubrication Technology, Trans. ASME Series F, Vol.95, n.2, April 1973
[2] Hirs G.G.:"A Systematic Study of Turbulent Film Flow". Journal of Lubrication Technology, Trans. ASME, January 1974
[3] Childs D.W.:"Rotordynamic Moment Coefficients for Finite Length Turbulent Seals". IFTOMM Conf.'Rotordynamics Problems in Power Plants', Rome, September 1982
[4] Nordmann R.,Dietzen F.J.:"Finite Difference Analysis of Rotordynamic Seal Coefficients for an Eccentric Shaft Position", Proc. Workshop Rotordynamic Instability Problems in High-Performance Turbomachinery, College Station, 1988
[5] Falco M., Mimmi G., Marenco G.: "Effects of Seals on Rotor Dynamics", Proc of the International Conference on Rotordynamics, Tokyo, 1986

A CONTRIBUTION TO THE DYNAMIC SIMULATION OF NONLINEAR ROTATING MACHINES WITH FLUID DYNAMICAL COUPLING ELEMENTS

D. WEBER[*], R. CARDINALI[**], R. NORDMANN[*]

[*] University of Kaiserslautern, Germany
[**] University of Campinas, Brasil

ABSTRACT

In this paper a simulation program for nonlinear rotordynamic analysis that allows to calculate the bending vibrations of systems with many degrees of freedom (DOF's) is presented. The program is based on the Finite Element Method in combination with a modal substructure technique to reduce the system's equations of motion. The solution of this nonlinear system is performed in time domain with a bimodal decoupling procedure in combination with a polygon method. The fluid dynamic coupling elements available in the program are seals (smooth and grooved) and journal bearings (circumferential, multi-lobe and tilting-pad). Whereas the seals may be modelled linearly, the journal bearings must be modelled nonlinearly, in the case of large excitation forces or of vertical machines. It is shown that the use of force tables to model the nonlinear bearing forces is a very fast method to realize parametric studies.

1. INTRODUCTION

The present paper deals with the nonlinear bending vibrations of rotating machines considering fluid dynamic coupling elements like seals or journal bearings.

In order to improve the availibility of high performance rotating machinery, it is essential to take into account the bending vibrations accurately. There exist different rotordynamic programs working with linearized models like that developed by DIEWALD [1]. All of them are valid for horizontal machines with conventional excitation forces. They proceed on the assumption that only small vibrations exist about a steady state point.

But in the case of horizontal machines with a rotational speed near the point of instability or in the case of large excitation forces, or in the case of vertical machines, the assumption of small vibrations is not valid. In these cases, the correlation between movements and forces for journal bearings shows a large nonlinearity. In contrast to the bearings, the seals can be modelled linearly, because their clearance is twice the clearance of journal bearings and previous research work shows that it is valid to work with linear models for seals up to 60% of their clearance [2].

The program presented here is based on the Finite Element formulation in combination with a modal substructure technique [3]. The following elements and effects, which are important for rotating machines, are taken into account: springs and dampers to model ball bearings; beams with the influence of gyroscopic effects, shear deformation, rotary inertia, clearance excitation, internal and external damping to model an elastic shaft; smooth or grooved seals; nonlinear journal bearings like cylindrical, multi-lobe or tilting-pad ones and discs also with the influence of gyroscopic effects to model the impeller.

The nonlinearities require a calculation in the time domain. It is thus possible to solve Reynolds' equation at each time step. Thus each simulation requires a large CPU-time and consequently is very time consuming and expensive.

A method is presented in this paper that relies on working with force tables for the bearings to reduce the CPU time required during the simulation. The paper is divided into two parts. In the theoretical part the analytical and numerical methods used are explained and in the second part some example simulations will be shown.

2. THEORETICAL PART - NONLINEAR JOURNAL FORCES

All necessary information about the substructure technique used here can be found in [3] or [4]. The most important issue of this paper is the handling of the nonlinear bearing forces.

Journal bearings are very important fluid dynamic coupling elements for the nonlinear simulation. Often tilting pad or multi lobe bearings are employed. At the bearings, the shaft is guided by a pressure field originating normally from a laminar circumferrential fluid flow. The magnitude of nonlinearity of the bearing forces depends on the relative displacements and velocities of the film. Fig. 1 shows the behaviour of a 360° cylindrical bearing.

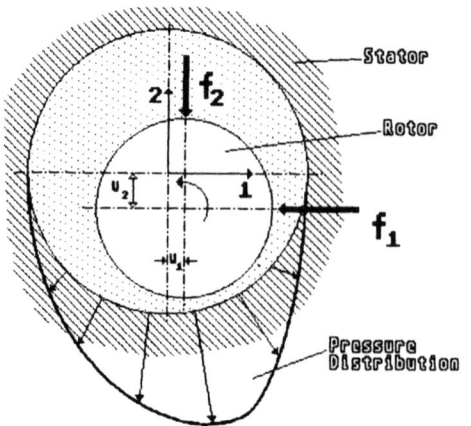

Fig. 1: Pressure distribution and forces on a 360° cylindrical bearing

To calculate the forces the Reynolds' equation, a partial differential equation of second order in the relative pressure Π has to be solved.

$$\frac{\partial}{\partial \varphi}\left\{H^3 \frac{\partial \Pi}{\partial \varphi}\right\} + \frac{1}{\beta^2}\frac{\partial}{\partial \zeta}\left\{H^3 \frac{\partial \Pi}{\partial \zeta}\right\} = 6\left\{\frac{\partial H}{\partial \varphi} + \frac{2\partial H}{\Omega \partial t}\right\} \tag{10}$$

To do so, axial and circumferential boundary conditions are necessary. In axial direction the pressure must be zero at the end of the bearing. In circumferential direction the pressure must be zero at the end of the pads too and it must always be greater than or equal to zero all over the surface of a pad. Here the physical conditions, which satisfy the continuity equation $\left(\left.\frac{\partial \Pi}{\partial \varphi}\right]_{\Pi(\varphi)=0} = 0\right)$ are used.

There are two different possibilities to handle the nonlinear journal forces. The first one is to solve the Reynolds equation for each time step and the second one is to work with force tables, which characterize the journal bearings.

Solving the Reynolds Equation at each Time Step: To do that, three classes of methods can be distinguished. The first class are the analytical methods for very short or infinitely wide bearings. The methods based on DUBOIS and OCVIRK /5/ works with a transformation of the Reynolds equation into an ordinary differential equation. After two integrations, the pressure Π is obtained.

The second class are the analytical approaches. The method based on FALKENHAGEN /6/ works with a correction of the infinite wide bearing solution. The method based on SPRINGER /7/ starts with the equivalent variation problem. To get the relative pressure Π, a setup with Tschebyscheff series is done.

The third class are the numerical solutions. With Finite Difference terms, the partial derivations are substituted. A one-dimensional method with a parabolic setup in axial direction and a two-dimensional method are available /8/.

The methods of the second and third classes are implemented in the presented program. They can be used for circumferential, multi-lobe and tilting-pad bearings. Working with the solution of the Reynolds equation in each time step may result in a large CPU-time.

Force Tables for Characterizing the Bearing Forces: The bearing forces f_1 and f_2 shown in Fig. 1 are dependent on the relative displacements u_1 and u_2 and the relative velocities $\dot{u}_1$ and $\dot{u}_2$ between the rotor and the stator.

$$f_1 = f_1(u_1, u_2, \dot{u}_1, \dot{u}_2) \qquad \text{and} \qquad f_2 = f_2(u_1, u_2, \dot{u}_1, \dot{u}_2) \tag{11}$$

To reduce the CPU-time necessary for the solution of the single DOF equations (9) the following idea is pursued. Solving the Reynolds equation (10) is seperated from the solution of the equation of motion. For all feasible states of motion of the shaft in the bearing, a table of the pertinent bearing forces f_1 and f_2 will be created before starting the rotordynamic simulation. To get this table the methods described in section 2.2.1 are available. For each kind of bearing and for each working condition a seperate table must be created. During the solution of the single DOF equations (9) the necessary bearing forces $\mathbf{f}_{nl}$ belonging to the momentary state of motion are chosen from the table.

To get the forces f_1 and f_2 of a special state of motion $\tilde{\mathbf{u}} = [\tilde{u}_1, \tilde{u}_2, \dot{\tilde{u}}_1, \dot{\tilde{u}}_2]$ it is necessary to find all neighbours of $\hat{\mathbf{u}}$ in the table. To obtain these $2^4 = 16$ neighbours it is not possible to make a direct search in the great force table, because that would lead to an unacceptable CPU-time. With the well known information about the number of variations of each parameter and their values it is possible to get all neighbours by an address calculation. With all these neighbours the forces f_1 and f_2 are obtained by a linear approximation. To do that, two linear sets of equations are solved by a Householder-algorithm.

The force table depends on 4 parameters. For each parameter, a lot of variations are necessary. This results in a very great table and thus in great CPU-time necessary to build up the table. To reduce the table and the calculation time the conditions for symmetry of the bearings are used. For tilting pad or multi lobe bearings with more than one pad, each state of motion can be transformed to the first pad. So the number of variations for u_1 and u_2 can be reduced drastically, dependent on the number of pads.

3. DEMONSTRATION EXAMPLES

3.1 Influence of the Solution Method of the Reynolds Equation on the CPU-time

With this example the influence of the method to calculate the bearing forces on the CPU-time is shown. Thus the mechanical part of this example should be very simple to reduce the influence of the substructure technique on the calculation time. A simple rigid rotor with only two DOF's is chosen, based on a publication of Merker /9/.

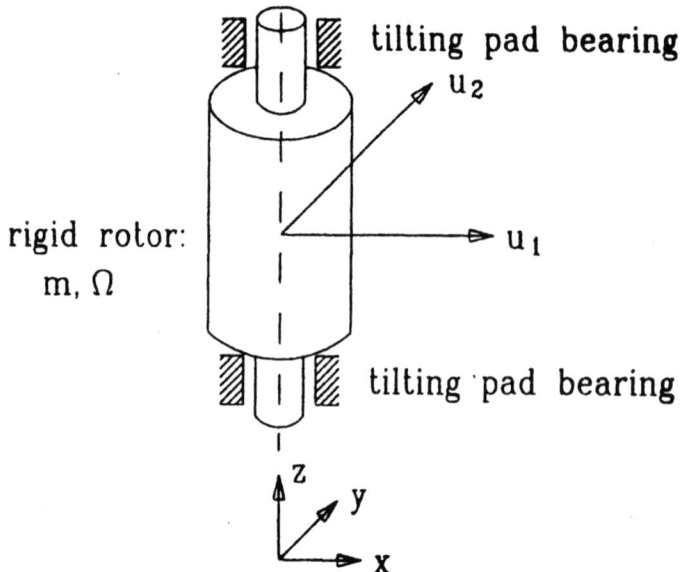

Fig. 2: Rigid rotor in two identical tilting pad bearings

As an excitation force an unbalance acts at the centre of gravity. Table 1 shows all necessary input data.

In version A the Reynolds equation is solved by a one dimensional Finite Difference Method in each time step. In version B the approximation based on Falkenhagen /6/ is used to calculate the bearing forces in each time step. In the last version C, the force tables are used. The tables were built up with the method based on Falkenhagen. All other input data are the same.

All orbits in Fig. 3 are nearly identical but in CPU-time (VAX-Station 3100) there are great differences:

$$
\begin{array}{lllll}
A & - & 7\ h & 30 & min \\
B & - & & 12 & min & 35\ sec \\
C & - & & 1 & min & 5\ sec
\end{array}
$$

This example shows, that working with force tables is a very fast method with regard to the calculation of bending vibration. But to build up the table, which has in this example 32856 combinations, a lot of CPU-time is required. Thus working with tables makes sense only if parametric studies must be done, however, parametric studies for the bearing parameters can be inhibitively time consuming because of the necessary recalculation of the bearing force table for each parameter change. Another possible application of force tables is to calculate the vibrations periodically within the scope of a diagnosis procedure at a machine in daily action. In these cases only one table for each bearing is necessary.

Table 1: Input data of the rigid rotor

rigid disc: mass = 1000 kg Ω = 7639,4 rpm

r [mm]	100,0
h_o [mm]	0,1
width [mm]	40,0
viscosity [kgm/s]	0,02
r_s [mm]	100,1
boundaries	Gümbel
number of pads	12
φ_{o1} [deg]	12,5
Φ_1 [deg]	0,0
φ_{s1} [deg]	0,0
Δ [mm]	0,0

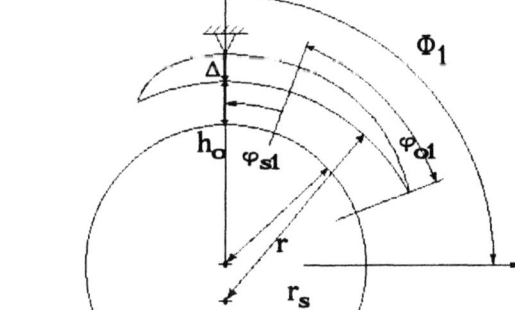

394

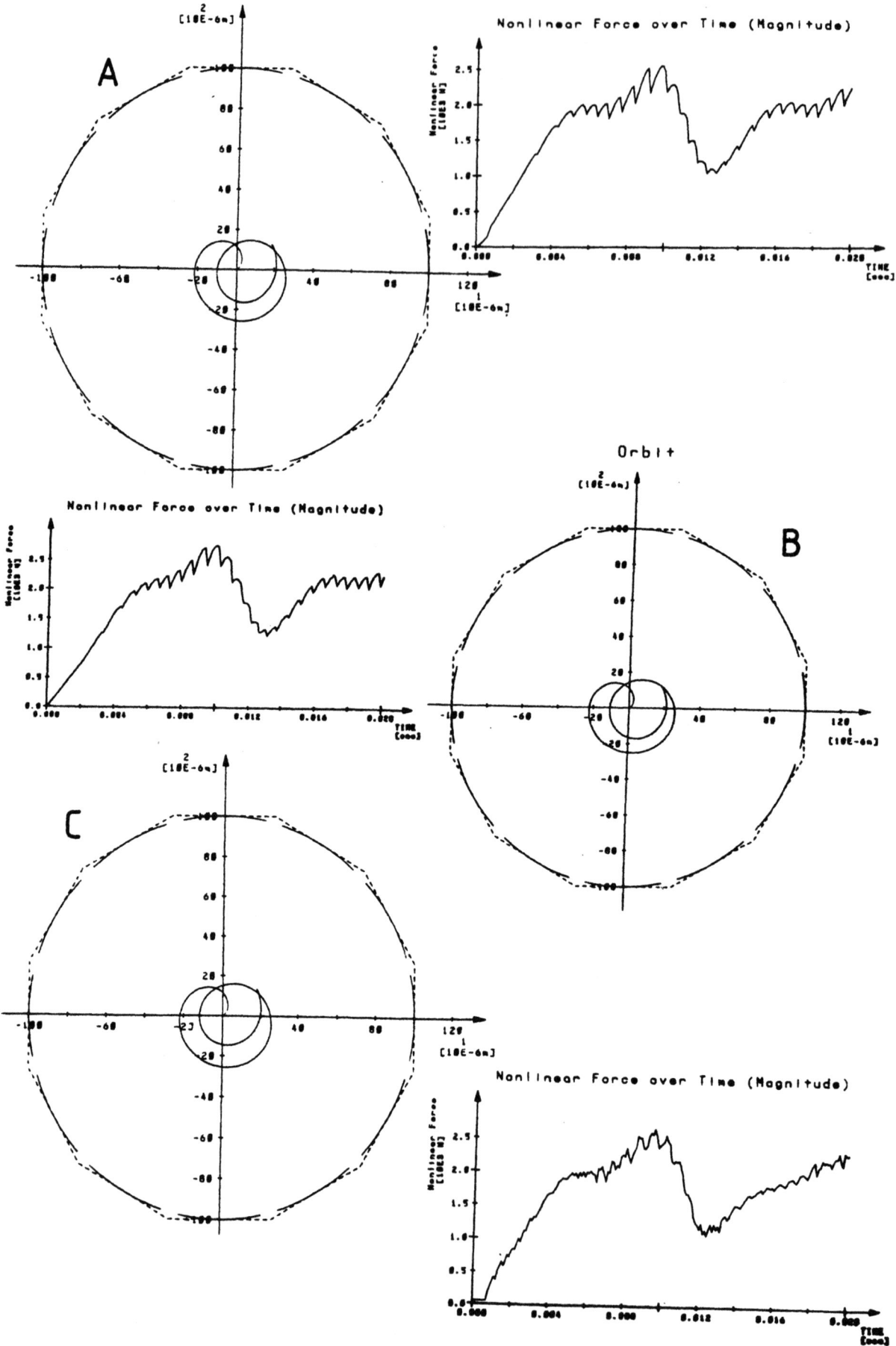

Fig. 3: Orbits of a rigid test-rotor

A - Finite Difference Method
B - Method of Falkenhagen
C - Force Tables

3.2 A Hydrogenerator as a Demonstration Example

A Francis type hydro-unit with vertical arrangement is modelled as a demonstration example. It has a nominal power of 160 MW at a rotational frequency of 1.43 Hz. Fig. 4 shows the Finite Element model. All necessary input data of the generator, turbine, shaft, bearings and seals can be found in /3/.

The expansion compared to the model of /3/ is the consideration of the influence of the bearing support. The masses M_1 and M_2 and the stiffnesses K_2 and K_4 of these parts of the casing were estimated with regard to some static FE calculations with ANSYS:

$$M_1 = 12400 \text{ kg} \qquad k_2 = 3,6\,E\,9 \text{ N/m}$$
$$M_2 = 8700 \text{ kg} \qquad k_4 = 2,5\,E\,9 \text{ N/m}$$

The excitation forces are the unbalanced masses of the generator and turbine as well as the stochastically fluctuating forces due to the turbulent flow inside the turbine /10/. Fig. 5 shows the orbits due to unbalanced forces at the generator (node 1, eccentricity = 0,3 mm) and at the turbine (node 7, eccentricity = 0,2 mm) plus stochastic forces in both directions at node 7 to model the hydraulic forces.

Fig. 5 shows the influence of the hydraulic forces very well. At the generator and at the generator bearing this influence can be neglected. Their orbits depend only on the unbalanced forces . But at node 4 and 5 as well as at the turbine bearing and the turbine the hydraulic forces are an essential part of the excitation.

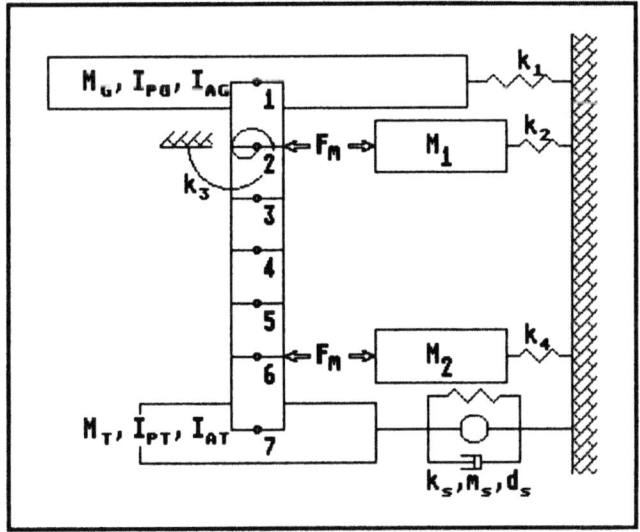

generator, magnetic pull
axial support, generator bearing
bearing support

turbine bearing, bearing support

turbine, seals

Fig.4: Model of a vertical hydro-unit

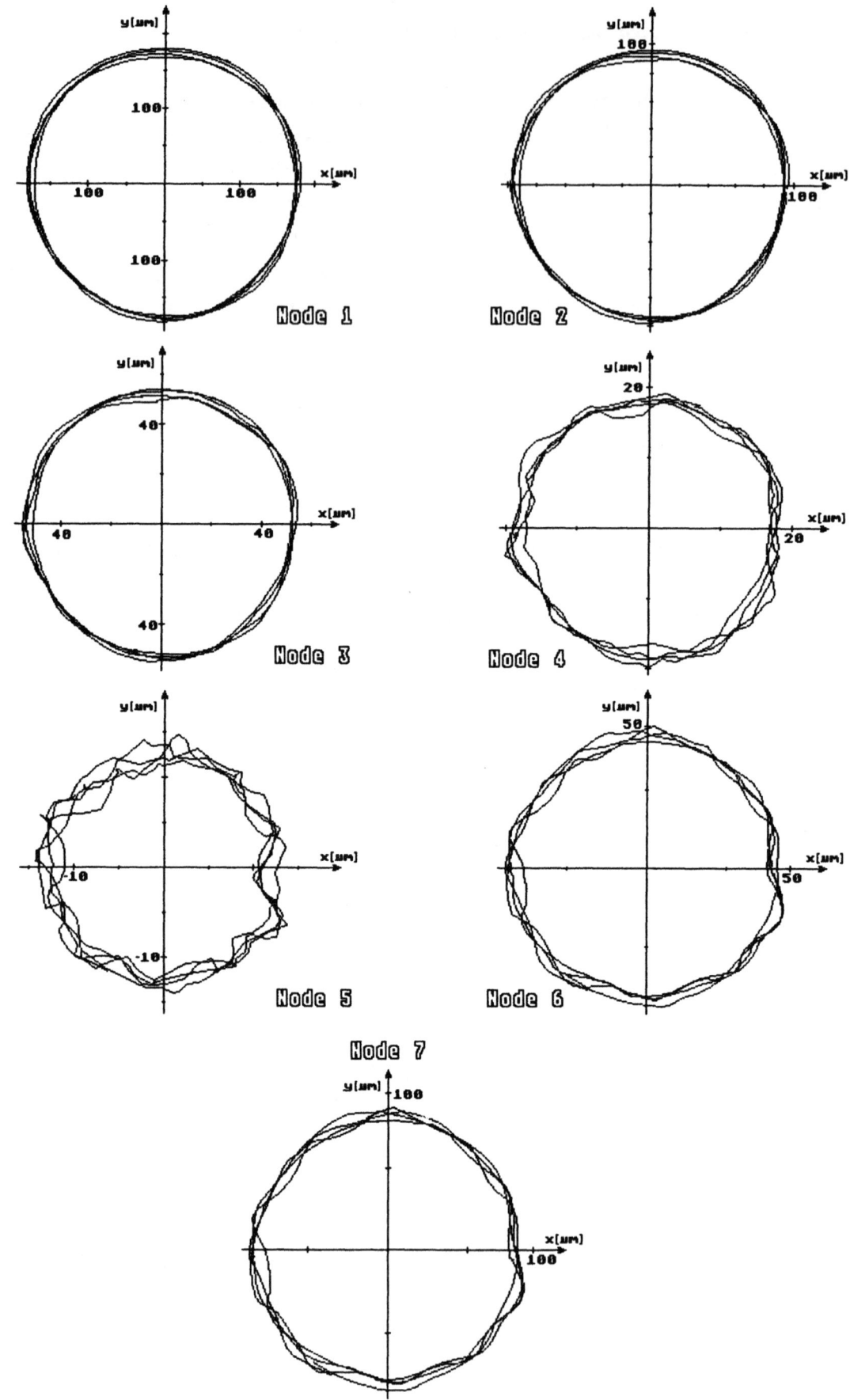

Fig. 5: Unbalanced and stochastic response of the hydrogenerator

4. CONCLUSIONS

The program presented here allows to calculate the bending vibrations of large rotor systems with regard to the nonlinear influence of journal bearings. The first example shown in section 3 demonstrates the influence of the calculation method to get the bearing forces on CPU-time. Working with force tables is a very fast possibility if parametric studies must be done or if the field of interest is the diagnosis of machines in daily action. If parametric studies should be done at the bearings the example given shows the importance of fast approximation methods to solve the Reynolds equation. With a method based on Falkenhagen and expanded on tilting pad bearings a drastic reduction of the CPU-time in comparison with the Finite Difference method is achieved. The second example shows the importance of the hydraulic forces at a hydrogenerator.

5. REFERENCES

[1]   W. Diewald and R. Nordmann, "Dynamic Analysis of Centrifugal Pump Rotors with Fluid-Mechanical Interactions", 11th Biennial Conference on Mechanical Vibrations and Noise, Boston/USA 1987.

[2]   F.J. Dietzen and R. Nordmann, "Calculating Rotordynamic Coefficients of Seals by Finite Difference Techniques", Journal of Tribology, 1987, Vol. 109, pp. 388-394.

[3]   D. Weber, R. Cardinali and R. Nordmann, "A Modal Substructure Technique for the Nonlinear Dynamic Simulation of Hydroelectric Machines", International Conference on Mechanics of Solids and Structures, Nanyang Technological University, Singapore 1991.

[4]   R. Gasch and R. Knothe, Strukturdynamik, Vol. II, Springer Verlag, New York, London, Paris, Tokyo, 1989.

[5]   G.B. Dubois and F.W.- Ocvirk, "Analytical Derivation and Experimental Evaluation of Short-Bearing Approximation for Full Journal Bearings", National Advisory Committee for Aeronautics, Report 1157, 1953.

[6]   G.L. Falkenhagen, E.J. Gunter, F.T. Schuller, "Stability and Transient Motion of a Vertical Three-Lobe Bearing System", 1971. ASME paper No 71-VIB-76.

[7]   H. Springer, "Zur Berechnung hydrodynamischer Lager mit Hilfe von Tschebeyscheff-Polynomen", Forschung im Ingenieurwesen, 1978, Vol. 4, pp. 126-134.

[8]   R. Klump, "Ein Beitrag zur Theorie der Kippsegmentlager", Diss. TH Karlsruhe 1975.

[9]   H.J. Merker, "Über den nichtlinearen Einfluß von Gleitlagern auf die Schwingungen von Rotoren", Diss. Universität Darmstadt, 1981.

[10]   R. Cardinali, R. Nordmann, A. Sperber, "Dynamic Simulation of Nonlinear Models of Hydroelectric Machinery", submittet to Journal of Mechanical Systems and Signal-Processing, 1991.

A CONSIDERATION OF DAMPING IN ROTORS SUPPORTED BY TILTING-PAD BEARINGS
M.F.White

Division of Marine Engineering
University of Trondheim
Trondheim, Norway

ABSTRACT

Tilting-pad journal bearings are often used in high performance machinery operating at speeds above the first critical speed, in order to improve bearing stability. In many cases such machinery may still be prone to subsynchronous or unstable vibration problems. The influence of the dynamic characteristics of the tilting-pad bearings on the response of the complete rotor system is therefore of great importance. This paper is primarily concerned with the damping provided by such bearings. A number of factors are considered including frequency dependence of the dynamic coefficients, influence of pad support flexibility, and a simple method for evaluation of effective damping required to maintain rotor stability.

DYNAMIC COEFFICIENTS

The dynamic coefficients for journal bearings are usually represented by eight coefficients, i.e. four stiffness and four damping coefficients. This type of linearised coefficient is useful for the analysis of rotor-bearing vibrational behaviour because it avoids the necessity of solving the rotor equation of motion and bearing lubrication equations simultaneously.

However, for tilting-pad bearings there are additional degrees of freedom due to pad rotation. In order to fully represent the dynamic characteristics of these bearings, (5N+4) stiffness and (5N+4) damping coefficients are needed where N is the number of pads. The fluid film forces can be expressed as follows:

$$F_x = \sum_{i=1}^{N} (k_{xx}x + c_{xx}\dot{x} + k_{xy}y + c_{xy}\dot{y} + k_{x\delta}\delta + c_{x\delta}\dot{\delta})$$

$$F_y = \sum_{i=1}^{N} (k_{yx}x + c_{yx}\dot{x} + k_{yy}y + c_{yy}\dot{y} + k_{y\delta}\delta + c_{y\delta}\dot{\delta}) \qquad (1)$$

$$I_i\ddot{\delta} = k_{\delta x}x + c_{\delta x}\dot{x} + k_{\delta y}y + c_{\delta y}\dot{y} + k_{\delta\delta}\delta + c_{\delta\delta}\dot{\delta})$$

A reduction process is often used to simplify the model and reduce the number of coefficients to eight. A feature of this reduction is that a frequency is assigned to the pad vibration and the dynamic behaviour becomes frequency dependent. Usually the so-called synchronous reduction is used, and the pad frequency is assumed to be synchronous. Published data for tilting-pad bearing coefficients is usually given in this form.

A detailed method for reduction of coefficients is shown by (Parsell et al 1983). First of all, stiffness coefficients are determined in the same way as for fixed pad bearings. The main difference is that not only stiffness coefficients k_{xx}, k_{xy}, k_{yx}, k_{yy} have to be calculated from the ratio of force change to displacement perturbation, but also coefficients like $k_{x\delta}$, $k_{y\delta}$, $k_{\delta x}$, $k_{\delta y}$, $k_{\delta\delta}$ are evaluated by applying small displacement perturbation of either translation (x,y) or pad rotation (δ) from the static equilibrium position and noting the corresponding change in force or moment. The static equilibrium is determined by the balance of forces and moments between the applied load and the oil film hydrodynamic reaction force determined using Reynolds equation. A method of solution is described more fully in (White & Chan 1991).

Similarly the damping coefficient can be computed by applying small velocity perturbations about the static equilibrium position. These coefficients are then combined according to the reduction method. An interesting feature of the reduction is that the reduced stiffness coefficients depend on pad damping coefficients whilst the reduced damping coefficients depend on the pad stiffness coefficients. This interconnection is the source of the frequency effect on the reduced coefficients.

When pad inertia is neglected and the pad motion is assumed to be at a frequency ω, the equations for calculating the reduced coefficients are given below:

$$K_{mn} = \sum_{i=1}^{N} (k_{mn} - \frac{e_1 d_1 + \omega^2 e_2 d_2}{d_1^2 + \omega^2 d_2^2}) \quad , \quad C_{mn} = \sum_{i=1}^{N} (c_{mn} - \frac{e_2 d_1 - e_1 d_2}{d_1^2 + \omega^2 d_2^2}) \qquad (2)$$

$$with \quad \begin{array}{ll} d_1 = k_{\delta\delta} & e_1 = k_{m\delta} k_{\delta n} - \omega^2 c_{m\delta} c_{\delta n} \\ d_2 = c_{\delta\delta} & e_2 = c_{m\delta} k_{\delta n} + c_{\delta n} k_{m\delta} \end{array}$$

where N is the number of pads, m,n are permutations of x,y, and k_{mn}, c_{mn} are stiffness and damping coefficients for individual pads and K_{mn}, C_{mn} are the reduced coefficients.

SUBSYNCHRONOUS DYNAMIC COEFFICIENTS

Rotordynamic instability problems are in many cases associated with subsyncronous vibration frequencies - typically at around half the rotational speed when the rotor-bearing system is at or near the threshold of instability. Some bearing dynamic coefficients have therefore been determined to illustrate the frequency effect for tilting-pad journal bearings with different preloads and offsets over a range of operating Sommerfeld numbers. Figures 1 to 6 compare the cases when the vibration response occurs at synchronous frequency and at half the rotational speed. It is clearly seen that zero preload and centrally pivoted bearings operating at high Sommerfeld numbers suffer from the greatest relative loss of damping at the subsynchronous whirl/speed ratio of 0.5, while the stiffness increases. An increase in either preload or offset can significantly reduce the frequency effect. The effect also falls off when the Sommerfeld number decreases, i.e. the speed is lower or the load larger. Care must therefore be exercised when using small preload tilting-pad journal bearings for high speed applications when the shaft static load is relatively light. In particular, when external excitation is present (e.g. aerodynamic cross-coupling) the bearings may not be able to provide sufficient damping to stabilize the total system even though they are themselves free from any cross coupling effect. This is discussed in a later section.

PAD SUPPORT FLEXIBILITY

It is worth noting that the influence of the pivot flexibility and the pad deformation can also lead to a decrease in the effective damping provided by the bearings (Lund 1987, Kirk & Reedy 1988) and this flexibility effect is also dependent on the frequency.

A very much simplified view can be obtained by considering the relative magnitudes of pivot stiffness and the oil film radial stiffness. Three different cases are given in Table 1. All of them assume the same stiffness expression of a line contact pivot (Kirk & Reedy 1988):

$$KHarris = 2.55 \times 10^6 \ W^{0.1} \ L^{0.8} \tag{3}$$

which is for use with British units, where L is pad length (in) and W is load (lbf).

Table 1: Stiffness of 5-pad, tilting-pad bearings, L=3.582 in (91 mm)	Case 1	Case 2	Case 3
	zero preload load between pads	zero preload load between pads	0.5 preload load on pad
W (lbf) N (rpm) Kxx (N/m) Kyy (N/m) Kp = KHarris (N/m)	379 (1686N) 1700 52. x 10^6 99. x 10^6 2242. x 10^6	3790 (16860N) 17000 520. x 10^6 990. x 10^6 2819. x 10^6	3890 (17303N) 1700 84. x 10^6 4500. 10^6 2837. x 10^6

More accurate estimates of pivot stiffness taking into account surface curvature give values of pivot stiffness, Kp, 30-40% higher than indicated in Table 1. However, the Harris expression (3) is useful for showing that the effect of pivot stiffness is most significant for heavily loaded machines because the influence of load on pivot stiffness is rather weak compared with the stronger influence of load on the oil film stiffness for the same Sommerfeld number (Cases 1 & 2). Preloaded bearings can generate larger oil film stiffnesses, and when they are operated at high eccentricity ratios (high loads) the stiffness can rise to a similar order of magnitude to that of the pivot. This is particularly the case for load-on-pad bearings (Case 3).

By looking at a simplified model of the oil film and pivot where the pivot damping is ignored and the pad inertia is neglected, the effect of the pivot can be regarded as simply putting a spring in series with the oil film spring and damper elements. The equivalent damping of the combined system will always be less than that of the bearing oil film alone. For Case 3 the relative motion across the oil film, and hence the damping force, will be considerably less than for the other two cases. Derivations of expressions for pad equivalent stiffness and damping may be found in (Kirk & Reedy 1988).

EXPERIMENTAL RESULTS

Measured oil film damping coefficients are often less than the predicted theoretical values (Someya 1989, Chan & White 1992). This may be due to the boundary conditions for the oil film extent and pressures at the film boundaries. Damping coefficients depend on both wedge film pressure and squeeze film pressure. For tilting-pad bearings the oil flow conditions at inlet and inside the bearings, i.e. whether they are completely flooded with oil or dependent on the supply pressure, are not completely known. Some discrepancies between theory and experiment are therefore expected. The greatest discrepancy occurred at high eccentricity ratio where the measured coefficients were as much as an order of magnitude less than the theoretical values.

Stiffness coefficients which only depend on wedge film pressure may be determined more accurately.

EFFECTIVE BEARING DAMPING TO MAINTAIN STABILITY

Unstable vibration phenomena in rotor-bearing systems have been found in many cases at or near to the first natural frequency. The first mode of vibration is therefore of great interest for stability analysis. The equations of motion for a shaft model may be written as:

$$
\begin{aligned}
m\ddot{x} + C\dot{x} + Kx + By &= F_x(t) \\
m\ddot{y} + C\dot{y} + Ky + Bx &= F_y(t)
\end{aligned}
\qquad (4)
$$

where m is the mass, C is the external bearing damping, K is the combined shaft and bearing stiffness, B is the aerodynamic cross coupling stiffness and F is the external applied force. For an axisymmetrical system the equations (4) may be combined into one complex equation,

$$
m\ddot{v} + C\dot{v} + (K-iB)\,v = F(t) \qquad (5)
$$

where the solution v is

$$
v = \frac{1}{2m}\{-C \pm \sqrt{C^2 - 4m(K-iB)}\} \qquad (6)
$$

The square root can be equated to a complex number a+ib. Equating real and imaginary parts,

$$
a^2 - b^2 = C^2 - 4mK \quad , \quad 2ab = 4mB
$$

and

$$a^2 - \frac{1}{2} [C^2 - 4mK + \sqrt{(C^2 - 4mK)^2 + 16m^2 B^2}]$$

For small damping and cross coupling, i.e. $C^2 \ll 4mK$ and $B^2 \ll K^2$, it can be shown that

$$a = Cs = \frac{B}{\omega_n} \quad , \quad b = 2m\omega_n \tag{7}$$

The stability condition

$$a^2 = C^2 - 4mK + \frac{4m^2 B^2}{a^2} < C^2 \tag{8}$$

or

$$B^2 < a^2 \frac{K}{m} < C^2 \frac{K}{m} \quad , \quad B < C\omega_n$$

Generalized coordinates may be used to analyse the first mode as described in (Clough & Penzien 1982), where

$v(x,t) = \phi(x) V(t)$, where $\phi(x)$ is the mode shape

$m = \displaystyle\int_{-l/2}^{l/2} m(x) \phi^2(x) \, dx + \sum_{i=1}^{N} M_i \phi^2(x_i)$ is generalized mass

$C = \displaystyle\sum_{i=1}^{N} C_i \phi^2(x_i)$ is generalized damping

$K = \displaystyle\int_{-l/2}^{l/2} EI(x) [\frac{d^2\phi(x)}{dx^2}]^2 dx + \sum_{i=1}^{N} K_i \phi^2(x_i)$ is generalized stiffness

and $B = \displaystyle\sum_{i=1}^{N} B_i \phi^2(x_i)$ is generalized cross coupling stiffness

A typical compressor with tilting-pad journal bearings will be considered as an example. The rotor has a nominal weight of 1306 kg and an axial span of 3.12 m. Two tilting-pad bearings are located at axial distances of 0.29 m and 0.22m from either end respectively.

If the bearings stiffnesses are assumed to be 350×10^6 N/m, then the calculated generalised mass and generalised stiffness for the first mode of vibration are 764 kg and 34.7×10^6 N/m. The corresponding first critical speed is 2036 rpm and the mode shape indicates that the vibration amplitude at the bearings is 0.06 relative to the maximum at mid span. From previous experience on compressors of this size, an aerodynamic cross coupling of 1.75×10^6 N/m is assumed at mid-shaft. Suppose now the bearing damping is 0.35×10^6 N-s/m, then the generalised damping becomes

$$C = 2 \cdot 0.35 \cdot 10^6 \cdot (0.06)^2 = 2520 N\text{-}s/m \qquad (9)$$

The stability condition requires that

$$B < C\omega_n = 2520 \cdot 2036 \cdot \frac{2\pi}{60} = 0.537 \cdot 10^6 N/m \qquad (10)$$

which is clearly not satisfied in this case.

Since the critical speed ω_n does not vary much if the damping is changed, the effective damping, Cs, needed to maintain stability can be estimated:

$$Cs > \frac{B}{\omega_n} = \frac{1.75 \cdot 10^6}{2036 \cdot 2\pi/60} = 8213 \quad N\text{-}s/m \qquad (11)$$

If the mode shape is taken into account as before, then the actual bearing damping, Ci, has to be at least

$$Ci = \frac{Cs}{2 \cdot (0.06)^2} = 1.14 \cdot 10^6 \quad N\text{-}s/m \qquad (12)$$

The equivalent negative damping due to cross coupling, denoted by a in equation (7), is approximately equal to the effective damping needed to maintain stability, i.e.

$$a = Cs = 8213 \quad N\text{-}s/m \qquad (13)$$

If the bearing stiffness is changed to 87.5×10^6 N/m, then the first critical speed is lowered to 1839 rpm. The modal mass is computed as 819 kg and the amplitudes at the bearings are about 0.2 of the maximum at mid span. In this case the effective damping required for stable rotor vibration becomes

$$Cs > \frac{B}{\omega_n} = \frac{1.75 \cdot 10^6}{1839 \cdot 2\pi/60} = 9089 \quad N\text{-}s/m$$

$$\qquad (14)$$

The corresponding tilting-pad bearing damping is then

$$Ci = \frac{Cs}{2 \cdot (0.2)^2} = 0.114 \cdot 10^6 \quad N\text{-}s/m \qquad (15$$

It can be seen that the actual damping necessary for stability is much smaller than for the previous case of higher bearing stiffness. These results are in agreement with computational results using much more complicated models to investigate the effects of bearing stiffness and damping. The advantage of this simplified method using generalised coordinates is that the model only needs to predict the frequency and mode shape of the first natural frequency. This can be estimated reasonably well from a knowledge of bearing stiffness and shaft properties. Normally when instability is expected the damping is small and should not significantly affect the mode shape or natural frequency. The more uncertain factors such as bearing damping and aerodynamic cross coupling can be studied parametrically in an easy way.

404

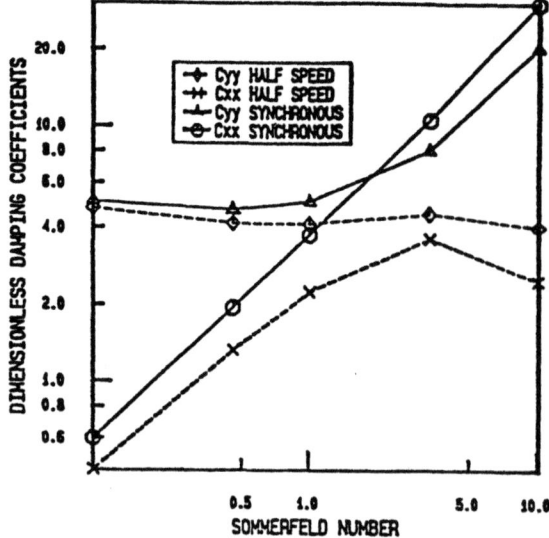

Fig. 1 Frequency dependence of damping coeff.
(5-pad, zero preload, load on pad, 0.5 offset)

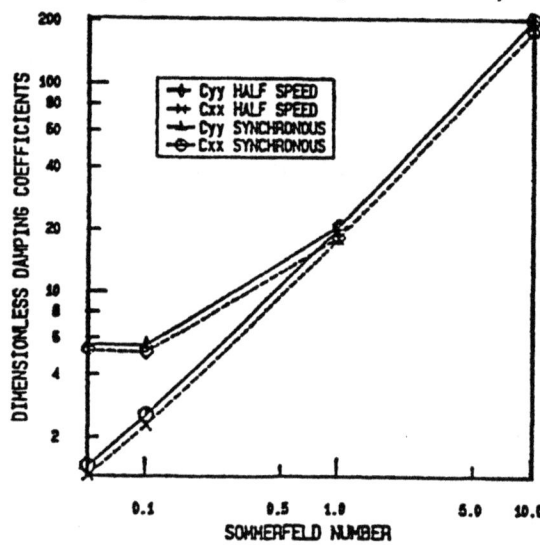

Fig. 2 Frequency dependence of damping coeff.
(5-pad, 0.5 preload, load on pad, 0.5 offset)

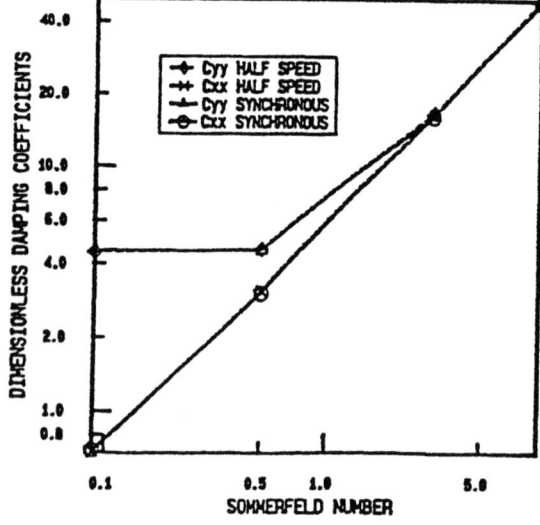

Fig. 3 Frequency dependence of damping coeff.
(5-pad, zero preload, load on pad, 0.59 offset)

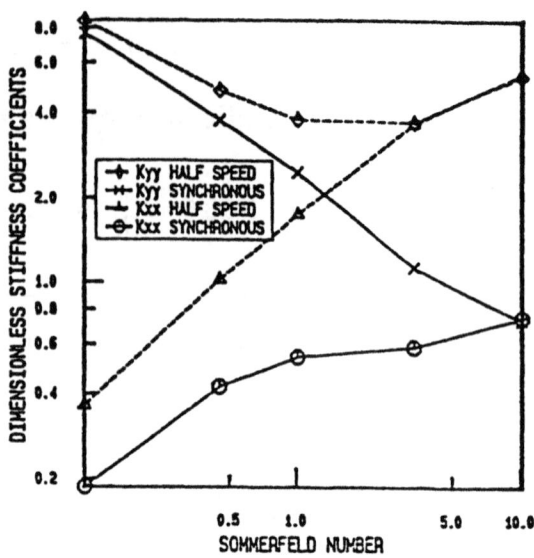

Fig. 4 Frequency dependence of stiffness coeff.
(5-pad, zero preload, load on pad, 0.5 offset)

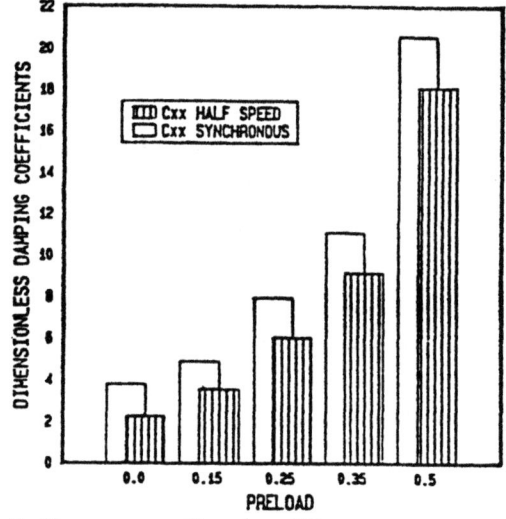

Fig. 5 Frequency effect for different preloads
(5-pad, load on pad, .5 offset, Sommerfeld no.=1)

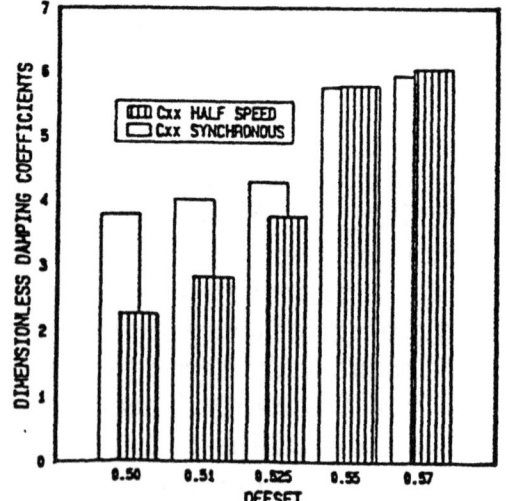

Fig. 6 Frequency effect for different offsets
(5-pad, zero preload, load on pad bearings,
Sommerfeld number = 1)

CONCLUSIONS

For tilting-pad journal bearings with small preloads operating at high Sommerfeld numbers the damping at subsynchronous frequencies is considerably lower than that predicted for synchronous vibration. This frequency effect is attenuated by increased bearing preloads and offsets.

For preloaded bearings and load-on-pad bearings at high eccentricity ratios the oil film stiffness can rise to the same order of magnitude as the pivot stiffness. Under these conditions the pad pivot flexibility can significantly reduce the effective bearing damping.

A simplified approach is presented for estimating the stability of the first mode of vibration of a shaft-bearing system using the method of generalised coordinates. The method only requires a knowledge of bearing stiffness and shaft properties. Other factors which are more difficult to estimate such as bearing damping or aerodynamic cross coupling stiffness can be studied parametrically.

Nomenclature:

B	Cross coupling stiffness	Kp	Pivot stiffness
C	Damping	F	Force
Ci	Bearing damping	m	Mass
Cp	Pad clearance (Rp - R)	N	Rotational speed (Hz)
Cs	Damping to maintain stability	R	Journal radius
D	Journal diameter	Rp	Pad radius
E	Modulus of elasticity	W	Load
I	Moment of area/pad inertia	δ	Pad rotation angle
k	Pad stiffness	μ	Dynamic viscosity
K	Stiffness	ω_n	Critical speed (rad/s)
ω	Rotational speed (rad/s)		

Dimensionless damping coefficient = $Cp \cdot \omega \ C/W$

Dimensionless stiffness coeffisient = $Cp \cdot K/W$

Sommerfeld number = $\dfrac{\mu \cdot N \cdot L \cdot D}{W} \cdot \dfrac{(R)^2}{Cp}$

REFERENCES

Chan SH, White MF (1991) Experimental Determination of Dynamic Characteristics of a Full Size Gas Turbine Tilting-pad Journal Bearing by an Impact Test Method, 13th ASME Conference on Mechanical Vibration and Noise, Miami.

Clough RW, Penzian J (1982) Dynamics of Structures, McGraw-Hill.

Kirk RG, Reedy SW (1988) Evaluation of pivot stiffness for typical tilting-pad journal bearings, ASME Journal of Vibration, Acoustics, Stress and Reliability in Design, 110:165-171.

Lund JW (1987) The influence of pad flexibility on the dynamic coefficients of a tilting-pad journal bearing, ASME Journal of Tribology, 109:65-70.

Parsell JK, Allaire PE, Barrett L (1983) Frequency effects in tilting-pad journal bearing dynamic coefficients, ASLE Transactions, 26,2:222-227.

White MF, Chan SH (1992) The subsynchronous dynamic behaviour of tilting-pad journal bearings, ASME Journal of Tribology, (accepted for publication).

The Response of a Spherical Journal Bearing to Rotor Unbalance

I.A. Craighead, Dept. of Mechanical Engineering, University of Strathclyde.
P.S. Leung, Dept. of Mech. Eng. & Manuf. Syst., Newcastle Polytechnic.

ABSTRACT

A spherical hydrodynamic journal bearing proposed by Leung et al (1) has two main advantages over traditional journal bearings:

1. It can accommodate large angular misalignment of the journal with little degradation of performance.
2. It has the capacity to resist axial forces as well as supporting a transverse load.

Earlier work (1) compared the static and dynamic performance of the spherical bearing with a similar cylindrical bearing and found only marginal differences in the performance of the two bearings. This paper goes on to study the performance of the spherical bearing when axial loading is present.
Steady state characteristics are established for various degrees of axial loading. With the journal displaced in the axial (z) direction, additional direct and cross coupled displacement and velocity coefficients are generated in the z direction. The resulting 18 linearised coefficients are evaluated for the spherical bearing under various loading conditions and the response of the system to rotor unbalance is assessed by determination of journal displacement and bearing forces over typical operating speeds.

INTRODUCTION

Spherical bearings have been in use for many years although they have usually been restricted to specialised applications due to high manufacturing costs. Examples include artificial human joints (2), a space station (3) and a water dam, (4). Early investigations of this type of bearing reflected the restricted applications being considered (5, 6, 7).
The use of a spherical journal bearing can now be contemplated in many conventional journal bearing applications as advances in CNC machining enable spherical surfaces to be produced almost as easily and economically as cylindrical surfaces. The use of a spherical journal bearing provides two significant improvements compared to a conventional cylindrical bearing:

a) Large misalignments between shaft and bearing can be accommodated without degradation in bearing performance. This is particularly beneficial where alignment is difficult or bearing supports/pedestals are subject to movement due to temperature, time etc. variations.

b) It locates the shaft axially and, within certain limits of operation, will eradicate the need for a separate axial location device such as a thrust bearing.

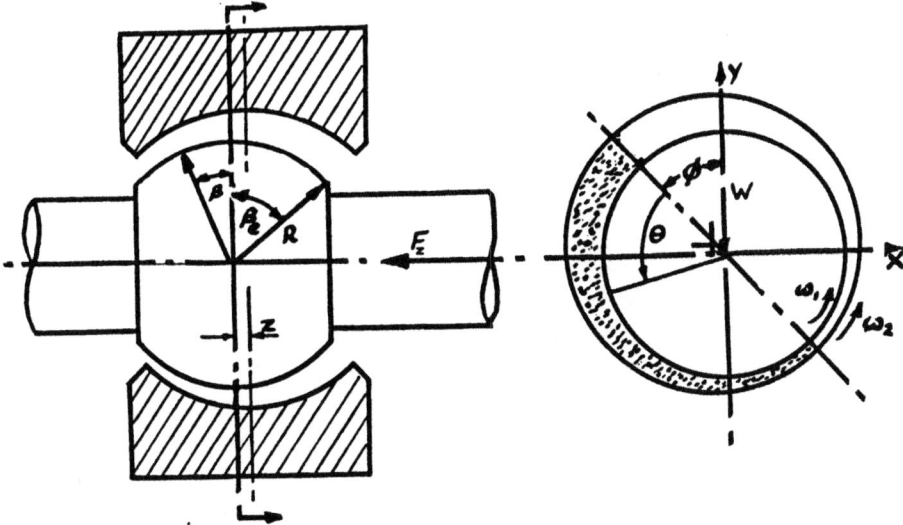

Figure 1   Configuration of the Spherical Bearing

To ensure satisfactory operation of a spherical bearing, its static and dynamic behaviour must be fully investigated.   In reference (1), Leung et al investigated the spherical bearing with no axial loading present and concluded that the bearing behaved in a similar manner to a conventional cylindrical journal bearing.   In this paper the linearised theory in (1) is extended to cover the situation when axial loading is present. The load vector becomes three dimensional and this results in cross–coupling in the x, y and z directions.   This in turn necessitates the use of 18 linearised dynamic coefficients to investigate the dynamic behaviour of the bearing.   The bearing response to rotor unbalance is considered here and the stability of the bearing is evaluated in (8).

STEADY STATE ANALYSIS

For a spherical bearing as shown in Figure 1, the Reynolds equation assuming incompressible flow and laminar conditions can be expressed as:

$$\frac{1}{R^2}\frac{1}{\cos^2\beta}\frac{\partial}{\partial\theta}\left(\frac{h^3}{12\mu}\frac{\partial P}{\partial\theta}\right) + \frac{1}{R^2}\frac{\partial}{\partial\beta}\left(\frac{h^3}{12\mu}\frac{\partial P}{\partial\beta}\right) + \frac{1}{R^2}\theta\tan\beta\frac{\partial}{\partial\theta}\left(\frac{h^3}{12\mu}\frac{\partial P}{\partial\beta}\right)$$

$$+ \frac{1}{R^2}\frac{\partial}{\partial\beta}\left(\frac{h^3}{12\mu}\theta\tan\beta\frac{\partial P}{\partial\theta}\right) + \frac{1}{R^2}\theta\tan\beta\frac{\partial}{\partial\theta}\left(\frac{h^3}{12\mu}\theta\tan\beta\frac{\partial P}{\partial\theta}\right) = \frac{(\omega_1+\omega_2)}{2}\frac{\partial h}{\partial\theta} + \frac{\partial h}{\partial t} \quad (1)$$

where P is the lubricant pressure h the film thickness, μ the lubricant viscosity and the other parameters are as shown in Figure 1.

In (1) it was shown that for values of β_e less than 45° the equation can be simplified to:–

$$\frac{1}{R^2\cos^2\beta}\frac{\partial}{\partial\theta}\left(\frac{h^3}{12\mu}\frac{\partial p}{\partial\theta}\right) + \frac{1}{R^2}\frac{\partial}{\partial\beta}\left(\frac{h^3}{12\mu}\frac{\partial p}{\partial\beta}\right) = \frac{(\omega_1+\omega^2)}{2}\frac{\partial h}{\partial\theta} + \frac{\partial h}{\partial t} \quad (2)$$

resulting in a marginal reduction in accuracy of solution but significant savings in computer time for a finite difference solution of the equation.

With an axial force present the shaft will move in the z direction and the fluid film thickness if given by

$$h = c(1 + \epsilon \cos\theta \cos\beta - z \tan\beta) \qquad (3)$$

and also

$$z_{max} = c(1 - \epsilon \cos\beta e)/\tan\beta e \qquad (4)$$

where c is the bearing radial clearance, ϵ is the eccentricity ratio and z is the nondimensional axial displacement $(\bar{z}/c)$

The pressure field generated by the bearing operating with an axial displacement results in a z component of force which under stable operating conditions will balance the applied force. A conventional finite difference approach (9) was used to solve the reduced Reynolds equation. The procedure adopted was to select a given eccentricity ratio with no axial loading present. The attitude angle and Sommerfeld number were then determined. An axial displacement was then introduced and the new Sommerfeld number, attitude angle and axial load were calculated. Iteration of the eccentricity ratio and axial displacement were then carried out to result in the bearing operating at the original Sommerfeld number with the required percentage of axial loading being carried. The iteration scheme adopted was

$$\epsilon_{new} = \epsilon_{old}(1 + 0.5(F_{old} - F_{new})/F_{old})$$

$$z_{new} = z_{old}(1 + 0.25(F_{zold} - F_{znew})/F_{zold}) \qquad (5)$$

and this was found to be stable and efficient under all circumstances.

Typical pressure plots are shown in Figure 2 indicating the peak pressure to be offset in the presence of axial loads.

The amount of axial loading that the bearing can sustain will depend largely on the value of β_e and the analysis showed that for $\beta_e = 30°$ approximately 10% of the lateral load could be tolerated and for $\beta_e = 45°$ approximately 25% of the lateral load could be resisted axially before

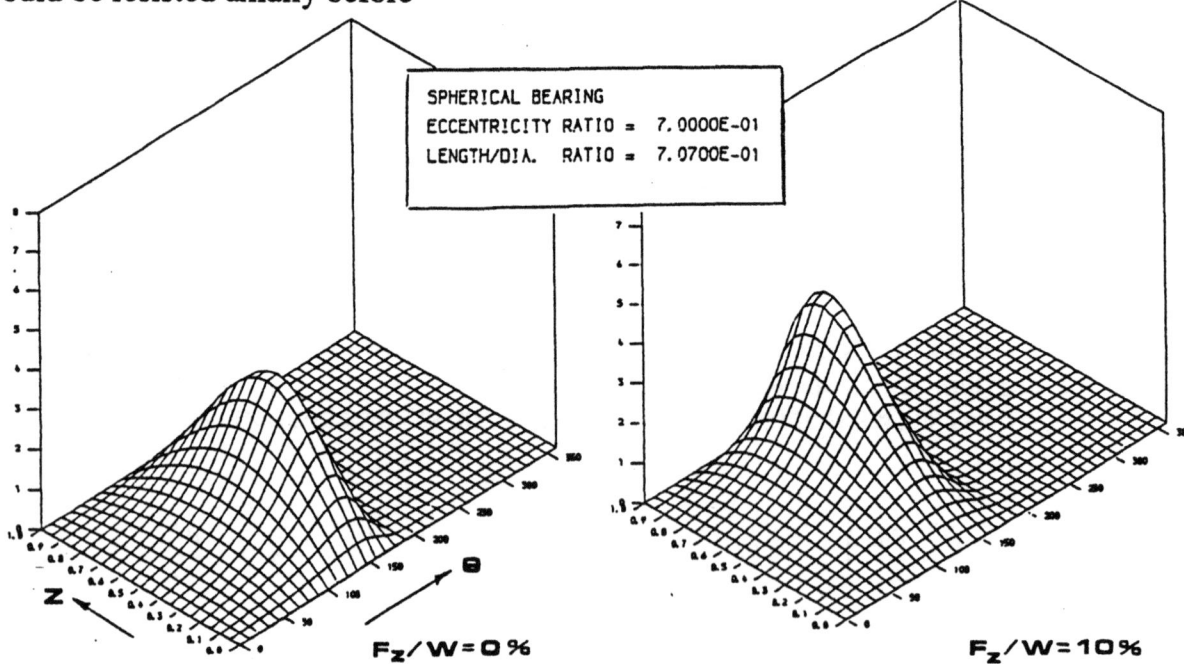

Figure 2 Pressure Fields for a Spherical Bearing

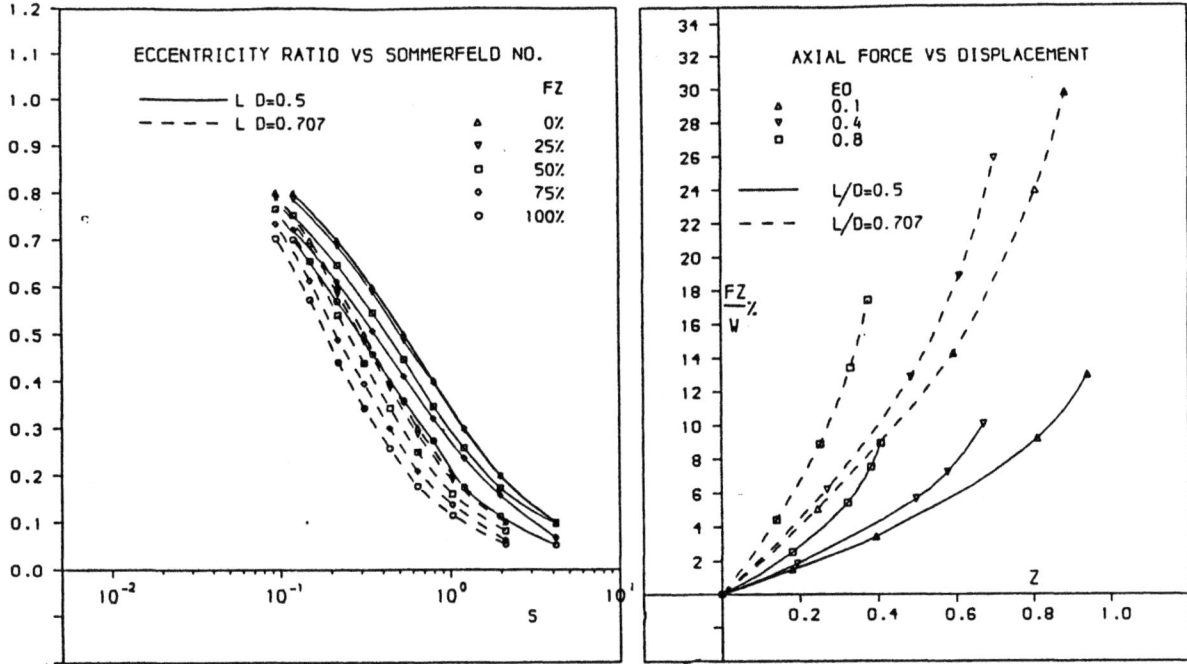

Figure 3   Change in eccentricity ratio and axial displacement with axial force

film clearances became too small (<5% of bearing radial clearance). Slightly higher percentages could be carried when operating at smaller eccentricity ratios. Figure 3 shows the modification in eccentricity ratio as the axial load is applied on the bearing with $\beta_e = 30°$ (L/D = 0.5) and $\beta_e = 45°$ (L/D = 0.707) and also the relationship between axial displacement and axial force.

DERIVATION OF DYNAMIC COEFFICIENTS

For each operating position the linearised behaviour of the system can be determined by use of dynamic coefficients. For 3 dimensional motion $(z \neq o)$ however, 9 displacement and 9 velocity coefficients are required and the equations for translational motion become:-

$$
\begin{bmatrix} M & 0 & 0 \\ O & M & O \\ O & O & M \end{bmatrix} \begin{Bmatrix} \ddot{x} \\ \ddot{y} \\ \ddot{z} \end{Bmatrix} + \begin{bmatrix} A_{xx} & A_{xy} & A_{xz} \\ A_{yz} & A_{yy} & A_{yz} \\ A_{zx} & A_{zy} & A_{zz} \end{bmatrix} \begin{Bmatrix} x \\ y \\ z \end{Bmatrix} + \begin{bmatrix} B_{xx} & B_{xy} & B_{xz} \\ B_{yx} & B_{yy} & B_{yz} \\ B_{zx} & B_{zy} & B_{zz} \end{bmatrix} \begin{Bmatrix} \dot{x} \\ \dot{y} \\ \dot{z} \end{Bmatrix} = \begin{Bmatrix} F_x \\ F_y \\ F_z \end{Bmatrix} \quad (6)
$$

The coefficients were determined for a range of eccentricity ratios (0.1 – 0.8) and a range of axial loads (0–10%) for the L/D = 0.5 bearing. Perturbations in displacement and velocity in the x, y and z directions about the steady state operating positions were used to obtain the coefficients. Examples of the variation in some of the coefficients are given in Figure 4.

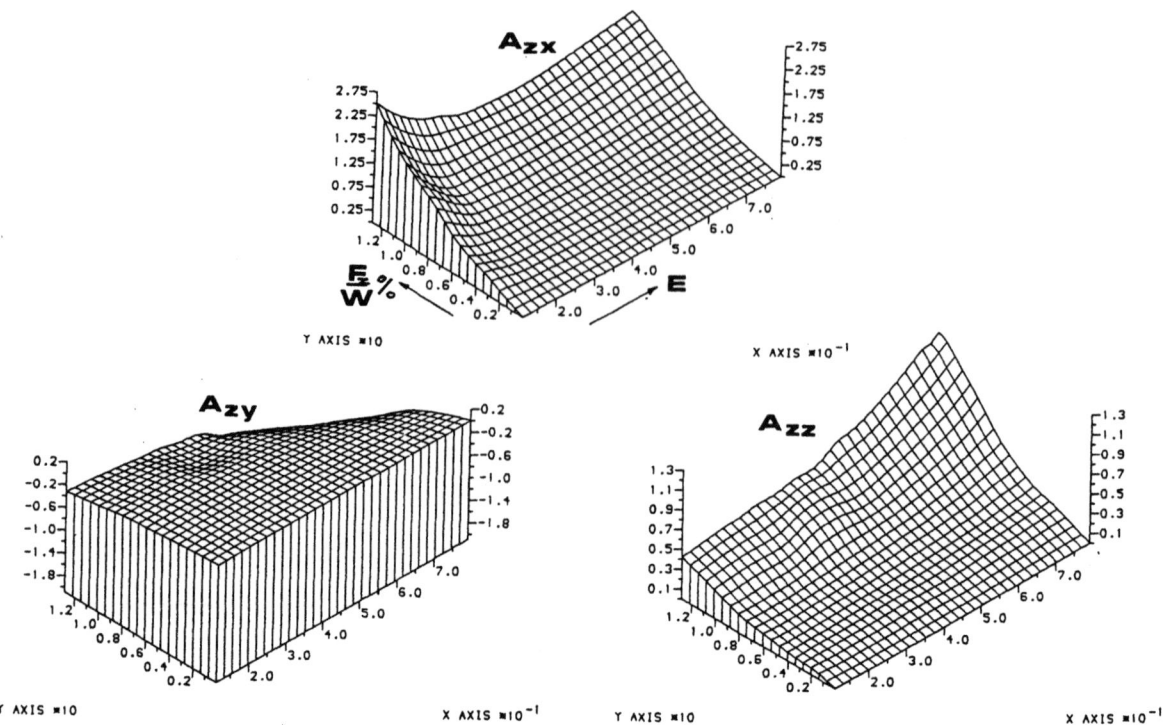

Figure 4 Sample displacement coefficients of the Spherical Bearing

ANALYSIS OF A RIGID ROTOR, TWO BEARING SYSTEM

Virtually all rotors possess some degree of unbalance at some state in their operational lifetime. This unbalance results in vibration in the x, y plane, ie. perpendicular to the axis of rotation. With the cross coupling present in an axially loaded spherical bearing this motion will become 3 dimensional and results in dynamic displacements and forces in the x, y and z directions. To investigate this behaviour, a two bearing rigid rotor system was considered (9). A rigid rotor was chosen to minimise any rotor influence on the bearing behaviour. Only one bearing was loaded axially, the second being assumed to be a conventional cylindrical bearing. The salient details of the rotor/bearing system are given in Table 1.

The analysis of a rigid rotor on two bearings has been widely documented (eg 9, 10) and results in a matrix equation

$$[A]\{D\} = \{F\} \tag{7}$$

Table 1 Rotor Bearing System Details

Rotor	Mass = 12500 kg	I_{zz} = 680 kgm^2	I_{xx} = 26340 kgm^2
	Length = 5 m	CoFG = 0.5 L	
	Unbalance 1 = 0.25 kg	Radius = 1 m	Axial pos = −1 m from CoFG
	Unbalance 2 = 0.25 kg	Radius = 1 m	
	Phase = 0		Axial pos = +1 m from CoFG
Bearings:	Length = 0.125 m	Diameter = 0.25	Radial Clearance = 0.16 mm
	Lubricant viscosity = 0.29 Poise		

where, in the present case, $[A]$ is a 10×10 matrix of terms based on inertias, dimensions, rotor speed and bearing coefficients, $\{D\}$ is the displacement matrix providing amplitude and phase details of x, y, z, θ and ψ displacements and $\{F\}$ is the forcing vector.

For each operating speed, the Sommerfeld number was calculated and used to interpolate the 18 dynamic coefficients. Solution of (7) was then achieved by Crouts method giving displacements and phase relationships for a given operating condition and out of balance. Knowledge of the dynamic displacements and bearing coefficients then allows calculation of the forces transmitted to the bearings. Figures 5 and 6 show the displacement and forces within the spherical bearing in the x, y and z directions.

DISCUSSION

The spherical bearing was shown to operate in a similar manner as an equivalent cylindrical bearing when there was no axial load (1). With the introduction of an axial load, the eccentricity ratio can be seen to reduce (the lateral load being kept constant) in Figure 3. This is to be expected as the journal surface is pushed "up" the bearing surface as z increases. The attitude angle is also affected slightly by the z displacement.

It has not been possible to present all of the coefficients in this paper due to space restrictions. However, Figure 4 includes the variation of Azx, Azy and Azz with eccentricity ratio and axial load and are typical of the 18 sets of coefficients obtained. Generally the direct stiffness and damping terms are found to increase with axial load whereas the cross coupling terms may increase or decrease depending on the particular coefficient and the eccentricity ratio considered. The effect of individual coefficient changes is difficult to ascertain as the coupled motions depend on all 18 coefficients.

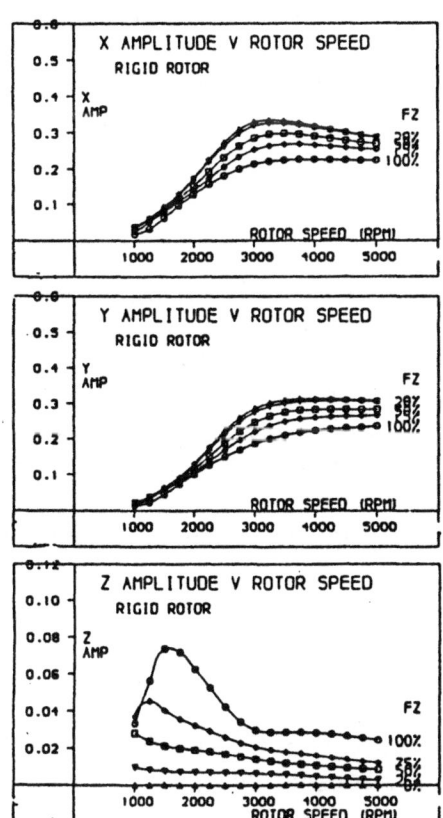

Figure 5 Bearing displacements.

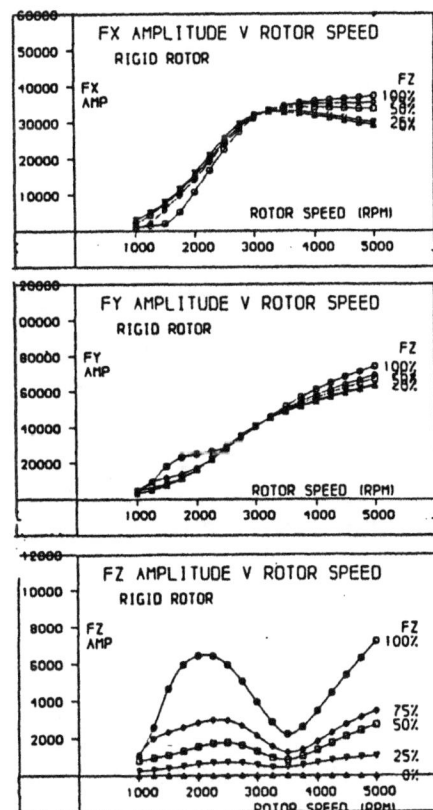

Figure 6 Bearing Forces

412

The displacements in the x and y directions (Figure 5) show a maximum amplitude at approximately 3200 rpm. From the Campbell diagram for the system (9) this corresponds to the closest approach of running speed (ie Forcing frequency) to a system (natural frequency). The maximum amplitude is just over 30% of the bearing radial clearance which is usually considered reliable for a linear analysis. With the application of axial load, the x and y amplitudes are reduced by up to approximately 1/3. This is to be expected as the displacement in the z direction effectively reduces the bearing clearance.

In the axial direction, distinct resonance peaks can be seen to increase from below 1250 rpm to 1600 rpm as the axial load is increased to a maximum. However, the amplitude is always less than 7.5% of the radial clearance and at a typical operating speed of 3000 rpm is no more than 3% of the bearing clearance.

Thus it appears that the dynamic axial displacement inherent in a spherical bearing supporting an unbalanced rotor should not present a problem to the safe operation of all but the most critical of machinery.

The bearing force plots (Figure 6) show that increasing the axial load raises the force transmitted in the x and y directions. This is probably largely due to increases in the direct stiffness coefficients. The force in the z direction shows a peak value at around 2200 rpm which then declines to a minimum at 3500 rpm before increasing in proportion to rotor speed. This would suggest some form of resonance. However, the peaks do not coincide with the displacement peaks indicating that the damping in the system plays a significant part in the transmission of force in the axial direction. In general terms, the dynamic force transmitted in the z direction is less than 10% of the magnitude of the force transmitted in the vertical or y direction. This implies that only minor modifications in bearing support design would be needed to accommodate axial forces.

Although the analysis shows no serious problems associated with the application of a spherical bearing to conventional situations it should not be overlooked that there is very little thrust capability without a lateral load. This is likely to preclude its application to lightly loaded rotors subject to large axial forces (for example vertical rotors).

CONCLUSION

A spherical journal bearing (L/D = 0.5) has been investigated and found to be capable of supporting an axial load of up to 10% of the lateral load. 18 linearised displacement and velocity coefficients have been determined for a range of loading conditions and then used to investigate the response of a rigid rotor supported by two bearings. It was found that for a typical system, axial amplitudes of approximately 20% of lateral vibration amplitudes were generated and dynamic forces in the axial direction were approximately 10% of the vertical dynamic force.

It is concluded that the bearing performs satisfactorily in the presence of typical rotor unbalance.

References

1. Leung PS, Craighead IA, Wilkinson TS (1989) An analysis of the steady state and dynamic characteristics of a spherical hydrodynamic journal bearing. Journ of Trib 111:459–467
2. Geonka PK, Booker JF (1980) Spherical bearing : static and dynamic analysis via the finite element method. Journ of Lubr Tech 102:308–319
3. Gross WA (1980) Fluid film lubrication. Wiley

4. Sautter S Spherical plain bearings in segment gates – An economic solution. Ball Bearing Journ. No. 230.
5. Shaw MC, Strang CD (1948) The hydrosphere – a new hydrodynamic bearing. Journ of App. Mechs 70:137–145
6. Wannier GH (1950) A contribution to the hydrodynamics of lubrication. Quat App Maths 8:1–32
7. Pan CHT (1968) Gas lubricated spherical bearings. Journ of Basic Eng 85:311–323
8. Craighead IA, Leung PS (Sept 1992) The stability of a spherical journal bearing with axial load. In preparation for IMechE Vibr in Rot Mach Conf Univ Bath
9. Craighead IA (1976) A study of the dynamics of rotor bearing systems and related fluid film bearing characteristics PhD Thesis, Univ of Leeds
10. Gunter EJ, De Choudhury P (1969) Rigid rotor dynamics. NASA Contract report CR – 1391

INFLUENCE OF BALL BEARING STIFFNESS ON SQUEEZE FILM BEHAVIOR

INCLUDING FLUID FLOW TURBULENCE AND INERTIA EFFECTS

B. BOU-SAID , D. NELIAS
Laboratoire de Mécanique des Contacts.
URA 856 . I.N.S.A Bât. 113
20 Avenue Albert Einstein 69621
Villeurbanne Cedex FRANCE

SUMMARY

The squeeze film damper is a fluid film bearing device to suppress vibration in high speed rotating machinery. The basic geometry is two concentric cylinders with a lubricating oil filling the annulus between them. The inner cylinder is fixed to a rolling element bearing outer race and usually performs a constant speed circular precession about the outer cylinder axis. The squeeze film bearings have found extensive use as a means of introducing damping into rotor-bearing assemblies and hence reducing vibration amplitudes to acceptable levels. Then a fundamental problem is to determine the squeeze film dynamic coefficients. We present in this study the influence of ball bearing stiffness on the squeeze film bearing. The two elements are first studied separatly and then gathered to obtain a dynamic representation of the assembly constituted of the ball bearing and the squeeze film damper.

The influence of flow characteristics as well as the ball bearing stiffness on the squeeze film stability map are studied.

I. INTRODUCTION

Squeeze film dampers have been the subject of numerous theoretical and experimental investigations since their development in the early 1960s to attenuate turborotor vibration in aircraft engine. They have found extensive use as a means of introducing damping into rotor bearing assemblies and hence reducing vibration amplitudes to acceptable levels. The basic geometry is two concentric cylinders with a lubricating oil filling the annulus between them . The outer cylinder is fixed to the stationnary housing. The inner cylinder is fixed to a rolling element bearing outer race and usually performs a constant speed circular precession about the outer cylinder axis. An important litterature exist concerning the squeeze film problem, and we can mention :

- the theoretical studies of John Tichy et al [1,2]
- the experimental works of I.A. San Andres et al [3], J. Ellis et al [4].

All these works have focused their analyses essentially on fluid inertia effects on squeeze film bearing performances. The rotational speed can rise up to 60 000 rmp and then the laminar assumption for the fluid flow is no more valid. It is necessary to take into account the fluid flow behavior and its incidence on bearing static and dynamic characteristics. In the litterature, the squeeze film bearing is always considered as an isolated element of a machine. As the inner cylinder is fixed to a rolling element, the ball bearing must have an influence on the squeeze film behavior. We present here the influence of cavitation, inertia and turbulence effects, ball bearing stiffness on static and dynamic characteristics of a squeeze film bearing. The squeeze film bearing and the ball bearing are first studied separately and then gathered to have a global dynamic representation.

II. THE SQUEEZE FILM BEARING

II.1 Basic equations

The basic geometry is presented in figure 1. The principal hypotheses are the following :
- no misalignment,
- the ratio between the radial clearance and the shaft radius is of the order of 10^{-3}.
- the fluid flow is newtonian, isoviscous and incompressible. The basic equations are [5]:
* the equation of continuity
* the equilibrium equations
* the equations of rheological behavior

The asociated boundary conditions are :
* for $y = o$ $u = v = w = o$
* for $y = h$ $u = w = o$ and $v = V = e\omega \sin \theta$
* and for $z = \pm$ L/2 $p = pa$ imposed pressure.

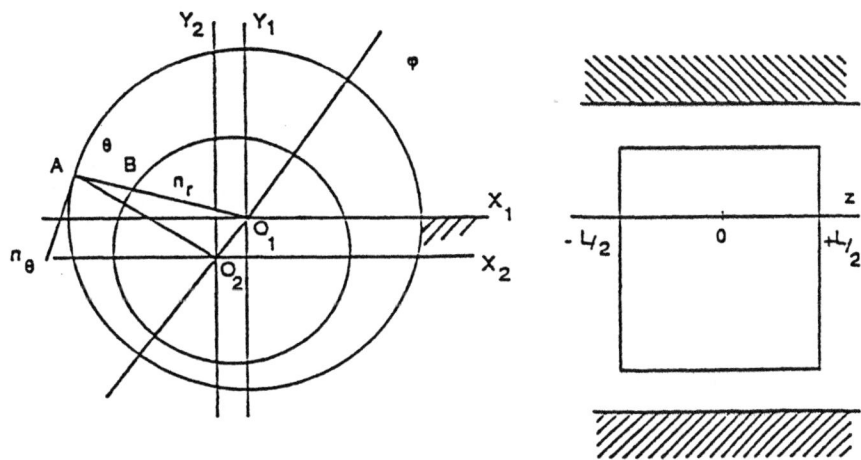

Figure 1 : Schematic of the bearing

II.2 Inertia and turbulence effects

The modeling of inertia and turbulence effects have been the subject of numerous investigations. We can mention principally Slezkin and Targ [6] and Constantinescu [7]. Starting from the work of J. Tichy [2] we can obtain a modified Reynolds equation which takes the following form [5] :

$$\frac{\partial}{\partial x}\left(\frac{h^3}{G_x}\frac{\partial p}{\partial x}\right) + \alpha^2 \frac{\partial}{\partial z}\left(h^3 G_z\frac{\partial p}{\partial z}\right) = 12\left(\frac{\partial h}{\partial t} - \frac{\partial h}{\partial \theta}\right)$$

$$+ R_e^* h^3 G(\theta, z) \qquad (1)$$

with :

$$G(\theta,z) = -\frac{1}{h}\left[\frac{\partial^2 I_{\theta\theta}}{\partial \theta^2} + 2\alpha^2 \frac{\partial^2 I_{\theta z}}{\partial\theta\,\partial z} + \alpha^4 \frac{\partial^2 I_{zz}}{\partial z^2} - \frac{\partial^2 h}{\partial t^2}\right]$$

$$-\frac{2}{h^2}\frac{\partial h}{\partial\theta}\left[\frac{\partial I_{\theta\theta}}{\partial\theta} + \alpha^2 \frac{\partial I_{\theta z}}{\partial z} + \frac{\partial q_\theta}{\partial t}\right] \qquad (2)$$

and : $q_\theta = \displaystyle\int_0^h u\,dy, \; q_z = \int_0^h w\,dy$

$$I_{\theta\theta} = \int_0^h u^2\,dy \; ; \; I_{\theta z} = \int_0^h u\,w\,dy \; ; \; I_{zz} = \int_0^h w^2\,dy$$

$$\alpha = R/L \; ; \; R_e^* = \frac{\omega\,C^2}{\nu}$$

$$G_x = \frac{1}{12 + 0,0136\,R_c^{0,9}}, G_z = \frac{1}{12 + 0,0043\,R_c^{0,96}}$$

where $R_c = \rho\dfrac{u\,h}{\mu}$ is the couette Reynolds number. If the poiseuille flow is dominant [5] we have :

$$G_x = G_z = G = \frac{6,8}{Re_p^{0,681}}$$

with $Re_p = \dfrac{\rho\,u_m\,h}{\mu}$ and u_m represents the mean velocity of the fluid due to the poiseuille flow.

II.3 Some results

To point out the importance of the different phenomena which govern the flow behavior the case of a squeeze film bearing of L/D ratio equals 0.5 has been treated. Figures 2.1 and 2.2 show the influence of inertia effects on the field pressure. For this case we take the Reynolds number: 20000 , the modified Reynolds number : 20

and the rotationnal speed : 60000 rpm. We observe a great difference in
the field pressure aspect between the two cases. Taking into account
the inertia effect increases the pressure. In Figure 3 we present the
evolution of the dimensionless critical mass versus the eccentricity
ratio. The evolution of this characteristic is largely modified if we
take into account the cavitation, the turbulence and the inertia
effects compared to the basic case.

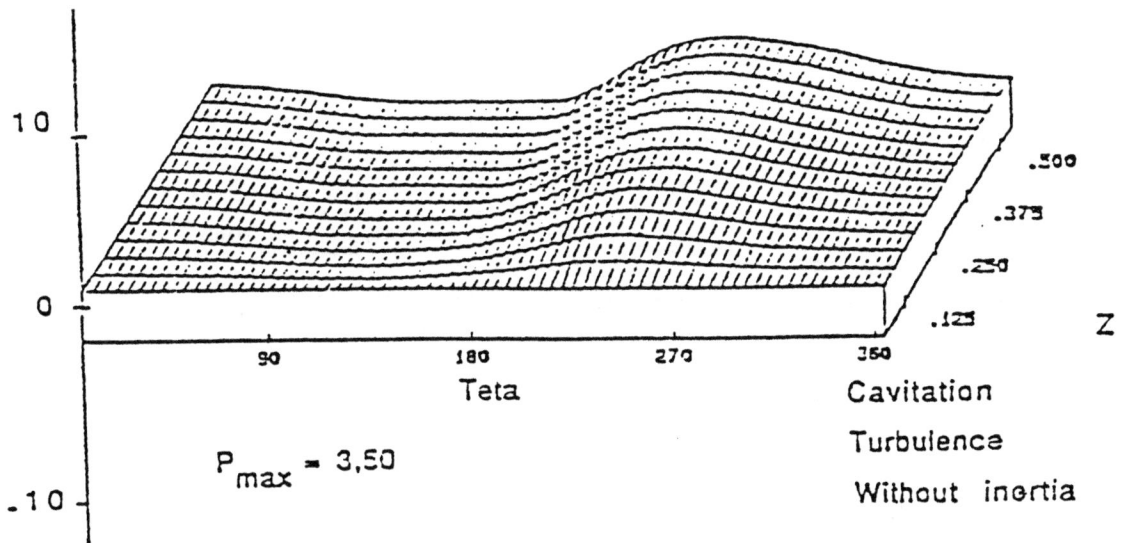

Figure 2.1 : Field pressure. Turbulence without inertia

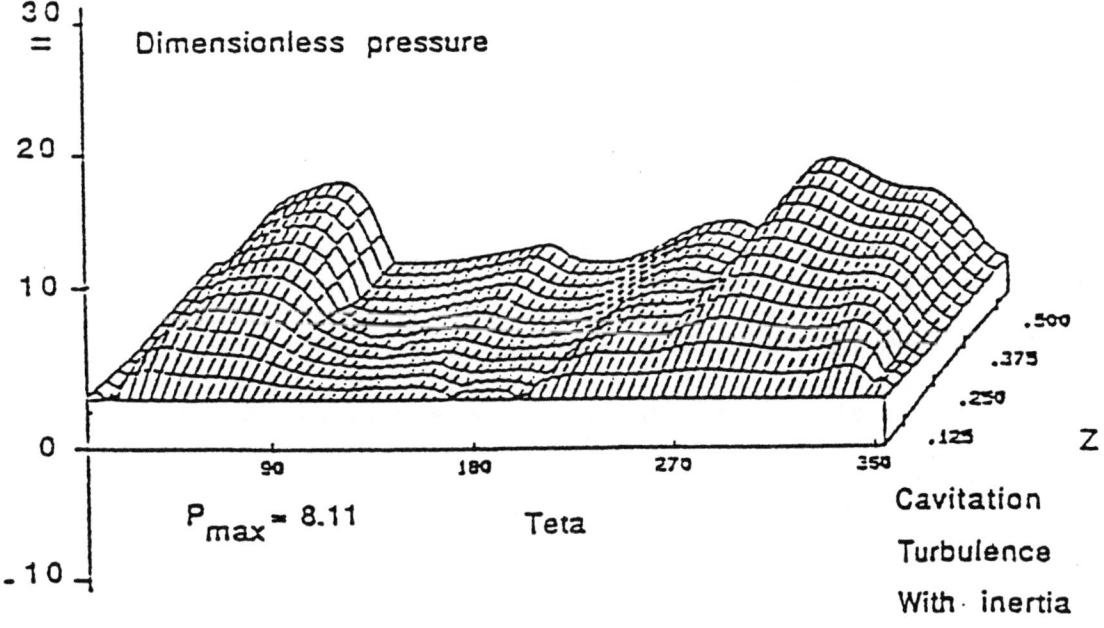

Figure 2.2 : Field pressure. Turbulence with inertia

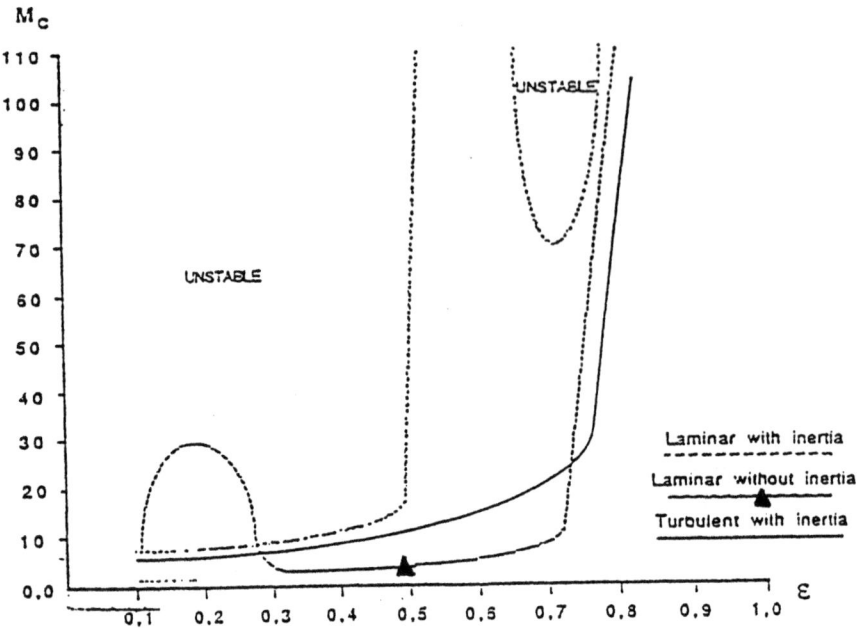

Figure 3 : Dimensionelss critical mass versus ε

III THE BALL BEARING

The development of a comprehensive quasi-static model which provides a realistic description of the ball bearing stiffness is briefly presented here. All details are given in [8,9]. The different assumptions are presented and the connection between the different forces are identified. The equilibrium of the bearing required some assumptions :

   - a quasi-static analysis and then inertial effects excepting centrifugal forces are neglected
   - no misalignment
   - the geometry is supposed to be perfect i.e rings and cage are cylindrical and balls spherical.
   - the ball behavior is symmetrical to the radial axis.
   - the cage is guided by the outer or the inner ring.

In this model, the kinematic of the cage and each ball is unknown. This kinematic is obtained from the equilibrium of the different bearing elements. Rings, cage and balls equilibrium equations are non-linear algebric equations, and can be solved simultaneously using an iterative numerical procedure. If N represents the number of balls we obtain 4N equations for the balls ,2 equations for the cage , 2 equations for the inner ring and 2(N-1) geometrical relations. For the details see [8]. Previous hydrodynamic and elastohydrodynamic models can be used to determine the normal and traction forces corresponding to the geometrical interactions between the different elements of a ball bearing [8]. The prediction of traction forces and moments is based on the lubricant constitutive equations in the high pressure region. The coefficients which define the behavior of a typical aeronautic lubricant have been estimated by an analysis of experimental traction data on a two disk machine [10].
From the analysis presented above we can see that the study of ball bearing consists of a rather sophisticated set of fairly simple to

extremely complicated interactions. Typical results obtained for a 50 mm pitch diameter ball bearing show the bearing stiffness versus the axial and normal loads (figures 4.1 and 4.2) in both lubricated and unlubricated cases.

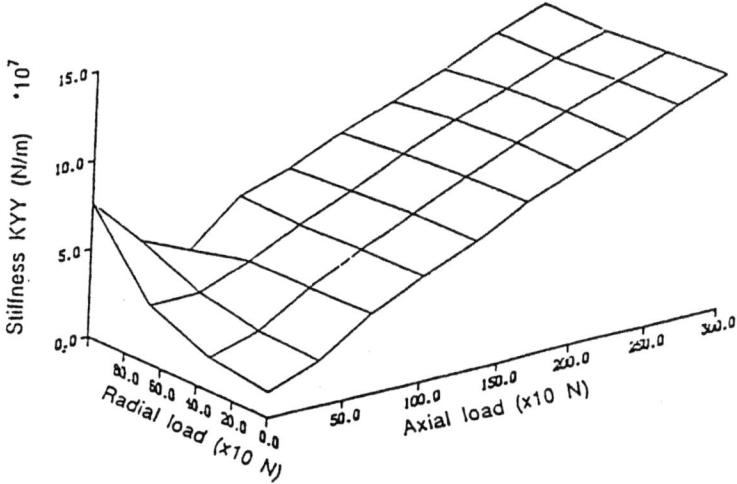

Figure 4.1 : Ball bearing stiffness. Lubricated case

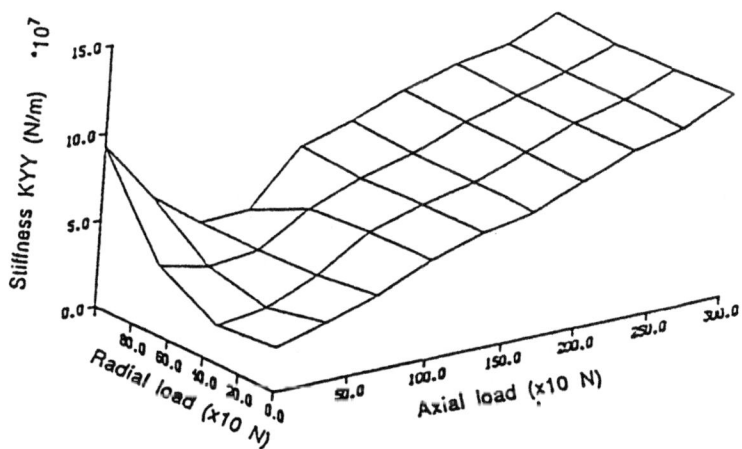

Figure 4.2 : Ball bearing stiffness. Unlubricated case

IV INFLUENCE OF BALL BEARING STIFFNESS ON SQUEEZE FILM STABILITY

The squeeze film bearing and the ball bearing are gathered to obtain a dynamic representation of the whole system. Figure 5 represents the influence of stiffness ball bearing on squeeze film stability map in both lubricated and unlubricated cases. We note an increase of the stability domain when we take into account the ball bearing stiffness especially when the ball bearing is lubricated. We note that an unconditional stability zone exists for small eccentricities which are the usual squeeze film bearings running conditions. These zones are not detected when using the only squeeze film theory.

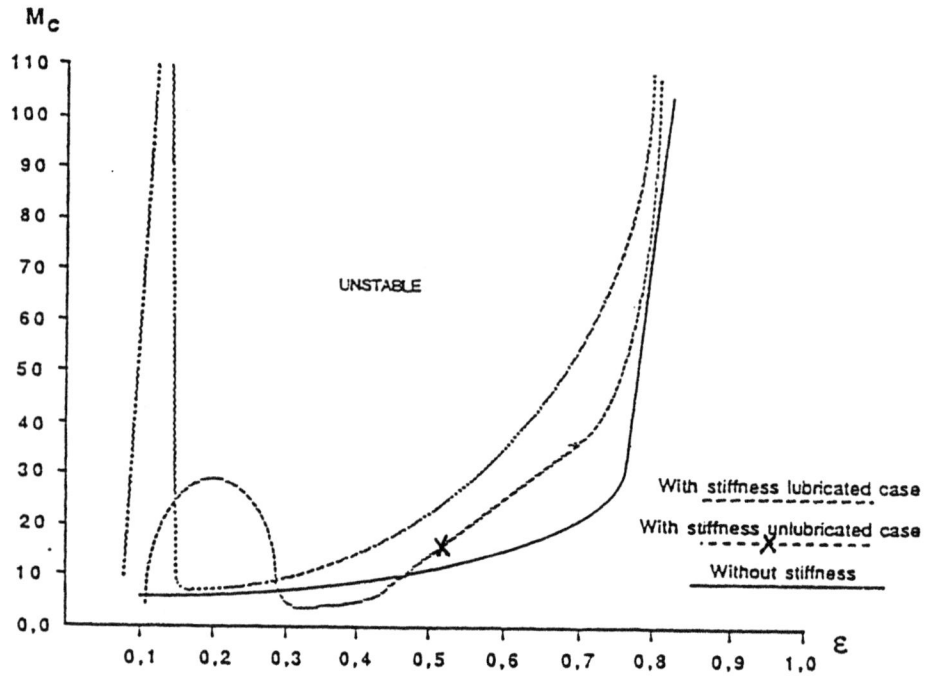

Figure 5 : Influence of stiffness ball bearing on squeeze film critical mass

V CONCLUSION

Taking into account the turbulence and the inertia effects point to show that :
    - the static characteristics are strongly modified
    - the domain of stability is completely different at small and high eccentricities compared to this obtained with the classical theory.

Gathering the ball bearing and the squeeze film bearing by the means of stiffness shows that an unconditional stability zone appears at small running eccentricities which is not predicted by the only squeeze film theory.

Nomenclature

C   : radial clearance
h   : filmthickness
L   : bearing lenght
M_c : dimensionless critical mass
p   : pressure
q_θ : circonférential flow

q_z : axial flow

R   : bearing radius
t   : time
u,v,w   : components of the fluid velocity
x,y,z   : cartésian coordinates
W   : load
ω : rotationnal speed
θ : angular coordinate
ρ : density
ε : eccentricity
μ : dynamic viscosity
ν : kinematic viscosity

REFERENCES

[1] J.A. TICHY, W.O. WINER "Inertial considerations in parallel circular squeeze film bearing" ASME, Journal of Lubrification Tribology. October 1970, p. 588-592.

[2] J.A. TICHY " A study of the effect of fluid inertia and end leakage in the finite squeeze film damper". ASME Journal of Lubrication Technology, January 1987, Vol. 109,p. 54-59.

[3] I.A. SAN ANDRES, J.M. VANCE " Experimental measurement of the dynamic pressure distribution in a squeeze film bearing damper executing circular centered orbit" ASLE preprint n° 86-TC-4D-2.

[4] J. ELLIS, B. ROBERTS, A. HOSSEINI SRANAKI " A comparison of identification methods of estimating squeeze film damper coefficients" ASME Journal of Tribology, Vol. 110 January 1988, p. 119-127.

[5] M. GUILLIEN, B. BOU-SAID, D. NELIAS "Influence de la turbulence et des effets d'inertie sur les caracteristiques statiques et dynamiques des paliers amortisseurs" Eighth World Congress on the Theory of Machines and Mechanisms. Prague Czechoslovakia, August 26-31, 1991.

[6] N.A. SLEZKIN, S.M. TARG "L'équation généralisée de Reynolds" C.R. Acad. Sc. de l'URSS, Vol 54, n°3, p. 205-208, 1946.

[7] V.N. CONSTANTINESCU "On turbulent lubrication". Proc. of the Inst. of Mech. Engrs, Vol. 173, n°38, p. 891-900, 1959.

[8] D. NELIAS " Etude du glissement dans les roulements à billes grande vitesse de turbomachine. Influence de la pollution du lubrifiant" Thèse de Doctorat. I.N.S.A de Lyon. 1989. 292 p.

[9] D. NELIAS, P. SAINSOT, B. BOU-SAID " Glissement dans les roulements à billes grande vitesse sous charges axiale et radiale combinées " Eighth World Congress on the Theory of Machines and Mechanisms. Prague Czechoslovakia, August 26-31, 1991.

[10] P.K. GUPTA, L. FLAMAND, D. BERTHE, M. GODET "On the traction behaviour of several lubricants" Trans. ASME Journal of Lubrication Technology. Vol. 103, n°1, 1981, p. 55-64.